Marine Enzymes

Marine Enzymes

Sources, Biochemistry and Bioprocesses for Marine Biotechnology

Editor

Antonio Trincone

MDPI • Basel • Beijing • Wuhan • Barcelona • Belgrade • Manchester • Tokyo • Cluj • Tianjin

Editor
Antonio Trincone
Istituto di Chimica
Biomolecolare,
Consiglio Nazionale delle
Ricerche,
Italy

Editorial Office
MDPI
St. Alban-Anlage 66
4052 Basel, Switzerland

This is a reprint of articles from the Special Issue published online in the open access journal *Marine Drugs* (ISSN 1660-3397) (available at: https://www.mdpi.com/journal/marinedrugs/special_issues/marine_enzymes).

For citation purposes, cite each article independently as indicated on the article page online and as indicated below:

LastName, A.A.; LastName, B.B.; LastName, C.C. Article Title. *Journal Name* **Year**, *Article Number*, Page Range.

ISBN 978-3-03943-024-6 (Hbk)
ISBN 978-3-03943-025-3 (PDF)

© 2020 by the authors. Articles in this book are Open Access and distributed under the Creative Commons Attribution (CC BY) license, which allows users to download, copy and build upon published articles, as long as the author and publisher are properly credited, which ensures maximum dissemination and a wider impact of our publications.

The book as a whole is distributed by MDPI under the terms and conditions of the Creative Commons license CC BY-NC-ND.

Contents

About the Editor ix

Preface to "Marine Enzymes" xi

Antonio Trincone
Enzymatic Processes in Marine Biotechnology
Reprinted from: *Mar. Drugs* **2017**, *15*, 93, doi:10.3390/md15040093 1

Giorgio Maria Vingiani, Pasquale De Luca, Adrianna Ianora, Alan D.W. Dobson and Chiara Lauritano
Microalgal Enzymes with Biotechnological Applications
Reprinted from: *Mar. Drugs*.**2019**, *17*, 459, doi:10.3390/md17080459 31

Qian Li, Fu Hu, Benwei Zhu, Yun Sun and Zhong Yao
Biochemical Characterization and Elucidation of Action Pattern of a Novel Polysaccharide Lyase 6 Family Alginate Lyase from Marine Bacterium *Flammeovirga* sp. NJ-04
Reprinted from: *Mar. Drugs* **2019**, *17*, 323, doi:10.3390/md17060323 51

Yanan Wang, Xuehong Chen, Xiaolin Bi, Yining Ren, Qi Han, Yu Zhou, Yantao Han, Ruyong Yao and Shangyong Li
Characterization of an Alkaline Alginate Lyase with pH-Stable and Thermo-Tolerance Property
Reprinted from: *Mar. Drugs* **2019**, *17*, 308, doi:10.3390/md17050308 63

Benwei Zhu, Limin Ning, Yucui Jiang and Lin Ge
Biochemical Characterization and Degradation Pattern of a Novel Endo-Type Bifunctional Alginate Lyase AlyA from Marine Bacterium *Isoptericola halotolerans*
Reprinted from: *Mar. Drugs* **2018**, *16*, 258, doi:10.3390/md16080258 77

Benwei Zhu, Fu Hu, Heng Yuan, Yun Sun and Zhong Yao
Biochemical Characterization and Degradation Pattern of a Unique pH-Stable PolyM-Specific Alginate Lyase from Newly Isolated *Serratia marcescens* NJ-07
Reprinted from: *Mar. Drugs* **2018**, *16*, 129, doi:10.3390/md16040129 91

Guiyuan Huang, Qiaozhen Wang, Mingqian Lu, Chao Xu, Fei Li, Rongcan Zhang, Wei Liao and Shushi Huang
AlgM4: A New Salt-Activated Alginate Lyase of the PL7 Family with Endolytic Activity
Reprinted from: *Mar. Drugs* **2018**, *16*, 120, doi:10.3390/md16040120 103

Peng Chen, Yueming Zhu, Yan Men, Yan Zeng and Yuanxia Sun
Purification and Characterization of a Novel Alginate Lyase from the Marine Bacterium *Bacillus* sp. Alg07
Reprinted from: *Mar. Drugs* **2018**, *16*, 86, doi:10.3390/md16030086 117

Irina Bakunina, Lubov Slepchenko, Stanislav Anastyuk, Vladimir Isakov, Galina Likhatskaya, Natalya Kim, Liudmila Tekutyeva, Oksana Son and Larissa Balabanova
Characterization of Properties and Transglycosylation Abilities of Recombinant α-Galactosidase from Cold-Adapted Marine Bacterium *Pseudoalteromonas* KMM 701 and Its C494N and D451A Mutants
Reprinted from: *Mar. Drugs* **2018**, *16*, 349, doi:10.3390/md16100349 131

Jingjing Sun, Congyu Yao, Wei Wang, Zhiwei Zhuang, Junzhong Liu, Fangqun Dai and Jianhua Hao
Cloning, Expression and Characterization of a Novel Cold-Adapted β-galactosidase from the Deep-sea Bacterium *Alteromonas* sp. ML52
Reprinted from: *Mar. Drugs* **2018**, *16*, 469, doi:10.3390/md16120469 153

Wei Ren, Ruanhong Cai, Wanli Yan, Mingsheng Lyu, Yaowei Fang and Shujun Wang
Purification and Characterization of a Biofilm-Degradable Dextranase from a Marine Bacterium
Reprinted from: *Mar. Drugs* **2018**, *16*, 51, doi:10.3390/md16020051 167

Ryuji Nishiyama, Akira Inoue and Takao Ojima
Identification of 2-keto-3-deoxy-D-Gluconate Kinase and 2-keto-3-deoxy-D-Phosphogluconate Aldolase in an Alginate-Assimilating Bacterium, *Flavobacterium* sp. Strain UMI-01
Reprinted from: *Mar. Drugs* **2017**, *15*, 37, doi:10.3390/md15020037 183

Maureen W. Ihua, Freddy Guihéneuf, Halimah Mohammed, Lekha M. Margassery, Stephen A. Jackson, Dagmar B. Stengel, David J. Clarke and Alan D. W. Dobson
Microbial Population Changes in Decaying *Ascophyllum nodosum* Result in Macroalgal-Polysaccharide-Degrading Bacteria with Potential Applicability in Enzyme-Assisted Extraction Technologies
Reprinted from: *Mar. Drugs* **2019**, *17*, 200, doi:10.3390/md17040200 201

Xian-Yu Zhu, Yong Zhao, Huai-Dong Zhang, Wen-Xia Wang, Hai-Hua Cong and Heng Yin
Characterization of the Specific Mode of Action of a Chitin Deacetylase and Separation of the Partially Acetylated Chitosan Oligosaccharides
Reprinted from: *Mar. Drugs* **2019**, *17*, 74, doi:10.3390/md17020074 221

Shangyong Li, Linna Wang, Xuehong Chen, Mi Sun and Yantao Han
Design and Synthesis of a Chitodisaccharide-Based Affinity Resin for Chitosanases Purification
Reprinted from: *Mar. Drugs* **2019**, *17*, 68, doi:10.3390/md17010068 237

Hang T. T. Cao, Maria D. Mikkelsen, Mateusz J. Lezyk, Ly M. Bui, Van T. T. Tran, Artem S. Silchenko, Mikhail I. Kusaykin, Thinh D. Pham, Bang H. Truong, Jesper Holck and Anne S. Meyer
Novel Enzyme Actions for Sulphated Galactofucan Depolymerisation and a New Engineering Strategy for Molecular Stabilisation of Fucoidan Degrading Enzymes
Reprinted from: *Mar. Drugs* **2018**, *16*, 422, doi:10.3390/md16110422 251

Hui Liu, Bi-Shuang Chen, Fayene Zeferino Ribeiro de Souza and Lan Liu
A Comparative Study on Asymmetric Reduction of Ketones Using the Growing and Resting Cells of Marine-Derived Fungi
Reprinted from: *Mar. Drugs* **2018**, *16*, 62, doi:10.3390/md16020062 269

Haiyan Jin, Yoshiko Hiraoka, Yurie Okuma, Elisabete Hiromi Hashimoto, Miki Kurita, Andrea Roxanne J. Anas, Hitoshi Uemura, Kiyomi Tsuji and Ken-Ichi Harada
Microbial Degradation of Amino Acid-Containing Compounds Using the Microcystin-Degrading Bacterial Strain B-9
Reprinted from: *Mar. Drugs* **2018**, *16*, 50, doi:10.3390/md16020050 285

Xinghao Yang, Xiao Xiao, Dan Liu, Ribang Wu, Cuiling Wu, Jiang Zhang, Jiafeng Huang, Binqiang Liao and Hailun He
Optimization of Collagenase Production by *Pseudoalteromonas* sp. SJN2 and Application of Collagenases in the Preparation of Antioxidative Hydrolysates
Reprinted from: *Mar. Drugs* **2017**, *15*, 377, doi:10.3390/md15120377 297

Shangyong Li, Linna Wang, Ximing Xu, Shengxiang Lin, Yuejun Wang, Jianhua Hao and Mi Sun
Structure-Based Design and Synthesis of a New Phenylboronic-Modified Affinity Medium for Metalloprotease Purification
Reprinted from: *Mar. Drugs* 2017, 15, 5, doi:10.3390/md15010005 313

Jiayi Wang, Jing Lin, Yunhui Zhang, Jingjing Zhang, Tao Feng, Hui Li, Xianghong Wang, Qingyang Sun, Xiaohua Zhang and Yan Wang
Activity Improvement and Vital Amino Acid Identification on the Marine-Derived Quorum Quenching Enzyme MomL by Protein Engineering
Reprinted from: *Mar. Drugs* 2019, 17, 300, doi:10.3390/md17050300 327

Yanan Li, Xue Kong and Haibin Zhang
Characteristics of a Novel Manganese Superoxide Dismutase of a Hadal Sea Cucumber (*Paelopatides* sp.) from the Mariana Trench
Reprinted from: *Mar. Drugs* 2019, , 84, doi:10.3390/md17020084 341

Yatong Wang, Yanhua Hou, Yifan Wang, Lu Zheng, Xianlei Xu, Kang Pan, Rongqi Li and Quanfu Wang
A Novel Cold-Adapted Leucine Dehydrogenase from Antarctic Sea-Ice Bacterium *Pseudoalteromonas* sp. ANT178
Reprinted from: *Mar. Drugs* 2018, 16, 359, doi:10.3390/md16100359 353

About the Editor

Antonio Trincone graduated with honors in Pharmacy at the University of Naples, discussing a thesis on the biogenesis of membrane lipids of archaebacteria. He is currently a senior researcher, and has been working at the Istituto di Chimica Biomolecolare belonging to Consiglio Nazionale delle Ricerche, Italy, in Naples, since 1983. He has been Professor of Organic Chemistry, and in charge for several years at the University of Salerno, Italy. Antonio Trincone is Specialty Chief Editor of Marine Biotechnology in Frontiers Marine Science. He has been the editor of a book entitled 'Biocatalysis: Chemistry and Biology' (Research Signpost, India); he is a reviewer for different scientific journals and research projects. He is part of the Editorial Board of *Marine Drugs*, and has guest-edited different Special Issues for this journal. Antonio Trincone was on the Scientific Committee for the 1st Symposium on Marine Enzymes and Polysaccharides (MEP'12), held in NhaTrang, Vietnam, from December 10th to 17th, 2012. He edited, among others, the book "Marine enzymes for biocatalysis: Sources, biocatalytic characteristics and bioprocesses of marine enzymes".

Preface to "Marine Enzymes"

Prominent conclusions by many scientists in the field of marine biotechnology emphasize that, due to marine biological diversity and to the specificity of biological marine metabolisms, the study of biocatalysts on a global scale from this environment is just starting, and possesses huge potential for the development of applications with industrial benefits. The oceans are the world's largest ecosystem with a biodiversity undescribed by science, at a degree more than 90%. Only a deep understanding of the complexity of this ecosystem will enable human beings to protect the oceans and organisms populating them, and pave the way for the sustainable exploitation of marine resources. This knowledge constitutes the core of marine biotechnology, and will certainly fuel various applications. Many fields are covered that are highly relevant to societal challenges: (i) in a biorefinery value-chain with marine enzymes for biochemical processes; (ii) in food industries for enzymatic procedures in seafood processing; (iii) in fields of fine chemicals—in pharmaceutical, cosmetics, agriculture and environmental sectors—enzymatic treatments are a tool to improve efficiency and selectivity for extraction/manipulation of structurally complex marine molecules, to gain access to bioactive compounds, and to provide complex core blocks for hemisynthesis; (iv) the field of marine biomarkers and applications in pollution monitoring (biosensor) and bioremediation could also be of high significance for the appreciation of marine sources for enzymes. However, many challenges remain, including a deep comprehension of the "marine biotechnology landscape" and a multidisciplinary approach, not only in education and training.

This book starts with two review articles—in the first, a refined search of literature is adopted with respect to previous surveys, centering generally on the enzymatic process more than on a single novel activity developing the analysis, according to the biotechnological field of applications; in the second, a more selective overview of the state of the art in marine and freshwater microalgal enzymes with potential biotechnological applications is provided, discussing future perspectives for this field. With recent advances, in fact, it is becoming easier to identify sequences encoding targeted enzymes, increasing the likelihood of the identification, heterologous expression, and characterization of these enzymes of interest, by genomics and transcriptomic approaches.

A list of six articles is dedicated to a specific enzymatic activity, alginate lyase, that is important in the production of alginate oligosaccharides for their recognized activities as anticoagulants, antioxidants, antineoplastics, plant growth accelerators and tumor inhibitors, in food, agricultural, and medical fields. The great biodiversity of marine source is important for the urgency to obtain alginate lyases with the optimal characteristics (e.g., pH-stability, thermo-tolerance, and single product distribution) needed for industrial applications.

A further batch of six articles are always dedicated to carbohydrate active marine enzymes, including two other additional reports on galactosidases, thus totalling more than half a book dedicated to this class of enzymes.

Among the interesting other-articles, the one reporting the asymmetric reduction of ketones using the growing and resting cells of 13 marine-derived fungi shows the importance of these organisms in this field and the possibility of adopting a simple system that offers alternative, highly enantioselective and minimally polluting properties to important enantiomeric pure alcohols.

Antonio Trincone
Editor

Review

Enzymatic Processes in Marine Biotechnology

Antonio Trincone

Istituto di Chimica Biomolecolare, Consiglio Nazionale delle Ricerche, Via Campi Flegrei, 34, 80078 Pozzuoli, Naples, Italy; antonio.trincone@icb.cnr.it; Tel.: +39-081-867-5095

Academic Editor: Keith B. Glaser
Received: 17 February 2017; Accepted: 20 March 2017; Published: 25 March 2017

Abstract: In previous review articles the attention of the biocatalytically oriented scientific community towards the marine environment as a source of biocatalysts focused on the habitat-related properties of marine enzymes. Updates have already appeared in the literature, including marine examples of oxidoreductases, hydrolases, transferases, isomerases, ligases, and lyases ready for food and pharmaceutical applications. Here a new approach for searching the literature and presenting a more refined analysis is adopted with respect to previous surveys, centering the attention on the enzymatic process rather than on a single novel activity. Fields of applications are easily individuated: (i) the biorefinery value-chain, where the provision of biomass is one of the most important aspects, with aquaculture as the prominent sector; (ii) the food industry, where the interest in the marine domain is similarly developed to deal with the enzymatic procedures adopted in food manipulation; (iii) the selective and easy extraction/modification of structurally complex marine molecules, where enzymatic treatments are a recognized tool to improve efficiency and selectivity; and (iv) marine biomarkers and derived applications (bioremediation) in pollution monitoring are also included in that these studies could be of high significance for the appreciation of marine bioprocesses.

Keywords: marine enzymes; biocatalysts; bioprocesses; biorefinery; seafood; marine biomarkers

1. Introduction

Currently there is enormous interest in marine biotechnology with the worldwide flourishing of editorial initiatives (journals, books, etc.) hosting important experimental results and surveys from several projects, especially those belonging to the FP7 program and to the topic "Blue growth" of H2020, potentially providing clues that will aid development of enabling technologies in the field. The oceans are the world's largest ecosystem, covering more than 70% of the earth's surface. They host the greatest diversity of life in unexplored habitats. The Census of Marine Life, evaluating marine biodiversity, ascertained that at least 50% and potentially more than 90% of marine species are undescribed by science [1]. Only a deep understanding of the complexity of the marine ecosystem will enable human beings to protect the oceans and organisms populating them, and pave the way for sustainable exploitation of marine resources. This knowledge will certainly fuel various applications and itself constitutes the core of marine biotechnology. Increasingly labeled "blue biotechnology", this wide field covers many aspects that are highly relevant to societal challenges, as is well established in the EU Framework Programme Horizon 2020. Several outlined emerging technologies are (i) robotics; (ii) miniaturized solutions for marine monitoring; (iii) biomimetics; (iv) acoustics; (v) nanobiotechnlogy; (vi) renewable energy harvesting (wave energy, algae biofuels); and (vii) high-performance computing. However, many challenges remain, including a deep comprehension of the "marine biotechnology landscape" and a multidisciplinary approach, not only in education and training [2].

Marine sources (microorganisms in general and symbionts in particular, extremophiles, fungi, plants and animals) are of scientific interest in that the origin of marine biomolecules (i.e., biocatalysts) features all marine bioprocesses. Knowledge of these biocatalysts and their habitat-related properties such as salt tolerance, hyperthermostability, barophilicity and cold adaptivity (of great interest for industry) is necessary for bioprocesses exploitation. One of the most explicative aspects is related to the stereo-chemical properties of a marine enzyme. Substrate specificity and affinity, as evolved properties that are linked to the metabolic functions of the enzymes, are key aspects. Two review articles already appeared in 2010 [3] and in 2011 [4] with different focuses on these topics. The first [3], to draw the attention of the biocatalytically oriented scientific community to the marine environment as a source of biocatalysts; in fact the discussion was mainly about the specific diversity of molecular assets of biocatalysis that were recognized with respect to terrestrial counterparts. The second review [4] spotlighted habitat-related properties from a biochemical point of view, also reporting on important examples in bioprocesses. Various updates of these former analyses of the literature have also recently been published, such as the one by Lima et al. 2016 [5] including marine examples of oxidoreductases, hydrolases, transferases, isomerases, ligases and lyases ready for food and pharmaceutical applications.

In the present review, a new approach for searching the literature and presenting a more refined analysis is adopted with respect to previous surveys. The focus of the literature search is centered on the enzymatic process more than on a single novel activity. This survey is developed according to the biotechnological field of applications where bioprocesses, based on marine enzymes and/or marine biomasses, are central. Focusing on enzymatic processes rather than on single activities helsd us to recognize the fields of application. For the first, a biorefinery value-chain, the provision of biomass is one of the most important aspects, with aquaculture as the prominent sector. In the food industry the interest in the marine domain is similarly developed to deal with the enzymatic procedures adopted in food manipulation. Moreover, as for the selective and easy extraction/modification of structurally complex marine molecules, enzymatic treatments are a recognized tool to improve efficiency and selectivity in fine chemistry processes to get access to bioactive compounds and provide complex core blocks ready for hemisynthesis. In closing, the field of marine biomarkers and derived applications (bioremediation) in pollution monitoring could be of high significance for the appreciation of marine bioprocesses.

In the fields indicated above, the selected primary articles are presented in tabulated form, picking up different aspects of importance in short explicative notes to avoid a huge amount of text. Selected modern review articles are listed under each paragraph to depict the present state of the art of the related field.

2. Literature Search

A survey of the literature has been conducted mainly by using the database ScienceDirect with access to 3800 scientific journals in major scientific disciplines. The search was based on two terms: (i) "enzymatic processing" (in abstract, title or keywords) and (ii) "marine" (in all fields). How to manipulate these queries to get an effective result in terms of the number of hits was first investigated using the search functions offered by the database. In particular, the W/in function (proximity operator) has been found useful for the first query, with the keywords used in the query sorted into phrases (low numbers), sentences (medium) and paragraphs (high) of the hit. This result is then refined by using the AND operator with the word "marine" in all fields. The resulting pattern of these searches is as follows: W/3 (137 hits), W/4 (153), W/5 (161), W/15 (256) and for W/50 (442). As a comparison, a more general coverage alternative search was also used, adopting the same keywords, in the Scopus database. It resulted in a similar score (478 hits), thus confirming the choice of ScienceDirect as the reference database when using W/50. This has been the value adopted for searching for articles for this review.

An interesting detail is the yearly distribution of the hits (shown in Figure 1 below). Two time intervals can be easily recognized; published results are doubled yearly from 2012 to 2016 with respect to the previous score in the interval 1993 to 2011, characterized by fewer than 20 hits per year. This picture reflects the strategic efforts of various funded programs created by the European Commission to support and foster research in the European Research Area and similar actions in other parts of the world. That logical reason for the variation of the results in the two intervals 1993–2011 and 2012–2016 further confirms the suitability of the keywords adopted.

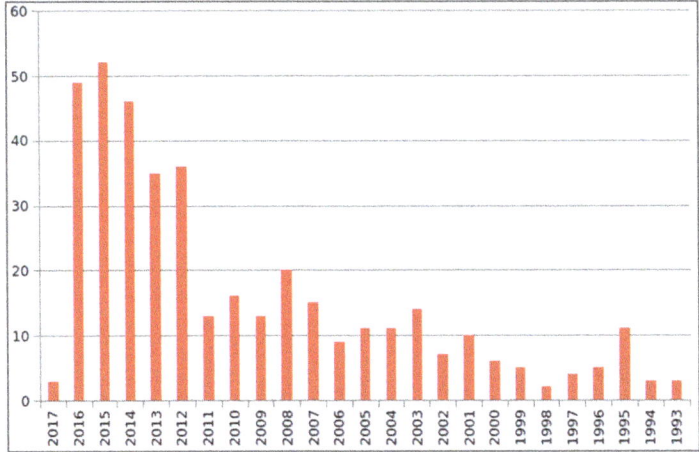

Figure 1. Yearly distribution of the number of hits in the search for articles in this review (see text for details). Fewer than 20 hits per year characterizes the interval 1993–2011; a doubled value is seen for 2012–2016.

3. Biorefinery

A future sustainable economy based on renewable resources is the main point of the concept of a biorefinery. Research and development studies have been underway for many years in different parts of the world to replace a large fraction of fossil resources. The most important aspect of a biorefinery value-chain is the provision of biomass with a consistent and regular supply of renewable carbon-based raw materials [6]. One of the sectors providing these feedstocks is aquaculture (algae and seaweeds), which, together with biochemical processes (marine enzymes in pre-treatment) adopted, is an important aspect of the domain of marine biotechnology. Needless to say, the focus in many review articles [7] is on the importance of extremophiles and thermostable enzymes to overcome the limitations of biocatalysts in current bioprocesses for lignocellulosic biomass conversion. In this context, it is also of interest to mention the features of marine biocatalysts related to the ecological features of the habitat in which marine organisms thrive. Generally the resulting enzymatic properties are very important from a biotechnological point of view [4,8].

The idea here is to depict both the state of the art about marine enzyme-based bioprocesses and the importance of marine-originating feedstocks in biorefinery. Therefore, biocatalysts and biomass are the two fundamental elements on which the analysis of primary articles in the literature is based here (Table 1 [9–42]). The selected articles inserted in Table 1 deal with (i) cellulases and other important carbohydrate-active enzymes; (ii) lipases, to manipulate feedstock oils for biodiesel production and (iii) other biocatalysts, including those commercially available. However, to depict the current state of interest when only marine feedstock exploitation is present, chemical treatments were also listed in these selected modern articles (entries 13–34, Table 1).

Table 1. Biorefinery.

Entry	Reference	Biocatalyst(s)	Biomass	Notes
		Cellulases and other important carbohydrate active hydrolases		
1/2015	[9]	Cellulase of marine fungus *Cladosporium sphaerospermum*	Seaweed biomass *Ulva fasciata*	Cellulases found are active and stable in different ionic liquids
2/2014	[10]	Lignocellulose depolymerizing multi-enzyme complex: lignin peroxidase, xylanase and cellulases		13 microbial marine isolates from seaweed were studied. They belong to the genera *Brachybacterium*, *Brevibacterium*, *Halomonas*, *Kokuria*, *Micrococcus*, *Nocardiopsis*, *Pseudomonas* and *Streptomyces*
3/2014	[11]	Cellulase from a marine bacterium, *Bacillus carboniphilus*		Optimization study of saccharification using marine microbial cellulase
4/2013	[12]	Cellulase from a marine *Bacillus* sp. H1666	*Ulva lactuca* macro algae is studied for cellulase treatment	Isolated enzyme has saccharification applicability on *Ulva lactuca*
5/2009	[13]	Cellulase isolated from a marine bacterium, *Bacillus subtilis* subsp. *subtilis* A-53		
6/2014	[14]	κ-carrageenase CgkA and λ-carrageenase CglA from *Pseudoalteromonas carrageenovora*	Carrageenan from red algae	Improvement of the process of degradation by the study of functional carrageenolytic complex
7/2016	[15]	Endo-type β-agarase AgaG1, screened from *Alteromonas* sp. GNUM1, exo-type β-agarase DagB from *S. coelicolor* A3 and a α-neoagarobiose hydrolase from *Alcanivorax* sp.	Agarose	Enzymatic agarose hydrolysis process without acid pretreatment
8/2012	[16]	Mimicked the natural agarolytic pathway using three microbial agarases (Aga16B, Aga50D and DagA) and NABH	Recalcitrant agar polysaccharide	
		Feedstock oils to biodiesel		
9/2016	[17]	-	Marine microalgae	Optimization study for disruption of thraustochytrid cell using bead mill for maximising lipid extraction yields and hydrolysis of oil extracted studied using commercial lipases
10/2013	[18]	Bacterial isolate *Flammeovirga yaeyamensis*	Oil-rich microalga (*Chlorella vulgaris* ESP-1)	Cell-wall destruction analyzed by SEM micrographs was associated with the activity of hydrolytic enzymes

Table 1. Cont.

Entry	Reference	Biocatalyst(s)	Biomass	Notes
		Feedstock oils to biodiesel		
11/2015	[19]	Chlorella protothecoides		Study of fermentations developed to produce lipid by heterotrophic C. protothecoides using cassava bagasse as the low-cost feedstock
12/2014	[20]		Nannochloropsis oceanica	Crude algal oils were extracted from the oleaginous microalga Nannochloropsis oceanica
		Marine feedstock valorization		
13/2015	[21]	Three enzymes used: alcalase, neutrase and esperase	Effluents obtained from chemical and enzymatic chitin production of Illex argentinus pen byproducts	Study for production of lactic acid bacteria, marine probiotic bacteria and two common gram (+) bacteria using effluents as substrates
14/2015	[22]	Chemical treatment	Mussel processing wastewaters	Laboratory optimization to transform mussel processing wastewater into a growth culture medium to produce microbial biomass. The lab-scale processes studied were upscaled to a pre-industrial level using a 70-L fermenter
15/2015	[23]	Sulfuric acid was seen as the best catalyst with a lipid conversion efficiency of 44.9%	Marine microalga Chlorella sp. BDUG 91771	-
16/2016	[24]	Pretreatment with hydrogen peroxide	Seaweed Ulva prolifera	Optimization study
17/1994	[25]	Commercial bacterial inoculum (Stabisil)	Crustacean shell waste from the world's fishing industry	Optimization study for recovery of protein, pigment and chitin from waste shell of prawn
18/2017	[26]	Cellulase and pectinase	Porphyridium cruentum, red microalgae	Evaluation of bioethanol production in response culture conditions of to Porphyridium cruemtum. Enzymatic hydrolysis resulted in high glucose conversion yields for both seawater and freshwater conditions
19/2016	[27]	Commercial cellulases and alginases	Brown algae Macrocystis pyrifera	Study of various pretreatments
20/2016	[28]	Mixture of commercial enzymes: Viscozyme® L, Cellic® CTec2, Cellic® HTec2	Macro-algae Gracilaria verrucosa	
21/2016	[29]	Chemical and enzymatic process with commercial cellulase and β-glucosidase	Red macroalgae Eucheuma cottonii	
22/2015	[30]	Celluclast and commercial alginate lyase (EC4.2.2.3) from Sphingobacterium spiritivorum were used	Brown seaweed Laminaria digitata	

Table 1. Cont.

Entry	Reference	Biocatalyst(s)	Biomass	Notes
		Feedstock oils to biodiesel		
23/2015	[31]	Commercial cellulase, xylanase and β-glucosidase	Nannochloropsis salina	Anaerobic digestion study
24/2015	[32]	Commercially available enzymes (pectinase) and enzyme mixtures (Accellerase 1500, Accellerase XC, and Accellerase XY) with multiple enzyme activities (exoglucanase, endoglucanase, hemi-cellulase, and β-glucosidases were used)	Nannochloropsis salina	Study of conversion of lipid-extracted biomass into fermentable sugars
25/2014	[33]	Commercial cellulase	Red algae Kappaphycus alvarezii	Optimization study
26/2014	[34]	Commercial Viscozyme L and Cellic CTec2	Marine green macro-algae Enteromorpha intestinalis	Study of hydrotermal method
27/2014	[35]	Free and immobilized yeast	Red alga Gracilaria sp.	A study for bioethanol production using hydrolisate of Gracilaria
28/2014	[36]	Yeast fermentation	Microalga Chlorella vulgaris	Pectinase enzyme was used for disrupting microalgal cells
29/2013	[37]	Saccharomyces fermentation	Red alga Gracilaria verrucosa	A study for the combined production of agar and bioethanol; the pulp was used after agar extraction
30/2013	[38]	Saccharomyces fermentation	Red algae Kappaphycus alvarezii	105 L of ethanol per ton of seaweed were obtained after a dilute acid pretreatment
31/2012	[39]	Yeast fermentation	Aquatic plant Zostera marina	Study of the potential of this plant as a source of bioactives and sugars for bioethanol production
32/2011	[40]	Saccharomyces fermentation	Byproduct from the alginate extraction process	Interesting study for exploitation of seaweed waste from alginate production
33/2015	[41]	-	Red alga Gracilaria verrucosa	Optimization study of this suitable feedstock for biosugar production
34/2015	[42]	β-1,3-glucanase, cellulase and β-glucosidase were studied	Red seaweed Grateloupia turuturu	Enzyme-assisted extraction of R-phycoerythrin with ultrasound technology

Non-conventional sources of cellulase enzymes have been sought for a long time. As mentioned in an old report, the identification of these biocatalysts in the marine fungus *Dendryphiella arenaria* [43] dates back 40 years. However, of recent interest are the complex biopolymers in microalgae cells subjected to breakdown in biological pretreatments, as reported in a more modern review [44]. The focus is not only on cellulases as the most-explored specific and efficient enzymes (entries 1–5 Table 1), but also on other hydrolytic enzymes, including hemicellulase, pectinase, protease and amylase, even in the form of an enzymatic cocktail seen as an effective tool with respect to single enzymes as well. Entire microbial communities associated with marine organisms are studied (entry 2 Table 1). Chitinases are very well represented in research, too, and new insights into the disruption of crystalline polysaccharides were gained, as seen in a recent review report [45]. Substrate-disrupting accessory non-hydrolytic proteins are novel tools to improve molecular accessibility to polymers with increased process efficiency. Other primary articles, dealing with different carbohydrate active hydrolases, are listed in Table 1 (entries 6–8).

Obviously lipases are also considered a convenient tool for converting a wide range of feedstock oils into biodiesel [46]. A study of 427 yeast strains from seawater, sediment, mud of salterns, guts of the marine fish and marine algae should also be mentioned here [47]. Industrial yeast *Yarrowia lipolytica* of marine origin is a biocatalyst of interest in metabolic engineering studies used to expand the substrate range [48]. Entries 9–12 in Table 1 are interesting primary articles along this line of research. Microalgae cultivation and macroalgae developed as pests due to eutrophication are generally seen as potential resources for biofuel production [49]. Another interesting aspect is the combination of macroalgae cultivation exploiting nutrients coming from marine aquaculture or other processes.

In industrial squid manufacturing for chitin production, a large volume of protein effluents containing peptones from alkaline and enzymatic hydrolysis of the pens are substrates used as a nitrogen source to reduce the cost of marine probiotic bacteria cultivation (entry 13, Table 1). A similar approach was also reported for effluents originating from the industrial thermal treatment of mussels (entry 14, Table 1); many other examples on these lines are reported here (entries 15–33, Table 1), where studies on different biomasses were conducted often by comparing chemical and enzymatic procedures and yeast fermentations adopted for bioethanol production. An enzymatic cocktail of glucanase, cellulase and glucosidase was studied for R-phycoerythrin extraction, assisted by ultrasound technology from the red seaweed *G. turuturu* (entry 34, Table 1) proliferating along the French coast.

Marine biomass-centered studies are also listed in the section about food industry development. Additionally, a particular and interesting aspect studied is the use of algicidal microorganisms to improve cell disruption during biotechnological processes aimed at producing biofuels. Secreted algicidal substances to be used as microalgae breakdown agents are reviewed [50].

4. Food Applications

Enzymatic procedures in food processing are mostly based on biocatalysts of terrestrial or microbial origin, with new enzymes currently obtained from these environments. For years the marine domain has been seen as a promising source of interesting biocatalysts [51] for modern applications. However, the enzymatic activities present in seafood or in byproducts were utilized for centuries in the traditional production of various cured and fermented seafood, allowing the preparation of numerous sauces and pastes from the time of the Greeks and Romans and also seafood processes in the Far East [52].

The recovery and processing of waste is generally a challenge in the food industry. Especially in the seafood sector, a more complete utilization of the raw material with minimization of the inherent problems of pollution and waste treatment is a current issue. Many reviews or chapters in books depicting the state of the art of this specific topic were found. One of the oldest reports found during our search dates back to 1978; interestingly, it already pointed out bioconversion tools for the processing and valorization of food waste in the conversion to useful products [53]. Upgrading of sea byproducts is, however, still central in more modern analysis [54], with more attention dedicated nowadays to therapeutic potential. Attention to antihypertensive and immunomodulatory agents (i.e., peptides obtained by enzymatic hydrolysis of fish proteins) is recognized more than the simple nutritional and biological properties of these materials. Both have recently been investigated in marine mussels [55]. While for mussel proteins the focus is on peptides obtained by bioprocesses, lipids (PUFAs) are also investigated for the prevention and treatment of rheumatoid arthritis. The use of all-natural stabilizers for food, in the form of (enzymatically) muscle-derived extracts, appears interesting, as well as the addition of plant extracts or pure phenolic compounds to combat oxidation in seafood [56]. Basic research related to the enzymatic processes occurring in seafood material has also been traced; a study of blackening processes in freeze-thawed prawns during storage is of interest. The respiratory pigment hemocyanin is converted into a phenoloxidase-like enzyme and acts as a potent inducer of post-harvest blackening; these discoveries are helpful for the development of anti-blackening treatments for these foods [57]. Biotechnologists are also interested in the large availability of seafood raw materials; byproducts from waste in processing (liver, skin, head, viscera, trimmings, etc.) amount to 60% or more [58] of total renewable raw material. Review articles were found for general aspects [59,60] and a particular focused on seaweeds [61].

The food and biorefinery sections overlap because byproducts, primarily used as feed with low returns, are also thought to be useful for biodiesel generation. Some articles could thus be listed in both sections; however, repetition is avoided here and also in this case tabulation is based on the same fundamental key for the analysis of literature adopted previously in Table 1; enzyme (marine or commercial)-based bioprocesses and types of marine-originating feedstocks are the two columns of Table 2, including a third one for notes [62–96]. Commercial or marine-originating proteases and lipases are most used for production in this field and a wide range of edible biomass for the valorization study is listed in Table 2. They are also an excellent, low-cost source for enzyme production [97].

An enzymatic approach can overcome the environmental impact of traditional processes and make the processes sustainable and cost-effective. In a very recent comprehensive review, it is reported that significant developments can be expected for enzyme applications in the fish and seafood industries [98] in the near future.

Table 2. Food applications.

Entry	Reference	Biocatalyst(s)	Biomass	Notes
1/2007	[62]	Antibacterial alkaline protease	Fish processing waste	Action exerted by cell lysis of pathogenic bacteria
2/2016	[63]	Alcalase for oil extraction	*Thunnus albacares* byproducts (heads)	Study for deodorization of fish oil
3/2016	[64]	Commercial alcalase	Shrimp waste	Response surface methodology study to grow hydrocarbon-degrading bacteria *Bacillus subtilis*
4/2016	[65]	Enzymatic deproteinization by commercial enzyme savinase	Norway lobster (*Nephrops norvegicus*) processing byproducts	Chitin and chitosan production
5/2016	[66]	Commercial enzymes used for the preparation of seaweed: Celluclast and Alcalase	Brown seaweed *Ecklonia radiata*	Study of extraction methods of the alga and potential in vitro prebiotic effect
6/2016	[67]	Commercial alcalase	Wastewater generated during shrimp cooking	Study for the production of enzymic hydrolysates with antioxidant capacity and production of essential amino acids
7/2016	[68]	Commercial alcalase	Head byproducts of *Prionace glauca*	Production of chondroitin sulphate from blue shark waste was studied after cartilage hydrolysis with alcalase
8/2016	[69]	Alcalase	Adhesive gum layer surrounding naturally fertilised ballan wrasse (*Labrus bergylta*) eggs	A study for the biological control (by cleaner fish *Labrus bergylta*) of sea lice in farming *Salmo salar*
9/2016	[70]	Commercial lipase B from *C. antarctica* (Lipozyme 435, immobilized lipase)	Sardine oil	Sardine oil was evaluated by glycerolysis using commercial lipase to produce monoacyl glycerols rich in omega-3 polyunsaturated fatty acids
10/2015	[71]	Proteases from *Bacillus subtilis* A26 (TRMH-A26), *Raja clavata* crude alkaline protease extract, alcalase and neutrase	Thornback ray (*Raja clavata*) muscle	Study of bioactivity of extracts after proteolytic hydrolysis with different enzyme preparations
11/2015	[72]	-	Brines marinated herring (*Clupea harengus*)	Brines from marinated herring processing used for recovery of useful material
12/2015	[73]	Commercial proteases	Atlantic salmon (*Salmo salar*) rest raw materials	Study of the production of different hydrolysates using commercial enzymes for the valorization of viscera-containing raw material from Atlantic salmon
13/2015	[74]	Marine proteases	Red scorpionfish (*Scorpaena scrofa*) viscera	Alkaline proteases of marine origin suggested for detergent formulations and deproteinization of shrimp shells
14/2015	[75]	Commercial alcalase, pepsin and trypsin	Common carp (*Cyprinus carpio*) egg	Hydrolysates improve the immune system with differential influences on the immune function. Interesting study for several applications in the health food, pharmaceutical, and nutraceutical industries

Table 2. Cont.

Entry	Reference	Biocatalyst(s)	Biomass	Notes
15/2014	[76]	Hydrolysis by bromelain	Protein byproducts of seaweed (*Gracilaria* sp.)	Set up of a flavouring agent with umami taste and seaweed odour
16/2014	[77]	Different commercial proteases	Fresh herring byproducts	Enzymatic hydrolysis to produce fish protein hydrolysates and separate oil
17/2014	[78]	Hydrolysis by commercial proteases	Cod (*Gadus morhua*) fillets	Study of influences of oxidative processes during protein hydrolysis using cod
18/2013	[79]	Proteolytic processing with commercial proteases	Fractions obtained from processing of Atlantic rock crab (*Cancer irroratus*) byproducts	Small peptides with biological activity recovered
19/2013	[80]	Commercial alcalase	Protein concentrates recovered from cuttlefish processing wastewater	Selective ultrafiltration methods under study for concentrating active components with antihypertensive and antioxidant activities
20/2013	[81]	Commercial alcalase	Tuna dark muscle	Basic study for fractionation of protein hydrolysates with ultrafiltration and nanofiltration
21/2012	[82]	Proteases and lipases from marine waste	Byproducts of Monterey sardine (*Sardinops sagax caerulea*) processing	Actions of enzymes from sardine byproduct (viscera and byproduct concentrate extracts) produced 3-fold greater hydrolysis than with the commercial enzyme
22/2012	[83]	Trypsin and alcalase	Waste byproducts of red seaweed *Porphyra columbina*	Study on protein water extracts wasted during traditional phycolloids extraction procedure from *P. columbina*. Interesting immunosuppressive effects and antihypertensive and antioxidant activities found
23/2012	[84]	Commercial alcalase	Fish byproducts	Comparison of methods including enzymatic extraction
24/2009	[85]	Proteolytic commercial enzyme mix	Snow crab (*Chionoecetes opilio*) byproduct fractions	Pilot scale enzymatic hydrolysis to entire snow crab byproducts followed by fractionation operations in order to recover enriched fractions of proteins, lipids and chitin
25/2009	[86]	Commercial alcalase preparations	*Gadus morhua* skin collagen	Optimization of parameters for the hydrolysis
26/2008	[87]	Three types of enzymes used: papain, trypsin and pepsin	Wastewater from the industrial processing of octopus	Marine peptones as promising alternatives to expensive commercial medium for growth of lactic acid bacteria
27/2008	[88]	Commerical proteases	*Dosidicus gigas* mantle	Tenderization of mantle for commercial use as substitute of *Illex argentinus*

Table 2. Cont.

Entry	Reference	Biocatalyst(s)	Biomass	Notes
28/2007	[89]	Alcalase	Shrimp processing discards	Isolation and characterisation of a natural antioxidant from shrimp waste
29/2005	[90]	Alcalase, Lecitase® a carboxylic ester hydrolase with inherent activity towards both phospholipid and triacylglycerol structures	Cod (*Gadus morhua*) byproducts	Study for protein and oil extractions
30/2005	[91]	Flavourzyme, a fungal protease/peptidase complex produced by *Aspergillus oryzae*, and Neutrase	Cod (*Gadus morhua*) byproducts	Composition of products generated by hydrolyses of byproducts of cod processing for optimization and design on desired product
31/2003	[92]	Crude papain was selected to perform the enzymatic extraction	Skate cartilage	Study for a low-cost process for glycosaminoglycan extraction from skate cartilage
32/2002	[93]	Umamizyme (commercial endo-peptidase activity from a strain of *A. oryzae*)	Tuna waste	Study for evaluation of activity of Umamizyme in comparison to other fungal enzymes
33/2001	[94]	Alcalase	Yellowfin tuna (*Thunnus albacares*) waste	Study of hydrolysis of tuna stomach proteins
34/2016	[95]	Six commercial enzyme mixtures and individual enzymes were used: Viscozyme® L, Celluclast® 1.5 L, Ultraflo® L and the three proteases Alcalase® 2.4 L FG, Neutrase® 0.8 L and Flavourzyme® 1000 L.	Brown alga *Ecklonia radiata*	Study of enzyme-assisted extraction of carbohydrates for the design and optimization of processes to obtain oligo- and polysaccharides
35/2014	[96]	Proteolytic preparations from *Bacillus mojavensis* A21, *Bacillus subtilis* A26, *Bacillus licheniformis* NH1, *B. licheniformis* MP1, *Vibrio metschnikovii* J, *Aspergillus clavatus* ES1 and crude alkaline protease extracts from Sardinelle (*Sardinella aurita*), Goby (*Zosterisessor ophiocephalus*) and Grey triggerfish (*Balistes capriscus*) prepared and characterized by the group	Shrimp processing byproducts	Enzymatic deproteinization for extraction of chitin

5. Fine Chemistry and Lab Techniques

Biocatalytic procedures using marine enzymes for the production of fine chemicals are an important aspect of this review. The production and manipulation of complex biomolecules benefited from important biotechnological features characterizing biocatalyzed reactions with respect to the use of purely chemical-based methods [99].

Technological improvements (metagenomics) applied to bioprospecting in understudied environments help to identify a greater repertoire of novel biocatalysts with complementarity about properties (stereochemistry, resistance, etc.). Marine enzymes offer hyperthermostability, salt tolerance, barophilicity, cold adaptability, chemoselectivity, regioselectivity and stereoselectivity [4], thus acting as useful and new alternatives to terrestrial biocatalysts in use. Particular importance is represented by enzymes showing resistance to organic solvents, with the examples from marine environments mostly related to halophilic proteins (salt reduces water activity, like organic solvent systems), as analyzed in a recent, comprehensive review article [100].

As for carbohydrate-active enzymes, selective and easy manipulation of structurally complex marine polysaccharides provides homogeneous core blocks (oligosaccharides) for analysis and hemisynthesis. This constitutes the core of a sustainable process when using renewable resources. Other preminent examples of enzymatic treatments as a tool to improve the extraction efficiency of specific bioactive compounds from seaweeds were recently reviewed [101] and laboratory techniques in the preparation of compounds for further research are discussed in a recent report [102] listing enzymes for the functionalization of chitosan such as polyphenoloxidases (PPO) (tyrosinases, laccases) and peroxidases (POD); examples from the marine environment are indicated.

Due to our specific design of search terms in querying literature databases, the coverage of articles dealing with the simple prospecting, isolation and identification of new marine biocatalyst(s) is partial; however, however those found are included in this section. Indeed, most articles listed in Table 3 [103–134] focused on bioprocesses for the synthesis of useful products and on enzymatic routes adopted for setting up laboratory techniques to study marine complex biomolecules (e.g., improving extraction, digestion of polysaccharides to simple components for structural determination, etc.). Various examples are found in the literature of the synthesis and hydrolysis of glycosidic bonds. Entry 8, Table 3 [110] is only one of the examples of synthetic strategies for the production of interesting products (enzymatically glycosylated natural lipophilic antioxidants). Polymers with synthetic carbohydrates have a wide range of applications in medical biotechnology as new biomaterials [135] and carbohydrate-active hydrolases can also be applied; moreover, these biocatalysts are important for the synthesis of a number of novel dietary carbohydrates in food technology, for the production of chromophoric oligosaccharides of strictly defined structure as valuable biochemical tools, etc. Other numerous applications oriented to vegetal waste treatment in recovering useful materials are reported in the section devoted to biorefinery.

Tabulation of primary articles in this section (Table 3) is based on biocatalyst(s) used, product(s) obtained with the biocatalyzed process, or evaluating of marine feedstock(s); comments about the contents of the article are also reported in the notes in Table 3. Entries 1–18 (Table 3) are related to carbohydrate-active hydrolases, while a few (entries 19–23) are listed for ester hydrolysis and proteolytic activities (entries 31 and 32).

Table 3. Fine chemistry and laboratory techniques.

Entry	Reference	Biocatalyst(s)	Product(s)	Feedstocks	Notes
		Carbohydrate active hydrolases			
1/2016	[103]	Alkaline β-agarase from marine bacterium *Stenotrophomonas* sp. NTa	From agarose as substrate neoagarobiose, neoagarotetraose and neoagarohexaose are the predominant products	-	First evidence of extracellular agarolytic activity in *Stenotrophomonas*, the enzyme exhibited stability across a wide pH range and resistance against some inhibitors, detergents and denaturants
2/2016	[104]	Cloned novel chitinase from a marine bacterium *Paenicibacillus barengoltzii* functionally expressed in *E. coli*	The chitinase hydrolyzed colloidal chitin to yield mainly N-acetyl chitobiose	Chitin (from crab shells)	Production of 21.6 mg·mL^{-1} of N-acetyl chitobiose from colloidal chitin with the highest conversion yield of 89.5% (w/w)
3/2015	[105]	Chitinase from the marine-derived *Pseudoalteromonas tunicata* CCUG 44952T	Active also on chromogenic substrate pNP-(GlcNAc) but not on pNP-(GlcNAc)$_2$ and pNP-(GlcNAc)$_3$	Colloidal and crystalline chitin	The recombinant enzyme exhibited antifungal activity against phytopathogenic and human pathogenic fungi, (biofungicide)
4/2014	[106]	Commercial pectinase or acidic hydrolysis	3-deoxy-D-*manno*-oct-2-ulosonic acid (Kdo): a sugar that is difficult to obtain by chemical synthesis and that has applications in medicinal chemistry	Marine microalgae, *Tetraselmis suecica*	Evaluation of *T. suecica* as feedstock for a KDO production
5/2014	[107]	α-amylase from marine *Nocardiopsis* sp. strain B2	-	-	Study for immobilization of a marine α-amylase by ionotropic gelation technique using gellan gum (GG)
6/2014	[108]	Endo- and exo-glucanases from marine sources: endo-1,3-β-D-glucanase (LIV) from *Pseudocardium sachalinensis* and the exo-1,3-β-D-glucanase from *Chaetomium indicum*	Different fractions of oligosaccharides	Laminaran from brown alga *Eisenia bicyclis*	Study for anticancer activity of the native laminaran and products of its enzymatic hydrolysis
7/2014	[109]	Amylolytic system in the digestive fluid of the sea hare, *Aplysia kurodai*	Maltotriose, maltose, and glucose	Sea lettuce (*Ulva pertusa*)	Enzymatic analysis of the amylolytic system in the digestive fluid of the sea hare *Aplysia kurodai* and efficient production of glucose from sea lettuce
8/2012	[110]	α-glucosidase from *Aplysia fasciata*	Glucosylated anti-oxidant derivatives of hydroxytyrosol	-	Biocatalytic production of mono- and disaccharide derivatives at final concentrations of 9.35 and 10.8 g/L of reaction
9/2006	[111]	Endo-1,3-β-D-glucanases (laminarinases) from marine mollusks *Spisula sachalinensis* and *Chlamys albidus*	Biologically active 1,3;1,6-β-D-glucan, called translam	Hydrolysis of laminaran	Study of immobilization
10/2006	[112]	Commercial enzymes	N-acetyl chitobiose	Various chitin substrates α-chitin from shrimp waste	Experimental conditions studied to achive 10% N-acetyl chitobiose

Table 3. Cont.

Entry	Reference	Biocatalyst(s)	Product(s)	Feedstocks	Notes
			Carbohydrate active hydrolases		
11/2004	[113]	1→3-β-D-glucanase LIV from marine mollusk *Spisula sachalinensis* and α-D-galactosidase from marine bacterium *Pseudoalteromonas* sp. KMM 701	Oligo- and polysaccharide derivatives possessing immunostimulating, antiviral, anticancer and/or radioprotective activity	Laminaran from the brown seaweeds *Laminaria cichorioides*	Immobilization study
12/1996	[114]	Chitin degrading enzymes from sea water bacterium strain identified as *Alteromonas*	β-(1→6)-(GlcNAc)$_2$	Chitin and chito-oligosaccharides	High transglycosylation activity of the enzyme preparation was also confirmed
13/2016	[115]	Endolytic alginate lyases	4-deoxy-L-erythro-5-hexoseulose uronic acid	Alginate and alginate oligosaccharides	In depth study of degradation process from alginate to unsaturated monosaccharides
14/2015	[116]	Ulvan-degrading bacterial β-lyase from a new *Alteromonas* species	Sulfated oligosaccharides from the seaweed *Ulva*	Ulvan	Fractions of molecular weight down to a 5 kDa of oligosaccharides mix are obtained
15/2014	[117]	Extracellular β-agarases from *Agarivorans albus* OAY2	Neoagarobiose NA$_2$, neoagarotetraose NA$_4$ and neoagarohexaose NA$_6$		Report about enzyme purification and oligosaccharides preparation
16/2007	[118]	Glycosyl hydrolases in crude extracts from extremophilic marine bacterium *Thermotoga neapolitana* (DSM 4359)	(β-1,4)-xylooligosaccharides of 1-hexanol, 9-fluorene methanol, 1,4-butanediol and geraniol		Transglycosylation reactions by xylose, galactose, fucose, glucose and mannose enzymatic transfers
17/2012	[119]	Commercial α-amylase	Carrageenan-derived oligosaccharide	Hydrolysis of κ-carrageenan	
18/2007	[120]	β-N-acetyl-D-glucosaminidase from prawn *Penaeus vannamei*			Mechanistic and inhibition studies
			Ester hydrolysis		
19/1995	[121]	Fungal deacetylase	Hexa-N-deacetylchitohexaose	Natural or artificial chitin substrates as well as N-acetylchito-oligosaccharides	Enzymatic deacetylation: methodological study
20/2012	[122]	Commercial immobilized lipase, lipozyme from *Thermomyces lanuginosa*	Diglycerides and monoglycerides containing polyunsaturated fatty acids	Menhaden oil	Enzymatic ethanolysis of menhaden oil
21/2010	[123]	Organism isolated from marine sediments	Fatty acid-based biopolymer	Triglycerides of sunflower, soybean, olive, sesame and peanut as substrates	Hydrolysis of triglycerides and dimerization of fatty acid to anhydrides and subsequent formation of a Fatty acid based biopolymer (FAbBP)
22/2007	[124]	Commercial enzymes	Acylglycerol synthesis	N-3 PUFA from tuna oil	-
23/2016	[125]	Novel marine microbial esterase PHE14	Asymmetric synthesis of D-methyl lactate by enzymatic kinetic resolution	Racemic methyl lactate commercially available	Esterase PHE14 exhibited very good tolerance to most organic solvents, surfactants and metal ions

Table 3. Cont.

Entry	Reference	Biocatalyst(s)	Product(s)	Feedstocks	Notes
			Oxidoreductases		
24/2014	[126]	Lipoxygenase/hydroperoxide lyase	Polyunsaturated aldehydes: 2,4,7-decatrienal and 2,4-decadienal	Macroalgal genus *Ulva* (Ulvales, Chlorophyta)	*Ulva mutabilis* is selected as cultivable for production
25/2012	[127]	Marine fungi *Aspergillus sclerotiorum* CBMAI 849 and *Penicillium citrinum* CBMAI 1186	Reduction of 1-(4-methoxyphenyl)-ethanone to its stereochemical pure alcohol (ee > 99%, yield = 95%)		Immobilization study
26/2003	[128]	Hydrogenase	Enzymatic production and regeneration of NADPH		6.2 g·L^{-1} NADPH produced with a total turnover number (ttn: mol produced NADPH/mol consumed enzyme) of 10,000
27/2003	[129]	Lipoxygenase–hydroperoxide lyase pathway	C6 and C9 unsaturated aldehydes	Brown alga *Laminaria angustata*	Study of biosynthetic pathway
28/2004	[130]	Cultures of the haptophyte microalga *Chrysotila lamellosa*	Alkanediones	Regiospecific oxygenation of alkenones	Biogenetic study
29/1996	[131]	Enzymatic extract of the marine gorgonian *Pseudopterogorgia americana*	9(11)-secosteroids	Cholesterol, stigmasterol and progesterone	Claimed as the first chemoenzymatic preparation of a natural product using the enzymatic machinery of a marine invertebrate
30/2012	[132]	Bromoperoxidase of brown alga *Ascophyllum nodosum*	4-bromopyrrole-2-carboxylate	Bromination of methyl pyrrole-2-carboxylate in bromoperoxidase II-catalyzed oxidation	Bromoperoxidase II mimics biosynthesis of methyl 4-bromopyrrole-2-carboxylate, a natural product isolated from the marine sponge *Axinella tenuidigitata*
			Proteolytic activities		
31/2004	[133]	Proteolytic enzymes	Products of proteolysis	Gastric fluid of the marine crab, *Cancer pagurus*	Influence of metal ions and organic solvents other than pH and temperature are analyzed, including long-term stability over a period of several months
32/2006	[134]	Alkaline serine protease		Marine gamma-Proteobacterium	Activity in presence of up to 30% NaCl. Water miscible and immiscible organic solvents like ethylene glycol, ethanol, butanol, acetone, DMSO, xylene and perchloroethylene enhance as well as stabilize the enzyme activity

Little is known about the distribution and diversity of *Candida* genus in marine environments, with only a few species isolated from these environments [136]. *C. rugosa* and *C. antarctica* are among the terrestrial yeasts, so biocatalysis-related articles dealing with these quite famous commercial lipases are not listed here. Interesting examples of oxidoreductases are also found in the literature, both for direct biocatalytic applications (including examples of immobilized enzymes) and for biogenetically related studies (see entries 24–30, Table 3).

Finally, it is worth mentioning in this section a remarkable review focused on the biosynthesis of oxylipins in non-mammals. Biocatalysts involved in pathways related to these biomolecules carry interesting and unusual catalytic properties for biocatalysis. A detailed biocatalytic knowledge of enzymatic catalysis in these reactions is needed to plan a possible direct in vitro biocatalytic lab-scale production of useful products [137]. With the similar aim of increasing knowledge about natural enzymes with interesting features, a review was compiled on the enzymatic breakage of dimethylsulfoniopropionate [138]. The compound, a zwitterionic osmolyte produced by corals, marine algae and some plants in massive amounts (ca. 10^9 tons per year), is transformed by marine microbes and bioprocesses involving this molecule and derivatives are of interest for assessing the ability of relevant enzymes to realize these transformations.

6. Sediments and Bioremediation

Among the studies on marine sources for enzymes, the fields of marine biomarkers and bioremediation applications are of high significance in this context [139].

In the list of parameters that are usually considered when assessing exposure to environmental pollutants in aquatic ecosystems, biotransformation enzymes (phase I and II), biotransformation products and stress proteins are of high interest for enzymatic processing. It is of value, for example, that a unique set of protein expression (signature) for exposure to different chemical compounds has been recognized for *Mytilus edulis* and the expressed proteins identified participate in α- and β-oxidation pathways, xenobiotic and amino acid metabolism, cell signalling and oxyradical metabolism [140]. Bioremediation as an enabling technology exploits naturally occurring organisms that with their metabolic ability are able to transform toxic substances in less hazardous compounds that in turn are included in biogeochemical cycles. Stereochemical aspects play an important role due to the fact that homochirality appears to be a requirement for the functioning of enzymes with specific (partial) incorporation of stereoforms. Enantioselective chromatographic separation of chiral environmental xenobiotics is covered in an interesting review. The study includes microbial transformation of chiral pollutants in aquatic ecosystems, enzymatic transformations, etc. [141]. An additional report states the effectiveness of enzymatic processes in bioremediation even though limited application is evidenced with respect to stability and the cost of biocatalysts. Marine enzymes are seen as a solution to this challenge. In particular, marine fungi and their laccases are used in the textile industry, mostly to deal with salty effluents [142].

A recent report of a European project in this field characterizes novel hydrocarbon degrading microbes isolated from the southern side of the Mediterranean Sea. Exploiting and managing the diversity and ecology of microorganisms thriving in these polluted sites is a major objective in terms of increasing knowledge of the bioremediation potential of these poorly investigated sites [143].

In Table 4 primary articles in the field are selected [144–168], indicating the biocatalyst(s) or organism(s) exploited and details of application, with notes about the results and importance of the work.

Table 4. Sediments and bioremediation.

Entry	Reference	Biocatalyst(s)/Organism(s)	Application(s)	Notes
1/2016	[144]	Marine microbial community for manganese oxidation	-	Several new genera associated with Mn(II) oxidation were found in metal-contaminated marine sediments and are seen as a solution for metal bioremediation
2/2013	[145]	Marine sedimentary bacterial communities	Anaerobic degradation of mixtures of isomeric pristenes and phytenes	Several bacterial products of transformation confirm the key role played by hydration in the metabolism of alkenes
3/2010	[146]	MAP kinase signaling pathway	-	Different pollutants generated different patterns of induction of the biomarker MAPK phosphorylation
4/2003	[147]	Isolate (isolate TKW) of sulfate-reducing bacteria	Reduction of chromate (CrO_4^{2-})	Soluble hexavalent chromium (Cr^{6+}) enzymatically transformed into less toxic and insoluble trivalent chromium (Cr^{3+}) with potential in bioremediation of sediments contaminated by metals
5/2013	[148]	Giant freshwater prawn Macrobrachium rosenbergii	Study for potential biomarkers of exposure to organophosphorus pollutants: molecular and immunological responses	Investigation on the effects of the pesticide trichlorfon used in aquaculture, on molecular and enzymatic processes related to the response of the giant freshwater prawn, Macrobrachium rosenbergii
6/1985	[149]	Five fish species: Salmo gairdneri, S. trutta, Galaxias maculatus, G. truttaceus and G. auratus	Study of detoxication enzyme activities for a fungicide, chlorothalonil	Metabolism of chlorothalonil
7/1986	[150]	-	Role of enzymatic processes in the metabolism of organic matter	Proteolysis and aerobic oxidation of organic material
8/1992	[151]	-	Study for distinction between enzymatic and non-enzymatic degradation pathways in marine ecosystems	Enzymatic degradation pathways for α-hexachlorocyclohexane
9/1995	[152]	Antioxidant enzymes in mussel, Mytilus galloprovincialis	Use of mussels as bioindicators in monitoring heavy metal pollution	Adaptation as a compensatory mechanism in chronically polluted organisms was found
10/1995	[153]	Dioxygenase	Dioxygenase pathway with subsequent conjugation and excretion	Metabolism of benzo[a]pyrene in a freshwater green alga, Selenastrum capricornutum
11/2000	[154]	Polychaete worms	Sulphide detoxification by polychaete worms Marenzelleria viridis (Verrill 1873) and Hediste diversicolor (O.F. Müller)	Detoxification end-product is thiosulphate
12/2001	[155]	-	-	Biodegradation of extracellular organic carbon by bacteria in sediments

Table 4. Cont.

Entry	Reference	Biocatalyst(s)/Organism(s)	Application(s)	Notes
13/2004	[156]	Antioxidant enzyme activities and lipid peroxidation in the gills of the hydrothermal vent mussel *Bathymodiolus azoricus*	Study of enzymatic defences (superoxide dismutase (SOD), catalase (CAT), total glutathione peroxidase (Total GPx) and selenium-dependent glutathione peroxidase (Se–GPx) and lipid peroxidation *against metals*	Assessment of physiological adaptation to continuous metal exposure in natural environment
14/2004	[157]	Marine glycosidases	Biostimulation of enzymatic activities (glycosidases) by oxygen supply	Enzymatic activity increased when oxygenation was increased and the supply of oxygen into the sediment enhanced enzymatic degradation rates
15/2005	[158]	Antioxidant enzymes in a model organism *Daphnia magna*	Study of age-related biochemical changes in aquatic organism	General evaluation of importance of oxidative stress in aging
16/2006	[159]	European eel (*Anguilla anguilla*) exposed to persistent organic pollutants	Detection of early warning responses to pollutant exposition	Metabolic responses including detoxification mechanisms (biotransformation, antioxidant process) in European eel (*Anguilla anguilla*) exposed to persistent organic pollutants
17/2007	[160]	Catalase, superoxide dismutase, glutathione peroxidase, glutathione reductase, glutathione S-transferases in wild populations of mussels (*Mytilus galloprovincialis*)	Study of biochemical response to to petrochemical environmental contamination	Environmental monitoring programmes to get data that could be used as a baseline reference during oil accidents
18/2007	[161]	Acetylcholinesterase, catalase, and glutathione-S-transferase (GST) of blue mussels (*Mytilus edulis*)	Study of specific reaction to exposure to nodularin	Acetylcholinesterase activity, catalase (CAT) activity and glutathione-S-transferase (GST) in blue mussels (*Mytilus edulis*) exposed to an extract made of natural cyanobacterial mixture containing toxic cyanobacterium *Nodularia spumigena*
19/2007	[162]	Induction of biotransformation enzymes in *Sparus aurata*	Relationship between specific molecular processes (induction of enzymes) and the behavioral performance of fish is of great interest in understanding the impact of PAHs at increasing levels of biological complexity	The study investigates biochemical response to phenantrene in *Sparus aurata*

Table 4. Cont.

Entry	Reference	Biocatalyst(s)/Organism(s)	Application(s)	Notes
20/2012	[163]	Tow deep-sea fish species, namely *Alepocephalus rostratus* and *Lepidion lepidion* and the decapod crustacean *Aristeus antennatus*	-	Study of hepatic biomarkers (ethoxyresorufin-O-deethylase, EROD, pentoxyresorufin-O-deethylase PROD, catalase CAT, carboxylesterase CbE, glutathione-S-transferase GST, total glutathione peroxidase GPX and glutathione reductase demonstrating seasonal variation despite constant temperature and salinity
21/2013	[164]	*Platichthys flesus*	Transcriptomic study	Assessment of hepatic transcriptional differences between fish exposed to mixture of brominated diphenyl ethers and controls
22/2015	[165]	Peptidase activities	A study of enzymes using labeled substrate in diverse regions of the ocean	Enzymatic capabilities differ in Pelagic–benthic environments, affecting the processing of marine organic matter
23/2015	[166]	Arginine kinase of the crustacean *Exopalaemon carinicauda*	Bioinformatic study. Insights on the role of Cr^{3+} on enyzme with respect to inhibition and aggregation with structural disruption	An investigation of the effect of Cr^{3+} on enzymes of seawater organisms providing information on the physiological role of metal pollution in marine environments
24/2015	[167]	Oxidoreductases and catalases in *Bacillus safensis*	Isolation and enzyme identification study	The organism is responsible for degradation of the petroleum aromatic fractions
25/2016	[168]	Dehydrogenase activity or peroxidase activity	A surface methodology study	Degradation of crude oil fitted linearly with increasing biomass and enzyme activities with growth

7. Others

Searching within records used in this review for the names of class of enzymes, oxidoreductases, transferases, hydrolases and lyases are equally represented, while isomerases and ligases are less often used as keywords. However, only 30% of the total records contain such names for classes, with common names of biocatalysts more often adopted in titles, abstracts and keywords. Thus, in this section a few other topics, difficult to insert in previous sections, are covered.

Enzymes acting on 1,4 glycosidic bonds between galacturonic acid residues in pectin were investigated for the improvement of banana fiber processing. At least one actinomycete strain of *Streptomyces lydicus* collected from estuarine and marine areas in India was found to be a potent producer of polygalacturonase [169].

In a study to reduce the cost of aquaculture of fish cobia by adding crustacean processing waste, an investigation of the endogenous chitinolytic enzymes in cobia was conducted. It suggested substantial endogenous production of enzymes of the chitinolytic system and that the activity from chitinolytic bacteria was not significant [170].

An interesting study based on enzymology tools was conducted for demonstrating the enhanced mixing processes between the sediment and the overlying waters of the Delaware Estuary. The authors used fluorescently labeled polysaccharides to determine the effects of suspended sediment transport on water column hydrolytic activities [171].

In another report, a rapid and easy-to-use set, composed of semi-quantitative kits, was adopted for the investigation of the heterotrophic bacterial community in meadows of *Posidonia oceanica* during environmental surveys. Although the set is composed of known kits (ApiZym galleries, Biolog microplates and BART™ tests) principal enzymatic activities, metabolic capabilities and benthic mineralisation processes were all studied [172].

A particular aspect is related to carbonic anhydrases; it is worth mentioning in this review since it was recently discussed in a book devoted to these enzymes from extremophiles and the possible biotechnological applications for which they can be used [173]. Among the CO_2 sequestration methods proposed in order to capture and concentrate CO_2 from combustion gases, the biomimetic approach [174] could indeed benefit from the diversity of marine carbonic anhydrases. However, these metallo-enzymes are also discussed for their potential as novel biomarkers in environmental monitoring and the development of biosensors for metals [175], and another interesting aspect, encompassing at least two of the fields listed in this review, is the investigation of other enzymatic activities possessed by carbonic anhydrases [176].

8. Conclusions

One of the recent transnational calls of ERA-NET (ERA-MBT), an action funded under the EU FP7 program, was focused on biorefinery and entitled "The development of biorefinery processes for marine biomaterials". The projects were required to develop the production of a large number of different products and novel processes through the application of biotechnological knowledge. Further technological developments to improve the integration and optimisation of the processing steps were required. In this realm seaweed biomass seems to have great potential as a raw material in a biobased economy, offering advantages such as no competion with food production, absence of fertilisers or pesticides in the value-chain and positive relapses removing an excess of nutrients from marine environments. However, a few key areas where development is still needed are technology for the improvement of large-scale cultivation and fractionation, and the identification of new marine microbial strains to break down macroalgal polysaccharide.

The use of macroalgae in food preparation is a classical topic in food applications found when inspecting the primary articles listed in Table 2. Macroalgae contain a high concentration of minerals, vitamins, trace elements and fibre and have low fat content. Different projects studying these aspects are in development. It is of interest that a close inspection of the affiliations of the corresponding authors of all articles in Table 2 reveals that Spain and France are the top two countries where research efforts were published; however, many countries produce interesting research results.

As for fine chemistry and laboratory techniques, different case studies recently illustrated the importance of biocatalysis, considering the specificities of marine enzymes with respect to their terrestrial counterparts [177]. Ketone reduction and epoxide hydrolases useful in organic synthesis appeared to be central for stereochemical aspects. Access to bioactive aldehydes with lipooxygenases and lipases actions and biodegradation of marine pollutants are covered, along with other lipolytic activities used for enantioselective hydrolysis. From the overall analysis and examples reported, the strategy regarding the potential of marine habitats is clear. It is important to report the note of two editors in a special issue dedicated to biocatalysis in *Current Opinion in Chemical Biology*; they stated that "At least a third of this planet's biomass resides in the oceans, and the rules of the marine biochemical game seem to be fundamentally different than those described in our biochemistry textbooks." [178].

In the bioremediation field, [141] is very stimulating, pointing out the importance of stereoselectivity in aquatic ecosystem biotransformations and reporting on the great impact that the use of cyclodextrins in chiral gas-chromatography had in widening this knowledge. All the information reported by the studies on processes is of great interest to biocatalysis practitioners, starting from pioneering investigations into the distribution of atropisomeric PCBs in the marine environment. Marine bacteria specialising in the degradation of hydrocarbons have been isolated from polluted seawater and some of these bacteria can grow on these substrates; they represent an extraordinary archive of mono- and dioxygenases, oxidases, dehydrogenases and other enzymatic activities that can be applied in regio- and stereoselective biocatalysis. There is a gap between the general knowledge from the studies in this section and the specificity/suitability required for preparative enzymatic processes; bridging this gap could shed more light on the useful features of the enzymes involved in this type of pollutant biotransformation, thus enabling more effective application.

Fewer than half of the results from our literature search are cited in the references list. About 20% of the hits are represented by books or reference works that were hardly used here. Additionally, only four scientific journals hosted more than 10 articles of the remaining corpus: *Bioresource Technology*, *Food Chemistry*, Process *Biochemistry* and *Chemosphere* together account fornot far off 100 articles. The top specialist marine-oriented journal was *Algal Research*, with only eight papers. Inserting representative primary articles in the tables and excluding the ones that do not belong resulted in the current total number of references. Therefore, scientific interest in marine enzymatic processing can be considered successfully published in non-specialized journals such as the ones cited above; the separation of fields, as adopted here, is only for ease of discussion. This has already been mentioned above for marine biomass-centered studies that could have been listed under food applications or biorefinery, or for enzymatic activities used for polysaccharide manipulation that could have been listed under fine chemistry and lab techniques instead of biorefinery.

In conclusion, all these aspects point to a final consideration of the importance of an interdisciplinary network in setting up successful research projects enabling the identification of an arsenal of enzymes and pathways greatly in demand for biotechnological applications. In continuing this research effort, further refining of the scientific literature could be of interest; exploration of the fields individuated above should be continued in depth, in specialized journals, in a manner that could help to reveal sub-fields along with more details pointing to a single process with room to discuss a single enzymatic activity.

Acknowledgments: The financial support for bibliographic search facilities is provided by CNR funding to Istituto di Chimica Biomolecolare.

Conflicts of Interest: The author declares no conflict of interest.

References

1. Census of Marine Life. Available online: http://www.coml.org (accessed on 23 March 2017).
2. Trincone, A. Increasing knowledge: The grand challenge in marine biotechnology. *Front. Mar. Sci.* **2014**, *1*. [CrossRef]
3. Trincone, A. Potential biocatalysts originating from sea environments. *J. Mol. Catal. B Enzym.* **2010**, *66*, 241–256. [CrossRef]
4. Trincone, A. Marine Biocatalysts: Enzymatic features and applications. *Mar. Drugs* **2011**, *9*, 478–499. [CrossRef] [PubMed]
5. Lima, R.N.; Porto, A.L.M. Recent advances in marine enzymes for biotechnological processes. In *Marine Enzymes Biotechnology: Production and Industrial Applications, Part I—Production of Enzymes*; Kim, S.-K., Toldra, F., Eds.; Academic Press Elsevier: New York, NY, USA, 2016; pp. 153–192.
6. Cherubini, F. The biorefinery concept: Using biomass instead of oil for producing energy and chemicals. *Energy Convers. Manag.* **2010**, *51*, 1412–1421. [CrossRef]
7. Bhalla, A.; Bansal, N.; Kumar, S.; Bischoff, K.; Sani, R. Improved lignocellulose conversion to biofuels with thermophilic bacteria and thermostable enzymes. *Bioresour. Technol.* **2013**, *128*, 751–759. [CrossRef] [PubMed]
8. Menon, V.; Rao, M. Trends in bioconversion of lignocellulose: Biofuels, platform chemicals & biorefinery concept. *Prog. Energy Combust. Sci.* **2012**, *38*, 522–550.
9. Trivedi, N.; Reddy, C.; Radulovich, R.; Jha, B. Solid state fermentation (SSF)-derived cellulase for saccharification of the green seaweed Ulva for bioethanol production. *Algal Res.* **2015**, *9*, 48–54. [CrossRef]
10. Satheeja Santhi, V.; Bhagat, A.; Saranya, S.; Govindarajan, G.; Jebakumar, S. Seaweed (*Eucheuma cottonii*) associated microorganisms, a versatile enzyme source for the lignocellulosic biomass processing. *Int. Biodeterior. Biodegrad.* **2014**, *96*, 144–151. [CrossRef]
11. Annamalai, N.; Rajeswari, M.; Balasubramanian, T. Enzymatic saccharification of pretreated rice straw by cellulase produced from *Bacillus carboniphilus* CAS 3 utilizing lignocellulosic wastes through statistical optimization. *Biomass Bioenergy* **2014**, *68*, 151–160. [CrossRef]
12. Harshvardhan, K.; Mishra, A.; Jha, B. Purification and characterization of cellulase from a marine *Bacillus* sp. H1666: A potential agent for single step saccharification of seaweed biomass. *J. Mol. Catal. B Enzym.* **2013**, *93*, 51–56. [CrossRef]
13. Kim, B.; Lee, B.; Lee, Y.; Jin, I.; Chung, C.; Lee, J. Purification and characterization of carboxymethylcellulase isolated from a marine bacterium, *Bacillus subtilis subsp. subtilis* A-53. *Enzyme Microb. Technol.* **2009**, *44*, 411–416. [CrossRef]
14. Kang, D.; Hyeon, J.; You, S.; Kim, S.; Han, S. Efficient enzymatic degradation process for hydrolysis activity of the carrageenan from red algae in marine biomass. *J. Biotechnol.* **2014**, *192*, 108–113. [CrossRef] [PubMed]
15. Seo, Y.B.; Park, J.; Huh, I.Y.; Hong, S.-K.; Chang, Y.K. Agarose hydrolysis by two-stage enzymatic process and bioethanol production from the hydrolysate. *Process Biochem.* **2016**, *51*, 759–764. [CrossRef]
16. Kim, H.; Lee, S.; Kim, K.; Choi, I. The complete enzymatic saccharification of agarose and its application to simultaneous saccharification and fermentation of agarose for ethanol production. *Bioresour. Technol.* **2012**, *107*, 301–306. [CrossRef] [PubMed]
17. Byreddy, A.; Barrow, C.; Puri, M. Bead milling for lipid recovery from thraustochytrid cells and selective hydrolysis of *Schizochytrium* DT3 oil using lipase. *Bioresour. Technol.* **2016**, *200*, 464–469. [CrossRef] [PubMed]
18. Chen, C.-Y.; Bai, M.-D.; Chang, J.-S. Improving microalgal oil collecting efficiency by pretreating the microalgal cell wall with destructive bacteria. *Biochem. Eng. J.* **2013**, *81*, 170–176. [CrossRef]
19. Chen, J.; Liu, X.; Wei, D.; Chen, G. High yields of fatty acid and neutral lipid production from cassava bagasse hydrolysate (CBH) by heterotrophic *Chlorella protothecoides*. *Bioresour. Technol.* **2015**, *191*, 281–290. [CrossRef] [PubMed]
20. Wang, Y.; Liu, J.; Gerken, H.; Zhang, C.; Hu, Q.; Li, Y. Highly-efficient enzymatic conversion of crude algal oils into biodiesel. *Bioresour. Technol.* **2014**, *172*, 143–149. [CrossRef] [PubMed]
21. Vázquez, J.; Caprioni, R.; Nogueira, M.; Menduíña, A.; Ramos, P.; Pérez-Martín, R. Valorisation of effluents obtained from chemical and enzymatic chitin production of *Illex argentinus* pen by-products as nutrient supplements for various bacterial fermentations. *Biochem. Eng. J.* **2015**, *116*, 34–44. [CrossRef]
22. Prieto, M.; Prieto, I.; Vázquez, J.; Ferreira, I. An environmental management industrial solution for the treatment and reuse of mussel wastewaters. *Sci. Total Environ.* **2015**, *538*, 117–128. [CrossRef] [PubMed]

23. Mathimani, T.; Uma, L.; Prabaharan, D. Homogeneous acid catalysed transesterification of marine microalga *Chlorella* sp. BDUG 91771 lipid–an efficient biodiesel yield and its characterization. *Renew. Energy* **2015**, *81*, 523–533. [CrossRef]
24. Li, Y.; Cui, J.; Zhang, G.; Liu, Z.; Guan, H.; Hwang, H.; Aker, W.; Wang, P. Optimization study on the hydrogen peroxide pretreatment and production of bioethanol from seaweed *Ulva prolifera* biomass. *Bioresour. Technol.* **2016**, *214*, 144–149. [CrossRef] [PubMed]
25. Healy, M.; Romo, C.; Bustos, R. Bioconversion of marine crustacean shell waste. *Resour. Conserv. Recycl.* **1994**, *11*, 139–147. [CrossRef]
26. Kim, H.M.; Oh, C.H.; Bae, H.-J. Comparison of red microalagae (*Porphyridium cruentum*) culture conditions for bioethanol production. *Biores. Technol.* **2017**, *233*, 44–50. [CrossRef] [PubMed]
27. Ravanal, M.; Pezoa-Conte, R.; von Schoultz, S.; Hemming, J.; Salazar, O.; Anugwom, I.; Jogunola, O.; Mäki-Arvela, P.; Willför, S.; Mikkola, J.; et al. Comparison of different types of pretreatment and enzymatic saccharification of *Macrocystis pyrifera* for the production of biofuel. *Algal Res.* **2016**, *13*, 141–147. [CrossRef]
28. Kwon, O.; Kim, D.; Kim, S.; Jeong, G. Production of sugars from macro-algae *Gracilaria verrucosa* using combined process of citric acid-catalyzed pretreatment and enzymatic hydrolysis. *Algal Res.* **2016**, *13*, 293–297. [CrossRef]
29. Tan, I.; Lee, K. Comparison of different process strategies for bioethanol production from *Eucheuma cottonii*: An economic study. *Bioresour. Technol.* **2016**, *199*, 336–346. [CrossRef] [PubMed]
30. Hou, X.; Hansen, J.; Bjerre, A. Integrated bioethanol and protein production from brown seaweed *Laminaria digitata*. *Bioresour. Technol.* **2015**, *197*, 310–317. [CrossRef] [PubMed]
31. Bohutskyi, P.; Chow, S.; Ketter, B.; Betenbaugh, M.; Bouwer, E. Prospects for methane production and nutrient recycling from lipid extracted residues and whole *Nannochloropsis salina* using anaerobic digestion. *Appl. Energy* **2015**, *154*, 718–731. [CrossRef]
32. Mirsiaghi, M.; Reardon, K. Conversion of lipid-extracted *Nannochloropsis salina* biomass into fermentable sugars. *Algal Res.* **2015**, *8*, 145–152. [CrossRef]
33. Abd-Rahim, F.; Wasoh, H.; Zakaria, M.; Ariff, A.; Kapri, R.; Ramli, N.; Siew-Ling, L. Production of high yield sugars from *Kappaphycus alvarezii* using combined methods of chemical and enzymatic hydrolysis. *Food Hydrocoll.* **2014**, *42*, 309–315. [CrossRef]
34. Kim, D.; Lee, S.; Jeong, G. Production of reducing sugar from *Enteromorpha intestinalis* by hydrothermal and enzymatic hydrolysis. *Bioresour. Technol.* **2014**, *161*, 348–353. [CrossRef] [PubMed]
35. Wu, F.; Wu, J.; Liao, Y.; Wang, M.; Shih, I. Sequential acid and enzymatic hydrolysis in situ and bioethanol production from *Gracilaria* biomass. *Bioresour. Technol.* **2014**, *156*, 123–131. [CrossRef] [PubMed]
36. Kim, K.; Choi, I.; Kim, H.; Wi, S.; Bae, H. Bioethanol production from the nutrient stress-induced microalga *Chlorella vulgaris* by enzymatic hydrolysis and immobilized yeast fermentation. *Bioresour. Technol.* **2014**, *153*, 47–54. [CrossRef] [PubMed]
37. Kumar, S.; Gupta, R.; Kumar, G.; Sahoo, D.; Kuhad, R. Bioethanol production from *Gracilaria verrucosa*, a red alga, in a biorefinery approach. *Bioresour. Technol.* **2013**, *135*, 150–156. [CrossRef] [PubMed]
38. Hargreaves, P.; Barcelos, C.; da Costa, A.; Pereira, N. Production of ethanol 3G from *Kappaphycus alvarezii*: Evaluation of different process strategies. *Bioresour. Technol.* **2013**, *134*, 257–263. [CrossRef] [PubMed]
39. Pilavtepe, M.; Sargin, S.; Celiktas, M.; Yesil-Celiktas, O. An integrated process for conversion of *Zostera marina* residues to bioethanol. *J. Supercrit. Fluids* **2012**, *68*, 117–122. [CrossRef]
40. Ge, L.; Wang, P.; Mou, H. Study on saccharification techniques of seaweed wastes for the transformation of ethanol. *Renew. Energy* **2011**, *36*, 84–89. [CrossRef]
41. Kim, S.; Hong, C.; Jeon, S.; Shin, H. High-yield production of biosugars from *Gracilaria verrucosa* by acid and enzymatic hydrolysis processes. *Bioresour. Technol.* **2015**, *196*, 634–641. [CrossRef] [PubMed]
42. Le Guillard, C.; Dumay, J.; Donnay-Moreno, C.; Bruzac, S.; Ragon, J.; Fleurence, J.; Bergé, J. Ultrasound-assisted extraction of R-phycoerythrin from *Grateloupia turuturu* with and without enzyme addition. *Algal Res.* **2015**, *12*, 522–528. [CrossRef]
43. Ladisch, M.; Lin, K.; Voloch, M.; Tsao, G. Process considerations in the enzymatic hydrolysis of biomass. *Enzyme Microb. Technol.* **1983**, *5*, 82–102. [CrossRef]
44. Carrillo-Reyes, J.; Barragán-Trinidad, M.; Buitrón, G. Biological pretreatments of microalgal biomass for gaseous biofuel production and the potential use of rumen microorganisms: A review. *Algal Res.* **2016**, *18*, 341–351. [CrossRef]

45. Eijsink, V.; Vaaje-Kolstad, G.; Vårum, K.; Horn, S. Towards new enzymes for biofuels: Lessons from chitinase research. *Trends Biotechnol.* **2008**, *26*, 228–235. [CrossRef] [PubMed]
46. Guldhe, A.; Singh, B.; Mutanda, T.; Permaul, K.; Bux, F. Advances in synthesis of biodiesel via enzyme catalysis: Novel and sustainable approaches. *Renew. Sustain. Energy Rev.* **2015**, *41*, 1447–1464. [CrossRef]
47. Wang, L.; Chi, Z.; Wang, X.; Liu, Z.; Li, J. Diversity of lipase-producing yeasts from marine environments and oil hydrolysis by their crude enzymes. *Ann. Microbiol.* **2007**, *57*, 495–501. [CrossRef]
48. Ledesma-Amaro, R.; Nicaud, J. Metabolic engineering for expanding the substrate range of *Yarrowia lipolytica*. *Trends Biotechnol.* **2016**, *34*, 798–809. [CrossRef] [PubMed]
49. Suganya, T.; Varman, M.; Masjuki, H.H.; Renganathan, S. Macroalgae and microalgae as a potential source for commercial applications along with biofuels production: A biorefinery approach. *Renew. Sust. Energy Rev.* **2016**, *55*, 909–941. [CrossRef]
50. Demuez, M.; González-Fernández, C.; Ballesteros, M. Algicidal microorganisms and secreted algicides: New tools to induce microalgal cell disruption. *Biotechnol. Adv.* **2015**, *33*, 1615–1625. [CrossRef] [PubMed]
51. Patel, A.; Singhania, R.; Pandey, A. Novel enzymatic processes applied to the food industry. *Curr. Opin. Food Sci.* **2016**, *7*, 64–72. [CrossRef]
52. Shahidi, F.; Janak Kamil, Y. Enzymes from fish and aquatic invertebrates and their application in the food industry. *Trends Food Sci. Technol.* **2001**, *12*, 435–464. [CrossRef]
53. Carroad, P.; Wilke, C. Enzymes and microorganisms in food industry waste processing and conversion to useful products: A review of the literature. *Resour. Recover. Conserv.* **1978**, *3*, 165–178. [CrossRef]
54. Cudennec, B.; Caradec, T.; Catiau, L.; Ravallec, R. Upgrading of sea by-products: Potential nutraceutical applications. In *Marine Medicinal Foods Implications and Applications—Animals and Microbes*; Kim, S.-K., Ed.; Academic Press: New York, NY, USA, 2012; Volume 65, pp. 479–494.
55. Grienke, U.; Silke, J.; Tasdemir, D. Bioactive compounds from marine mussels and their effects on human health. *Food Chem.* **2014**, *142*, 48–60. [CrossRef] [PubMed]
56. Undeland, I. Oxidative stability of seafood. In *Oxidative Stability and Shelf Life of Foods Containing Oils and Fats*; Hu, M., Jacobsen, C., Eds.; AOCS Press: Champaign, IL, USA, 2016; pp. 391–460.
57. Adachi, K.; Hirata, T.; Nagai, K.; Sakaguchi, M. Hemocyanin a most likely inducer of black spots in Kuruma prawn *Penaeus japonicus* during storage. *J. Food Sci.* **2008**, *66*, 1130–1136. [CrossRef]
58. Chalamaiah, M.; Dinesh kumar, B.; Hemalatha, R.; Jyothirmayi, T. Fish protein hydrolysates: Proximate composition, amino acid composition, antioxidant activities and applications: A review. *Food Chem.* **2012**, *135*, 3020–3038. [CrossRef] [PubMed]
59. Guerard, F. Enzymatic methods for marine by-products recovery. In *Maximising the Value of Marine By-Products Woodhead Publishing Series in Food Science, Technology and Nutrition*; Shahidi, F., Ed.; Woodhead Publishing: Cambridge, UK, 2007; pp. 107–143.
60. Sila, A.; Bougatef, A. Antioxidant peptides from marine by-products: Isolation, identification and application in food systems. A review. *J. Funct. Foods* **2016**, *21*, 10–26. [CrossRef]
61. Fleurence, J. Seaweeds as food. In *Seaweed in Health and Disease Prevention*; Fleurence, J., Levine, I., Eds.; Academic Press: San Diego, CA, USA, 2016; pp. 149–167.
62. Bhaskar, N.; Sudepa, E.; Rashimi, H.; Tamilselvi, A. Partial purification and characterization of protease of *Bacillus proteolyticus* CFR3001 isolated from fish processing waste and its antibacterial activities. *Bioresour. Technol.* **2007**, *98*, 2758–2764. [CrossRef] [PubMed]
63. De Oliveira, D.A.S.B.; Minozzo, M.G.; Licodiedoff, S.; Waszczynskyj, N. Physicochemical and sensory characterization of refined and deodorized tuna (*Thunnus albacares*) by-product oil obtained by enzymatic hydrolysis. *Food Chem.* **2016**, *207*, 187–194. [CrossRef] [PubMed]
64. Zhang, K.; Zhang, B.; Chen, B.; Jing, L.; Zhu, Z.; Kazemi, K. Modeling and optimization of Newfoundland shrimp waste hydrolysis for microbial growth using response surface methodology and artificial neural networks. *Mar. Pollut. Bull.* **2016**, *109*, 245–252. [CrossRef] [PubMed]
65. Sayari, N.; Sila, A.; Abdelmalek, B.E.; Abdallah, R.B.; Ellouz-Chaabouni, S.; Bougatef, A.; Balti, R. Chitin and chitosan from the Norway lobster by-products: Antimicrobial and anti-proliferative activities. *Int. J. Biol. Macromol.* **2016**, *87*, 163–171. [CrossRef] [PubMed]
66. Charoensiddhi, S.; Conlon, M.A.; Vuaran, M.S.; Franco, C.M.M.; Zhang, W. Impact of extraction processes on prebiotic potential of the brown seaweed *Ecklonia radiata* by in vitro human gut bacteria fermentation. *J. Funct. Foods* **2016**, *24*, 221–230. [CrossRef]

67. Tonon, R.V.; dos Santos, B.A.; Couto, C.C.; Mellinger-Silva, C.; Brígida, A.I.S.; Cabral, L.M.C. Coupling of ultrafiltration and enzymatic hydrolysis aiming at valorizing shrimp wastewater. *Food Chem.* **2016**, *198*, 20–27. [CrossRef] [PubMed]
68. Vázquez, J.A.; Blanco, M.; Fraguas, J.; Pastrana, L.; Pérez-Martín, R. Optimisation of the extraction and purification of chondroitin sulphate from head by-products of *Prionace glauca* by environmental friendly processes. *Food Chem.* **2016**, *198*, 28–35. [CrossRef] [PubMed]
69. Grant, B.; Picchi, N.; Davie, A.; Leclercq, E.; Migaud, H. Removal of the adhesive gum layer surrounding naturally fertilised ballan wrasse (*Labrus bergylta*) eggs. *Aquaculture* **2016**, *456*, 44–49. [CrossRef]
70. Solaesa, Á.G.; Sanz, M.T.; Falkeborg, M.; Beltrán, S.; Guo, Z. Production and concentration of monoacylglycerols rich in omega-3 polyunsaturated fatty acids by enzymatic glycerolysis and molecular distillation. *Food Chem.* **2016**, *190*, 960–967. [CrossRef] [PubMed]
71. Lassoued, I.; Mora, L.; Nasri, R.; Aydi, M.; Toldrá, F.; Aristoy, M.-C.; Barkia, A.; Nasri, M. Characterization, antioxidative and ACE inhibitory properties of hydrolysates obtained from thornback ray (*Raja clavata*) muscle. *J. Proteom.* **2015**, *128*, 458–468. [CrossRef] [PubMed]
72. Gringer, N.; Hosseini, S.V.; Svendsen, T.; Undeland, I.; Christensen, M.L.; Baron, C.P. Recovery of biomolecules from marinated herring (*Clupea harengus*) brine using ultrafiltration through ceramic membranes. *LWT Food Sci. Technol.* **2015**, *63*, 423–429. [CrossRef]
73. Opheim, M.; Šližytė, R.; Sterten, H.; Provan, F.; Larssen, E.; Kjos, N.P. Hydrolysis of Atlantic salmon (*Salmo salar*) rest raw materials—Effect of raw material and processing on composition, nutritional value, and potential bioactive peptides in the hydrolysates. *Process Biochem.* **2015**, *50*, 1247–1257. [CrossRef]
74. Younes, I.; Nasri, R.; Bkhairia, I.; Jellouli, K.; Nasri, M. New proteases extracted from red scorpionfish (*Scorpaena scrofa*) viscera: Characterization and application as a detergent additive and for shrimp waste deproteinization. *Food Bioprod. Process.* **2015**, *94*, 453–462. [CrossRef]
75. Chalamaiah, M.; Hemalatha, R.; Jyothirmayi, T.; Diwan, P.V.; Bhaskarachary, K.; Vajreswari, A.; Ramesh Kumar, R.; Dinesh Kumar, B. Chemical composition and immunomodulatory effects of enzymatic protein hydrolysates from common carp (*Cyprinus carpio*) egg. *Nutrition* **2015**, *31*, 388–398. [CrossRef] [PubMed]
76. Laohakunjit, N.; Selamassakul, O.; Kerdchoechuen, O. Seafood-like flavour obtained from the enzymatic hydrolysis of the protein by-products of seaweed (*Gracilaria* sp.). *Food Chem.* **2014**, *158*, 162–170. [CrossRef] [PubMed]
77. Šližytė, R.; Carvajal, A.K.; Mozuraityte, R.; Aursand, M.; Storrø, I. Nutritionally rich marine proteins from fresh herring by-products for human consumption. *Process Biochem.* **2014**, *49*, 1205–1215. [CrossRef]
78. Halldorsdottir, S.M.; Sveinsdottir, H.; Freysdottir, J.; Kristinsson, H.G. Oxidative processes during enzymatic hydrolysis of cod protein and their influence on antioxidant and immunomodulating ability. *Food Chem.* **2014**, *142*, 201–209. [CrossRef] [PubMed]
79. Beaulieu, L.; Thibodeau, J.; Bonnet, C.; Bryl, P.; Carbonneau, M.-É. Detection of antibacterial activity in an enzymatic hydrolysate fraction obtained from processing of Atlantic rock crab (*Cancer irroratus*) by-products. *Pharmanutrition* **2013**, *1*, 149–157. [CrossRef]
80. Amado, I.R.; Vázquez, J.A.; González, M.P.; Murado, M.A. Production of antihypertensive and antioxidant activities by enzymatic hydrolysis of protein concentrates recovered by ultrafiltration from cuttlefish processing wastewaters. *Biochem. Eng. J.* **2013**, *76*, 43–54. [CrossRef]
81. Saidi, S.; Deratani, A.; Ben Amar, R.; Belleville, M.-P. Fractionation of a tuna dark muscle hydrolysate by a two-step membrane process. *Sep. Purif. Technol.* **2013**, *108*, 28–36. [CrossRef]
82. Castro-Ceseña, A.B.; del Pilar Sánchez-Saavedra, M.; Márquez-Rocha, F.J. Characterisation and partial purification of proteolytic enzymes from sardine by-products to obtain concentrated hydrolysates. *Food Chem.* **2012**, *135*, 583–589. [CrossRef] [PubMed]
83. Cian, R.E.; Martínez-Augustin, O.; Drago, S.R. Bioactive properties of peptides obtained by enzymatic hydrolysis from protein byproducts of *Porphyra columbina*. *Food Res. Int.* **2012**, *49*, 364–372. [CrossRef]
84. Rubio-Rodríguez, N.; de Diego, S.M.; Beltrán, S.; Jaime, I.; Sanz, M.T.; Rovira, J. Supercritical fluid extraction of fish oil from fish by-products: A comparison with other extraction methods. *J. Food Eng.* **2012**, *109*, 238–248. [CrossRef]
85. Beaulieu, L.; Thibodeau, J.; Bryl, P.; Carbonneau, M.-É. Characterization of enzymatic hydrolyzed snow crab (*Chionoecetes opilio*) by-product fractions: A source of high-valued biomolecules. *Bioresour. Technol.* **2009**, *100*, 3332–3342. [CrossRef] [PubMed]

86. Huo, J.; Zhao, Z. Study on enzymatic hydrolysis of *Gadus morrhua* skin collagen and molecular weight distribution of hydrolysates. *Agric. Sci. China* **2009**, *8*, 723–729. [CrossRef]
87. Vázquez, J.A.; Murado, M.A. Enzymatic hydrolysates from food wastewater as a source of peptones for lactic acid bacteria productions. *Enzyme Microb. Technol.* **2008**, *43*, 66–72. [CrossRef]
88. Barcia, I.; Sánchez-Purriños, M.L.; Novo, M.; Novás, A.; Maroto, J.F.; Barcia, R. Optimisation of *Dosidicus gigas* mantle proteolysis at industrial scale. *Food Chem.* **2008**, *107*, 869–875. [CrossRef]
89. Guerard, F.; Sumaya-Martinez, M.T.; Laroque, D.; Chabeaud, A.; Dufossé, L. Optimization of free radical scavenging activity by response surface methodology in the hydrolysis of shrimp processing discards. *Process Biochem.* **2007**, *42*, 1486–1491. [CrossRef]
90. Šližytė, R.; Rustad, T.; Storrø, I. Enzymatic hydrolysis of cod (*Gadus morhua*) by-products. *Process Biochem.* **2005**, *40*, 3680–3692. [CrossRef]
91. Daukšas, E.; Falch, E.; Šližytė, R.; Rustad, T. Composition of fatty acids and lipid classes in bulk products generated during enzymic hydrolysis of cod (*Gadus morhua*) by-products. *Process Biochem.* **2005**, *40*, 2659–2670. [CrossRef]
92. Lignot, B.; Lahogue, V.; Bourseau, P. Enzymatic extraction of chondroitin sulfate from skate cartilage and concentration-desalting by ultrafiltration. *J. Biotechnol.* **2003**, *103*, 281–284. [CrossRef]
93. Guerard, F.; Guimas, L.; Binet, A. Production of tuna waste hydrolysates by a commercial neutral protease preparation. *J. Mol. Catal. B Enzym.* **2002**, *19–20*, 489–498. [CrossRef]
94. Guérard, F.; Dufossé, L.; De La Broise, D.; Binet, A. Enzymatic hydrolysis of proteins from yellowfin tuna (*Thunnus albacares*) wastes using alcalase. *J. Mol. Catal. B Enzym.* **2001**, *11*, 1051–1059. [CrossRef]
95. Charoensiddhi, S.; Lorbeer, A.J.; Lahnstein, J.; Bulone, V.; Franco, C.M.M.; Zhang, W. Enzyme-assisted extraction of carbohydrates from the brown alga *Ecklonia radiata*: Effect of enzyme type, pH and buffer on sugar yield and molecular weight profiles. *Process Biochem.* **2016**, *10*, 1503–1510. [CrossRef]
96. Younes, I.; Hajji, S.; Frachet, V.; Rinaudo, M.; Jellouli, K.; Nasri, M. Chitin extraction from shrimp shell using enzymatic treatment. Antitumor, antioxidant and antimicrobial activities of chitosan. *Int. J. Biol. Macromol.* **2014**, *69*, 489–498. [CrossRef] [PubMed]
97. Sovik, S.L.; Rustad, T. Effect of season and fishing ground on the activity of lipases in byproducts from cod (*Gadus morhua*). *LWT Food Sci. Technol.* **2005**, *38*, 867–876. [CrossRef]
98. Fernandes, P. Enzymes in fish and seafood processing. *Front. Bioeng. Biotechnol.* **2016**, *4*, 59. [CrossRef] [PubMed]
99. Blamey, J.M.; Fischer, F.; Meyer, H.P.; Sarmiento, F.; Zinn, M. Enzymatic biocatalysis in chemical transformations: A promising and emerging field in green chemistry practice. In *Biotechnology of Microbial Enzymes*; Brahmachari, G., Ed.; Academic Press: San Diego, CA, USA, 2017; pp. 347–403.
100. Doukyu, N.; Ogino, H. Organic solvent-tolerant enzymes. *Biochem. Eng. J.* **2010**, *48*, 270–282. [CrossRef]
101. Hardouin, K.; Bedoux, G.; Burlot, A.S.; Nyvall-Collén, P.; Bourgougnon, N. Enzymatic recovery of metabolites from seaweeds: Potential applications. In *Advances in Botanical Research*; Nathalie Bourgougnon, N., Ed.; Academic Press: San Diego, CA, USA, 2014; Volume 71, pp. 279–320.
102. Aljawish, A.; Chevalot, I.; Jasniewski, J.; Scher, J.; Muniglia, L. Enzymatic synthesis of chitosan derivatives and their potential applications. *J. Mol. Catal. B Enzym.* **2015**, *112*, 25–39. [CrossRef]
103. Zhu, Y.; Zhao, R.; Xiao, A.; Li, L.; Jiang, Z.; Chen, F.; Ni, H. Characterization of an alkaline β-agarase from *Stenotrophomonas* sp. NTa and the enzymatic hydrolysates. *Int. J. Biol. Macromol.* **2016**, *86*, 525–534. [CrossRef] [PubMed]
104. Yang, S.; Fu, X.; Yan, Q.; Guo, Y.; Liu, Z.; Jiang, Z. Cloning, expression, purification and application of a novel chitinase from a thermophilic marine bacterium *Paenibacillus barengoltzi*. *Food Chem.* **2016**, *192*, 1041–1048. [CrossRef] [PubMed]
105. García-Fraga, B.; da Silva, A.; López-Seijas, J.; Sieiro, C. A novel family 19 chitinase from the marine-derived *Pseudoalteromonas tunicata* CCUG 44952T: Heterologous expression, characterization and antifungal activity. *Biochem. Eng. J.* **2015**, *93*, 84–93. [CrossRef]
106. Kermanshahi-pour, A.; Sommer, T.; Anastas, P.; Zimmerman, J. Enzymatic and acid hydrolysis of *Tetraselmis suecica* for polysaccharide characterization. *Bioresour. Technol.* **2014**, *173*, 415–421. [CrossRef] [PubMed]

107. Chakraborty, S.; Jana, S.; Gandhi, A.; Sen, K.; Zhiang, W.; Kokare, C. Gellan gum microspheres containing a novel α-amylase from marine *Nocardiopsis* sp. strain B2 for immobilization. *Int. J. Biol. Macromol.* **2014**, *70*, 292–299. [CrossRef] [PubMed]
108. Menshova, R.; Ermakova, S.; Anastyuk, S.; Isakov, V.; Dubrovskaya, Y.; Kusaykin, M.; Um, B.; Zvyagintseva, T. Structure, enzymatic transformation and anticancer activity of branched high molecular weight laminaran from brown alga *Eisenia bicyclis*. *Carbohydr. Polym.* **2014**, *99*, 101–109. [CrossRef] [PubMed]
109. Tsuji, A.; Nishiyama, N.; Ohshima, M.; Maniwa, S.; Kuwamura, S.; Shiraishi, M.; Yuasa, K. Comprehensive enzymatic analysis of the amylolytic system in the digestive fluid of the sea hare, *Aplysia kurodai*: Unique properties of two α-amylases and two α-glucosidases. *FEBS Open Bio* **2014**, *4*, 560–570. [CrossRef] [PubMed]
110. Trincone, A.; Pagnotta, E.; Tramice, A. Enzymatic routes for the production of mono- and di-glucosylated derivatives of hydroxytyrosol. *Bioresour. Technol.* **2012**, *115*, 79–83. [CrossRef] [PubMed]
111. Shchipunov, Y.; Burtseva, Y.; Karpenko, T.; Shevchenko, N.; Zvyagintseva, T. Highly efficient immobilization of endo-1,3-β-d-glucanases (laminarinases) from marine mollusks in novel hybrid polysaccharide-silica nanocomposites with regulated composition. *J. Mol. Catal. B Enzym.* **2006**, *40*, 16–23. [CrossRef]
112. Ilankovan, P.; Hein, S.; Ng, C.; Trung, T.; Stevens, W. Production of N-acetyl chitobiose from various chitin substrates using commercial enzymes. *Carbohydr. Polym.* **2006**, *63*, 245–250. [CrossRef]
113. Shchipunov, Y.; Karpenko, T.; Bakunina, I.; Burtseva, Y.; Zvyagintseva, T. A new precursor for the immobilization of enzymes inside sol–gel-derived hybrid silica nanocomposites containing polysaccharides. *J. Biochem. Biophys. Methods* **2004**, *58*, 25–38. [CrossRef]
114. Shimoda, K. Efficient preparation of β-(1→6)-(GlcNAc)$_2$ by enzymatic conversion of chitin and chito-oligosaccharides. *Carbohydr. Polym.* **1996**, *29*, 149–154. [CrossRef]
115. Inoue, A.; Nishiyama, R.; Ojima, T. The alginate lyases FlAlyA, FlAlyB, FlAlyC, and FlAlex from *Flavobacterium* sp. UMI-01 have distinct roles in the complete degradation of alginate. *Algal Res.* **2016**, *19*, 355–362. [CrossRef]
116. Coste, O.; Malta, E.; López, J.; Fernández-Díaz, C. Production of sulfated oligosaccharides from the seaweed *Ulva* sp. using a new ulvan-degrading enzymatic bacterial crude extract. *Algal Res.* **2015**, *10*, 224–231. [CrossRef]
117. Yang, M.; Mao, X.; Liu, N.; Qiu, Y.; Xue, C. Purification and characterization of two agarases from *Agarivorans albus* OAY02. *Process Biochem.* **2014**, *49*, 905–912. [CrossRef]
118. Tramice, A.; Pagnotta, E.; Romano, I.; Gambacorta, A.; Trincone, A. Transglycosylation reactions using glycosyl hydrolases from *Thermotoga neapolitana*, a marine hydrogen-producing bacterium. *J. Mol. Catal. B Enzym.* **2007**, *47*, 21–27. [CrossRef]
119. Wu, S. Degradation of κ-carrageenan by hydrolysis with commercial α-amylase. *Carbohydr. Polym.* **2012**, *89*, 394–396. [CrossRef] [PubMed]
120. Xie, X.; Du, J.; Huang, Q.; Shi, Y.; Chen, Q. Inhibitory kinetics of bromacetic acid on β-N-acetyl-D-glucosaminidase from prawn (*Penaeus vannamei*). *Int. J. Biol. Macromol.* **2007**, *41*, 308–313. [CrossRef] [PubMed]
121. Martinou, A.; Kafetzopoulos, D.; Bouriotis, V. Chitin deacetylation by enzymatic means: Monitoring of deacetylation processes. *Carbohydr. Res.* **1995**, *273*, 235–242. [CrossRef]
122. Shin, S.; Sim, J.; Kishimura, H.; Chun, B. Characteristics of menhaden oil ethanolysis by immobilized lipase in supercritical carbon dioxide. *J. Ind. Eng. Chem.* **2012**, *18*, 546–550. [CrossRef]
123. Kavitha, V.; Radhakrishnan, N.; Madhavacharyulu, E.; Sailakshmi, G.; Sekaran, G.; Reddy, B.; Rajkumar, G.; Gnanamani, A. Biopolymer from microbial assisted in situ hydrolysis of triglycerides and dimerization of fatty acids. *Bioresour. Technol.* **2010**, *101*, 337–343. [CrossRef] [PubMed]
124. Liu, S.; Zhang, C.; Hong, P.; Ji, H. Lipase-catalysed acylglycerol synthesis of glycerol and n−3 PUFA from tuna oil: Optimisation of process parameters. *Food Chem.* **2007**, *103*, 1009–1015. [CrossRef]
125. Wang, Y.; Zhang, Y.; Sun, A.; Hu, Y. Characterization of a novel marine microbial esterase and its use to make D-methyl lactate. *Chin. J. Catal.* **2016**, *37*, 1396–1402. [CrossRef]
126. Alsufyani, T.; Engelen, A.; Diekmann, O.; Kuegler, S.; Wichard, T. Prevalence and mechanism of polyunsaturated aldehydes production in the green tide forming macroalgal genus *Ulva* (Ulvales, Chlorophyta). *Chem. Phys. Lipids* **2014**, *183*, 100–109. [CrossRef] [PubMed]

127. Rocha, L.; de Souza, A.; Rodrigues Filho, U.; Campana Filho, S.; Sette, L.; Porto, A. Immobilization of marine fungi on silica gel, silica xerogel and chitosan for biocatalytic reduction of ketones. *J. Mol. Catal. B Enzym.* **2012**, *84*, 160–165. [CrossRef]
128. Mertens, R.; Greiner, L.; van den Ban, E.; Haaker, H.; Liese, A. Practical applications of hydrogenase I from *Pyrococcus furiosus* for NADPH generation and regeneration. *J. Mol. Catal. B Enzym.* **2003**, *24–25*, 39–52. [CrossRef]
129. Boonprab, K.; Matsui, K.; Akakabe, Y.; Yotsukura, N.; Kajiwara, T. Hydroperoxy-arachidonic acid mediated n-hexanal and (Z)-3- and (E)-2-nonenal formation in *Laminaria angustata*. *Phytochemistry* **2003**, *63*, 669–678. [CrossRef]
130. Rontani, J.; Beker, B.; Volkman, J. Regiospecific oxygenation of alkenones in the benthic haptophyte Anand HAP 17. *Phytochemistry* **2004**, *65*, 3269–3278. [CrossRef] [PubMed]
131. Kerr, R.; Rodriguez, L.; Keliman, J. A chemoenzymatic synthesis of 9(11)-secosteroids using an enzyme extract of the marine gorgonian *Pseudopterogorgia americana*. *Tetrahedron Lett.* **1996**, *37*, 8301–8304. [CrossRef]
132. Wischang, D.; Radlow, M.; Schulz, H.; Vilter, H.; Viehweger, L.; Altmeyer, M.; Kegler, C.; Herrmann, J.; Müller, R.; Gaillard, F.; et al. Molecular cloning, structure, and reactivity of the second bromoperoxidase from *Ascophyllum nodosum*. *Bioorg. Chem.* **2012**, *44*, 25–34. [CrossRef] [PubMed]
133. Saborowski, R.; Sahling, G.; del Toro, M.A.N.; Walter, I.; García-Carreño, F. Stability and effects of organic solvents on endopeptidases from the gastric fluid of the marine crab *Cancer pagurus*. *J. Mol. Catal. B Enzym.* **2004**, *30*, 109–118. [CrossRef]
134. Sana, B.; Ghosh, D.; Saha, M.; Mukherjee, J. Purification and characterization of a salt, solvent, detergent and bleach tolerant protease from a new gamma-Proteobacterium isolated from the marine environment of the Sundarbans. *Process Biochem.* **2006**, *41*, 208–215. [CrossRef]
135. Wang, Q.; Dordick, J.S.; Linhardt, R.J. Synthesis and application of carbohydrate-containing polymers. *Chem. Mater.* **2002**, *14*, 3232–3244. [CrossRef]
136. Wang, L.; Chi, Z.; Yue, L.; Chi, Z.; Zhang, D. Occurrence and diversity of *Candida* genus in marine environments. *J. Ocean Univ. China* **2008**, *7*, 416–420. [CrossRef]
137. Andreou, A.; Brodhun, F.; Feussner, I. Biosynthesis of oxylipins in non-mammals. *Prog. Lipid Res.* **2009**, *48*, 148–170. [CrossRef] [PubMed]
138. Johnston, A.; Green, R.; Todd, J. Enzymatic breakage of dimethylsulfoniopropionate—A signature molecule for life at sea. *Curr. Opin. Chem. Biol.* **2016**, *31*, 58–65. [CrossRef] [PubMed]
139. Van der Oost, R.; Beyer, J.; Vermeulen, N. Fish bioaccumulation and biomarkers in environmental risk assessment: A review. *Environ. Toxicol. Pharmacol.* **2003**, *13*, 57–149. [CrossRef]
140. Apraiz, I. Identification of proteomic signatures of exposure to marine pollutants in mussels (*Mytilus edulis*). *Mol. Cell. Proteom.* **2006**, *5*, 1274–1285. [CrossRef] [PubMed]
141. Hühnerfuss, H. Chromatographic enantiomer separation of chiral xenobiotics and their metabolites–A versatile tool for process studies in marine and terrestrial ecosystems. *Chemosphere* **2000**, *40*, 913–919. [CrossRef]
142. Demarche, P.; Junghanns, C.; Nair, R.; Agathos, S. Harnessing the power of enzymes for environmental stewardship. *Biotechnol. Adv.* **2012**, *30*, 933–953. [CrossRef] [PubMed]
143. Daffonchio, D.; Ferrer, M.; Mapelli, F.; Cherif, A.; Lafraya, Á.; Malkawi, H.; Yakimov, M.; Abdel-Fattah, Y.; Blaghen, M.; Golyshin, P.; et al. Bioremediation of Southern Mediterranean oil polluted sites comes of age. *New Biotechnol.* **2013**, *30*, 743–748. [CrossRef] [PubMed]
144. Zhou, H.; Pan, H.; Xu, J.; Xu, W.; Liu, L. Acclimation of a marine microbial consortium for efficient Mn(II) oxidation and manganese containing particle production. *J. Hazard. Mater.* **2016**, *304*, 434–440. [CrossRef] [PubMed]
145. Rontani, J.-F.; Bonin, P.; Vaultier, F.; Guasco, S.; Volkman, J.K. Anaerobic bacterial degradation of pristenes and phytenes in marine sediments does not lead to pristane and phytane during early diagenesis. *Org. Geochem.* **2013**, *58*, 43–55. [CrossRef]
146. Châtel, A.; Hamer, B.; Talarmin, H.; Dorange, G.; Schröder, H.C.; Müller, W.E.G. Activation of MAP kinase signaling pathway in the mussel *Mytilus galloprovincialis* as biomarker of environmental pollution. *Aquat. Toxicol.* **2010**, *96*, 247–255. [CrossRef] [PubMed]
147. Cheung, K.H.; Gu, J.-D. Reduction of chromate (CrO_4^{2-}) by an enrichment consortium and an isolate of marine sulfate-reducing bacteria. *Chemosphere* **2003**, *52*, 1523–1529. [CrossRef]

148. Chang, C.-C.; Rahmawaty, A.; Chang, Z.-W. Molecular and immunological responses of the giant freshwater prawn, *Macrobrachium rosenbergii*, to the organophosphorus insecticide, trichlorfon. *Aquat. Toxicol.* **2013**, *130–131*, 18–26. [CrossRef] [PubMed]
149. Davies, P.E. The toxicology and metabolism of chlorothalonil in fish. III. Metabolism, enzymatics and detoxication in *Salmo* spp. and *Galaxias* spp. *Aquat. Toxicol.* **1985**, *7*, 277–299. [CrossRef]
150. Agatova, A.I.; Andreeva, N.M.; Kucheryavenko, A.V.; Torgunova, N.I. Transformation of organic matter in areas inhabited by natural and artificially cultured populations of marine invertebrates in the bay of Pos'et (sea of Japan). *Aquaculture* **1986**, *53*, 49–66. [CrossRef]
151. Pfaffenberger, B.; Hühnerfuss, H.; Kallenborn, R.; Köhler-Günther, A.; König, W.A.; Krüner, G. Chromatographic separation of the enantiomers of marine pollutants. Part 6: Comparison of the enantioselective degradation of α-hexachlorocyclohexane in marine biota and water. *Chemosphere* **1992**, *25*, 719–725. [CrossRef]
152. Regoli, F.; Principato, G. Glutathione, glutathione-dependent and antioxidant enzymes in mussel, *Mytilus galloprovincialis*, exposed to metals under field and laboratory conditions: Implications for the use of biochemical biomarkers. *Aquat. Toxicol.* **1995**, *31*, 143–164. [CrossRef]
153. Warshawsky, D.; Cody, T.; Radike, M.; Reilman, R.; Schumann, B.; LaDow, K.; Schneider, J. Biotransformation of benzo[a]pyrene and other polycyclic aromatic hydrocarbons and heterocyclic analogs by several green algae and other algal species under gold and white light. *Chem. Biol. Interact.* **1995**, *97*, 131–148. [CrossRef]
154. Hahlbeck, E.; Arndt, C.; Schiedek, D. Sulphide detoxification in *Hediste diversicolor* and *Marenzelleria viridis*, two dominant polychaete worms within the shallow coastal waters of the southern Baltic Sea. *Comp. Biochem. Physiol. B Biochem. Mol. Biol.* **2000**, *125*, 457–471. [CrossRef]
155. Goto, N.; Mitamura, O.; Terai, H. Biodegradation of photosynthetically produced extracellular organic carbon from intertidal benthic algae. *J. Exp. Mar. Biol. Ecol.* **2001**, *257*, 73–86. [CrossRef]
156. Company, R.; Serafim, A.; Bebianno, M.J.; Cosson, R.; Shillito, B.; Fiala-Médioni, A. Effect of cadmium, copper and mercury on antioxidant enzyme activities and lipid peroxidation in the gills of the hydrothermal vent mussel *Bathymodiolus azoricus*. *Mar. Environ. Res.* **2004**, *58*, 377–381. [CrossRef] [PubMed]
157. Gallizia, I.; Vezzulli, L.; Fabiano, M. Oxygen supply for biostimulation of enzymatic activity in organic-rich marine ecosystems. *Soil Biol. Biochem.* **2004**, *36*, 1645–1652. [CrossRef]
158. Barata, C.; Carlos Navarro, J.; Varo, I.; Carmen Riva, M.; Arun, S.; Porte, C. Changes in antioxidant enzyme activities, fatty acid composition and lipid peroxidation in *Daphnia magna* during the aging process. *Comp. Biochem. Physiol. B Biochem. Mol. Biol.* **2005**, *140*, 81–90. [CrossRef] [PubMed]
159. Buet, A.; Banas, D.; Vollaire, Y.; Coulet, E.; Roche, H. Biomarker responses in European eel (*Anguilla anguilla*) exposed to persistent organic pollutants. A field study in the Vaccarès lagoon (Camargue, France). *Chemosphere* **2006**, *65*, 1846–1858. [PubMed]
160. Lima, I.; Moreira, S.M.; Osten, J.R.-V.; Soares, A.M.V.M.; Guilhermino, L. Biochemical responses of the marine mussel Mytilus galloprovincialis to petrochemical environmental contamination along the north-western coast of Portugal. *Chemosphere* **2007**, *66*, 1230–1242. [CrossRef] [PubMed]
161. Kankaanpää, H.; Leiniö, S.; Olin, M.; Sjövall, O.; Meriluoto, J.; Lehtonen, K.K. Accumulation and depuration of cyanobacterial toxin nodularin and biomarker responses in the mussel *Mytilus edulis*. *Chemosphere* **2007**, *68*, 1210–1217. [CrossRef] [PubMed]
162. Correia, A.D.; Gonçalves, R.; Scholze, M.; Ferreira, M.; Henriques, M.A.R. Biochemical and behavioral responses in gilthead seabream (*Sparus aurata*) to phenanthrene. *J. Exp. Mar. Biol. Ecol.* **2007**, *347*, 109–122. [CrossRef]
163. Koenig, S.; Solé, M. Natural variability of hepatic biomarkers in Mediterranean deep-sea organisms. *Mar. Environ. Res.* **2012**, *79*, 122–131. [CrossRef] [PubMed]
164. Williams, T.D.; Diab, A.M.; Gubbins, M.; Collins, C.; Matejusova, I.; Kerr, R.; Chipman, J.K.; Kuiper, R.; Vethaak, A.D.; George, S.G. Transcriptomic responses of European flounder (*Platichthys flesus*) liver to a brominated flame retardant mixture. *Aquat. Toxicol.* **2013**, *142–143*, 45–52. [CrossRef] [PubMed]
165. Arnosti, C. Contrasting patterns of peptidase activities in seawater and sediments: An example from arctic fjords of Svalbard. *Mar. Chem.* **2015**, *168*, 151–156. [CrossRef]
166. Si, Y.-X.; Gu, X.-X.; Cai, Y.; Yin, S.-J.; Yang, J.-M.; Park, Y.-D.; Lee, J.; Qian, G.-Y. Molecular dynamics simulation integrating study for Cr^{3+}-binding to arginine kinase. *Process Biochem.* **2015**, *50*, 1363–1371. [CrossRef]

167. Da Fonseca, F.S.A.; Angolini, C.F.F.; Arruda, M.A.Z.; Junior, C.A.L.; Santos, C.A.; Saraiva, A.M.; Pilau, E.; Souza, A.P.; Laborda, P.R.; de Oliveira, P.F.L.; et al. Identification of oxidoreductases from the petroleum *Bacillus safensis* strain. *Biotechnol. Rep.* **2015**, *8*, 152–159. [CrossRef]
168. Pi, Y.; Meng, L.; Bao, M.; Sun, P.; Lu, J. Degradation of crude oil and relationship with bacteria and enzymatic activities in laboratory testing. *Int. Biodeter. Biodegrad.* **2016**, *106*, 106–116. [CrossRef]
169. Nicemol, J.; Niladevi, K.N.; Anisha, G.S.; Prema, P. Hydrolysis of pectin: An enzymatic approach and its application in banana fiber processing. *Microbiol. Res.* **2008**, *163*, 538–544.
170. Fines, B.C.; Holt, G.J. Chitinase and apparent digestibility of chitin in the digestive tract of juvenile cobia. *Rachycentron canadum. Aquaculture* **2010**, *303*, 34–39. [CrossRef]
171. Ziervogel, K.; Arnosti, C. Enzyme activities in the Delaware Estuary affected by elevated suspended sediment load. *Estuar. Coast. Shelf Sci.* **2009**, *84*, 253–258. [CrossRef]
172. Richir, J.; Velimirov, B.; Poulicek, M.; Gobert, S. Use of semi-quantitative kit methods to study the heterotrophic bacterial community of *Posidonia oceanica* meadows: Limits and possible applications. *Estuar. Coast. Shelf Sci.* **2012**, *109*, 20–29. [CrossRef]
173. Capasso, C.; Supuran, C.T. Carbonic Anhydrases from Extremophiles and Their Biotechnological Applications. In *Carbonic Anhydrases as Biocatalysts*; Supuran, C.T., De Simone, G., Eds.; Elsevier: Amsterdam, The Netherlands, 2015; pp. 311–324.
174. Migliardini, F.; De Luca, V.; Carginale, V.; Rossi, M.; Corbo, P.; Supuran, C.T.; Capasso, C. Biomimetic CO_2 capture using a highly thermostable bacterial alpha-carbonic anhydrase immobilized on a polyurethane foam. *J. Enzyme Inhib. Med. Chem.* **2013**, *29*, 146–150. [CrossRef] [PubMed]
175. Lionetto, M.G.; Caricato, R.; Giordano, M.E.; Erroi, E.; Schettino, T. Carbonic anhydrase as pollution biomarker: An ancient enzyme with a new use. *Int. J. Environ. Res. Public Health* **2012**, *9*, 3965–3977. [CrossRef] [PubMed]
176. Innocenti, A.; Scozzafava, S.; Parkkila, L.; Puccetti, G.; de Simone, G.; Supuran, C.T. Investigations of the esterase, phosphatase, and sulfatase activities of the cytosolic mammalian carbonic anhydrase isoforms I, II, and XIII with 4-nitrophenyl esters as substrates. *Bioorg. Med. Chem. Lett.* **2008**, *18*, 2267–2271. [CrossRef] [PubMed]
177. Trincone, A. Biocatalytic processes using marine biocatalysts: Ten cases in point. *Curr. Org. Chem.* **2013**, *17*, 1058–1066. [CrossRef]
178. Dawfik, D.S.; van der Donk, W.A. Editorial overvies: Biocatalysis and biotransformation: Esoteric, niche enzymology. *Curr. Opin. Chem. Biol.* **2016**, *31*, v–vii. [CrossRef]

© 2017 by the author. Licensee MDPI, Basel, Switzerland. This article is an open access article distributed under the terms and conditions of the Creative Commons Attribution (CC BY) license (http://creativecommons.org/licenses/by/4.0/).

Review

Microalgal Enzymes with Biotechnological Applications

Giorgio Maria Vingiani [1], Pasquale De Luca [2], Adrianna Ianora [1], Alan D.W. Dobson [3,4] and Chiara Lauritano [1,*]

1. Marine Biotechnology Department, Stazione Zoologica Anton Dohrn, CAP80121 (NA) Villa Comunale, Italy
2. Research Infrastructure for Marine Biological Resources Department, Stazione Zoologica Anton Dohrn, CAP80121 (NA) Villa Comunale, Italy
3. School of Microbiology, University College Cork, College Road, T12 YN60 Cork, Ireland
4. Environmental Research Institute, University College Cork, Lee Road, T23XE10 Cork, Ireland
* Correspondence: chiara.lauritano@szn.it; Tel.: +39-081-5833221

Received: 4 July 2019; Accepted: 1 August 2019; Published: 5 August 2019

Abstract: Enzymes are essential components of biological reactions and play important roles in the scaling and optimization of many industrial processes. Due to the growing commercial demand for new and more efficient enzymes to help further optimize these processes, many studies are now focusing their attention on more renewable and environmentally sustainable sources for the production of these enzymes. Microalgae are very promising from this perspective since they can be cultivated in photobioreactors, allowing the production of high biomass levels in a cost-efficient manner. This is reflected in the increased number of publications in this area, especially in the use of microalgae as a source of novel enzymes. In particular, various microalgal enzymes with different industrial applications (e.g., lipids and biofuel production, healthcare, and bioremediation) have been studied to date, and the modification of enzymatic sequences involved in lipid and carotenoid production has resulted in promising results. However, the entire biosynthetic pathways/systems leading to synthesis of potentially important bioactive compounds have in many cases yet to be fully characterized (e.g., for the synthesis of polyketides). Nonetheless, with recent advances in microalgal genomics and transcriptomic approaches, it is becoming easier to identify sequences encoding targeted enzymes, increasing the likelihood of the identification, heterologous expression, and characterization of these enzymes of interest. This review provides an overview of the state of the art in marine and freshwater microalgal enzymes with potential biotechnological applications and provides future perspectives for this field.

Keywords: microalgae; enzymes; marine biotechnology; -omics technologies; heterologous expression; homologous expression

---

1. Introduction

Water covers around 71% of the Earth's surface, with salt water responsible for 96.5% of this percentage [1]. Due to its molecular structure and chemical properties, water includes (and often participates in) every chemical reaction that is biologically relevant [2]. In such reactions, enzymes cover a fundamental role: They are organic macromolecules that catalyze biological reactions (so-called "biocatalysts" [3]). Due to their substrate-specificity, enzymes are commonly used in several sectors (such as food processing, detergent, pharmaceuticals, biofuel, and paper production) to improve, scale, and optimize industrial production. For example, hydrolases, which are enzymes that catalyze the hydrolysis of chemical bonds, have applications in several fields. Examples of industrially relevant hydrolases are cellulases for biofuel production [4], amylases for syrup production [5], papain, phytases and galactosidases for food processing [6], and other hydrolases which have various pharmaceutical applications [7]. The demand for new enzymes is growing every year, and many financial reports expect the global enzyme market value to surpass the $10 billion mark by 2024 (Allied Market Research, 2018, https://www.alliedmarketresearch.com/enzymes-market;ResearchandMarket.com, 2018, https://www.researchandmarkets.com/research/6zpvw9/industrial?w=4), of which $7 billion alone will be for industrial applications (BCC Research, 2018, https://www.bccresearch.com/market-research/biotechnology/global-markets-for-enzymes-in-industrial-applications.html).

Microalgae are photosynthetic unicellular organisms that can be massively cultivated under controlled conditions in photobioreactors with relatively small quantities of micro- and macro-nutrients [8], and can thus fit perfectly into this market sector. Microalgae continue to be used in a number of biotechnological applications. Searching the available literature in the PubMed database, this trend is clearly visible (search filters used were the word "microalgae" in the Title/Abstract field and the word "biotechnolog*" in the Text Word field, using the asterisk wildcard to expand the term selection; Figure 1). Considering the full 20-year interval between "1999–2018", it is clear that as of 2012, there has been a rapid increase in the number of publications involving both "microalgae" and "biotechnology", reaching a peak in the years 2015–2016.

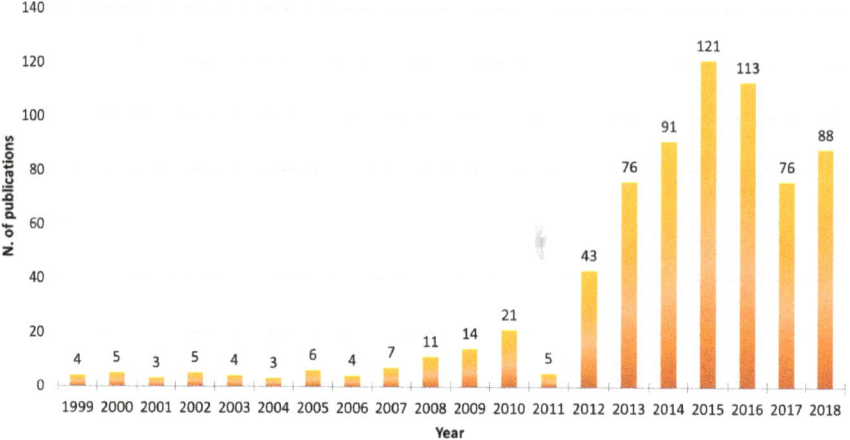

Figure 1. Microalgae Biotechnology PubMed Search Results 1999–2018. Using PubMed database search in the 20-years interval 1999–2018, the following search filters were set: The word "microalgae" in the [Title/Abstract] field and the word "biotechnolog*" in the [Text Word] field, using the asterisk (*) wildcard to expand the term selection (such as biotechnology, biotechnological, and biotechnologies).

The literature regarding the biotechnological applications of microalgae is dominated by four main research sectors: (1) Direct use of microalgal cells, for bioremediation applications and as food supplements [9]; (2) Extraction of bioactives for different applications (e.g., cosmeceutical, nutraceutical, and pharmaceutical applications, and for biofuel production [10,11]); (3) Use of microalgae as platforms for heterologous expression or endogenous gene editing and overexpression [12]; (4) Use of microalgae as sources of enzymes for industrial applications [13]. The latter field appears to be less well-studied compared to the others, due to the high costs currently involved in enzyme extraction and characterization, as well as the scarcity of annotated microalgal genomes.

Recent projects, such as those funded under the European Union Seventh Framework 2007–2017 (EU FP-7), e.g., BIOFAT (https://cordis.europa.eu/project/rcn/100477/factsheet/en) and GIAVAP (https://cordis.europa.eu/project/rcn/97420/factsheet/en), together with Horizon 2020 programs, e.g., ALGAE4A-B (http://www.algae4ab.eu/project.html) and VALUEMAG (https://www.valuemag.eu/), have resulted in an increase in –omics data (i.e., genomics, transcriptomics, proteomics and metabolomics data) available for microalgae, improving the possibility of finding new enzymes from both marine and freshwater species [14]. Mogharabi and Faramarzi recently reported the isolation of some enzymes from algae and highlighted their potential as cell factories [15]. This review aims to provide a summary of the current literature on microalgal enzymes with potential biotechnological applications with a particular focus on enzymes involved in the production of high-value added lipids and biodiesel, healthcare applications, and bioremediation.

2. Enzymes from Microalgae

2.1. Enzymes for High-Value Added Lipids and Biodiesel Production

Microalgae are known to accumulate large amounts of lipids [16], with triglycerides (TAGs) and poly-unsaturated fatty acids (PUFA) being the most studied from a biotechnological application standpoint, particularly for the production of biodiesel and nutraceuticals [9,16–18]. TAGs, esters derived from glycerol and three chained fatty acids (FA) which are usually stored in cytosol-located lipid droplets [19], can be used to produce biodiesel following acid- or base-catalyzed transesterification reactions [20]. PUFAs, for their part, have well-proven beneficial health effects [21,22], especially Ω-3 fatty acids such as docosahexaenoic acid (DHA) and eicosapentaenoic acid (EPA) (Figure 2).

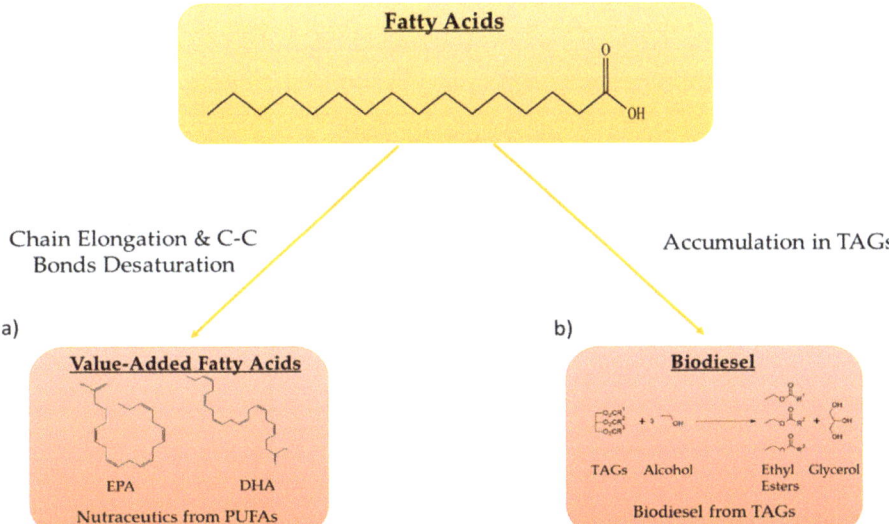

Figure 2. Examples of fatty acids of biotechnological interest. (**a**) Through various reactions of elongation and formation of double C-C bonds, poly-unsaturated fatty acids (PUFA) can be synthetized, such as eicosapentaenoic acid (EPA) and docosahexaenoic acid (DHA) with nutraceutical or food applications; (**b**) Accumulation in triglycerides (TAGs) and biodiesel formation via chemical transesterification.

The most frequently studied enzyme involved in lipid synthesis is acyl-CoA diacylglycerol acyltransferase (DGAT), involved in the final reaction of the TAG biosynthetic pathway [23,24]. Three independent groups of enzymes, referred to as acyl-CoA diacylglycerol acyltransferases type 1, 2, and 3 (DGATs 1-2-3), take part in the acyl-CoA-dependent formation of TAGs from its precursor sn-1,2-diacylglycerol (DAG) [25]. The individual contribution of each DGAT isoenzyme to the fatty acid profile of TAG differs between species [24,26].

A gene encoding DGAT1 was initially discovered in the green alga *Chlorella ellipsoidea* by Guo et al. [27], and an experiment involving overexpression of DGAT1 was subsequently performed in the oleaginous microalgae *Nannochloropsis oceanica* [28]. The first DGAT2 sequence was obtained from the green alga *Ostreococcus tauri* [29], and different studies involving overexpression of DGAT2 were performed. In particular, DGAT2 overexpression led to an increase in TAG production in the diatoms *Phaeodactylum tricornutum* [30] and *Thalassiosira pseudonana* [31], and in the oleaginous microalgae *Neochloris oleoabundans* [32] and *N. oceanica* [33]. Different isoforms of DGAT2 (NoDGAT2A, 2C, 2D) have successively been identified in *N. oceanica* and different combinations of either overexpression or under-expression have been analyzed. These combinations gave different fatty acid-production profiles, with some optimized for nutritional applications and others for biofuel purposes [34]. Even if the green alga *Chlamydomonas reinhardtii* is considered a common biofuel feedstock, it showed no clear trends following overexpression of different DGAT2 isoforms, with increased levels of TAG in some reports [35], while levels were not increased in others [36]. Recently, Cui and coworkers [37] characterized a dual-function wax ester synthase (WS)/DGAT enzyme in *P. tricornutum*, whose overexpression led to an accumulation of both TAGs and wax esters. This was the first report of this particular enzyme in a microalga, and a patent involving the enzyme was subsequently filed (Patent Code: CN107299090A, 2017).

In addition to DGAT, other genes have been targeted in order to increase high-value added lipid production, including glucose-6-phosphate dehydrogenase (G6PD), Δ6-desaturase, 6-phosphogluconate dehydrogenase (6PGD), glycerol-3-phosphate acyltransferase (GPAT1-GPAT2), and acetyl-CoA synthetase 2 (ACS2). Overexpression of these enzymes resulted in increased lipid contents [38–42]. In particular, two patents for desaturases have been filed. One covers a Δ6-desaturase from *Nannochloropsis* spp., which converts linoleic acid to γ-linolenic acid (GLA) and α-linolenic acid (ALA) to stearidnoic acid (Patent Code: CN101289659A, 2010). The other covers a Ω6-desaturase from *Arctic chlamydomonas* sp. *ArF0006*, which converts oleic acid to linoleic acid (Patent Code: KR101829048B1, 2018).

Other approaches to increase lipid production and/or alter lipid profiles via gene disruption have been employed. Examples include the knock-out of a phospholipase A2 (PLA2) gene via CRISPR/Cas9 ribonucleoproteins in *C. reinhardtii* [43], microRNA silencing of the stearoyl-ACP desaturase (that forms oleic acid via addition of a double-bond in a lipid chain [44]) in *C. reinhardtii* [45], and meganuclease and TALE nuclease genome modification in *P. tricornutum* [46]. This last approach involved modifying the expression of seven genes, potentially affecting the lipid content (UDP-glucose pyrophosphorylase, glycerol-3-phosphate dehydrogenase, and enoyl-ACP reductase), the acyl chain length (long chain acyl-CoA elongase and a putative palmitoyl-protein thioesterase), and the degree of fatty acid saturation (Ω-3 fatty acid desaturase and Δ-12-fatty acid desaturase). In particular, a mutant for UDP-glucose pyrophosphorylase showed a 45-fold increase in TAG accumulation under nitrogen starvation conditions. Figure 3 provides an overview of the subcellular localization of metabolic pathways and engineered enzymes in the aforementioned examples.

Finally, Sorigué and coworkers [47] reported, for the first time, the presence of a photoenzyme named fatty acid photodecarboxylase (FAP) in *Chlorella variabilis* str microalgae. NC64A. FAP converts fatty acids to hydrocarbons and may be useful in light-driven production of hydrocarbons. It is worth mentioning that Misra et al. [48] have developed a database to catalogue the enzymes which have been identified as being responsible for lipid synthesis from available microalgal genomes (e.g., *C. reinhardtii*, *P. tricornutum*, *Volvox carteri*), called dEMBF (website: http://bbprof.immt.res.in/embf/). To date, the database has collected 316 entries from 16 organisms, while providing different browsing options (Search by: "Enzyme Classification", "Organism", and "Enzyme Class") and different web-based tools (NCBI's Blast software integrated, sequence comparison, Motif prediction via the MEME software). The enzymes discussed in this section are reported in Table 1.

Table 1. Enzymes from Microalgae for Lipid and Biodiesel Production. Marine and freshwater ecological strain sources are abbreviated as M or F, respectively. Algal classes of *Bacillariophyceae*, *Chlorophyceae*, *Trebouxiophyceae*, *Eustigmatophyceae*, *Mamiellophyceae*, *Coscinodiscophyceae*, and *Cyanidiophyceae* are abbreviated as BA, CH, TR, EU, MA, CO, and CY, respectively.

Ref.	Enzymes	Microalgae	Strain Source	Microalgal Class	Main Results
[39]	Δ6-Desaturase	*Phaeodactylum tricornutum*	M	BA	Neutral lipid production enhanced and increase of EPA content
[41]	acetyl-CoA synthetase	*Chlamydomonas reinhardtii*	F	CH	Increase in neutral lipid production
[27]	acyl-CoA diacylglycerol acyltransferase 1	*Chlorella ellipsoidea*	F	TR	Sequence identification and function of TAG accumultation characterized
[28]	acyl-CoA diacylglycerol acyltransferase 1A	*Nannochloropsis oceanica*	M	EU	Increase in TAGs production both in nitrogen-replete and -deplete conditions
[36]	acyl-CoA diacylglycerol acyltransferase 2	*Chlamydomonas reinhardtii*	F	CH	No TAGs overproduction
[35]	acyl-CoA diacylglycerol acyltransferase 2	*Chlamydomonas reinhardtii*	F	CH	Five DGAT2 homologous genes identification and the overexpression of CrDGAT2-1 and CrDGAT2-5 resulting in a significant increase in lipid production
[33]	acyl-CoA diacylglycerol acyltransferase 2	*Nannochloropsis oceanica*	M	EU	Increase in neutral lipid production
[32]	acyl-CoA diacylglycerol acyltransferase 2	*Neochloris oleoabundans*	F	CH	Change of lipid profile
[29]	acyl-CoA diacylglycerol acyltransferase 2	*Ostreococcus tauri*	M	MA	Gene identification and enzyme characterization in heterologous systems
[30]	acyl-CoA diacylglycerol acyltransferase 2	*Phaeodactylum tricornutum*	M	BA	Increase in neutral lipid production with enrichment EPA-PUFAs content
[31]	acyl-CoA diacylglycerol acyltransferase 2	*Thalassiosira pseudonana*	M	CO	Increase in TAGs production with focus on the intracellular enzyme localization
[34]	acyl-CoA diacylglycerol acyltransferase 2A, 2C, 2D	*Nannochloropsis oceanica*	M	EU	Differential DGAT2 isoforms expression in different engineered strains with individual specialized lipid profiles

Table 1. *Cont.*

Ref.	Enzymes	Microalgae	Strain Source	Microalgal Class	Main Results
[47]	fatty acid photodecarboxylase	*Chlorella variabilis*	F	TR	Enzyme identification and alkane synthase activity tested
[38]	glucose-6-phosphate dehydrogenase	*Phaeodactylum tricornutum*	M	BA	Modest increase in neutral lipid production with a lipid composition switch from polyunsaturated to monounsaturated
[42]	glucose-6-phosphate dehydrogenase; phosphogluconate dehydrogenase	*Fistulifera solaris*	M	BA	Slight increase in TAGs production
[40]	glycerol-3-phosphate acyltransferase 1, 2	*Cyanidioschyzon merolae*	F	CY	Significant increase in TAGs production
[43]	phospholipase A2	*Chlamydomonas reinhardtii*	F	CH	Increase in TAGs production
[45]	stearoyl-ACP desaturase	*Chlamydomonas reinhardtii*	F	CH	Production of TAGs enriched in stearic acid
[46]	UDP-glucose pyrophosphorylase, glycerol-3-phosphate dehydrogenase, enoyl-ACP reductase, long chain acyl-CoA elongase, putative palmitoyl-protein thioesterase, Ω-3 fatty acid desaturase and Δ-12-fatty acid desaturase	*Phaeodactylum tricornutum*	M	BA	Significant increase in TAGs production (45-fold increase for UDP-glucose pyrophosphorylase mutant)
[37]	wax esther synthase/acyl-CoA diacylglycerol acyltransferase	*Phaeodactylum tricornutum*	M	BA	Increase in neutral lipids and wax esters production
Patent Code (Year)	**Enzymes**	**Microalgae**	**Strain Source**	**Microalgal Class**	**Notes**
CN107299090A (2017)	wax esther synthase/acyl-CoA diacylglycerol acyltransferase	*Phaeodactylum tricornutum*	M	BA	Neutral lipids and wax esters production enhanced
CN101289659A (2010)	Δ6-Desaturase	*Nannochloropsis* spp.	M	EU	The enzyme sequence was identified and the enzyme characterized in bacterial systems
KR101829048B1 (2018)	Ω6-Desaturase	Arctic *Chlamydomonas* sp. ArF0006	F	CH	The enzyme sequence was identified and the enzyme characterized in bacterial systems

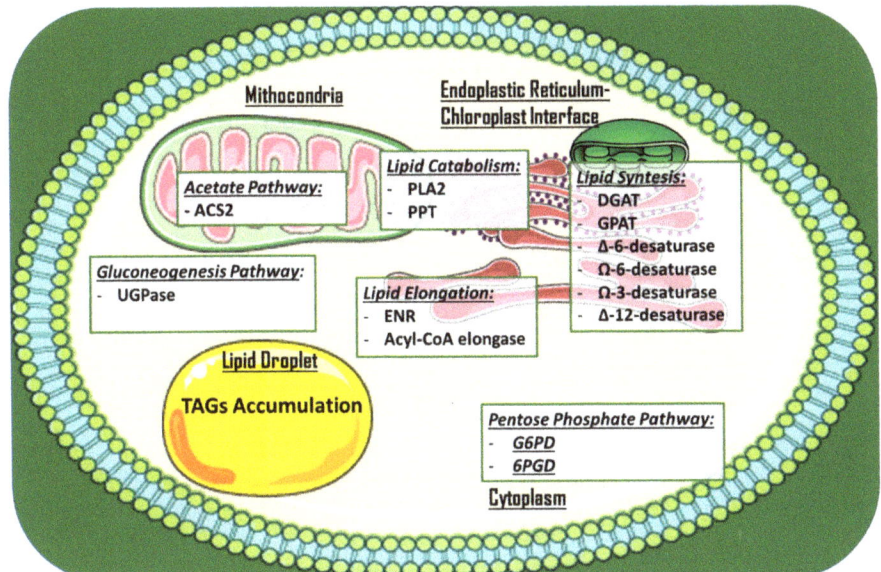

Figure 3. Main studied and engineered enzymes for TAGs and PUFAs in microalgae for the production of high value-added lipids. Enzymes are roughly divided in subcellular compartments. A single lipid droplet where TAGs are accumulated is added. Abbreviations: DGAT: Acyl-CoA diacylglycerol acyltransferase; G6PD: Glucose-6-phosphate dehydrogenase; 6PGD: 6-phosphogluconate dehydrogenase; GPAT: Glycerol-3-phosphate acyltransferase; ACS2: acetyl-CoA synthetase 2; PLA2: Phospholipase A2; Δ-6/Δ-12-Desaturase: delta-6/delta-12 fatty acid desaturase; Ω-3/Ω-6-desaturase: omega-2/omega-6 fatty acid desaturase; ENR: Enoyl-acyl carrier protein reductase; UGPase: UDP-glucose pyrophosphorylase; TAG: Triglyceride.

2.2. Enzymes for Healthcare Application

Enzymes for healthcare applications can include: (1) Enzymes used directly as "drugs", or (2) enzymes involved in the biosynthetic pathway of bioactive compounds (Figure 4). Regarding the first group, the most studied enzyme is L-asparaginase. L-asparaginase is an L-asparagine amidohydrolase enzyme used for the treatment of acute lymphoblastic leukemia, acute myeloid leukemia, and non-Hodgkin's lymphoma [49]. Its hydrolytic effect reduces asparagine availability for cancer cells that are unable to synthesize L-asparaginase autonomously [50] L-asparaginase was historically first discovered and then produced in bacteria (e.g., *Escherichia coli*, *Erwinia aroideae*, *Bacillus cereus*) [51–53]. However, in order to overcome some of the economical and safety limits associated with marketing the enzyme [54,55], increased efforts began to focus on the identification and characterization of the enzyme in microalgae strains.

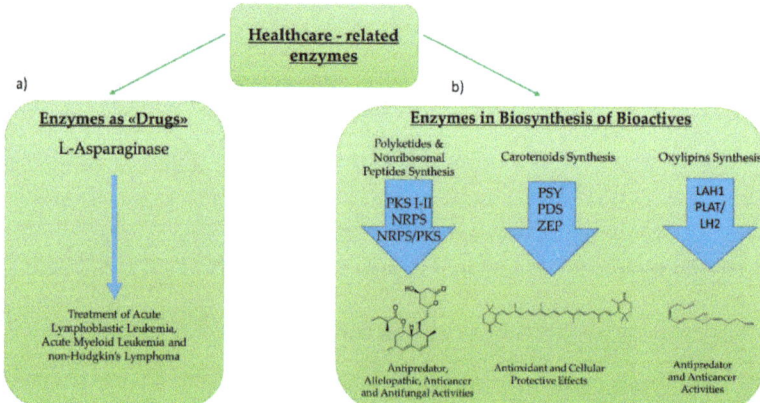

Figure 4. Enzymes for Healthcare Applications. Enzymes for healthcare applications can include: (**a**) Enzymes used directly as "drugs", such as the L-asparaginase (**b**) enzymes involved in the biosynthetic pathway of active compounds, such as polyketides, carotenoids, or oxylipins. In the synthesis of polyketides, the enzymes studied are polyketide synthases and nonribosomal peptide synthases. For the synthesis of carotenoids, the most studied enzymes are phytoene synthase (PSY), phytoene decarboxylase (PDS) and zeaxanthin epoxidase (ZEP). For the synthesis of oxylipins the studied enzymes are lipoic acid hydrolases (LAH) and PLAT (Polycystin-1, Lipoxygenase, Alpha-Toxin)/LH2 (Lipoxygenase homology). An example of molecules and their roles for each pathway is also outlined.

Paul [56] first purified an L-asparaginase in *Chlamidomonas* spp. with limited anticancer activity, and tested it in an in vivo anti-lymphoma assay. Ebrahiminezhad and coworkers screened 40 microalgal isolates via activity assays and reported on *Chlorella vulgaris* as a novel potential feedstock for L-asparaginase production [57].

Regarding enzymatic pathways involved in the synthesis of bioactive compounds, many studies have focused on polyketide synthases (PKS) and nonribosomal peptide synthetases (NRPS). PKS produce polyketides, while NRPS produce nonribosomal peptides. Both classes of secondary metabolites are formed by sequential reactions operated by these "megasynthase" enzymes [58,59]. Polyketides and nonribosomal peptides have been reported to have antipredator, allelopathic, anticancer, and antifungal activities [58,60–62]. PKS can be multi-domain enzymes (Type I PKS), large enzyme complexes (Type II), or homodimeric complexes (Type III) [63]. Genes potentially encoding these first two types' of PKSs have been identified in several microalgae (e.g., *Amphidinium carterae*, *Azadinium spinosum*, *Gambierdiscus* spp., *Karenia brevis* [64–67]). Similarly, NRPSs have a modular organization similar to type I PKSs, and genes potentially encoding NRPSs have been found in different microalgae [68]. Moreover, metabolites that are likely to derive from hybrid NRPS/PKS gene clusters have been reported from *Karenia brevis* [69]. However, to our knowledge, there are no studies reporting the direct correlation of a PKS or NRPS gene from a microalga with the production of a bioactive compound.

Other microalgal enzymes which have been widely studied are those involved in the synthesis of compounds with nutraceutical and cosmeceutical applications, such as those involved in carotenoid synthesis (e.g., astaxanthin, β-carotene, lutein, and canthaxanthin). Carotenoids are isoprenoid pigments, which have many cellular protective effects, such as antioxidant effects occurring via the chemical quenching of O_2 and other reactive oxygen species [70–72]. Their antioxidant properties can potentially protect humans from a compromised immune response, premature aging, arthritis, cardiovascular diseases, and/or certain cancers [72]. Among microalgae, the most studied for the industrial production of carotenoids are the halophile microalga *Dunaliella salina* and the green alga *Haematococcus pluvialis*, which naturally produce high amounts of carotenoids [73]. Moreover, *D.*

salina is a particularly versatile feedstock, and many researchers have focused on obtaining maximum carotenoid yields without impeding its growth [74–76]. In addition, *D. salina* has been successfully transformed via different approaches, such as microparticle bombardment [77] or via *Agrobacterium tumefaciens* [78], increasing the feasibility of its use for biotechnological applications.

The most studied enzymes involved in carotenoid synthesis are: β-carotene oxygenase, lycopene-β-cyclase, phytoene synthase, phytoene desaturase, β-carotene hydroxylase, and zeaxanthin epoxidase [79]. In order to improve the production of carotenoids, different metabolic engineering approaches have been employed. The initial method used was to induce random or site directed mutations in an attempt to improve the activity of enzymes involved in the carotenoid metabolic pathway. Increased production of carotenoids can also be achieved by changing culturing conditions or by employing genetic modifications [79]. For example, mRNA levels of β-carotene oxygenase, involved in the biosynthesis of ketocarotenoids [80], increased in *Chlorella zofengiensis* under combined nitrogen starvation and high-light irradiation, and an increase canthaxanthin, zeaxanthin, and astaxanthin was observed [81]. Couso et al. [82] reported an upregulation in lycopene-β-cyclase, which converts lycopene to β-carotene [83] in *C. reinhardtii* under conditions of high light.

Regarding genetic modifications, Cordero [84] transformed the green microalga *C. reinhardtii* by overexpressing a phytoene synthase (which converts geranylgeranyl pyrophosphate to phytoene) isolated from *Chlorella zofingiensis*, resulting in a 2.0- and 2.2-fold increase in violaxanthin and lutein production, respectively. A phytoene desaturase, which transforms the colorless phytoene into the red-colored lycopene [85], was mutated in *H. pluvialis* by Steinbrenner and Sandmann [86], resulting in the upregulation of the enzyme and an increase in astaxanthin production. Galarza and colleagues expressed a nuclear phytoene desaturase in the plastidial genome of *H. pluvialis*, resulting in a 67% higher astaxanthin accumulation when the strain was grown under stressful conditions [87]. The insertion of a β-carotene hydroxylase from *C. reinhardtii* in *Dunaliella salina* resulted in a 3-fold increase of violaxanthin and a 2-fold increase of zeaxanthin [78]. The inhibition of *D. salina* phytoene desaturase using RNAi technology [88] resulted in an increase in phytoene content, but also a decrease in photosynthetic efficiency and growth rate.

More modern methods which have been used include the use of CRISPR/Cas9 (clustered regularly interspaced short palindromic repeats/CRISPR-associated protein 9) for precise and highly efficient "knock-out" of key genes [89]. For example, Baek et al. have used CRISPR/Cas9 to knock-out the zeaxanthin epoxidase (ZEP) gene in *C. reinhardtii* [90]. This enzyme is involved in the conversion of zeaxantin to violaxantin [91], and with its knock-out they obtained a 47-fold increase in zeaxanthin productivity. The current state-of-art involved in metabolic engineering for carotenoid production in microalgae is further discussed in other reviews [72,92].

Other studies have focused on enzymes involved in the synthesis of oxylipins, which are secondary metabolites that have previously been shown to have antipredator and anticancer activities [93–95]. Although oxylipin chemistry and putative biosynthetic pathways have been extensively studied in both plants and microalgae [96–98], the related enzymes and genes have only recently been identified and characterized in microalgae. Adelfi and coworkers have studied genes involved in the biosynthesis of oxylipins in *Pseudo-nitzchia multistriata* and performed transcriptome analysis on these genes in *Pseudo-nitzchia arenysensis* [99]. In diatoms, they characterized, for the first time, two patatin-like lypolitic acid hydrolases (LAH1) involved in the release of the fatty acid precursors of oxylipins and tested their galactolipase activity in vitro. Transcriptomic analysis also revealed three of seven putative patatin genes (g9879, g2582, and g3354) in *N. oceanica* and demonstrated that they were u-regulated under nitrogen-starvation conditions [100]. Similarly, Lauritano and coworkers analyzed the transcriptome of the green alga *Tetraselmis suecica* and reported three PLAT (Polycystin-1, Lipoxygenase, Alpha-Toxin)/LH2 (Lipoxygenase homology) domain transcripts [68]. The group also performed in silico domain assessment and structure predictions. The enzymes discussed in this section are described in Table 2.

Table 2. Enzymes from Microalgae for Healthcare Applications. Marine, freshwater, and soil strain sources are abbreviated as M, F, or S, respectively. Algal classes of *Chlorophyceae, Trebouxiophyceae, Bacillariophyceae, Dinophyceae,* and *Chlorodendrophyceae*, are abbreviated as CH, TR, BA, DY, and CR respectively.

Reference	Enzymes	Microalgae	Strain Source	Microalgal Class	Main Results
[78]	β-carotene hydroxylase	*Dunaliella salina*	M	CH	Increase in violaxanthin and zeaxanthin production
[81]	β-carotene oxygenase	*Chlorella zofingiensis*	S	TR	Increase in canthaxanthin, zeaxanthin and astaxanthin production under combined nitrogen starvation and high light stress
[56]	L-asparaginase	*Chlamydomonas* spp.	F	CH	Enzyme purified and tested
[57]	L-asparaginase	*Chlorella vulgaris*	F, S	TR	Screening of 40 microalgal isolates searching for new L-asparaginase sources
[82]	lycopene-β-cyclase	*Chlamydomonas reinhardtii*	F	CH	Increased gene expression under high light stress
[99]	lypolitic acid hydrolase 1	*Pseudo-nitzschia multistrata, Pseudo-nitzschia arenysensis*	M	BA	Enzyme finding, characterization and retrieval of homologous sequences in other diatoms
[69]	non-ribosomal peptide synthase	*Karenia brevis*	M	DY	Gene cluster identification and chloroplastic localization identification
[68]	polycystin-1, Lipoxygenase, Alpha-Toxin/ lipoxygenase homology 2	*Tetraselmis suecica*	M	CR	Three putative enzyme sequences identification and in silico domain assessment and structure prediction
[88]	phytoene desaturase	*Dunaliella salina*	M	CH	Increase in phytoene production
[84]	phytoene synthase	*Chlamydomonas reinhardtii*	F	CH	Increase in violaxanthin (2.0 fold) and lutein (2.2-fold) production
[86]	phytoene desaturase	*Haematococcus pluvialis*	F	CH	Increase in astaxanthin production
[87]	phytoene desaturase	*Haematococcus pluvialis*	F	CH	Increase in astaxanthin production
[64]	polyketide synthase	*Amphidinium carterae*	M	DY	Identification of a transcript coding for type I PKS β-ketosynthase domain

Table 2. Cont.

Reference	Enzymes	Microalgae	Strain Source	Microalgal Class	Main Results
[65]	polyketide synthase	Azadinium spinosum	M	DY	Identification of type I PKS domains using a combination of genomic and transcriptomic anayses
[66]	polyketide synthase	Gamberdiscus polynesiensis, Gamberdiscus excentricus	M	CH	Identification of transcripts coding for type I and type II PKS domains
[67]	polyketide synthase	Karenia brevis	M	DY	Identification of eight transcripts, six of which coding for type I PKS catalytic domains
[90]	zeaxanthin epoxidase	Chlamydomonas reinhardtii	F	CH	Increase in zeaxanthin production of 47-fold

2.3. Enzymes for Bioremediation

Bioremediation is the use of microorganisms and their enzymes for the degradation and/or transformation of toxic pollutants into less dangerous metabolites/moieties. The potential, which microalgae possess to proliferate in environments that are rich in nutrients (e.g., eutrophic environments) and to biosequestrate heavy metal ions, makes them ideal candidate organisms for bioremediation strategies [101,102]. The optimal goal in this area is to combine bioremediation activities with the possibility of extracting lipids and other high-value added compounds from the biomass that is produced [103–106] in order to reduce overall costs and to recycle materials. In this section, the focus will be on enzymatic bioremediation, which is a novel approach involving the direct use of purified or partially purified enzymes from microorganisms, and in this case, from microalgae, in order to detoxify a specific toxicant/pollutant [107]. This method has recently started to demonstrate promising results through the use of bacterial enzymes [108,109]. Examples are the use of enzymes for the bioremediation of industrial waste and, in particular, the recent use of chromate reductases found in chromium resistant bacteria, known to detoxify the highly toxic chromium Cr(VI) to the less-toxic Cr(III) [110].

In microalgae, a recent study focused on Cr(VI) reduction involving *C. vulgaris* [111]. This activity was suggested to involve both a biological route, through the putative enzyme chromium reductase, and a nonbiological route: Using the scavenger molecule glutathione (GSH). With respect to chromium removal, several strains of microalgae have been reported to be capable of achieving Cr(IV) removal from water bodies, including *Scenedesmus* and *Chlorella* species [112–114]. In the aforementioned transcriptome study on the green algae *Tetraselmis suecica*, a transcript for a putative nitrilase was reported [68]. Given that nitrilases are enzymes that catalyze the hydrolysis of nitriles to carboxylic acids and ammonia [115] and that this enzyme has recently been used for cyanide bioremediation in wastewaters [116], this nitrilase in *T. suecica* may prove to be useful in the treatment of cyanide contaminated water bodies.

Other enzymes have been reported to be overexpressed in microalgae when they are exposed to contaminants, but it is not clear whether or not they are directly involved in their degradation or whether they are produced as a stress defensive response in the cell in order to help balance cellular homeostasis (e.g., to detoxify reactive oxygen/nitrogen species produced after exposure to contaminants). Examples of these enzymes include peroxidases (Px), superoxide dismutase (SOD), catalase (CAT), and glutathione reductase (GR). SOD, Px, and CAT typically function in helping detoxify the cell from oxygen reactive species [117,118], while GR replenishes bioavailable glutathione, catalyzing the reduction of glutathione disulfide (GSSG) to the sulfhydryl form (GSH) [119]. Regarding the detoxification of reactive nitrogen species, the most studied enzymes in microalgae are the nitrate and nitrite reductases. The first enzyme reduces nitrate (NO_3^-) to nitrite (NO_2^-), while the second subsequently reduces nitrite to ammonia (NH_4^+). NH_4^+ is then assimilated into amino acids via the glutamine synthetase/glutamine-2-oxoglutarate amino-transferase cycle [120] (Figure 5).

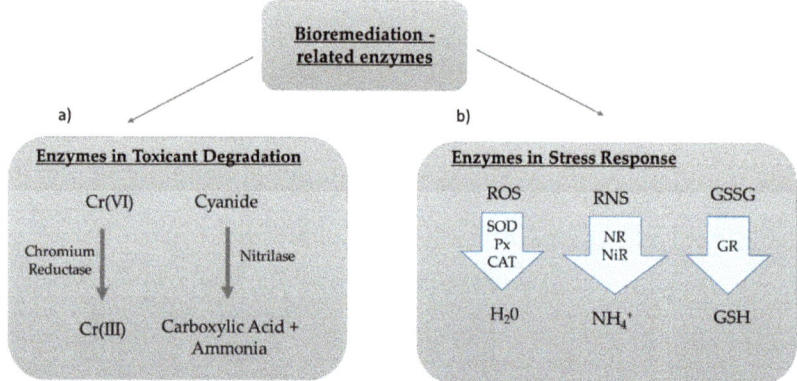

Figure 5. Enzymes for Bioremediation. Enzymes for Bioremediation can be: (**a**) Enzymes directly used for the degradation of toxicant compounds to less or non toxic versions (e.g., the hexavalent Chromium is converted to the less toxic trivalent Chromium due to the activity of Chromium Reductase); (**b**) Enzymes involved in cellular stress response mechanisms, such as peroxidases (Px), superoxide dismutase (SOD), and catalase (CAT) that detoxify reactive oxygen species (ROS), nitrate reductase (NR), and nitrite reductase (NiR) that detoxify reactive nitrogen species (RNS) in ammonium, and GR, that catalyzes the reduction of glutathione disulfide (GSSG) to glutathione (GSH).

For example, peroxidase activity has been reported in extracts from the green alga *Selenastrum capricornutum* (now named *Raphidocelis subcapitata* [121]), which was highly sensitive to very small concentrations of copper (Cu) (0.1 mM), and the authors proposed that the enzyme could be employed as a sensitive bioindicator of copper contamination in fresh waters [122]. Levels of Px, SOD, CAT, and GR have been reported to be upregulated following Cu contamination in *P. tricornutum* and following lead (Pb) contamination in two lichenic microalgal strains from the *Trebouxia* genus (prov. names, TR1 and TR9) [123,124]. In Morelli's work, an increase of 200% in CAT activity indicated its important role in Cu detoxification. In contrast, Alvarez and coworkers reported that Px, SOD, CAT, and GR activity was higher in TR1 than in TR9 under control conditions (with the exception of CAT), while prolonged exposure to Pb resulted in the enzymatic activities of the two microalgae changing to similar levels, reflecting the different physiological and anatomical adaptations of the two organisms. TR1 possesses a thinner cell wall, thereby requiring it to have a more efficient basal enzymatic defence system, while TR9 has a thicker cell wall and induces the expression of intracellular defense mechanisms when the contaminant concentrations are high and physical barriers are no longer effective. Further studies will be required to assess whether these TR1 enzymes are more efficient than enzymes from other microalgal sources and the potential applications that these enzymes may have. All of the enzymes discussed in this section are reported in Table 3.

Table 3. Enzymes from Microalgae with utility in Bioremediation applications Marine, freshwater, and lichenic strain sources are abbreviated as M, F, and L respectively. Algal classes of *Trebouxiophyceae*, *Chlorodendrophyceae*, *Chlorophyceae*, and *Bacillariophyceae* are abbreviated as TR, CR, CH, and BA, respectively.

Reference	Enzymes	Microalgae	Strain Source	Microalgal Class	Main Results
[111]	Putative Cr Reductase	*Chlorella vulgaris*	F	TR	Enzymatic Cr conversion (from Cr(VI) to Cr(III)) detected
[68]	Nitrilase	*Tetraselmis suecica*	M	CR	Putative enzyme sequence identification
[122]	Putative ascorbate peroxidase	*Selenastrum capricornutum*	F	CH	High sensitivity to Cu concentration activity
[123]	superoxide-dismutase, catalase, glutathione reductase	*Phaeodactylum tricornutum*	M	BA	Higher detected enzymatic activity after Cu accumulation
[124]	superoxide-dismutase, catalase, glutathione reductase, ascorbate peroxidase	Trebouxia 1 (TR1), Trebouxia 9 (TR9)	L	TR	Constitutive higher enzymatic activity detected in TR1, while exposed to Pb brings TR1 and TR9 enzymatic activities to comparable levels

3. Conclusions and Future Perspectives

Among aquatic organisms that have recently received attention as potential sources of industrially relevant enzymes [125,126], microalgae, in particular, stand out as a new sustainable and ecofriendly source of biological products (e.g., lipids, carotenoids, oxylipins, and polyketides). This review summarized the available information on enzymes from microalgae with possible biotechnological applications, with a particular focus on value-added lipid production, together with healthcare and bioremediation applications.

The promise of microalgae as potential sources of novel enzymes of interest is reflected in the abundance of recent reports in the literature in this area. However, the biotechnological exploitation of their enzymes in comparison to other potential sources has only become more feasible quite recently, primarily due to the implementation of novel isolation and culturing procedures, together with an increase in the availability of -omics data. This data has facilitated the use of a broader array of approaches, such as site-specific mutagenesis, bioinformatics-based searches for genes of interest, and/or the use of genome editing tools (e.g., CRISPR/Cas9 and TILLING), resulting in promising results particularly with respect to high-performance lipid [46] and carotenoid [89] production in different microalgae.

The majority of studies to date have focused on enzymes involved in pathways for lipid synthesis in order to increase their total production or to direct cellular production to lipid classes with applications as nutraceuticals, cosmeceuticals, or as a feedstock for biodiesel production. For this reason, several recent studies have focused on the improvement of lipid production in oleaginous microalgae. In addition, algal biomass is often used for the extraction of both lipids and other value-added products, such as pigments and proteins, in order to maximize the production of useful products such as these at the lowest possible cost [127–129].

Future approaches to maximize the enzymatic potential of microalgae are likely to focus on three different approaches: (1) The use of ever-increasing amounts of available -omics data to optimize microalgal strains for the production of valuable products, through the overexpression of one or more enzymes through the use of genome editing tools; (2) identification and subsequent characterization of metabolic pathways involved in the production of specific bioactives (e.g., polyketides), many of which are still poorly characterized; (3) the search for genes with direct biotechnological applications (e.g., L-asparaginase, chromate reductase, nitrilase) in microalgal genomes and transcriptomes datasets. A common element in all three approaches is the potential use of next generation sequencing based approaches (NGS) [130], the price of which is declining rapidly [131].

The feasibility of employing any of the aforementioned three approaches will be directly influenced by progress in methods to decrease the costs of growth and genetic manipulation of microalgae. The ultimate aim would be to mimic what has happened in the area of bacterial enzymology, where robust pipelines for enzyme discovery have been established. If this could be achieved, then it is clear that microalgae are likely to meet our expectations as a promising source of novel enzymes with utility in a variety of different biotechnological applications.

Author Contributions: G.M.V., P.D.L., A.I., A.D.W.D. and C.L. co-wrote the review.

Funding: G.M.V. was supported by a Stazione Zoologica Ph.D. fellowship via the Open University.

Acknowledgments: Authors thank Servier Medical Art (SMART) website (https://smart.servier.com/) by Servier for the elements of Figure 3. SMART is licensed under a Creative Commons Attribution 3.0 Unported License.

Conflicts of Interest: The authors declare no conflict of interest.

References

1. Schneider, S.H.; Root, T.L.; Mastrandrea, M.D. *Encyclopedia of climate and weather*; Oxford University Press: Oxford, UK, 2011.
2. Bagchi, B. *Water in Biological and Chemical Processes*; Cambridge University Press: Cambridge, UK, 2013.
3. Faber, K. *Biotransformations in Organic Chemistry*; Springer: Heidelberg, Germany, 2011.
4. Cao, Y.; Tan, H. Effects of cellulase on the modification of cellulose. *Carbohydr. Res.* **2002**, *337*, 1291–1296. [CrossRef]
5. Nigam, P.S. Microbial enzymes with special characteristics for biotechnological applications. *Biomolecules* **2013**, *3*, 597–611. [CrossRef] [PubMed]
6. Fernandes, P. Enzymes in food processing: a condensed overview on strategies for better biocatalysts. *Enzyme Res.* **2010**, *2010*, 1–19. [CrossRef] [PubMed]
7. Vellard, M. The enzyme as drug: application of enzymes as pharmaceuticals. *Curr. Opin. Biotechnol.* **2003**, *14*, 444–450. [CrossRef]
8. Andersen, R.A. *Algal culturing techniques*; Academic Press: Cambridge, MA, USA, 2005.
9. Khan, M.I.; Shin, J.H.; Kim, J.D. The promising future of microalgae: current status, challenges, and optimization of a sustainable and renewable industry for biofuels, feed, and other products. *Microb. Cell Fact.* **2018**, *17*, 36. [CrossRef] [PubMed]
10. Martínez Andrade, K.; Lauritano, C.; Romano, G.; Ianora, A. Marine Microalgae with Anti-Cancer Properties. *Mar. Drugs* **2018**, *16*, 165. [CrossRef]
11. Bhalamurugan, G.L.; Valerie, O.; Mark, L. Valuable bioproducts obtained from microalgal biomass and their commercial applications: A review. *Environ. Eng. Res.* **2018**, *23*, 229–241. [CrossRef]
12. Doron, L.; Segal, N.; Shapira, M. Transgene Expression in Microalgae-From Tools to Applications. *Front. Plant Sci.* **2016**, *7*, 505. [CrossRef]
13. Brasil, B.d.S.A.F.; de Siqueira, F.G.; Salum, T.F.C.; Zanette, C.M.; Spier, M.R. Microalgae and cyanobacteria as enzyme biofactories. *Algal Res.* **2017**, *25*, 76–89. [CrossRef]
14. Lauritano, C.; Ianora, A. Grand Challenges in Marine Biotechnology: Overview of Recent EU-Funded Projects. In *Grand Challenges in Marine Biotechnology*; Rampelotto, P.H., Trincone, A., Eds.; Springer: Heidelberg, Germany, 2018; pp. 425–449.
15. Mogharabi, M.; Faramarzi, M.A. Are Algae the Future Source of Enzymes? *Trends Pept. Protein Sci.* **2016**, *1*, 1–6.
16. Chisti, Y. Biodiesel from microalgae. *Biotechnol. Adv.* **2007**, *25*, 294–306. [CrossRef]
17. Sanghvi, A.M.; Martin Lo, Y. Present and Potential Industrial Applications of Macro- and Microalgae. *Recent Patents Food, Nutr. Agric.* **2010**, *2*, 187–194.
18. Bellou, S.; Baeshen, M.N.; Elazzazy, A.M.; Aggeli, D.; Sayegh, F.; Aggelis, G. Microalgal lipids biochemistry and biotechnological perspectives. *Biotechnol. Adv.* **2014**, *32*, 1476–1493. [CrossRef]

19. Moriyama, T.; Toyoshima, M.; Saito, M.; Wada, H.; Sato, N. Revisiting the Algal "Chloroplast Lipid Droplet": The Absence of an Entity That Is Unlikely to Exist. *Plant Physiol.* **2018**, *176*, 1519–1530. [CrossRef]
20. Fukuda, H.; Kondo, A.; Noda, H. Biodiesel fuel production by transesterification of oils. *J. Biosci. Bioeng.* **2001**, *92*, 405–416. [CrossRef]
21. Wells, M.L.; Potin, P.; Craigie, J.S.; Raven, J.A.; Merchant, S.S.; Helliwell, K.E.; Smith, A.G.; Camire, M.E.; Brawley, S.H. Algae as nutritional and functional food sources: revisiting our understanding. *J. Appl. Phycol.* **2017**, *29*, 949–982. [CrossRef]
22. Caporgno, M.P.; Mathys, A. Trends in microalgae incorporation into innovative food products with potential health benefits. *Front. Nutr.* **2018**, *5*, 1–10. [CrossRef]
23. Merchant, S.S.; Kropat, J.; Liu, B.; Shaw, J.; Warakanont, J. TAG, You're it! *Chlamydomonas* as a reference organism for understanding algal triacylglycerol accumulation. *Curr. Opin. Biotechnol.* **2012**, *23*, 352–363. [CrossRef]
24. Xu, Y.; Caldo, K.M.P.; Pal-Nath, D.; Ozga, J.; Lemieux, M.J.; Weselake, R.J.; Chen, G. Properties and Biotechnological Applications of Acyl-CoA:diacylglycerol Acyltransferase and Phospholipid:diacylglycerol Acyltransferase from Terrestrial Plants and Microalgae. *Lipids* **2018**, *53*, 663–688. [CrossRef]
25. Lung, S.-C.; Weselake, R.J. Diacylglycerol acyltransferase: a key mediator of plant triacylglycerol synthesis. *Lipids* **2006**, *41*, 1073–1088. [CrossRef]
26. Shockey, J.M.; Gidda, S.K.; Chapital, D.C.; Kuan, J.-C.; Dhanoa, P.K.; Bland, J.M.; Rothstein, S.J.; Mullen, R.T.; Dyer, J.M. Tung tree DGAT1 and DGAT2 have nonredundant functions in triacylglycerol biosynthesis and are localized to different subdomains of the endoplasmic reticulum. *Plant Cell* **2006**, *18*, 2294–2313. [CrossRef]
27. Guo, X.; Fan, C.; Chen, Y.; Wang, J.; Yin, W.; Wang, R.R.C.; Hu, Z. Identification and characterization of an efficient acyl-CoA: Diacylglycerol acyltransferase 1 (DGAT1) gene from the microalga *Chlorella ellipsoidea*. *BMC Plant Biol.* **2017**, *17*, 1–16. [CrossRef]
28. Wei, H.; Shi, Y.; Ma, X.; Pan, Y.; Hu, H.; Li, Y.; Luo, M.; Gerken, H.; Liu, J. A type-I diacylglycerol acyltransferase modulates triacylglycerol biosynthesis and fatty acid composition in the oleaginous microalga, *Nannochloropsis oceanica*. *Biotechnol. Biofuels* **2017**, *10*, 1–18. [CrossRef]
29. Wagner, M.; Hoppe, K.; Czabany, T.; Heilmann, M.; Daum, G.; Feussner, I.; Fulda, M. Identification and characterization of an acyl-CoA:diacylglycerol acyltransferase 2 (DGAT2) gene from the microalga *Ostreococcus tauri*. *Plant Physiol. Biochem.* **2010**, *48*, 407–416. [CrossRef]
30. Niu, Y.-F.; Zhang, M.-H.; Li, D.-W.; Yang, W.-D.; Liu, J.-S.; Bai, W.-B.; Li, H.-Y. Improvement of Neutral Lipid and Polyunsaturated Fatty Acid Biosynthesis by Overexpressing a Type 2 Diacylglycerol Acyltransferase in Marine Diatom *Phaeodactylum tricornutum*. *Mar. Drugs* **2013**, *11*, 4558–4569. [CrossRef]
31. Manandhar-Shrestha, K.; Hildebrand, M. Characterization and manipulation of a DGAT2 from the diatom *Thalassiosira pseudonana*: Improved TAG accumulation without detriment to growth, and implications for chloroplast TAG accumulation. *Algal Res.* **2015**, *12*, 239–248. [CrossRef]
32. Klaitong, P.; Fa-aroonsawat, S.; Chungjatupornchai, W. Accelerated triacylglycerol production and altered fatty acid composition in oleaginous microalga *Neochloris oleoabundans* by overexpression of diacylglycerol acyltransferase 2. *Microb. Cell Fact.* **2017**, *16*, 1–10. [CrossRef]
33. Li, D.-W.; Cen, S.-Y.; Liu, Y.-H.; Balamurugan, S.; Zheng, X.-Y.; Alimujiang, A.; Yang, W.-D.; Liu, J.-S.; Li, H.-Y. A type 2 diacylglycerol acyltransferase accelerates the triacylglycerol biosynthesis in heterokont oleaginous microalga *Nannochloropsis oceanica*. *J. Biotechnol.* **2016**, *229*, 65–71. [CrossRef]
34. Xin, Y.; Lu, Y.; Lee, Y.-Y.; Wei, L.; Jia, J.; Wang, Q.; Wang, D.; Bai, F.; Hu, H.; Hu, Q.; et al. Producing Designer Oils in Industrial Microalgae by Rational Modulation of Co-evolving Type-2 Diacylglycerol Acyltransferases. *Mol. Plant* **2017**, *10*, 1523–1539. [CrossRef]
35. Deng, X.-D.; Gu, B.; Li, Y.-J.; Hu, X.-W.; Guo, J.-C.; Fei, X.-W. The roles of acyl-CoA: diacylglycerol acyltransferase 2 genes in the biosynthesis of triacylglycerols by the green algae *Chlamydomonas reinhardtii*. *Mol. Plant* **2012**, *5*, 945–947. [CrossRef]
36. La Russa, M.; Bogen, C.; Uhmeyer, A.; Doebbe, A.; Filippone, E.; Kruse, O.; Mussgnug, J.H. Functional analysis of three type-2 DGAT homologue genes for triacylglycerol production in the green microalga *Chlamydomonas reinhardtii*. *J. Biotechnol.* **2012**, *162*, 13–20. [CrossRef]
37. Cui, Y.; Zhao, J.; Wang, Y.; Qin, S.; Lu, Y. Characterization and engineering of a dual-function diacylglycerol acyltransferase in the oleaginous marine diatom *Phaeodactylum tricornutum*. *Biotechnol. Biofuels* **2018**, *11*, 32. [CrossRef]

38. Xue, J.; Balamurugan, S.; Li, D.-W.; Liu, Y.-H.; Zeng, H.; Wang, L.; Yang, W.-D.; Liu, J.-S.; Li, H.-Y. Glucose-6-phosphate dehydrogenase as a target for highly efficient fatty acid biosynthesis in microalgae by enhancing NADPH supply. *Metab. Eng.* **2017**, *41*, 212–221. [CrossRef]
39. Zhu, B.-H.; Tu, C.-C.; Shi, H.-P.; Yang, G.-P.; Pan, K.-H. Overexpression of endogenous delta-6 fatty acid desaturase gene enhances eicosapentaenoic acid accumulation in *Phaeodactylum tricornutum*. *Process Biochem.* **2017**, *57*, 43–49. [CrossRef]
40. Fukuda, S.; Hirasawa, E.; Takemura, T.; Takahashi, S.; Chokshi, K.; Pancha, I.; Tanaka, K.; Imamura, S. Accelerated triacylglycerol production without growth inhibition by overexpression of a glycerol-3-phosphate acyltransferase in the unicellular red alga *Cyanidioschyzon merolae*. *Sci. Rep.* **2018**, *8*, 1–12. [CrossRef]
41. Rengel, R.; Smith, R.T.; Haslam, R.P.; Sayanova, O.; Vila, M.; León, R. Overexpression of acetyl-CoA synthetase (ACS) enhances the biosynthesis of neutral lipids and starch in the green microalga *Chlamydomonas reinhardtii*. *Algal Res.* **2018**, *31*, 183–193. [CrossRef]
42. Osada, K.; Maeda, Y.; Yoshino, T.; Nojima, D.; Bowler, C.; Tanaka, T. Enhanced NADPH production in the pentose phosphate pathway accelerates lipid accumulation in the oleaginous diatom *Fistulifera solaris*. *Algal Res.* **2017**, *23*, 126–134. [CrossRef]
43. Shin, Y.S.; Jeong, J.; Nguyen, T.H.T.; Kim, J.Y.H.; Jin, E.; Sim, S.J. Targeted knockout of phospholipase A2 to increase lipid productivity in *Chlamydomonas reinhardtii* for biodiesel production. *Bioresour. Technol.* **2019**, *271*, 368–374. [CrossRef]
44. Los, D.A.; Murata, N. Structure and expression of fatty acid desaturases. *Biochim. Biophys. Acta Lipids Lipid Metab.* **1998**, *1394*, 3–15. [CrossRef]
45. De Jaeger, L.; Springer, J.; Wolbert, E.J.H.; Martens, D.E.; Eggink, G.; Wijffels, R.H. Gene silencing of stearoyl-ACP desaturase enhances the stearic acid content in *Chlamydomonas reinhardtii*. *Bioresour. Technol.* **2017**, *245*, 1616–1626. [CrossRef]
46. Daboussi, F.; Leduc, S.; Maréchal, A.; Dubois, G.; Guyot, V.; Perez-Michaut, C.; Amato, A.; Falciatore, A.; Juillerat, A.; Beurdeley, M.; et al. Genome engineering empowers the diatom *Phaeodactylum tricornutum* for biotechnology. *Nat. Commun.* **2014**, *5*, 3831. [CrossRef]
47. Sorigué, D.; Légeret, B.; Cuiné, S.; Blangy, S.; Moulin, S.; Billon, E.; Richaud, P.; Brugière, S.; Couté, Y.; Nurizzo, D.; et al. An algal photoenzyme converts fatty acids to hydrocarbons. *Science* **2017**, *357*, 903–907. [CrossRef]
48. Misra, N.; Panda, P.K.; Parida, B.K.; Mishra, B.K. dEMBF: A comprehensive database of enzymes of microalgal biofuel feedstock. *PLoS One* **2016**, *11*, 146–158. [CrossRef]
49. Batool, T.; Makky, E.A.; Jalal, M.; Yusoff, M.M. A comprehensive review on L-asparaginase and its applications. *Appl. Biochem. Biotechnol.* **2016**, *178*, 900–923. [CrossRef]
50. Ali, U.; Naveed, M.; Ullah, A.; Ali, K.; Shah, S.A.; Fahad, S.; Mumtaz, A.S. L-asparaginase as a critical component to combat Acute Lymphoblastic Leukemia (ALL): A novel approach to target ALL. *Eur. J. Pharmacol.* **2016**, *771*, 199–210. [CrossRef]
51. Roberts, J.; Prager, M.D.; Bachynsky, N. The antitumor activity of *Escherichia coli* L-asparaginase. *Cancer Res.* **1966**, *26*, 2213–2217.
52. Peterson, R.E.; Ciegler, A. L-asparaginase production by various bacteria. *Appl. Microbiol.* **1969**, *17*, 929–930.
53. Thenmozhi, C.; Sankar, R.; Karuppiah, V.; Sampathkumar, P. L-asparaginase production by mangrove derived *Bacillus cereus* MAB5: Optimization by response surface methodology. *Asian Pac. J. Trop. Med.* **2011**, *4*, 486–491. [CrossRef]
54. Ahmad, N.; Pandit, N.; Maheshwari, S. L-asparaginase gene-a therapeutic approach towards drugs for cancer cell. *Int. J. Biosci.* **2012**, *2*, 1–11.
55. Vidya, J.; Sajitha, S.; Ushasree, V.; Sindhu, R.; Binod, P.; Madhavan, A.; Pandey, A. Genetic and metabolic engineering approaches for the production and delivery of L-asparaginases: An overview. *Bioresour. Technol.* **2017**, *245*, 1775–1781. [CrossRef]
56. Paul, J.H. Isolation and characterization of a *Chlamydomonas* L-asparaginase. *Biochem. J.* **1982**, *203*, 109–115. [CrossRef]
57. Ebrahiminezhad, A.; Rasoul-Amini, S.; Ghoshoon, M.B.; Ghasemi, Y. *Chlorella vulgaris*, a novel microalgal source for L-asparaginase production. *Biocatal. Agric. Biotechnol.* **2014**, *3*, 214–217. [CrossRef]
58. Sasso, S.; Pohnert, G.; Lohr, M.; Mittag, M.; Hertweck, C. Microalgae in the postgenomic era: a blooming reservoir for new natural products. *FEMS Microbiol. Rev.* **2012**, *36*, 761–785. [CrossRef]

59. Berry, J. Marine and freshwater microalgae as a potential source of novel herbicides. In *Herbicides and Environment*, Kortekamp A.; InTechOpen: London, UK, 2011.
60. Kobayashi, J. Amphidinolides and its related macrolides from marine dinoflagellates. *J. Antibiot. (Tokyo)* **2008**, *61*, 271–284. [CrossRef]
61. Kellmann, R.; Stüken, A.; Orr, R.J.S.; Svendsen, H.M.; Jakobsen, K.S. Biosynthesis and molecular genetics of polyketides in marine dinoflagellates. *Mar. Drugs* **2010**, *8*, 1011–1048. [CrossRef]
62. Kohli, G.S.; John, U.; Van Dolah, F.M.; Murray, S.A. Evolutionary distinctiveness of fatty acid and polyketide synthesis in eukaryotes. *ISME J.* **2016**, *10*, 1877–1890. [CrossRef]
63. Jenke-Kodama, H.; Sandmann, A.; Müller, R.; Dittmann, E. Evolutionary implications of bacterial polyketide synthases. *Mol. Biol. Evol.* **2005**, *22*, 2027–2039. [CrossRef]
64. Lauritano, C.; De Luca, D.; Ferrarini, A.; Avanzato, C.; Minio, A.; Esposito, F.; Ianora, A. De novo transcriptome of the cosmopolitan dinoflagellate *Amphidinium carterae* to identify enzymes with biotechnological potential. *Sci. Rep.* **2017**, *7*, 11701. [CrossRef]
65. Meyer, J.M.; Rödelsperger, C.; Eichholz, K.; Tillmann, U.; Cembella, A.; McGaughran, A.; John, U. Transcriptomic characterisation and genomic glimps into the toxigenic dinoflagellate *Azadinium spinosum*, with emphasis on polykeitde synthase genes. *BMC Genomics* **2015**, *16*, 27. [CrossRef]
66. Kohli, G.S.; Campbell, K.; John, U.; Smith, K.F.; Fraga, S.; Rhodes, L.L.; Murray, S.A. Role of modular polyketide synthases in the production of polyether ladder compounds in ciguatoxin-producing *Gambierdiscus polynesiensis* and *G. excentricus* (Dinophyceae). *J. Eukaryot. Microbiol.* **2017**, *64*, 691–706. [CrossRef]
67. Monroe, E.A.; Van Dolah, F.M. The toxic dinoflagellate *Karenia brevis* encodes novel Type I-like polyketide synthases containing discrete catalytic domains. *Protist* **2008**, *159*, 471–482. [CrossRef]
68. Lauritano, C.; De Luca, D.; Amoroso, M.; Benfatto, S.; Maestri, S.; Racioppi, C.; Esposito, F.; Ianora, A. New molecular insights on the response of the green alga *Tetraselmis suecica* to nitrogen starvation. *Sci. Rep.* **2019**, *9*, 3336. [CrossRef]
69. López-Legentil, S.; Song, B.; DeTure, M.; Baden, D.G. Characterization and localization of a hybrid non-ribosomal peptide synthetase and polyketide synthase gene from the toxic dinoflagellate *Karenia brevis*. *Mar. Biotechnol.* **2010**, *12*, 32–41. [CrossRef]
70. Fiedor, J.; Burda, K. Potential role of carotenoids as antioxidants in human health and disease. *Nutrients* **2014**, *6*, 466–488. [CrossRef]
71. Musser, A.J.; Maiuri, M.; Brida, D.; Cerullo, G.; Friend, R.H.; Clark, J. The nature of singlet exciton fission in carotenoid aggregates. *J. Am. Chem. Soc.* **2015**, *137*, 5130–5139. [CrossRef]
72. Gong, M.; Bassi, A. Carotenoids from microalgae: A review of recent developments. *Biotechnol. Adv.* **2016**, *34*, 1396–1412. [CrossRef]
73. Rammuni, M.N.; Ariyadasa, T.U.; Nimarshana, P.H.V.; Attalage, R.A. Comparative assessment on the extraction of carotenoids from microalgal sources: Astaxanthin from *H. pluvialis* and β-carotene from *D. salina*. *Food Chem.* **2019**, *277*, 128–134. [CrossRef]
74. Lamers, P.P.; Janssen, M.; De Vos, R.C.H.; Bino, R.J.; Wijffels, R.H. Exploring and exploiting carotenoid accumulation in *Dunaliella salina* for cell-factory applications. *Trends Biotechnol.* **2008**, *26*, 631–638. [CrossRef]
75. Prieto, A.; Pedro Cañavate, J.; García-González, M. Assessment of carotenoid production by *Dunaliella salina* in different culture systems and operation regimes. *J. Biotechnol.* **2011**, *151*, 180–185. [CrossRef]
76. Besson, A.; Formosa-Dague, C.; Guiraud, P. Flocculation-flotation harvesting mechanism of *Dunaliella salina*: From nanoscale interpretation to industrial optimization. *Water Res.* **2019**, *155*, 352–361. [CrossRef]
77. Tan, C.; Qin, S.; Zhang, Q.; Jiang, P.; Zhao, F. Establishment of a micro-particle bombardment transformation system for *Dunaliella salina*. *J. Microbiol.* **2005**, *43*, 361–365.
78. Simon, D.P.; Anila, N.; Gayathri, K.; Sarada, R. Heterologous expression of β-carotene hydroxylase in *Dunaliella salina* by *Agrobacterium*-mediated genetic transformation. *Algal Res.* **2016**, *18*, 257–265. [CrossRef]
79. Saini, D.K.; Chakdar, H.; Pabbi, S.; Shukla, P. Enhancing production of microalgal biopigments through metabolic and genetic engineering. *Crit. Rev. Food Sci. Nutr.* **2019**, 1–15. [CrossRef]
80. Huang, J.C.; Wang, Y.; Sandmann, G.; Chen, F. Isolation and characterization of a carotenoid oxygenase gene from *Chlorella zofingiensis* (Chlorophyta). *Appl. Microbiol. Biotechnol.* **2006**, *71*, 473–479. [CrossRef]
81. Cordero, B.F.; Couso, I.; Leon, R.; Rodriguez, H.; Vargas, M.A. Isolation and characterization of a lycopene ε-cyclase gene of *Chlorella (Chromochloris) zofingiensis*. Regulation of the carotenogenic pathway by nitrogen and light. *Mar. Drugs* **2012**, *10*, 2069–2088. [CrossRef]

82. Couso, I.; Vila, M.; Vigara, J.; Cordero, B.F.; Vargas, M.Á.; Rodríguez, H.; León, R. Synthesis of carotenoids and regulation of the carotenoid biosynthesis pathway in response to high light stress in the unicellular microalga *Chlamydomonas reinhardtii*. *Eur. J. Phycol.* **2012**, *47*, 223–232. [CrossRef]
83. Cunningham, F.X.; Pogson, B.; Sun, Z.; McDonald, K.A.; DellaPenna, D.; Gantt, E.; Gantt, E. Functional analysis of the beta and epsilon lycopene cyclase enzymes of *Arabidopsis* reveals a mechanism for control of cyclic carotenoid formation. *Plant Cell* **1996**, *8*, 1613–1626. [CrossRef]
84. Cordero, B.F.; Couso, I.; León, R.; Rodríguez, H.; Vargas, M.Á. Enhancement of carotenoids biosynthesis in *Chlamydomonas reinhardtii* by nuclear transformation using a phytoene synthase gene isolated from *Chlorella zofingiensis*. *Appl. Microbiol. Biotechnol.* **2011**, *91*, 341–351. [CrossRef]
85. Fraser, P.D.; Misawa, N.; Linden, H.; Yamano, S.; Kobayashi, K.; Sandmann, G. Expression in *Escherichia coli*, purification, and reactivation of the recombinant *Erwinia uredovora* phytoene desaturase. *J. Biol. Chem.* **1992**, *267*, 19891–19895.
86. Steinbrenner, J.; Sandmann, G. Transformation of the green alga *Haematococcus pluvialis* with a phytoene desaturase for accelerated astaxanthin biosynthesis. *Appl. Environ. Microbiol.* **2006**, *72*, 7477–7484. [CrossRef]
87. Galarza, J.I.; Gimpel, J.A.; Rojas, V.; Arredondo-Vega, B.O.; Henríquez, V. Over-accumulation of astaxanthin in *Haematococcus pluvialis* through chloroplast genetic engineering. *Algal Res.* **2018**, *31*, 291–297. [CrossRef]
88. Srinivasan, R.; Babu, S.; Gothandam, K.M. Accumulation of phytoene, a colorless carotenoid by inhibition of phytoene desaturase (PDS) gene in *Dunaliella salina* V-101. *Bioresour. Technol.* **2017**, *242*, 311–318. [CrossRef]
89. Cong, L.; Ran, F.A.; Cox, D.; Lin, S.; Barretto, R.; Habib, N.; Hsu, P.D.; Wu, X.; Jiang, W.; Marraffini, L.A.; et al. Multiplex genome engineering using CRISPR/Cas systems. *Science.* **2013**, *339*, 819–823. [CrossRef]
90. Baek, K.; Yu, J.; Jeong, J.; Sim, S.J.; Bae, S.; Jin, E. Photoautotrophic production of macular pigment in a *Chlamydomonas reinhardtii* strain generated by using DNA-free CRISPR-Cas9 RNP-mediated mutagenesis. *Biotechnol. Bioeng.* **2018**, *115*, 719–728. [CrossRef]
91. Frommolt, R.; Goss, R.; Wilhelm, C. The de-epoxidase and epoxidase reactions of *Mantoniella squamata* (Prasinophyceae) exhibit different substrate-specific reaction kinetics compared to spinach. *Planta* **2001**, *213*, 446–456. [CrossRef]
92. Gimpel, J.A.; Henríquez, V.; Mayfield, S.P. In Metabolic Engineering of Eukaryotic Microalgae: Potential and Challenges Come with Great Diversity. *Front. Microbiol.* **2015**, *6*, 1376. [CrossRef]
93. De los Reyes, C.; Ávila-Román, J.; Ortega, M.J.; de la Jara, A.; García-Mauriño, S.; Motilva, V.; Zubía, E. Oxylipins from the microalgae *Chlamydomonas debaryana* and *Nannochloropsis gaditana* and their activity as TNF-α inhibitors. *Phytochemistry* **2014**, *102*, 152–161. [CrossRef]
94. Lauritano, C.; Romano, G.; Roncalli, V.; Amoresano, A.; Fontanarosa, C.; Bastianini, M.; Braga, F.; Carotenuto, Y.; Ianora, A. New oxylipins produced at the end of a diatom bloom and their effects on copepod reproductive success and gene expression levels. *Harmful Algae* **2016**, *55*, 221–229. [CrossRef]
95. Ávila-Román, J.; Talero, E.; de los Reyes, C.; García-Mauriño, S.; Motilva, V. Microalgae-derived oxylipins decrease inflammatory mediators by regulating the subcellular location of NFκB and PPAR-γ. *Pharmacol. Res.* **2018**, *128*, 220–230. [CrossRef]
96. Pohnert, G. Phospholipase A2 activity triggers the wound-activated chemical defense in the diatom *Thalassiosira rotula*. *Plant Physiol.* **2002**, *129*, 103–111. [CrossRef]
97. Matos, A.R.; Pham-Thi, A.T. Lipid deacylating enzymes in plants: Old activities, new genes. *Plant Physiol. Biochem.* **2009**, *47*, 491–503. [CrossRef]
98. Cutignano, A.; Lamari, N.; D'ippolito, G.; Manzo, E.; Cimino, G.; Fontana, A. Lipoxygenase products in marine diatoms: A concise analytical method to explore the functional potential of oxylipins. *J. Phycol.* **2011**, *47*, 233–243. [CrossRef]
99. Adelfi, M.G.; Vitale, R.M.; d'Ippolito, G.; Nuzzo, G.; Gallo, C.; Amodeo, P.; Manzo, E.; Pagano, D.; Landi, S.; Picariello, G.; et al. Patatin-like lipolytic acyl hydrolases and galactolipid metabolism in marine diatoms of the genus *Pseudo-nitzschia*. *Biochim. Biophys. Acta Mol. Cell Biol. Lipids* **2019**, *1864*, 181–190. [CrossRef]
100. Li, J.; Han, D.; Wang, D.; Ning, K.; Jia, J.; Wei, L.; Jing, X.; Huang, S.; Chen, J.; Li, Y.; et al. Choreography of Transcriptomes and Lipidomes of *Nannochloropsis* Reveals the Mechanisms of Oil Synthesis in Microalgae. *Plant Cell* **2014**, *26*, 1645–1665. [CrossRef]
101. De la Noüe, J.; Laliberté, G.; Proulx, D. Algae and waste water. *J. Appl. Phycol.* **1992**, *4*, 247–254. [CrossRef]
102. Mathimani, T.; Pugazhendhi, A. Utilization of algae for biofuel, bio-products and bio-remediation. *Biocatal. Agric. Biotechnol.* **2019**, *17*, 326–330. [CrossRef]

103. Kuo, C.-M.; Chen, T.-Y.; Lin, T.-H.; Kao, C.-Y.; Lai, J.-T.; Chang, J.-S.; Lin, C.-S. Cultivation of *Chlorella* sp. GD using piggery wastewater for biomass and lipid production. *Bioresour. Technol.* **2015**, *194*, 326–333. [CrossRef]
104. Kim, H.-C.; Choi, W.J.; Chae, A.N.; Park, J.; Kim, H.J.; Song, K.G. Evaluating integrated strategies for robust treatment of high saline piggery wastewater. *Water Res.* **2016**, *89*, 222–231. [CrossRef]
105. Hemalatha, M.; Sravan, J.S.; Yeruva, D.K.; Venkata Mohan, S. Integrated ecotechnology approach towards treatment of complex wastewater with simultaneous bioenergy production. *Bioresour. Technol.* **2017**, *242*, 60–67. [CrossRef]
106. Rugnini, L.; Costa, G.; Congestri, R.; Antonaroli, S.; Sanità di Toppi, L.; Bruno, L. Phosphorus and metal removal combined with lipid production by the green microalga *Desmodesmus sp.*: An integrated approach. *Plant Physiol. Biochem.* **2018**, *125*, 45–51. [CrossRef]
107. Sharma, B.; Dangi, A.K.; Shukla, P. Contemporary enzyme based technologies for bioremediation: a review. *J. Environ. Manage.* **2018**, *210*, 10–22. [CrossRef]
108. Thatoi, H.; Das, S.; Mishra, J.; Rath, B.P.; Das, N. Bacterial chromate reductase, a potential enzyme for bioremediation of hexavalent chromium: a review. *J. Environ. Manage.* **2014**, *146*, 383–399. [CrossRef]
109. Sivaperumal, P.; Kamala, K.; Rajaram, R. Bioremediation of industrial waste through enzyme producing marine microorganisms. In *Advances in Food and Nutrition Research*; Academic Press: Cambridge, MA, USA, 2017; Volume 80, pp. 165–179, ISBN 9780128095874.
110. Joutey, N.T.; Sayel, H.; Bahafid, W.; El Ghachtouli, N. Mechanisms of hexavalent chromium resistance and removal by microorganisms. *Rev. Environ. Contam. Toxicol.* **2015**, *233*, 45–69.
111. Yen, H.-W.; Chen, P.-W.; Hsu, C.-Y.; Lee, L. The use of autotrophic *Chlorella vulgaris* in chromium (VI) reduction under different reduction conditions. *J. Taiwan Inst. Chem. Eng.* **2017**, *74*, 1–6. [CrossRef]
112. Han, X.; Wong, Y.S.; Wong, M.H.; Tam, N.F.Y. Biosorption and bioreduction of Cr(VI) by a microalgal isolate, *Chlorella miniata*. *J. Hazard. Mater.* **2007**, *146*, 65–72. [CrossRef]
113. Jácome-Pilco, C.R.; Cristiani-Urbina, E.; Flores-Cotera, L.B.; Velasco-García, R.; Ponce-Noyola, T.; Cañizares-Villanueva, R.O. Continuous Cr(VI) removal by *Scenedesmus incrassatulus* in an airlift photobioreactor. *Bioresour. Technol.* **2009**, *100*, 2388–2391. [CrossRef]
114. Pradhan, D.; Sukla, L.B.; Mishra, B.B.; Devi, N. Biosorption for removal of hexavalent chromium using microalgae *Scenedesmus sp*. *J. Clean. Prod.* **2019**, *209*, 617–629. [CrossRef]
115. Raczynska, J.E.; Vorgias, C.E.; Antranikian, G.; Rypniewski, W. Crystallographic analysis of a thermoactive nitrilase. *J. Struct. Biol.* **2011**, *173*, 294–302. [CrossRef]
116. Park, J.M.; Trevor Sewell, B.; Benedik, M.J. Cyanide bioremediation: the potential of engineered nitrilases. *Appl. Microbiol. Biotechnol.* **2017**, *101*, 3029–3042. [CrossRef]
117. Cirulis, J.T.; Scott, J.A.; Ross, G.M. Management of oxidative stress by microalgae. *Can. J. Physiol. Pharmacol.* **2013**, *91*, 15–21. [CrossRef]
118. Lauritano, C.; Orefice, I.; Procaccini, G.; Romano, G.; Ianora, A. Key genes as stress indicators in the ubiquitous diatom *Skeletonema marinoi*. *BMC Genomics* **2015**, *16*, 411. [CrossRef]
119. Couto, N.; Wood, J.; Barber, J. The role of glutathione reductase and related enzymes on cellular redox homoeostasis network. *Free Radic. Biol. Med.* **2016**, *95*, 27–42. [CrossRef]
120. Rogato, A.; Amato, A.; Iudicone, D.; Chiurazzi, M.; Ferrante, M.I.; d'Alcalà, M.R. The diatom molecular toolkit to handle nitrogen uptake. *Mar. Genomics* **2015**, *24*, 95–108. [CrossRef]
121. Suzuki, S.; Yamaguchi, H.; Nakajima, N.; Kawachi, M. *Raphidocelis subcapitata* (=*Pseudokirchneriella subcapitata*) provides an insight into genome evolution and environmental adaptations in the *Sphaeropleales*. *Sci. Rep.* **2018**, *8*, 8058. [CrossRef]
122. Sauser, K.R.; Liu, J.K.; Wong, T.-Y. Identification of a copper-sensitive ascorbate peroxidase in the unicellular green alga *Selenastrum capricornutum*. *Biometals* **1997**, *10*, 163–168. [CrossRef]
123. Morelli, E.; Scarano, G. Copper-induced changes of non-protein thiols and antioxidant enzymes in the marine microalga *Phaeodactylum tricornutum*. *Plant Sci.* **2004**, *167*, 289–296. [CrossRef]
124. Álvarez, R.; del Hoyo, A.; García-Breijo, F.; Reig-Armiñana, J.; del Campo, E.M.; Guéra, A.; Barreno, E.; Casano, L.M. Different strategies to achieve Pb-tolerance by the two *Trebouxia* algae coexisting in the lichen *Ramalina farinacea*. *J. Plant Physiol.* **2012**, *169*, 1797–1806. [CrossRef]

125. Kennedy, J.; Margassery, L.M.; Morrissey, J.P.; O'Gara, F.; Dobson, A.D.W. Metagenomic strategies for the discovery of novel enzymes with biotechnological application from marine ecosystems. *Mar. Enzym. Biocatal.* **2013**, 109–130.
126. Trincone, A. Enzymatic processes in marine biotechnology. *Mar. Drugs* **2017**, *15*, 93. [CrossRef]
127. Raut, N.; Al-Balushi, T.; Panwar, S.; Vaidya, R.S.; Shinde, G.B. Microalgal Biofuel. In *Biofuels-Status and Perspective*; InTech: London, UK, 2015.
128. Ruiz, J.; Olivieri, G.; de Vree, J.; Bosma, R.; Willems, P.; Reith, J.H.; Eppink, M.H.M.; Kleinegris, D.M.M.; Wijffels, R.H.; Barbosa, M.J. Towards industrial products from microalgae. *Energy Environ. Sci.* **2016**, *9*, 3036–3043. [CrossRef]
129. Gifuni, I.; Pollio, A.; Safi, C.; Marzocchella, A.; Olivieri, G. Current Bottlenecks and Challenges of the Microalgal Biorefinery. *Trends Biotechnol.* **2018**, *37*, 242–252. [CrossRef]
130. Lauritano, C.; Ferrante, M.I.; Rogato, A. Marine Natural Products from Microalgae: An -Omics Overview. *Mar. Drugs* **2019**, *17*, 269. [CrossRef]
131. Mardis, E.R. A decade's perspective on DNA sequencing technology. *Nature* **2011**, *470*, 198–203. [CrossRef]

© 2019 by the authors. Licensee MDPI, Basel, Switzerland. This article is an open access article distributed under the terms and conditions of the Creative Commons Attribution (CC BY) license (http://creativecommons.org/licenses/by/4.0/).

Article

Biochemical Characterization and Elucidation of Action Pattern of a Novel Polysaccharide Lyase 6 Family Alginate Lyase from Marine Bacterium *Flammeovirga* sp. NJ-04

Qian Li, Fu Hu, Benwei Zhu *, Yun Sun and Zhong Yao

College of Food Science and Light Industry, Nanjing Tech University, Nanjing 211816, China; njlq@njtech.edu.cn (Q.L.); hufu@njtech.edu.cn (F.H.); sunyun_food@njtech.edu.cn (Y.S.); yaozhong@njtech.edu.cn (Z.Y.)
* Correspondence: zhubenwei@njtech.edu.cn; Tel.: +86-25-5813-9419

Received: 7 May 2019; Accepted: 24 May 2019; Published: 31 May 2019

Abstract: Alginate lyases have been widely used to prepare alginate oligosaccharides in food, agricultural, and medical industries. Therefore, discovering and characterizing novel alginate lyases with excellent properties has drawn increasing attention. Herein, a novel alginate lyase FsAlyPL6 of Polysaccharide Lyase (PL) 6 family is identified and biochemically characterized from *Flammeovirga* sp. NJ-04. It shows highest activity at 45 °C and could retain 50% of activity after being incubated at 45 °C for 1 h. The Thin-Layer Chromatography (TLC) and Electrospray Ionization Mass Spectrometry (ESI-MS) analysis indicates that FsAlyPL6 endolytically degrades alginate polysaccharide into oligosaccharides ranging from monosaccharides to pentasaccharides. In addition, the action pattern of the enzyme is also elucidated and the result suggests that FsAlyPL6 could recognize tetrasaccharide as the minimal substrate and cleave the glycosidic bonds between the subsites of −1 and +3. The research provides extended insights into the substrate recognition and degradation pattern of PL6 alginate lyases, which may further expand the application of alginate lyases.

Keywords: alginate lyase; polysaccharide lyase of family 6; characterization; degradation pattern

1. Introduction

Alginate is a linear acidic polysaccharide that constitutes the cell wall of brown algae [1]. It consists of two uronic acids, namely the β-D-mannuronate (M) and the α-L-guluronate (G), which are randomly arranged into different blocks [2]. The alginate has been widely used in food, agricultural and medical industries due to its favorable properties and versatile activities. However, the applications of alginate have been greatly limited by its disadvantages such as high molecular weight, low solubility, and poor bioavailability. In addition, the alginate molecule could not get into the circulation system due to its huge molecular structure. Therefore, it could not exhibit its physiological activities. Alginate oligosaccharides, as the degrading products of alginate, are smaller with excellent solubility and bioavailability than the polysaccharides. In addition, the physiological effects, such as anticoagulant, antioxidant, and antineoplastic activities, can also be retained after degradation. Therefore, they have been widely used as anticoagulants, plant growth accelerators and tumor inhibitors in food, agricultural, and medical fields [3–5]. Therefore, it holds great promise to degrade the alginate to prepare functional alginate oligosaccharides [6].

Alginate lyases could degrade alginate to oligosaccharides by β-elimination mechanism and therefore they belong to the Polysaccharides Lyase (PL) family [7]. Recently, alginate lyases have drawn increasing attention for preparing alginate oligosaccharides with the advantages such as high efficiency and specificity and mild degrading conditions [8]. Up to now, numerous alginate lyases have been isolated, identified, and characterized [9]. Unfortunately, only a few show high activity and thermal stability, which are essential properties for industrial applications [10,11]. Previously, two alginate lyases with excellent characteristics have been identified from the *Flammeovirga* sp. NJ-04. In this study, a novel alginate lyase of PL 6 family has been cloned and characterized from the strain. The biochemical properties and degrading pattern of the enzyme have been investigated and this research would further expand the applications of alginate lyases in related fields.

2. Results and Discussion

2.1. Sequence Analysis of FsAlyPL6

The gene of FsAlyPL6 was cloned and analyzed from *Flammeovirga* sp. NJ-04. The open reading frame (ORF) consisted of 2238 bps and encoded a putative alginate lyase of 745 amino acid residues with a theoretical molecular mass of 83.09 kDa. According to the conserved domain analysis, the FsAlyPL6 contained an N-terminal catalytic domain (Met^1-Asn^{366}) and a C-terminal domain (Gln^{367}-Lys^{745}). Based on the sequence alignments shown in Figure 1, FsAlyPL6 shared the highest identity (45%) with AlyGC (BAEM00000000.1) from *Glaciecola chathamensis* S18K6T, which indicated FsAlyPL6 is a new member of family PL6. In addition, FsAlyPL6 contained three conserved regions "NG(G/A)E", "KS", and "R(H/S)G" (marked in Figure 1), which are involved in substrate binding and catalytic activity [12]. The alginate lyases of PL6 family can be divided into three subfamilies, namely subfamilies 1, 2, and 3. In order to confirm the subfamilies of FsAlyPL6, the phylogenetic tree was used to compare the sequence homology with alginate lyases from diverse subfamilies. As is shown in Figure 2, FsAlyPL6 clustered with representative enzymes of subfamily 1, which indicated FsAlyPL6 is a new member of the subfamily 1 alginate lyase.

2.2. Expression and Purification of FsAlyPL6

The gene of FsAlyPL6 was ligated into pET-21a(+) and then the recombinant plasmid was transformed into *E. coli* BL21 (DE3) for heterologously expression. The recombinant FsAlyPL6 was then purified by Ni-NTA sepharose affinity chromatography and analyzed by SDS-PAGE (Figure 3). A clear band (about 80 kDa) can be observed in gel, which was consistent with the theoretical molecular mass of 83.09 kDa. Three kinds of substrates (sodium alginate, polyM, and polyG) were employed to determine the substrate specificity of FsAlyPL6. As shown in Table 1, FsAlyPL6 exhibited higher activity towards sodium alginate (483.95 U/mg) and it showed lower activity towards to polyM (221.5 U/mg). However, it showed the lowest activity towards to polyG (19.35 U/mg). Accordingly, FsAlyPL6 is a polyMG-preferred lyase like most of PL6 family alginate lyases with the exceptions of Patl3640 from *Pseudoalteromonas atlantica* T6c and Pedsa0631 from *Pedobacter saltans* [13]. Both of them preferred polyG to polyMG blocks. In addition, TsAly6A from *Thalassomonas* sp. LD5 [14], OalS6 from *Shewanella* sp. Kz7 [15], OalC6 from *Cellulophaga* sp. SY116 [16], and AlyF from *Vibrio* sp. OU02 [17] are all characterized as polyG-preferred alginate lyases. The kinetic parameters of FsAlyPL6 towards sodium alginate, polyM, and polyG were calculated based on the hyper regression analysis. As shown in Table 1, the K_m values of FsAlyPL6 towards sodium alginate, polyM, and polyG were 0.50 mg/mL, 1.52 mg/mL, and 1.62 mg/mL, respectively. FsAlyPL6 had a lower K_m value towards sodium alginate than to polyM and polyG. Accordingly, FsAlyPL6 exhibited higher affinity towards MG-block than to M-block and G-block. The k_{cat} values of FsAlyPL6 towards sodium alginate, polyM and polyG were 33.98 s^{-1}, 17.66 s^{-1}, and 4.98 s^{-1}, respectively. It indicated that FsAlyPL6 had higher catalytic efficiency towards MG-block than to the other two blocks.

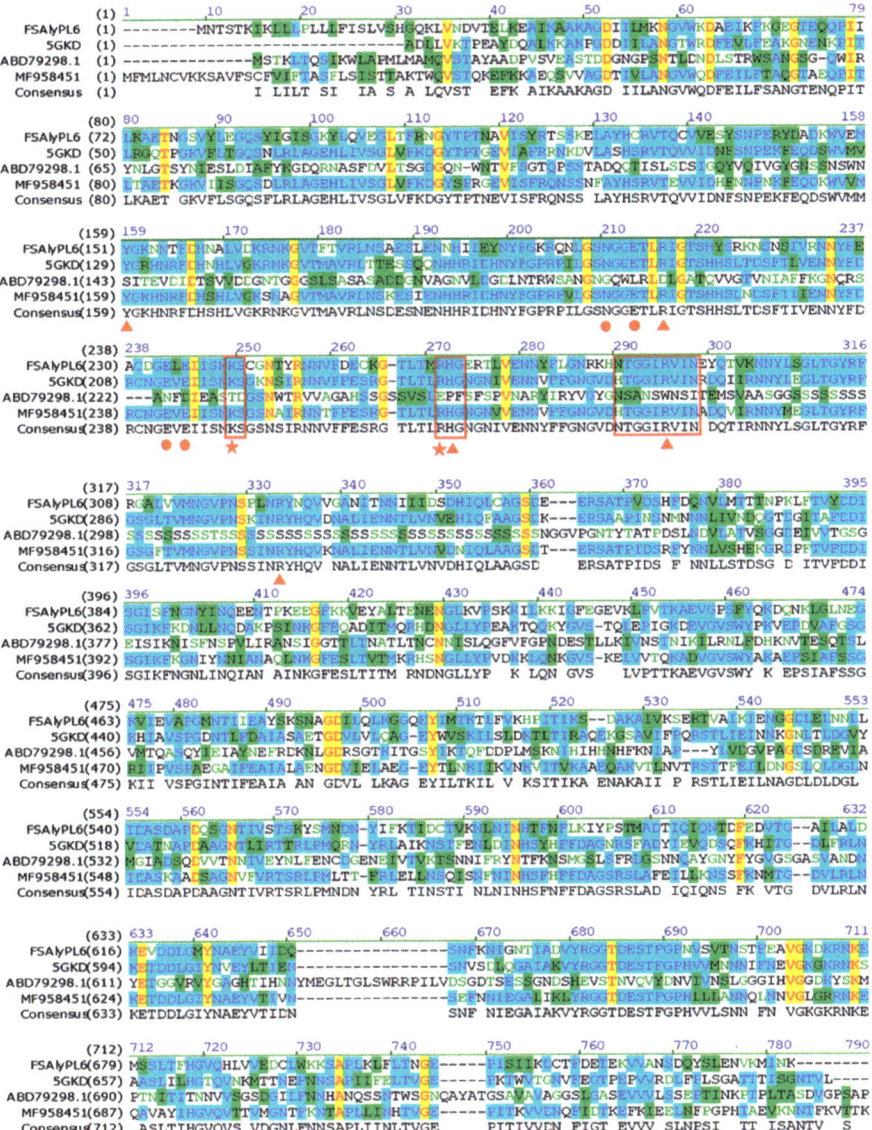

Figure 1. Multiple amino acid sequences alignment of AlyPL6 and other alginate lyases of PL6 family: AlyGC (BAEM00000000.1) from *Glaciecola chathamensis* S18K6T, polysaccharide lyase (ABD79298.1) from *Saccharophagus degradans* 2–40, and TsAly6A (MF958451) from *Thalassomonas* sp. LD5. Three boxes enclose conserved regions. Residues in FsAlyPL6, which are responsible for the enzymatic activity Ca^{2+} binding and catalysis, are marked in triangle, dots, and stars, respectively.

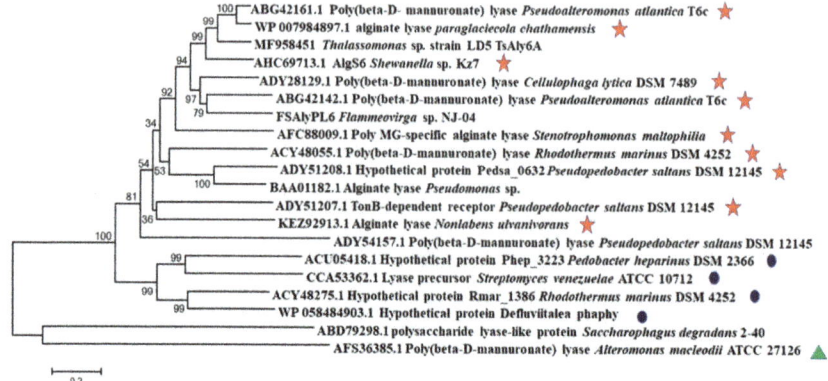

Figure 2. Phylogenetic analysis of FsAlyPL6 with other alginate lyases of PL6 family based on amino acid sequence comparisons. The species names are indicated along with accession numbers of corresponding alginate lyase sequences. Bootstrap values of 1000 trials are presented in the branching points. The subfamilies 1, 2, and 3 are marked with stars, dots, and triangle, respectively.

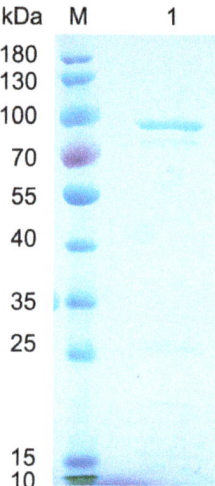

Figure 3. Sodium dodecyl sulfate polyacrylamide gel electrophoresis (SDS-PAGE) analysis of purified FsAlyPL6. Lane M protein: restrained marker (Thermo Scientific, Waltham, MA, USA); lane 1: purified FsAlyPL6.

Table 1. Specificity and kinetics of FsAlyPL6.

Substrate	Sodium Alginate	PolyM	PolyG
Activity (U/mg)	483.95	221.5	19.35
K_m (mg/mL)	0.50	1.52	1.62
V_{max} (nmol/s)	1.36	0.71	0.20
k_{cat} (s^{-1})	33.98	17.66	4.98
k_{cat}/K_m ($mL \cdot s^{-1} \cdot mg^{-1}$)	62.91	11.58	3.08

2.3. Biochemical Characterization of FsAlyPL6

The optimal temperature of FsAlyPL6 is 45 °C and it retains more than 90% of maximal activity after being incubated at 45 °C for 1 h (Figure 4A). Compared with other PL6 family alginate lyases,

FsAlyPL6 exhibits preferable thermal characteristics than most PL6 family alginate lyases. For example, AlyF of *Vibrio* OU02 showed the maximal activity at 30 °C [17] and AlyGC from *G. chathamensis* S18K6T has an optimal temperature of 30 °C [12]. OalC6 of *Cellulophaga* sp. SY116 exhibits highest activity at 40 °C and retains about 80% of highest activity after being incubated at 40 °C for 1 h [16]. In addition, FsAlyPL6 retains 95% activity after being incubated at 35 °C for 60 min and inactivated gradually with temperature increased (Figure 4B). This remarkable characteristic indicated FsAlyPL6 possesses great potential in industrial applications for preparation alginate oligosaccharides. The optimal pH of FsAlyPL6 is 9.0 and it retains about 90% activity incubated at pH 9.0–10.0 for 12 h (Figure 4C,D), which indicated FsAlyPL6 is an alkaline-stable lyase. To the best of our knowledge, few alginate lyases of PL6 family are alkaline-stable lyases, and most of them exhibit the maximal activities around neutral pH values such as OalC6 of *Cellulophaga* sp. SY116 has an optimal pH of 6.6 and it retains only 60% of its maximal activity after being incubated at pH 6.0 for 6 h [16]. The OalS6 from *Shewanella* sp. Kz7 exhibits maximal activity at pH 7.2 and retains 80% after being hatched at pH 6.0–8.0 for 24 h [15]. The influences of metal ions on enzyme activity were also investigated. As shown in Table 2, like TsAly6A from *Thalassomonas* sp. LD5 [14], the activity of FsAlyPL6 can be activated by Ca^{2+} and Mg^{2+}. FsAlyPL6 is inhibited by various divalent metal ions such as Cu^{2+}, Zn^{2+} and Ni^{2+}, which is similar to OalS6 from *Shewanella* sp. Kz7 [15].

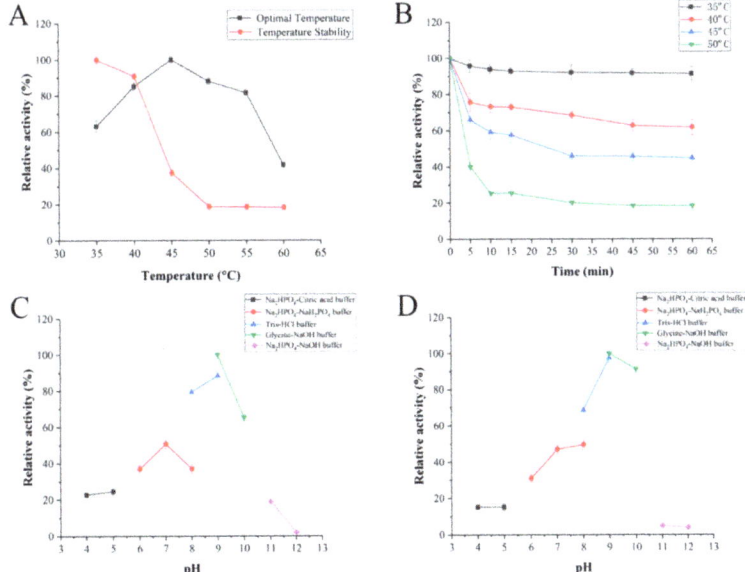

Figure 4. Biochemical characterization of FsAlyPL6: (**A**) The optimal temperature and thermal stability of FsAlyPL6; (**B**) the thermal-induced denaturation of FsAlyPL6; (**C**) the optimal pH of the FsAlyPL6; (**D**) the pH stability of FsAlyPL6.

Table 2. Cont.

Reagent	Relative Activity (%)
Control	100.00 ± 2.97
K^+	93.26 ± 2.23
Na^+	118.57 ± 1.08
Ca^{2+}	104.33 ± 1.12
Mg^{2+}	102.31 ± 2.78
Co^{2+}	22.14 ± 1.32

Table 2. Effects of metal ions on activity of FsAlyPL6.

Reagent	Relative Activity (%)
Zn^{2+}	24.88 ± 3.57
Cu^{2+}	15.28 ± 1.20
Ni^{2+}	50.19 ± 3.93
Mn^{2+}	6.46 ± 0.60
Fe^{3+}	26.55 ± 1.21

2.4. Action Pattern and Substrate Docking of FsAlyPL6 Product Analysis

To elucidate the action mode of FsAlyPL6, the degradation products of three substrates for different times (0–48 h) were analyzed by TLC (Figure 5). As the degrading process continues, three kinds of substrates are degraded into oligosaccharides with lower degrees of polymerization (DPs) (2–5) and monosaccharide, which indicated that FsAlyPL6 can cleave the glycosidic bonds within the substrates in an endolytic manner. The ESI-MS results indicated that degradation products of FsAlyPL6 towards the three different substrates include monosaccharide, and oligosaccharides with different DPs (2–5) can be detected (Figure 6A–C). Most of PL6 family enzymes are endo-type alginate lyases, which produce oligosaccharides with DPs (2–4). However, the Patl3640 from *Pseudoalteromonas atlantica* T6c [13], Pedsa0631 from *Pedobacter saltans* [13], OalS6 from *Shewanella* sp. Kz7 [15], and OalC6 from *Cellulophaga* sp. SY116 degrade the substrates into monosaccharides in an exolytic manner [16].

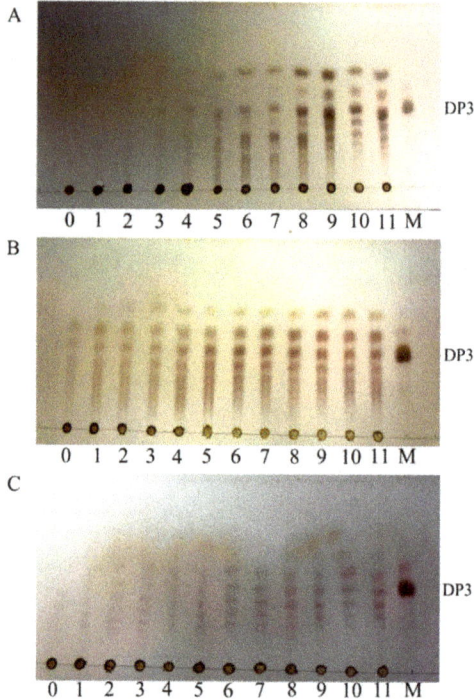

Figure 5. TLC analysis of degrading products of FsAlyPL6 towards alginate (**A**), polyM (**B**), and polyG (**C**). Lane M, the oligosaccharide standard; lanes 0–11, the samples taken by 0 min, 5 min, 10 min, 15 min, 30 min, 60 min, 2 h, 6 h, 12 h, 24 h, and 48 h, respectively.

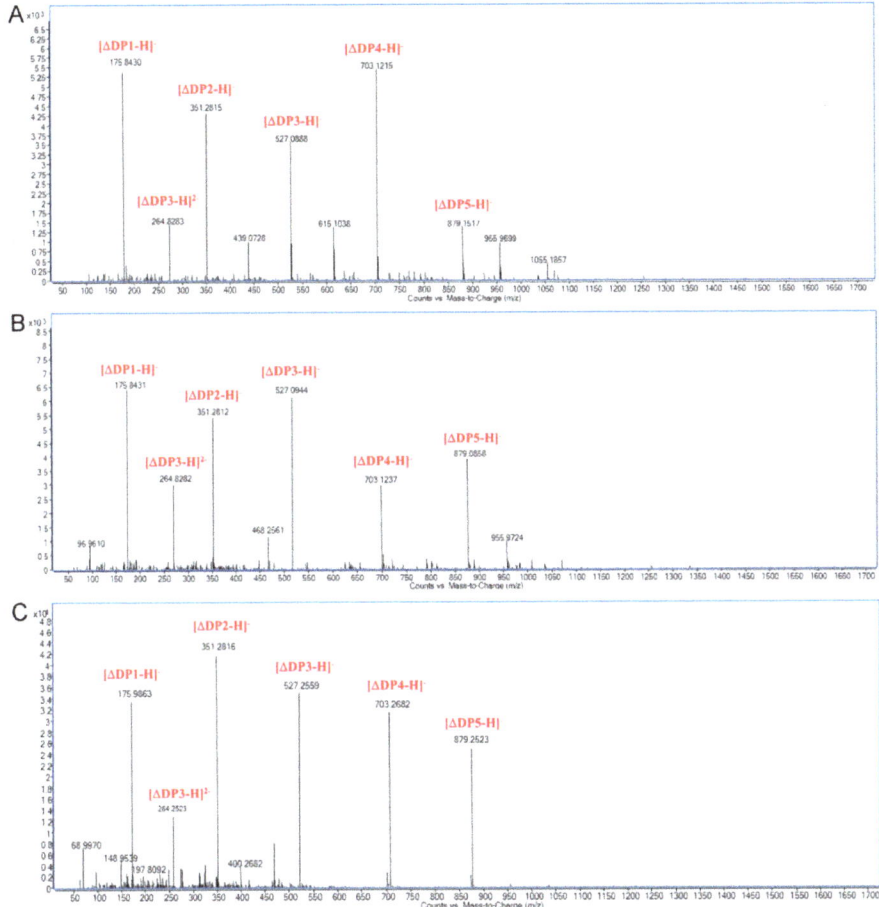

Figure 6. ESI-MS analysis of products of FsAlyPL6 towards alginate (**A**), polyM (**B**), and polyG (**C**).

The three-dimensional model of the FsAlyPL6 was constructed by PHYRE2 and the tetrasaccharide (MMMM) was docked into the FsAlyPL6. Because the sequence similarity between FsAlyPL6 and AlyGC was high (45%), the protein model was successfully constructed with 100% confidence. As shown in Figure 7A, the overall structure of FsAlyPL6 was predicted to fold into a "twin tower-like" structure (Figure 7A), which is similar to the structure of AlyGC (Figure 7B). However, AlyGC is an exo-type alginate lyase and FsAlyPL6 degrade alginate into oligosaccharide in an endolytic manner. The key residues for substrate recognition were identified by the sequence alignment and protein–substrate interactions. As shown in Figure 7C, the residues R_{239}, R_{263}, K_{218}, E_{213}, and Y_{332} are were highly conserved and involved in the interaction between the protein and substrates in subsites −1, +1, +2 and +3, respectively (Figure 8A,B). Based on the docking and β-elimination mechanism, the residues K_{218} and R_{239} acted as the Brønsted base and Brønsted acid, respectively, in the cleavage reaction of FsAlyPL6 on alginate, which is consistent with the residues of AlyGC (Figure 8B).

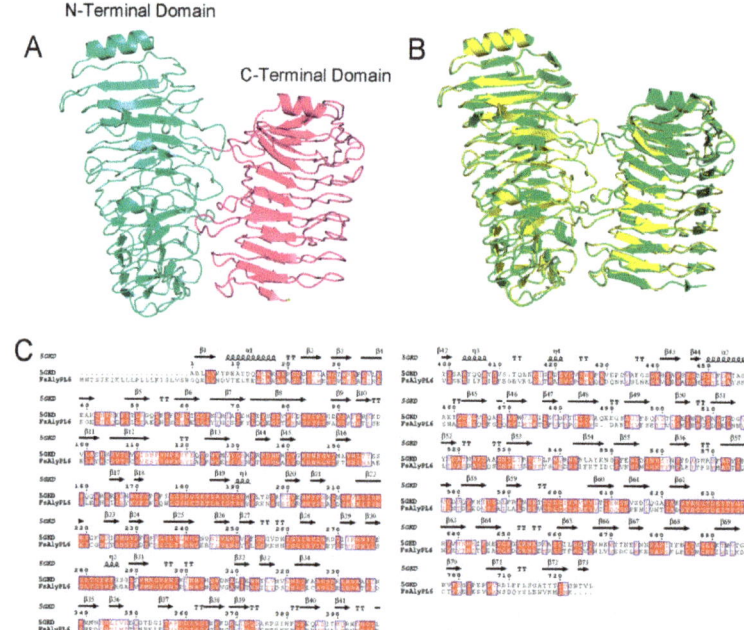

Figure 7. (**A**) Overall structure of FsAlyPL6; (**B**) the structural comparison of FsAlyPL6 (green) and AlyGC (yellow); (**C**) sequence alignments of FsAlyPL6 and AlyGC.

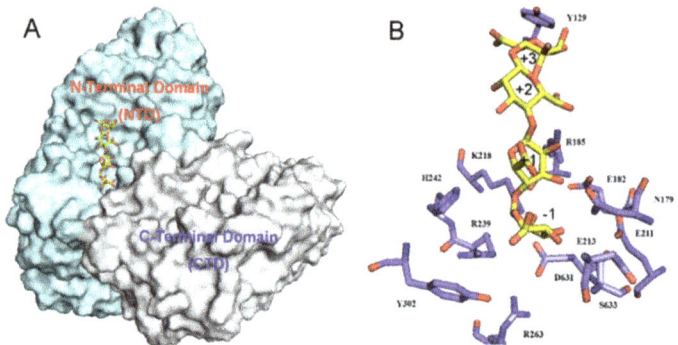

Figure 8. (**A**) Stereo view of the alginate tetrasaccharide (MMMM) bound to the tunnel-shaped active site of FsAlyPL6. (**B**) The presentation of catalytic residues responsible for binding and catalyzing the substrate.

3. Materials and Methods

3.1. Materials and Strains

Sodium alginate (M/G ratio: 77/23) was purchased from Sigma-Aldrich (St. Louis, MO, USA). PolyG and polyM (purity: about 95%; M/G ratio: 3/97 and 97/3, respectively) were purchased from Qingdao BZ Oligo Biotech Co., Ltd. (Qingdao, China). *Flammeovirga* sp. NJ-04 was isolated from the South China Sea and conserved in our laboratory. It was cultured at 35 °C in 2216E medium (Difoc). *Escherichia coli* DH5α and *E. coli* BL21 (DE3) were used for plasmid construction and as the hosts for gene expression, respectively. These strains were cultured at 37 °C in Luria-Bertani (LB) broth or on LB broth agar plates (LB broth was supplemented with 1.5% agar and contained 100 µg/mL ampicillin).

3.2. Cloning and Sequence Analysis of Alginate Lyase

As previously reported, a gene cluster for degrading alginate has been identified within the genome of the strain *Flammeovirga* sp. NJ-04 [10]. According to the sequence of the putative alginate lyase gene sequence (WP_044204792.1), a pair of special primers was designed as described in Supplementary Materials. For gene expression, the alginate lyase gene *FsAlyPL6* was subcloned and then ligated into pET-21a(+) expression vector. The theoretical molecular (Mw) and isoelectric point (pI) were calculated using Compute pI/Mw tool (https://web.expasy.org/compute_pi/). Molecular Evolutionary Genetics Analysis (MEGA) Program version 6.0 (Center for Evolutionary Medicine and Informatics, The Biodesign Institute, Tempe, AZ, USA) was applied to construct a phylogenetic tree through a neighbor-joining method based on alginate lyase protein sequences of PL6 family. The Vector NTI (Invitrogen, Thermo Scientific, Waltham, MA, USA) was used to obtain multiple sequence alignment. The homology modeling and docking was built by Protein Homology/analogY Recognition Engine V 2.0 (Structural Bioinformatics Group, Imperial College, London, Britain).

3.3. Hereologous Expression and Purification of the Recombinant Enzyme

The recombinant plasmid pET-21a(+)–*FsAlyPL6* was transformed into *E. coli* BL21 (DE3). It was then cultured in an LB medium (containing 100 µg/mL of ampicillin) at 37 °C by shaking at 200 rpm for 5 h, followed by being induced with 0.1 mM IPTG at 25 °C for 36 h when OD_{600} reached 0.6. The purification of FsAlyPL6 was performed as follows. The cells were harvested by centrifugation and then sonicated in lysis buffer (50 mM Tris-HCl with 300 mM NaCl, pH 8.0). The cell homogenate that contained recombinant protein were purified by using a His-trap column (GE Healthcare, Uppsala, Sweden). SDS on 12% (w/v) resolving gel was applied to detect the purity of the recombinant protein.

3.4. Substrate Specificity and Enzymatic Kinetics

The reaction was performed using 20 µL FsAlyPL6 (4 µg) mixed with 180 µL 0.8% alginate, polyM, and polyG respectively. The enzyme activity was measured using the ultraviolet absorption method [11]. One unit was defined as the amounts of enzyme required to increase absorbance at 235 nm (extinction coefficient: 6150 $M^{-1} \cdot cm^{-1}$) by 0.1 per min. The kinetic parameters of the FsAlyPL6 towards alginate, polyM, and polyG were investigated by measuring the enzyme activity with these substrates at different concentrations (0.4–10 mg/mL). Velocity (V), K_m, and V_{max} values were calculated as previously reported [10]. The radio of V_{max} versus enzyme concentration ([E]) was used to calculate the turnover number (k_{cat}) of the enzyme.

3.5. Biochemical Characterization of the Recombinant Enzyme FsAlyPL6

The effects of temperature on the enzyme activity were determined by testing the activity at different temperatures (35 °C to 60 °C). The thermal stability was characterized by measuring the residual activity after the purified FsAlyPL6 was incubated at 35–60 °C for 1 h. Furthermore, the thermally induced denaturation was also determined by measuring the residual activity after incubating the enzyme at 35–50 °C for 0–60 min. To investigated the optimal pH of the FsAlyPL6, 1% alginate mixed with different buffers at 45 °C (50 mM phosphate–citrate (pH 4.0–5.0), 50 mM NaH_2PO_4–Na_2HPO_4 (pH 6.0–8.0), 50 mM Tris–HCl (pH 7.0–9.0), and glycine–NaOH (pH 9.0–12.0)) were used as the substrates and the purified enzyme incubated with these substrates under standard conditions. Moreover, the pH stability was evaluated based on the residual activity after being incubated with indifferent buffers (pH 4.0–12.0) for 20 h. The effects of metal ions on the enzymatic activity were performed by incubating the FsAlyPL6 with substrates that contained various metal compounds with a final concentration of 1 mM. The reaction performed under standard tested conditions and the substrates blend without any metal ion was taken as the control.

3.6. Action Pattern and Degradation Product Analysis

In order to elucidate the action pattern of the FsAlyPL6, the thin-layer chromatography (TLC) was applied to analyze the degrading products of FsAlyPL6 towards sodium alginate, polyM and polyG. The reaction and treatment of the samples were performed as previously reported [10]. In order to investigate the composition of the degrading products, ESI-MS was employed as follows: The supernatant (2 µL) was loop-injected to an LTQ XL linear ion trap mass spectrometer (Thermo Fisher Scientific, Waltham, MA, USA) after centrifugation. The oligosaccharides were detected in a negative-ion mode using the following settings: ion source voltage, 4.5 kV; capillary temperature, 275–300 °C; tube lens, 250 V; sheath gas, 30 arbitrary units (AU); and scanning the mass range, 150–2000 *m/z*.

3.7. Molecular Modeling and Docking Analysis

Protein Homology/analogY Recognition Engine V 2.0 was applied to construct the three-dimensional structure of FsAlyPL6 according to the known structure of alginate lyase AlyGC from *Glaciecola chathamensis* S18K6T (PDB: 5GKD) with a sequence identity of 45%. The molecular docking of the FsAlyPL6 and MMMM was performed using Molecular Operating Environment (MOE, Chemical Computing Group Inc., Montreal, QC, Canada). The ligand-binding sites were defined using the bound ligand in the homology models. PyMOL (http://www.pymol.org) was used to visualize and analyze the modeled structure and to construct graphical presentations and illustrative figures.

4. Conclusions

In this study, we reported a new PL family alginate lyase FsAlyPL6 from the marine *Flammeovirga* sp. NJ-04. It preferred to degrade the polyMG block and showed highest activity at 45 °C and could retain 50% of activity after being incubated at 45 °C for 1 h. The FsAlyPL6 endolytically degraded alginate polysaccharide and released oligosaccharides with DPs of 1–5. In addition, it could recognize tetrasaccharide as the minimal substrate and cleave the glycosidic bonds between the subsites of −1 and +3 to release oligosaccharides. The research provides extended insights into the degradation pattern of PL6 alginate lyases and further expands the application of alginate lyases.

Supplementary Materials: The following are available online at http://www.mdpi.com/1660-3397/17/6/323/s1, Table S1: The primers for cloning the gene of FsAlyPL6.

Author Contributions: Q.L. and F.H. conceived and designed the experiments; B.Z., Q.L., and F.H. performed the experiments; Y.S., Y.S., and Z.Y. analyzed the data; B.Z. wrote the paper. All authors reviewed the manuscript.

Funding: This research was funded by the National Natural Science Foundation of China (grant number: 31601410 and 21776137).

Acknowledgments: The work was supported by the National Natural Science Foundation of China (grant numbers: 31601410 and 21776137).

Conflicts of Interest: The authors declare no conflicts of interest.

References

1. Gacesa, P. Enzymic degradation of alginates. *Int. J. Biochem.* **1992**, *24*, 545–552. [CrossRef]
2. Lee, K.Y.; Mooney, D.J. Alginate: Properties and biomedical applications. *Prog. Polym. Sci.* **2012**, *37*, 106–126. [CrossRef] [PubMed]
3. An, Q.D.; Zhang, G.H.; Zhang, Z.C.; Zheng, G.S.; Luan, L.; Murata, Y.; Li, X. Alginate-deriving oligosaccharide production by alginase from newly isolated Flavobacterium sp. LXA and its potential application in protection against pathogens. *J. Appl. Microbiol.* **2010**, *106*, 161–170. [CrossRef] [PubMed]
4. Iwamoto, M.; Kurachi, M.; Nakashima, T.; Kim, D.; Yamaguchi, K.; Oda, T.; Iwamoto, Y.; Muramatsu, T. Structure-activity relationship of alginate oligosaccharides in the induction of cytokine production from RAW264.7 cells. *FEBS Lett.* **2005**, *579*, 4423–4429. [CrossRef] [PubMed]

5. Tusi, S.K.; Khalaj, L.; Ashabi, G.; Kiaei, M.; Khodagholi, F. Alginate oligosaccharide protects against endoplasmic reticulum- and mitochondrial-mediated apoptotic cell death and oxidative stress. *Biomaterials* **2011**, *32*, 5438–5458. [CrossRef] [PubMed]
6. Falkeborg, M.; Cheong, L.Z.; Gianfico, C.; Sztukiel, K.M.; Kristensen, K.; Glasius, M.; Xu, X.; Guo, Z. Alginate oligosaccharides: Enzymatic preparation and antioxidant property evaluation. *Food Chem.* **2014**, *164*, 185–194. [CrossRef] [PubMed]
7. Wong, T.Y.; And, L.; Schiller, N.L. Alginate Lyase: Review of Major Sources and Enzyme Characteristics, Structure-Function Analysis, Biological Roles, and Applications. *Annu. Rev. Microbiol.* **2000**, *54*, 289–340. [CrossRef] [PubMed]
8. Zhu, B.; Chen, M.; Yin, H.; Du, Y.; Ning, L. Enzymatic Hydrolysis of Alginate to Produce Oligosaccharides by a New Purified Endo-Type Alginate Lyase. *Mar. Drugs* **2016**, *14*, 108. [CrossRef] [PubMed]
9. Zhu, B.; Yin, H. Alginate lyase: Review of major sources and classification, properties, structure-function analysis and applications. *Bioengineered* **2015**, *6*, 125–131. [CrossRef] [PubMed]
10. Zhu, B.; Ni, F.; Sun, Y.; Yao, Z. Expression and characterization of a new heat-stable endo-type alginate lyase from deep-sea bacterium Flammeovirga sp. NJ-04. *Extremophiles* **2017**, *21*, 1027–1036. [CrossRef] [PubMed]
11. Inoue, A.; Anraku, M.; Nakagawa, S.; Ojima, T. Discovery of a Novel Alginate Lyase from Nitratiruptor sp. SB155-2 Thriving at Deep-sea Hydrothermal Vents and Identification of the Residues Responsible for Its Heat Stability. *J. Biol. Chem.* **2016**, *291*, 15551–15563. [CrossRef] [PubMed]
12. Xu, F.; Dong, F.; Wang, P.; Cao, H.Y.; Li, C.Y.; Li, P.Y.; Pang, X.H.; Zhang, Y.Z.; Chen, X.L. Novel Molecular Insights into the Catalytic Mechanism of Marine Bacterial Alginate Lyase AlyGC from Polysaccharide Lyase Family 6. *J. Biol. Chem.* **2017**, *292*, 4457–4468. [CrossRef] [PubMed]
13. Mathieu, S.; Henrissat, B.; Labre, F.; Skjak-Braek, G.; Helbert, W. Functional Exploration of the Polysaccharide Lyase Family PL6. *PLoS ONE* **2016**, *11*, e0159415. [CrossRef] [PubMed]
14. Gao, S.; Zhang, Z.L.; Li, S.Y.; Su, H.; Tang, L.Y.; Tan, Y.L.; Yu, W.G.; Han, F. Characterization of a new endo-type polysaccharide lyase (PL) family 6 alginate lyase with cold-adapted and metal ions-resisted property. *Int. J. Biol. Macromol.* **2018**, *120*, 729–735. [CrossRef] [PubMed]
15. Li, S.Y.; Wang, L.N.; Han, F.; Gong, Q.H.; Yu, W.G. Cloning and characterization of the first polysaccharide lyase family 6 oligoalginate lyase from marine *Shewanella* sp Kz7. *J. Biochem.* **2016**, *159*, 77–86. [CrossRef] [PubMed]
16. Li, S.Y.; Wang, L.N.; Chen, X.H.; Zhao, W.W.; Sun, M.; Han, Y.T. Cloning, Expression, and Biochemical Characterization of Two New Oligoalginate Lyases with Synergistic Degradation Capability. *Mar. Biotechnol.* **2018**, *20*, 75–86. [CrossRef] [PubMed]
17. Lyu, Q.; Zhang, K.; Shi, Y.; Li, W.; Diao, X.; Liu, W. Structural insights into a novel Ca^{2+}-independent PL-6 alginate lyase from *Vibrio* OU02 identify the possible subsites responsible for product distribution. *BBA Gen. Subjects* **2019**, *1863*, 1167–1176. [CrossRef] [PubMed]

© 2019 by the authors. Licensee MDPI, Basel, Switzerland. This article is an open access article distributed under the terms and conditions of the Creative Commons Attribution (CC BY) license (http://creativecommons.org/licenses/by/4.0/).

Article

Characterization of an Alkaline Alginate Lyase with pH-Stable and Thermo-Tolerance Property

Yanan Wang [1,†], Xuehong Chen [1,†], Xiaolin Bi [2], Yining Ren [1], Qi Han [1], Yu Zhou [1], Yantao Han [1,*], Ruyong Yao [3] and Shangyong Li [1,*]

1. Department of Pharmacology, College of Basic Medicine, Qingdao University, Qingdao 266071, China; sunshine4581@163.com (Y.W.); chen-xuehong@163.com (X.C.); Renyn796@163.com (Y.R.); xiaoyu19990727@163.com (Q.H.); zy18339956716@163.com (Y.Z.)
2. Department of Rehabilitation Medicine, Qingdao University, Qingdao 266071, China; 18661809159@163.com
3. Central Laboratory of Medicine, Qingdao University, Qingdao 266071, China; yry0303@163.com
* Correspondence: hanyt@qdu.edu.cn (Y.H.); lisy@qdu.edu.cn (S.L.); Tel.: +86-0532-8378-0027 (Y.H. & S.L.)
† These authors contributed equally to this paper.

Received: 23 April 2019; Accepted: 14 May 2019; Published: 24 May 2019

Abstract: Alginate oligosaccharides (AOS) show versatile bioactivities. Although various alginate lyases have been characterized, enzymes with special characteristics are still rare. In this study, a polysaccharide lyase family 7 (PL7) alginate lyase-encoding gene, *aly08*, was cloned from the marine bacterium *Vibrio* sp. SY01 and expressed in *Escherichia coli*. The purified alginate lyase Aly08, with a molecular weight of 35 kDa, showed a specific activity of 841 U/mg at its optimal pH (pH 8.35) and temperature (45 °C). Aly08 showed good pH-stability, as it remained more than 80% of its initial activity in a wide pH range (4.0–10.0). Aly08 was also a thermo-tolerant enzyme that recovered 70.8% of its initial activity following heat shock treatment for 5 min. This study also demonstrated that Aly08 is a polyG-preferred enzyme. Furthermore, Aly08 degraded alginates into disaccharides and trisaccharides in an endo-manner. Its thermo-tolerance and pH-stable properties make Aly08 a good candidate for further applications.

Keywords: Alginate lyase; Thermo-tolerant; pH-stability; Endo-manner; *Vibrio* sp. SY01

1. Introduction

Alginate is an acidic hetero-polysaccharide extracted from brown algae, which accounting for 22–44% of its dry weight [1–3]. Alginate mainly contains two different uronic acids, including α-L-guluronic acid (G) and β-D-mannuronic acid (M). They are arranged into three different kinds of blocks by (1→4)-linked monosaccharides: homopolymeric G blocks, polyguluronate (PolyG); homopolymeric M blocks, polymannuronate (PolyM); and random or heteropolymeric blocks of alternating M and G units (PolyMG) [4,5].

Alginate lyase (E.C. 4.2.2.3 and E.C. 4.2.2.11) is a kind of polysaccharide lyase that degrades alginate by β-eliminating the glycoside 1-4 O-bonds between C4 and C5 at the non-reducing end, thus producing unsaturated alginate oligosaccharides (UAOS) as main products [6,7]. Due to its high efficiency, specificity and mild degradation function, alginate lyases have attracted widespread attention in industrial applications, especially in the preparation of alginate oligosaccharides [8,9].

According to the Carbohydrate-Active enZYmes (CAZy) databases, alginate lyases belong to PL families 5, 6, 7, 14, 15, 17, and 18 based on the analysis of their amino acid sequences [10–12]. Based on the substrate specificity, alginate lyases can be further classified into two types, one type is the G block-specific lyases (polyG lyases, EC 4.2.2.11), and the other type is the M block-specific lyases (polyM lyases, EC 4.2.2.3) [13,14]. In PL families 5, 7, 14, 17, and 18, most of the reported alginate lyases are polyM lyases. Only the alginate lyase reported in PL family 6 is mainly comprised of polyG lyases. Thus far, hundreds of alginate lyases have been purified, cloned, and characterized from marine microorganisms, brown seaweeds, and mollusks [15–18]. However, these reported enzymes with characteristics specific for commercial use are rare. Cold-adapted alginate lyases can run biocatalytic processes at low temperature and reduce the danger of contamination. Thermo-tolerant enzymes persist at high temperatures, thereby not only improving degradation efficiency but also reducing production costs. Meanwhile, high proportion product in a mixture of products will be propitious to the purification of oligosaccharide. Therefore, there is an urgency to obtain an alginate lyase with the optimal characteristics (e.g., pH-stability, thermo-tolerance, and single product distribution) needed for industrial applications.

In this study, a new alginate lyase-encoding gene, *aly08*, was cloned from *Vibrio* sp. SY01, and expressed in *Escherichia coli* BL21 (DE3). The recombinant enzyme Aly08 degraded alginate, yielding alginate disaccharides and trisaccharides as main products. This study also revealed that Aly08 was a polyG-preferred enzyme with special characteristics, such as wide pH-stability, thermo-tolerance, and single product distribution. These special features suggest that Aly08 may play essential roles in saccharification processes of alginate and carbon cycling.

2. Results and Discussion

2.1. Sequence Analysis of Aly08

The marine bacterium *Vibrio* sp. SY01 was isolated from Yellow sea sediment, China. It grew rapidly in the alginate sole-carbon medium and efficiently degraded brown seaweed with a high alginate lyase activity (more than 50 U/mL). The genome sequence analysis of *Vibrio* sp. SY01 showed that it contained the putative alginate lyase-encoding gene, *aly08*, consisting of 897 bp of an open reading frame (ORF). The identified alginate lyase, Aly08, contained 299 amino acid residues. Signal peptide analysis showed that Aly08 predicted a putative signal peptide (Met^1 to Phe^{22}) in its N-terminal. Furthermore, the theoretically isoelectric point (pI) and theoretical molecular weight (Mw) of mature Aly08 were 4.57 and 32.89 kDa, respectively. According to a search of Conserved Domain Database (CDD) of NCBI, Aly08 is a new alginate lyase with a single-domain belonging to the alginate lyase superfamily 2.

Based on the sequences of Aly08 and other reported PL family 7 alginate lyases, phylogenetic trees were created. Pectate lyase (Genbank number CAD56882) from *Bacillus licheniformis* 14A was included as a control (Figure 1). Among all of the reported alginate lyase, Aly08 had the highest identity sequence of amino acids (78%) with a PL family 7 alginate lyase, AlyL2 (Genbank number MH791447), from *Agarivorans* sp. L11 [19].

A deeply branched cluster was formed in the phylogenetic tree among the enzymes Aly08 (Genbank number MH791447), AlyL2 (Genbank number AJO61885), AlgMsp (Genbank number BAJ62034), and Alg7D (Genbank number ABD81807). According to the multiple sequence alignment (Figure 2), the enzyme contains three conserved regions "RTELREMLR", "QIH", and "MYFKAG" which are related to substrate binding and catalytic activity in PL family 7 [20]. These results identified Aly08 as a new member of the PL family 7. Thus far, several alginate lyases have been identified from various bacteria, such as *Pseudoalteromona*, *Flavobacterium*, *Nitratiruptor*, *Agarivorans*, and *Vibrio* [7,17,21–23]. After determining their various properties, most of the alginate lyases showed a preference towards polyM blocks. In this study, the purified alginate lyase Aly08 is a polyG-preferred alginate lyase containing the "QIH" conserved region, according to their sequence analysis (Figure 2). The conserved region, "QIH" or "QVH", plays a key role in the substrate preferences of alginate lyases. Aly08, along with AlgMsp, A1-II and Alg2A containing the "QIH" conserved region showed preferences for polyG [24–26]. In addition, other alginate lyases derived from PL family 7, such as AlxM from *Photobacterium* sp. ATCC 43367, and A9mT from *Vibrio* sp. A9m, possess "QVH" regions show preference for degrading polyM blocks (Table 1) [27,28]. Through further sequence screening, it was found that "QIH" or "QVH" may be an indicator for substrate-preferred analysis. Furthermore,

Aly08 can be used for the next part of combining a polyM-preferred alginate lyase for synergetic degradation alginate or brown seaweeds.

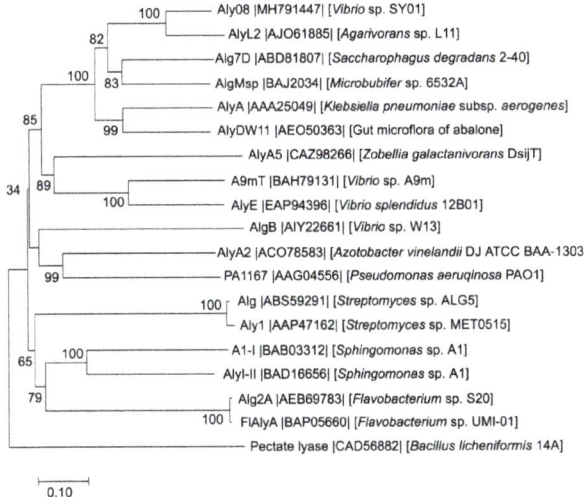

Figure 1. Phylogenetic analysis of Aly08 with other reported alginate lyases. The reliability of the phylogenetic reconstructions was determained by boot-strapping values (1000 replicates). Branch-related numbers are bootstrap values (confidence limits) representing the substitution frequency of each amino acid residue. A pectate lyase (CAD56882) from *Bacillus licheniformis* 14A was taken as control.

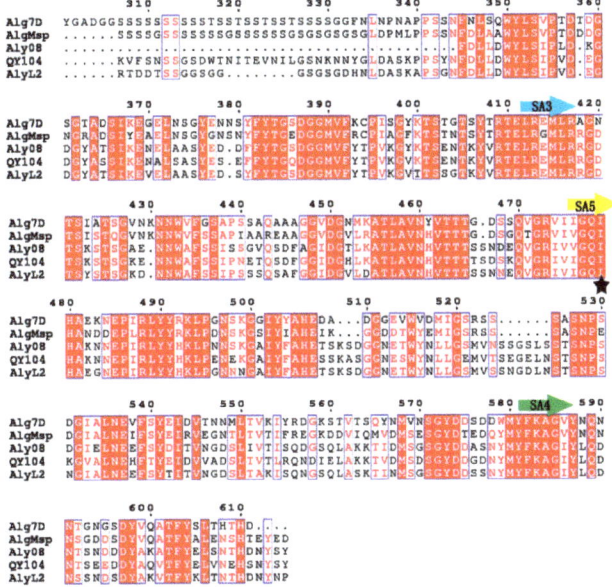

Figure 2. Sequence comparison of Aly08 with related alginate lyases from PL family 7: Alg7D (ABD81807) from *Saccharophagus degradans* 2–40, AlgMsp (BAJ62034) from *Microbulbifer* sp. 6532A, AlyV4 (AGL7859) from *Vibrio* sp. QY104, and AlyL2 (AJO61885) from *Agarivorans* sp. L11. The conserved regions and identical residues are marked with bands and black star, respectively.

Table 1. Comparison of the properties of Aly08 with other alginate lyases.

Protein Name	Optimal pH/ Temperature (°C)	Conserved Region QIH/QVH	Substrate Specificity	Products (DP)	Source	References
Aly08	8.35/45	QIH	PolyG	2,3	*Vibrio* sp. SY01	This study
AlgNJU-03	7.0/30	QIH	PolyG, polyM,alginate	2,3,4	*Vibrio* sp. NJU-03	[20]
AlgMsp	8.0/40	QIH	PolyG	2–5	*Microbulbifer* sp. 6532A	[24]
A1-II′	7.5/40	QIH	polyG,polyM	3,4	*Sphingomonas* sp.A1	[29]
Aly2	6.0/40	QIH	polyG	2,3,4	*Flammeovirga* sp. strain MY04	[30]
AlyPI	7.0/40	QIH	polyG,polyM	-	*Pseudoalteromonas* sp. CY24.	[31]
Alg2A	8.3/40	QIH	polyG	5,6,7	*Flavobacterium* sp. S20	[26]
AlxM	-	QVH	polyM	-	*Photobacterium* sp. ATCC 43367	[32]
A1m	9.0/30	QIH	polyG	-	*Agarivorans* sp. JAM-A1m	[33]
A9mT	7.5/30	QVH	polyM	-	*Vibrio* sp. A9m	[27]
FlAlyA	7.7/55	QIH	polyM, polyG	2–5	*Flavobacterium* sp. strain UMI-01	[34]
AlyH1	7.5/40	QIH	polyG, alginate	2,3,4	*Vibrio furnissii* H1	[35]

2.2. Expression, Purification, and Characterization of Aly08

The expression strain *E. coli* BL21-pET22b-Aly08 was grown in LB broth and Aly08 was purified by a Ni-NTA affinity column. The specific activity of purified Aly08 was 841.1 U/mg with high viscosity sodium alginate as substrate. Moreover, the purified enzyme Aly08 was analyzed by sodium dodecyl sulfate polyacrylamide gel electrophoresis (SDS-PAGE) and observed as a single band on the gel with an approximate Mw of 35 kDa (Figure 3), which was corresponding to the theoretical molecular mass of 32.89 kDa.

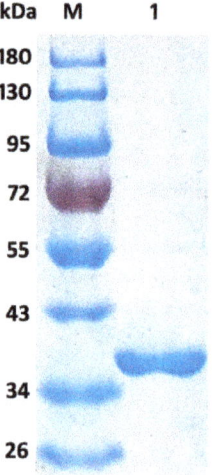

Figure 3. SDS-PAGE analysis of the recombinant enzyme Aly08. Lane M, protein marker; Lane 1, the purified Aly08.

Then, the characterization of purified Aly08 was analyzed as follows. The enzyme Aly08 showed maximum activity at 45 °C, and maintained activities of 82.8% and 48.7% when the enzyme was incubated at a low temperature for 1 h, 10 °C and 20 °C, respectively (Figure 4A,B). The optimal pH for Aly08 was found to be 8.35 (Figure 4C). In addition, Aly08 holds more than 60% of activity in a wide pH range from 7.0–11.0 after incubation in different buffers at 4 °C for 12 h, and was particularly stable under alkaline conditions. (Figure 4D). As previous study, most of the alginate lyases showed optimal pH and stability close to a neutral environment, such as AlyH1 from *Vibrio furnissii* H1 shows high activity at the optimal pH 7.5 and it was stable at pH 6.5–8.5. In addition, AlyH1 only retains about 20%

of residual activity when incubated at pH 9.5 for 12 h [35]. AlgNJU-03 from *Vibrio* sp. NJU-03 possessed a neutral optimal pH at 7.0 and its pH-stability range from 6.0 to 8.0 [20]. Those alginate lyases prefer neutral pH and they only show high activities within a narrow pH range after incubating for several hours and always exhibit instability under alkaline conditions. Another reported high-alkaline alginate lyase A1m from marine bacteria *Agarivorans* sp. JAM-A1m, exhibited high activity at pH 9.0 under glycine-NaOH buffer with 0.2 M NaCl added to the reaction mixture. However, A1m was only stable and maintain more than 60% of residual activity in a short period of 1 h over a narrow pH range of 7.0–9.0 [33]. Comparing with A1m, Aly08 was stable over 12 h with its 80% residual activity even at pH 9.0–11.0, while another enzyme derived from *Vibrio* sp. NJ-04 maintains good stability at pH 4.0–10.0, but the enzyme only maintains its maximum activity under neutral conditions (pH 7.0) [36]. Thus, Aly08 is an alkaline-stable lyase with industrial application potential as it has been proven to conduct catalysis reactions and maintain activity in a broader pH range.

In particular, the substrate preferred by Aly08 was determined by experimenting with three polymeric substrates (sodium alginate, polyM block and polyG block). Aly08 was found to prefer polyG blocks (1078.2 U/mg) rather than polyM blocks (297.1 U/mg) and native alginate (841.1 U/mg).

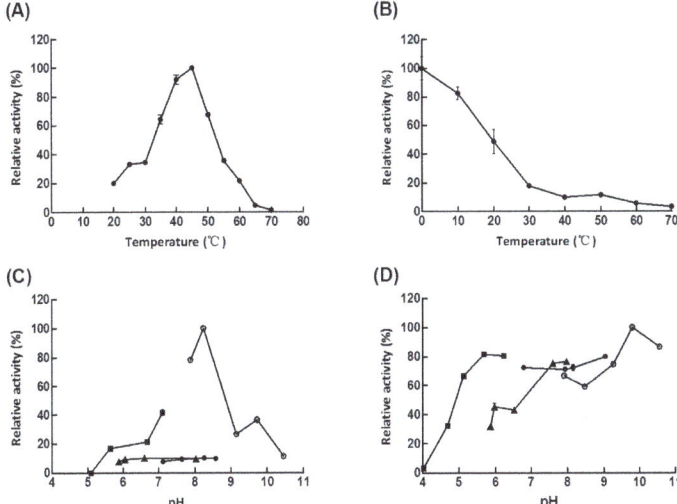

Figure 4. The biochemical characteristics of Aly08. (**A**) The optimal temperature of Aly08. (**B**) The thermal-stability of Aly08. (**C**) Optimal pH for the relative activity of Aly08 was determined in 20 mM Tris-HCl buffer (solid circle), 20 mM phosphate buffer (solid triangle), 20 mM citic-Na$_2$HPO$_4$ (solid square), or 20 mM glycine-NaOH buffer (hollow circle). (**D**) pH stability of Aly08 in 20 mM Tris-HCl buffer (solid circle), 20 mM phosphate buffer (solid triangle), 20 mM citic-Na$_2$HPO$_4$ (solid square), or 20 mM glycine-NaOH buffer (hollow circle).

Moreover, activities of Aly08 were enhanced by NaCl (different concentrations from 10 mM to 3 M), and the activity reached its maximum at 300 mM NaCl, at which point the activity was about eight times higher than the activity in the absence of NaCl (Table 2). Similarly, the activity of AlgM4 from *V. weizhoudaoensis* M0101 was increased about seven times at 1 M NaCl and activated by a concentration range of NaCl at 0–1 M [37]. For these alginates from marine bacteria, a certain level of NaCl concentration is essential for strain survival and enzyme activation. The effects of other metal ions on the activity of Aly08 were also shown in Table 2. Aly08 showed no obvious activated effect in the presence of NH_4^+, Li^+, Zn^{2+}, Ba^{2+}, and Co^{2+}, while Al^{3+} and Ni^{2+} showed no obvious inhibiting effects on relative enzymatic activity. Enzyme activity was activated by divalent ions, such as Ca^{2+} and Mn^{2+}. Different concentrations of KCl had little effect on the activities of Aly08 (Table S2). Interestingly, the enzyme activity of the reaction system containing divalent ions of Ca^{2+} was about twice as high as that of the reaction system without any ions. However, other reagents such as SDS and EDTA showed significant inactivation effects wherein relative activity was reduced to 50% and 55.9%, respectively.

Table 2. Effects of metal ions, EDTA and SDS on the activity of Aly08. Notes: Activity without addition of chemicals was defined as 100%. Data are shown as means ± SD ($n = 3$).

Reagent Added	Concentration (mM)	Relative Activity (%)
None	-	100.00 ± 0.24
NaCl	10	382.22 ± 2.64
	50	580.89 ± 4.36
	300	865.96 ± 26.46
	800	647.33 ± 11.25
	3000	361.97 ± 10.74
SDS	1	50.00 ± 3.32
EDTA	1	55.88 ± 8.60
$Al_2(SO_4)_3$	1	85.33 ± 10.63
KCl	1	99.67 ± 0.86
KCl	100	103.54 ± 0.16
$NiCl_2$	1	97.24 ± 2.20
$(NH_4)_2SO_4$	1	105.56 ± 1.37
$MnSO_4$	1	100.64 ± 1.78
Li_2SO_4	1	103.91 ± 0.93
$ZnCl_2$	1	114.27 ± 2.48
$BaCl_2$	1	120.74 ± 17.14
$CoCl_2$	1	130.86 ± 4.03
$MnCl_2$	1	166.39 ± 5.47
$CaCl_2$	1	281.18 ± 29.18

2.3. Thermo-Tolerance and Heat Recovery of Aly08

When we sought to determine the thermostability of Aly08, we found an interesting phenomenon. After ice-bath, the residual activity of heat-treatment enzymes was always higher than that of enzymes without an ice-bath (Figure 5A,B). After incubation at 30 °C and 40 °C for 1 h, Aly08 retained only 17.9% and 9.9% of its initial activity when directly assayed its activities. However, when the enzyme incubation at 0 °C for 30 min, the residual activity could recover to 43.8% and 39.4% in the same heat treatment condition (Figure 5A). Moreover, even the enzyme was boiled for 5 min, Aly08 was able to recover 78.3% of its initial activity after 30 min incubating in the ice-bath (Figure 5B).

To determine the optimal incubation temperature that contributed to the recovery of activity after boiling for 5 min, the enzyme was incubated for 30 min at various temperatures (0–80 °C). The recovered activity of the enzyme reached levels of 76.3%, 63.9%, and 30.1% after incubation at 0 °C, 10 °C, and 20 °C for 30 min, respectively. When the incubation conditions were above 30 °C, the recovery of activity was measured at less than 10% (Figure 5C).

To further determine the optimal incubation time that contributed to the recovery of activity, the enzyme was incubated at 0 °C for different times. The activity of Aly08 gradually increased with prolonged culture time at 0 °C. Aly08 was rapidly re-activated approximately 56.7% and 71.3% of its activities after incubation at 0 °C for 5 min and 10 min, respectively. After incubation for 20 min, the activity was restored to 77.9%, after which the activity recovery rate began to decrease (Figure 5D).

The thermo-stability experiment indicated that low temperature may contribute to the recovery of Aly08. The thermo-tolerance of Aly08 could promote effective storage and transportation as the inactivated enzyme with heat treatment is able to successfully restore most of its activity after incubation at 0 °C.

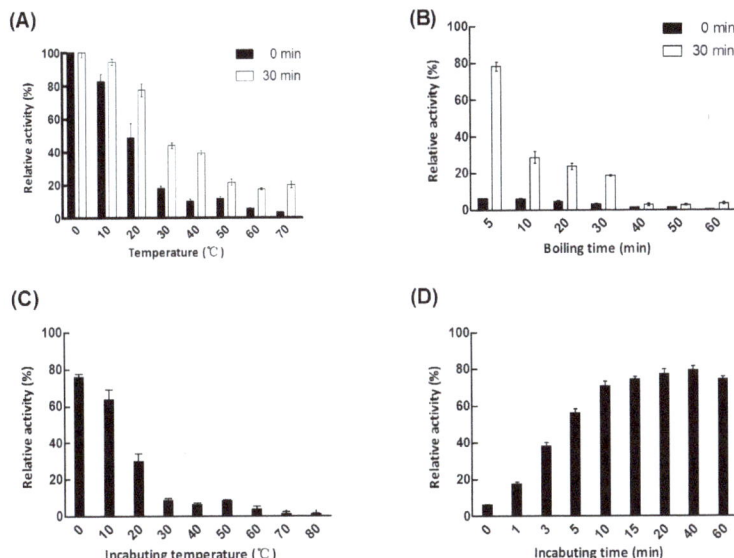

Figure 5. Thermo-tolerance and heat recovery of Aly08. (**A**) The difference of thermostability of enzymes incubation at ice-bath for 0 min (black columnar) and 30 min (white columnar). (**B**) Effects of boiling times on enzyme Aly08. Black and white columns indicate the activity of the heat-inactivated enzyme following ice-bath for 0 min and 30 min, respectively. (**C**) Effects of different incubation temperatures on the activity recovery of Aly08 under 5 min heat-inactivated conditions. (**D**) Effects of incubation time at 0 °C on the activity recovery of heat-inactivated Aly08. The enzyme activity without any treatment was 100%.

2.4. Action Pattern and Final Product Analysis

The action mode of Aly08 was determined by size-exclusion chromatography with a Superdex™ peptide 10/300 column (General Electric Company, Boston, MA, USA) using high-performance liquid chromatography (HPLC) platform (Figure 6). The hydrolysis pattern of Aly08 works as an endo-type because of the rapid depolymerization of substrates, the rise in polydispersity, and the production of intermediate oligosaccharides. Meanwhile, the action mode of Aly08 was further monitored by viscosity analysis (Figure S1). The viscosity of the alginate solution decreased rapidly during the first 5 min following the addition of Aly08 but changed little during subsequent time periods. During the whole observation period, the oligosaccharide content which was tested by A235 increased steadily. It can be further suggested that Aly08 is an endo-type enzyme in accordance with this finding.

The hydrolytic degradations were analyzed by thin-layer chromatography (TLC) method after the alginate was completely degraded (Figure 7A). In the hydrolysis proceeds, there was a gradual decrease of alginate polysaccharide and an accumulation of oligosaccharides with various DPs. And two clear spots of end product (2 h) appeared on the TLC plate, indicating that the migration rate was in good agreement with the alginate disaccharide (DP2) and trisaccharide (DP3) marker. The final degradation product was also determined by negative-ion electrospray ionization mass spectrometry (ESI-MS) (Figure 7B). Two main spectra were 351.1 m/z [ΔDP2-H]$^-$ and 527.2 m/z [ΔDP3-H]$^-$, corresponding to the molecular mass of the unsaturated alginate disaccharides and trisaccharides, respectively [38].

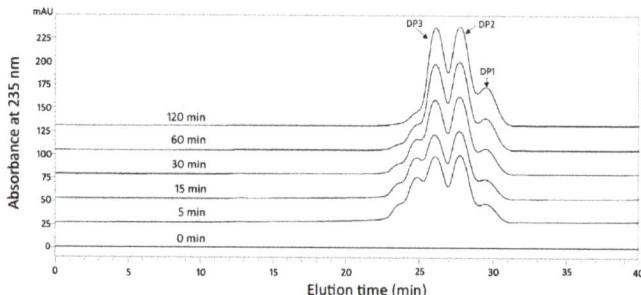

Figure 6. Degradation patterns of Aly08 toward sodium alginate. The elution positions of the unsaturated oligosaccharide product fractions with different degrees of polymerization are shown with arrows: DP1 represents unsaturated monosaccharide, DP2 represents unsaturated disaccharide, DP3 represents unsaturated trisaccharide.

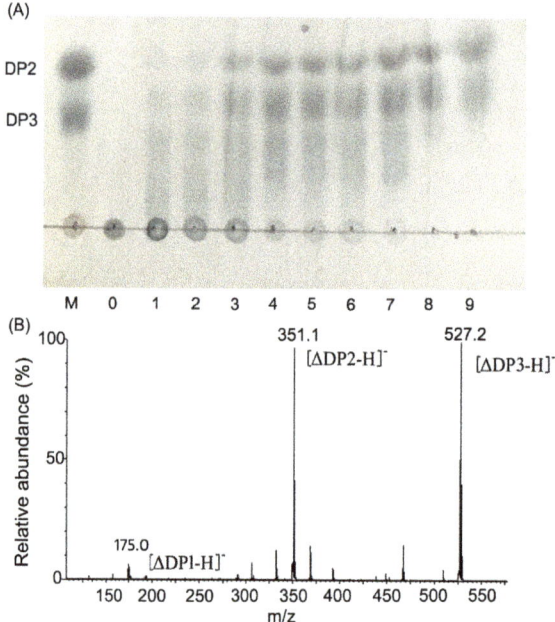

Figure 7. The hydrolytic products of Aly08. (**A**) TLC analysis of the hydrolytic products of Aly08. Lane M, standard alginate oligosaccharides (DP2-3); Line 0, alginate; Lane 1–9, hydrolytic products of Aly08 for different times (1, 2, 5, 10, 15, 30, 60, 90, and 120 min) toward 0.3% (w/v) high viscosity sodium alginate. DP2 and DP3 indicate alginate disaccharide and trisaccharide, respectively. (**B**) ESI-MS analysis of the end products of Aly08.

Through HPLC, viscosity, TLC and ESI-MS analysis, Aly08 was shown to degrade alginate polymer as an endo-type manner, eventually degrading alginate into disaccharides and trisaccharides. Previous studies have reported that enzymatic oligosaccharide products have a variety of specific biological activities and possess broad potential application prospects in many fields, such as antioxidant activities, regulation of plant root growth, anti-inflammatory activities, and regulation of lipoprotein metabolism [39–42]. The single homogeneous products in the progress of enzymatic production of oligosaccharides is conducive to the oligosaccharide purification and application. Thus far, most of the reported products of alginate lyase are mixtures of DP2–DP5, such as AlyA-OU02 from

V. spiendidus OU02, appear to take disaccharides, trisaccharides, and tetrasaccharides as the main hydrolytic products [28]. Additionally, the final degradation products of the alginate lyase FlAlyA from *Flavobacterium* sp. UMI-01 are DP2–DP5 [34]. Compared with those alginate lyase, the end products of Aly08 are a disaccharide and trisaccharide, which are advantageous for further separation and industrial high-efficiency production. Aly08 may have potential as a tool for the preparation of single homogeneous products of monosaccharides which have wide pharmaceutical applications.

3. Materials and Methods

3.1. Materials

High viscosity sodium alginate (20–50 kDa, 100–260 monosaccharide in polymer, M/G ratio: 1.66) and low viscosity sodium alginate (1–5 kDa, 5–26 monosaccharide in polymer the, M/G ratio: 1.66) was purchased from Bright Moon Seaweed Group (Qingdao, China), PolyM and PolyG blocks (purity: 95%) were purchased from Qingdao BZ Oligo Biotech Co., Ltd. (Qingdao, China). Standard alginate disaccharide and trisaccharide were also purchased from Qingdao BZ Oligo Biotech Co., Ltd. Standard monosaccharide (glucuronic acid) was purchased from Sigma. In addition, *E. coli* strains DH5α and BL21 (DE3) (Solarbio, Beijing, China) were grown in Luria–Bertani (LB) medium and used for plasmid construction and as a host for gene expression, respectively. LB broth supplemented with ampicillin (50 μg/mL) was used to grow both strains at 37 °C. The expression vectors used for gene cloning were pET-22b(+) (Novagen, Madison, WI, USA) plasmids. Oligonucleotides used for the cloning and expression of *aly08* were shown in Table S1.

3.2. Strains and Nucleotides

The sea mud samples were isolated from the sediment surface layer of Yellow Sea bottom (depth 36 m, E 120.13° N 35.76°, collected in May, 2017) and then immersed, diluted and spread on alginate sole-carbon selective medium plates [2 g $(NH_4)_2SO_4$, 30 g NaCl, 0.1 g $MgSO_4 \cdot 7H_2O$, 7 g K_2HPO_4, 3 g KH_2PO_4, 0.1 g $FeSO_4$, 5 g sodium alginate, dissolved in 1 L distilled water with 10 g agar, pH 7.0)]. At least 100 strains were isolated from the detectable colonies after incubation at 25 °C for 4 days, and then the strains were inoculated into agar-free selective medium for the purpose of identifying the activities of alginate lyases. A higher activity strain was screened out, 27F and 1492R primers were used to amplify the 16S rDNA gene of this strain. The 16S rDNA gene sequence of this strain was blasted to obtain the closely related sequence using the BLASTn algorithm program, the sequence was aligned with its closely related sequences using MEGA 6.0. According to the sequence alignment, this strain was classified as *Vibrio* sp. SY01. This strain has been preserved at the China Center for Type Culture Collection (CCTCC) under no. M2018769.

3.3. Sequence Analysis

In our previous study, genomic sequence analysis of *Vibrio.* sp SY01 showed a putative alginate lyase-encoding gene, *aly08*. The gene was cloned from the genome of strain SY01 and deposited in the Genbank database (accession number MH791447). The open reading frame (ORF) was identified with the program ORF finder (https://www.ncbi.nlm.nih.gov/orffinder/) and the signal peptide was analyzed using the SignalP 4.0 server (http://www.cbs.dtu.dk/services/SignalP/). The domain analysis of *aly08*, and its family analysis, is based on a comparison with the CDD (https://www.ncbi.nlm.nih.gov/cdd). The theoretically pI and Mw of *aly08* was determined by pI/Mw Tool (http://web.expasy.org/compute_pi/). Afterwards, the BLAST algorithm program on NCBI was used to search for similar sequences of *aly08*. Multiple sequence alignment was constructed using ClustalX 2.1 (National Center for Biotechnology Information, Bethesda, MD, USA), and the phylogenetic tree was created using the bootstrapping neighbor-joining method of MEGA 6.0 (Center for Evolutionary Medicine and Informatics, The Biodesign Institute, Tempe, AZ, USA).

3.4. Heterologous Expression and Purification of Recombinant Aly08

In order to express Aly08, the primers (PyAly08-F and PyAly08-R) were used to amplify the genomic DNA of *Vibrio* sp. SY01 without a signal sequence or stop codon. The *aly08* gene was then ligated into the expression vector pET-22b(+) with recognition sites *Nde* I and *Xho* I. In addition, the recombinant plasmid pET22b-Aly08 with a C-terminal 6 × His-tag was transformed into the *E. coli* BL21 (DE3) and grown on media (50 μg/mL ampicillin). Single colonies of *E. coli* BL21-pET22b-Aly08 were picked and cultured in LB medium (50 μg/mL ampicillin) at 37 °C with shaking at 200 rpm until OD_{600} reached 0.6–0.8. In order to induce the expression of protein, and the incubation was continued for 20 h at 20 °C (containing 0.1 mM isopropyl β-D-thiogalactoside (IPTG)). The cultured supernatant was harvested by high-speed refrigerated centrifuge (Hitachi, Tokyo, Japan) system (12,000 rpm, 5 min, 4 °C) and loaded onto a Ni-NTA sepharose column (GE Healthcare, Little Chalfont, Buckinghamshire, UK) using the AKTA150 automatic intelligent protein purification system (GE Healthcare, Little Chalfont, Buckinghamshire, UK) which had been equilibrated with wash buffer (20 mM phosphate buffer (pH 7.6), 500 mM NaCl). The Ni-NTA sepharose column was then eluted with a linear gradient of imidazole (25-500 mM imidazole, 20 mM phosphate buffer, 500 mM NaCl, pH7.6) in order to collect the active fractions. The active fraction was further analyzed by 12% SDS-PAGE, and the PageRuler Prest Protein Ladder (Thermo Scientific, USA) was used as a protein standard marker. Afterwards, the protein concentration of purified Aly08 was determined by a BCA protein assay kit (Beyotime Biotechnology, Shanghai China), bovine serum albumin (BSA) was used as a standard.

3.5. Alginate Lyase Activity Assay

Absorption at 235 nm (A235) was used to measure the activity of Aly08 as previously described [43–45]. The appropriately diluted enzyme (100 μL) was mixed with 900 μL of 0.3% (*w/v*) sodium alginate solution (10 mM glycine-NaOH buffer and 100 mM NaCl, pH 8.35). Then, the reaction system was incubated at 45 °C for 10 min and terminated by boiling for 10 min, and its absorbance was measured on a NanoPhotometer Pearl-360 spectrophotometer (IMPLEN, Munich, Germany). Alginate lyase activity was determined by increasing A235 as the production of unsaturated double bonds occurred as the alginate lyase cleaved glycosidic bonds at the non-reducing end of the polymer chain. One unit (U) of enzyme activity was defined as the amount of enzyme required to increase A235 by 0.1 per minute, under the above conditions.

3.6. Biochemical Characterization of the Recombinant Enzyme

The enzyme and substrate was incubated under 10 mM glycine-NaOH buffer (pH 8.35) at various temperatures (10–70 °C) to obtain the optimal temperature for Aly08. The thermal stability of Aly08 was then assayed by measuring its activity after pre-incubation at various temperatures (10–70 °C) for 1 h. The influence of pH values on Aly08 was calculated by measuring the residual activities in different buffers. The following buffers were used: 50 mM Na_2HPO_4-citric acid (pH 4.6–7.0), Tris-HCl (pH 7.6–8.6), glycine-NaOH (pH 8.6–10.6), and Na_2HPO_4-NaH_2PO_4 (pH 6.6–7.6). The highest activities represent 100% enzyme activity. The pH stability of Aly08 was determined by measuring the residual activity after incubating the purified enzyme with various pH buffers (pH 4.6–10.6) at 4 °C for 12 h. Substrate solution (10 mM glycine-NaOH buffer and 100 mM NaCl, pH 8.35) with three different substrates [0.3% (*w/v*) sodium alginate, polyM block and polyG block] were prepared and used for assaying the activities of the purified enzyme Aly08 in order to determine the preferred substrate of Aly08. Afterwards, based on the protein concentration measured by the BCA protein kit, the specific activity of Aly08 with the different substrates was calculated. To measure the effects of chemical compounds and metal ions on enzymatic activity, different metal ions and chemical compounds were added to the reaction system with a final concentration of 1 mM. A reaction mixture containing no metal ion or chemical compounds was used as a control. All reactions were performed in triplicate and the reaction parameters were expressed as the mean ± standard deviation.

3.7. Thermo-Tolerance Properties of Aly08

Purified Aly08 was placed at different temperatures for 1 h and then divided into two parts. One part was directly measured for its activity, while the other part was measured for its activity following incubation in an ice-bath for 30 min. In order to further observe whether the difference in temperature stability is related to boiling time, different boiling times (5, 10, 20, 30, 40, 50, and 60 min) were selected to evaluate the residual activity of the purified enzyme. After that the heat-treatment group was divided into two parts, one part was directly measured for its activity, while the other part was measured for its activity following incubation in an ice-bath for 30 min. To determine the effects of different temperatures on the activity recovery of heat-inactivated Aly08, the enzyme was immediately incubated for 30 min between 0 °C and 80 °C after boiling 5 min. Moreover, the enzyme was boiled for 5 min and further incubated at 0 °C for different times (0–60 min) to evaluate the effect of the time of low temperature treatment on the recovery of enzyme activity after it was heat treated.

3.8. Analysis of Reaction Products and Hydrolytic Pattern

Low viscosity sodium alginate was used for analysis the reaction product of Aly08. In order to examine the hydrolysis pattern of Aly08, the products from different incubation times (0, 5, 15, 30, 60, and 120 min) were monitored at A235 using gel filtration chromatography with a Superdex™ peptide 10/300 GL column with 0.2 M NH_4HCO_3 (flow rate: 0.6 mL/min) as an eluent on HPLC platform (LC-20A, Shimadzu, Japan) [46]. The mode of action was also analyzed using an Ostwald viscometer (No.1; Shibata Scientific Technology) with high viscosity sodium alginate as substrate. The equal component products (0.5 mL), which were degraded by Aly08 at 1, 5, 10, 15, and 30 min, were removed to characterize the viscosity and degradation products.

The hydrolytic degradation products were monitored using TLC method, wherein a reaction system containing 0.3% of high viscosity sodium alginate and Aly08 was constructed and the samples selected at different times (1, 2, 5, 10, 15, 30, 60, 90, and 120 min). The reaction products were analyzed using a TLC plate (TLC silica gel 60 F254, Merck KGaA, 64271 Darmstadt, Germany) with butanol/acetic-acid/water (2:1:1, by vol.) and color-developed with sulfuric acid/ethanol reagent (1:4, by vol.) after heating the TLC plate at 80 °C for 30 min. ESI-MS system (Thermo Fisher Scientific™ Q Exactive™ Hybrid Quadrupole-Orbitrap™, Waltham, MA, USA) was employed to further investigate the composition and degree of polymerization (DP) of the end products.

4. Conclusions

In conclusion, we purified and characterized a new alginate lyase, Aly08, from marine bacterium *Vibrio* sp. SY01. Its special characteristics (such as: thermo-tolerance and pH stability) make Aly08 a superior candidate for industrial applications. Further analysis will focus on analyzing the three-dimensional structure of Aly08 and exploring its molecular mechanisms.

Supplementary Materials: The following are available online at http://www.mdpi.com/1660-3397/17/5/308/s1, **Figure S1.** Viscosity measurement of Aly08; **Table S1.** Primers used in this study.

Author Contributions: S.L. and Y.H. designed the experiments. Y.W., L.B., Y.R., Q.H. and Z.Y. conducted the experiments. S.L., X.C. and R.Y. analyzed the data. Y.W., S.L. and Y.H. wrote the main manuscript. All authors reviewed the manuscript.

Funding: This research was funded by Shandong Provincial Natural Science Foundation grant number ZR2019BD027; National Nature Science Foundation of China, grant numbers 81573451 and 81602621; the Starting Research Funding of Qingdao University, grant number 41118010106 and 41118010250; the Key Lab of Marine Bioactive Substance and Modern Analytical Techniques, SOA, grant number MBSMAT-2018-06; and the Key Research and Development Program of Shandong Province, grant number 2018YYSP024.

Conflicts of Interest: The authors state that there is no conflict of interest.

References

1. Scieszka, S.; Klewicka, E. Algae in food: A general review. *Crit. Rev. Food Sci. Nutr.* **2018**, 1–10. [CrossRef] [PubMed]
2. Senturk Parreidt, T.; Muller, K.; Schmid, M. Alginate-based edible films and coatings for food packaging applications. *Foods* **2018**, *7*, 710. [CrossRef] [PubMed]
3. Wargacki, A.J.; Leonard, E.; Win, M.N.; Regitsky, D.D.; Santos, C.N.; Kim, P.B.; Cooper, S.R.; Raisner, R.M.; Herman, A.; Sivitz, A.B.; et al. An engineered microbial platform for direct biofuel production from brown macroalgae. *Science* **2012**, *335*, 308–313. [CrossRef] [PubMed]
4. Pawar, S.N.; Edgar, K.J. Alginate derivatization: A review of chemistry, properties and applications. *Biomaterials* **2012**, *33*, 3279–3305. [CrossRef]
5. Yamasaki, M.; Moriwaki, S.; Miyake, O.; Hashimoto, W.; Murata, K.; Mikami, B. Structure and function of a hypothetical *Pseudomonas aeruginosa* protein PA1167 classified into family PL-7: A novel alginate lyase with a beta-sandwich fold. *J. Biol. Chem.* **2004**, *279*, 31863–31872. [CrossRef] [PubMed]
6. Ertesvag, H. Alginate-modifying enzymes: Biological roles and biotechnological uses. *Front. Microbiol.* **2015**, *6*, 523.
7. Wong, T.Y.; Preston, L.A.; Schiller, N.L. Alginate lyase: Review of major sources and enzyme characteristics, structure-function analysis, biological roles, and applications. *Annu. Rev. Microbiol.* **2000**, *54*, 289–340. [CrossRef] [PubMed]
8. Li, S.Y.; Wang, Z.P.; Wang, L.N.; Peng, J.X.; Wang, Y.N.; Han, Y.T.; Zhao, S.F. Combined enzymatic hydrolysis and selective fermentation for green production of alginate oligosaccharides from Laminaria japonica. *Bioresour. Technol.* **2019**, *281*, 84–89. [CrossRef]
9. Sharma, S.; Horn, S.J. Enzymatic saccharification of brown seaweed for production of fermentable sugars. *Bioresour. Technol.* **2016**, *213*, 155–161. [CrossRef]
10. Kim, H.T.; Chung, J.H.; Wang, D.; Lee, J.; Woo, H.C.; Choi, I.G.; Kim, K.H. Depolymerization of alginate into a monomeric sugar acid using Alg17C, an exo-oligoalginate lyase cloned from *Saccharophagus degradans* 2-40. *Appl. Microbiol. Biotechnol.* **2012**, *93*, 2233–2239. [CrossRef] [PubMed]
11. Xu, F.; Dong, F.; Wang, P.; Cao, H.Y.; Li, C.Y.; Li, P.Y.; Pang, X.H.; Zhang, Y.Z.; Chen, X.L. Novel molecular insights into the catalytic mechanism of marine bacterial alginate lyase AlyGC from polysaccharide lyase Family 6. *J. Biol. Chem.* **2017**, *292*, 4457–4468. [CrossRef] [PubMed]
12. Garron, M.L.; Cygler, M. Uronic polysaccharide degrading enzymes. *Curr. Opin. Struct. Biol.* **2014**, *28*, 87–95. [CrossRef] [PubMed]
13. Kam, N.; Park, Y.J.; Lee, E.Y.; Kim, H.S. Molecular identification of a polyM-specific alginate lyase from *Pseudomonas* sp. strain KS-408 for degradation of glycosidic linkages between two mannuronates or mannuronate and guluronate in alginate. *Can. J. Microbiol.* **2011**, *57*, 1032–1041. [CrossRef] [PubMed]
14. Yang, M.; Yu, Y.; Yang, S.; Shi, X.; Mou, H.; Li, L. Expression and characterization of a new PolyG-specific alginate lyase from marine bacterium *Microbulbifer* sp. Q7. *Front. Microbiol.* **2018**, *9*, 2894. [CrossRef] [PubMed]
15. Chen, X.L.; Dong, S.; Xu, F.; Dong, F.; Li, P.Y.; Zhang, X.Y.; Zhou, B.C.; Zhang, Y.Z.; Xie, B.B. Characterization of a new cold-adapted and salt-activated polysaccharide lyase family 7 alginate lyase from *Pseudoalteromonas* sp. SM0524. *Front. Microbiol.* **2016**, *7*, 1120. [CrossRef]
16. Inoue, A.; Anraku, M.; Nakagawa, S.; Ojima, T. Discovery of a novel alginate lyase from *Nitratiruptor* sp. SB155-2 thriving at deep-sea hydrothermal vents and identification of the residues responsible for its heat stability. *J. Biol. Chem.* **2016**, *291*, 15551–15563. [CrossRef]
17. Jagtap, S.S.; Hehemann, J.H.; Polz, M.F.; Lee, J.K.; Zhao, H. Comparative biochemical characterization of three exolytic oligoalginate lyases from *Vibrio splendidus* reveals complementary substrate scope, temperature, and pH adaptations. *Appl. Environ. Microbiol.* **2014**, *80*, 4207–4214. [CrossRef]
18. Bonugli-Santos, R.C.; Dos Santos Vasconcelos, M.R.; Passarini, M.R.; Vieira, G.A.; Lopes, V.C.; Mainardi, P.H.; Dos Santos, J.A.; de Azevedo Duarte, L.; Otero, I.V.; da Silva Yoshida, A.M.; et al. Marine-derived fungi: Diversity of enzymes and biotechnological applications. *Front. Microbiol.* **2015**, *6*, 269. [CrossRef]
19. Li, S.Y.; Yang, X.M.; Bao, M.M.; Wu, Y.; Yu, W.G.; Han, F. Family 13 carbohydrate-binding module of alginate lyase from *Agarivorans* sp. L11 enhances its catalytic efficiency and thermostability, and alters its substrate preference and product distribution. *FEMS Microbiol. Lett.* **2015**, *362*. [CrossRef]

20. Zhu, B.W.; Sun, Y.; Ni, F.; Ning, L.M.; Yao, Z. Characterization of a new endo-type alginate lyase from *Vibrio* sp. NJU-03. *Int. J. Biol. Macromol.* **2018**, *108*, 1140–1147. [CrossRef]
21. Ogura, K.; Yamasaki, M.; Mikami, B.; Hashimoto, W.; Murata, K. Substrate recognition by family 7 alginate lyase from *Sphingomonas* sp. A1. *J. Mol. Biol.* **2008**, *380*, 373–385. [CrossRef]
22. Schiller, N.L.; Monday, S.R.; Boyd, C.M.; Keen, N.T.; Ohman, D.E. Characterization of the *Pseudomonas aeruginosa* alginate lyase gene (algL): Cloning, sequencing, and expression in *Escherichia coli*. *J. Bacteriol.* **1993**, *175*, 4780–4789. [CrossRef]
23. Thomas, F.; Lundqvist, L.C.; Jam, M.; Jeudy, A.; Barbeyron, T.; Sandstrom, C.; Michel, G.; Czjzek, M. Comparative characterization of two marine alginate lyases from *Zobellia galactanivorans* reveals distinct modes of action and exquisite adaptation to their natural substrate. *J. Biol. Chem.* **2013**, *288*, 23021–23037. [CrossRef]
24. Swift, S.M.; Hudgens, J.W.; Heselpoth, R.D.; Bales, P.M.; Nelson, D.C. Characterization of AlgMsp, an alginate lyase from *Microbulbifer* sp. 6532A. *PLoS ONE* **2014**, *9*, e112939. [CrossRef]
25. Yoon, H.J.; Hashimoto, W.; Miyake, O.; Okamoto, M.; Mikami, B.; Murata, K. Overexpression in *Escherichia coli*, purification, and characterization of *Sphingomonas* sp. A1 alginate lyases. *Protein Expr. Purif.* **2000**, *19*, 84–90. [CrossRef] [PubMed]
26. Huang, L.S.; Zhou, J.G.; Li, X.; Peng, Q.; Lu, H.; Du, Y.G. Characterization of a new alginate lyase from newly isolated *Flavobacterium* sp. S20. *J. Ind. Microbiol. Biotechnol.* **2013**, *40*, 113–122. [CrossRef] [PubMed]
27. Uchimura, K.; Miyazaki, M.; Nogi, Y.; Kobayashi, T.; Horikoshi, K. Cloning and sequencing of alginate lyase genes from deep-sea strains of *Vibrio* and *Agarivorans* and characterization of a new *Vibrio* enzyme. *Mar. Biotechnol.* **2010**, *12*, 526–533. [CrossRef]
28. Zhuang, J.J.; Zhang, K.K.; Liu, X.H.; Liu, W.Z.; Lyu, Q.Q.; Ji, A.G. Characterization of a novel polyM-preferred alginate lyase from marine *Vibrio splendidus* OU02. *Mar. Drugs* **2018**, *16*, 295. [CrossRef]
29. Miyake, O.; Ochiai, A.; Hashimoto, W.; Murata, K. Origin and diversity of alginate lyases of families PL-5 and -7 in *Sphingomonas* sp. strain A1. *J. Bacteriol.* **2004**, *186*, 2891–2896. [CrossRef] [PubMed]
30. Peng, C.N.; Wang, Q.B.; Lu, D.R.; Han, W.J.; Li, F.C. A novel bifunctional endolytic alginate lyase with variable alginate-degrading modes and versatile monosaccharide-producing properties. *Front. Microbiol.* **2018**, *9*, 167. [CrossRef] [PubMed]
31. Duan, G.F.; Han, F.; Yu, W.G. Cloning, sequence analysis, and expression of gene alyPI encoding an alginate lyase from marine bacterium *Pseudoalteromonas* sp. CY24. *Can. J. Microbiol.* **2009**, *55*, 1113–1118. [CrossRef] [PubMed]
32. Brown, B.J.; Preston, J.F.; Ingram, L.O. Cloning of alginate lyase gene (alxM) and expression in *Escherichia coli*. *Appl. Environ. Microbiol.* **1991**, *57*, 1870–1872. [PubMed]
33. Kobayashi, T.; Uchimura, K.; Miyazaki, M.; Nogi, Y.; Horikoshi, K. A new high-alkaline alginate lyase from a deep-sea bacterium *Agarivorans* sp. *Extremophiles* **2009**, *13*, 121–129. [CrossRef] [PubMed]
34. Inoue, A.; Takadono, K.; Nishiyama, R.; Tajima, K.; Kobayashi, T.; Ojima, T. Characterization of an alginate lyase, FlAlyA, from *Flavobacterium* sp. strain UMI-01 and its expression in *Escherichia coli*. *Mar. Drugs* **2014**, *12*, 4693–4712. [CrossRef]
35. Zhu, X.Y.; Li, X.Q.; Shi, H.; Zhou, J.; Tan, Z.B.; Yuan, M.D.; Yao, P.; Liu, X.Y. Characterization of a novel alginate lyase from marine bacterium *Vibrio furnissii* H1. *Mar. Drugs* **2018**, *16*, 30. [CrossRef]
36. Zhu, B.W.; Ni, F.; Ning, L.M.; Sun, Y.; Yao, Z. Cloning and characterization of a new pH-stable alginate lyase with high salt tolerance from marine *Vibrio* sp. NJ-04. *Int. J. Biol. Macromol.* **2018**, *115*, 1063–1070. [CrossRef] [PubMed]
37. Huang, G.Y.; Wang, Q.Z.; Lu, M.Q.; Xu, C.; Li, F.; Zhang, R.C.; Liao, W.; Huang, S.S. AlgM4: A new salt-activated alginate lyase of the PL7 family with endolytic activity. *Mar. Drugs* **2018**, *16*, 120. [CrossRef] [PubMed]
38. Zhu, B.W.; Ni, F.; Ning, L.M.; Yao, Z. Elucidation of degrading pattern and substrate recognition of a novel bifunctional alginate lyase from *Flammeovirga* sp. NJ-04 and its use for preparation alginate oligosaccharides. *Biotechnol. Biofuels* **2019**, *12*, 13. [CrossRef] [PubMed]
39. Yang, J.H.; Bang, M.A.; Jang, C.H.; Jo, G.H.; Jung, S.K.; Ki, S.H. Alginate oligosaccharide enhances LDL uptake via regulation of LDLR and PCSK9 expression. *J. Nutr. Biochem.* **2015**, *26*, 1393–1400. [CrossRef]

40. Falkeborg, M.; Cheong, L.Z.; Gianfico, C.; Sztukiel, K.M.; Kristensen, K.; Glasius, M.; Xu, X.; Guo, Z. Alginate oligosaccharides: Enzymatic preparation and antioxidant property evaluation. *Food Chem.* **2014**, *164*, 185–194. [CrossRef]
41. Zhang, Y.H.; Yin, H.; Zhao, X.M.; Wang, W.X.; Du, Y.G.; He, A.; Sun, K.G. The promoting effects of alginate oligosaccharides on root development in Oryza sativa L. mediated by auxin signaling. *Carbohydr. Polym.* **2014**, *113*, 446–454. [CrossRef] [PubMed]
42. Qu, Y.; Wang, Z.M.; Zhou, H.H.; Kang, M.Y.; Dong, R.P.; Zhao, J.W. Oligosaccharide nanomedicine of alginate sodium improves therapeutic results of posterior lumbar interbody fusion with cages for degenerative lumbar disease in osteoporosis patients by downregulating serum miR-155. *Int. J. Nanomed.* **2017**, *12*, 8459–8469. [CrossRef] [PubMed]
43. Qin, H.M.; Miyakawa, T.; Inoue, A.; Nishiyama, R.; Nakamura, A.; Asano, A.; Ojima, T.; Tanokura, M. Structural basis for controlling the enzymatic properties of polymannuronate preferred alginate lyase FlAlyA from the PL-7 family. *Chem Commun. (Camb)* **2018**, *54*, 555–558. [CrossRef]
44. Doi, H.; Tokura, Y.; Mori, Y.; Mori, K.; Asakura, Y.; Usuda, Y.; Fukuda, H.; Chinen, A. Identification of enzymes responsible for extracellular alginate depolymerization and alginate metabolism in *Vibrio algivorus*. *Appl. Microbial. Biotechnol.* **2017**, *101*, 1581–1592. [CrossRef] [PubMed]
45. Li, S.Y.; Wang, L.N.; Hao, J.H.; Xing, M.X.; Sun, J.J.; Sun, M. Purification and characterization of a new alginate lyase from marine bacterium *Vibrio* sp. SY08. *Mar. Drugs* **2016**, *15*, 1. [CrossRef] [PubMed]
46. Chen, P.; Zhu, Y.M.; Men, Y.; Zeng, Y.; Sun, Y.X. Purification and characterization of a novel alginate lyase from the marine bacterium *Bacillus* sp. Alg07. *Mar. Drugs* **2018**, *16*, 86. [CrossRef] [PubMed]

© 2019 by the authors. Licensee MDPI, Basel, Switzerland. This article is an open access article distributed under the terms and conditions of the Creative Commons Attribution (CC BY) license (http://creativecommons.org/licenses/by/4.0/).

Article

Biochemical Characterization and Degradation Pattern of a Novel Endo-Type Bifunctional Alginate Lyase AlyA from Marine Bacterium *Isoptericola halotolerans*

Benwei Zhu [1,*], Limin Ning [2,*], Yucui Jiang [2] and Lin Ge [3]

1 College of Food Science and Light Industry, Nanjing Tech University, Nanjing 211816, Jiangsu, China
2 College of Medicine and Life Science, Nanjing University of Chinese Medicine, Nanjing 210023, Jiangsu, China; jiangyucuinju@163.com
3 Technology Transfer Center, Nanjing Forest University, Nanjing 210037, Jiangsu, China; njfuelin@126.com
* Correspondence: zhubenwei@njtech.edu.cn (B.Z.); ninglimin@njucm.edu.cn (L.N.);
Tel.: +86-25-8581-1558 (B.Z.); +86-25-5813-9419 (L.N.)

Received: 28 June 2018; Accepted: 27 July 2018; Published: 31 July 2018

Abstract: Alginate lyases are important tools to prepare oligosaccharides with various physiological activities by degrading alginate. Particularly, the bifunctional alginate lyase can efficiently hydrolyze the polysaccharide into oligosaccharides. Herein, we cloned and identified a novel bifunctional alginate lyase, AlyA, with a high activity and broad substrate specificity from bacterium *Isoptericola halotolerans* NJ-05 for oligosaccharides preparation. For further applications in industry, the enzyme has been characterized and its action mode has been also elucidated. It exhibited the highest activity (7984.82 U/mg) at pH 7.5 and 55 °C. Additionally, it possessed a broad substrate specificity, showing high activities towards not only polyM (polyβ-D-mannuronate) (7658.63 U/mg), but also polyG (poly α-L-guluronate) (8643.29 U/mg). Furthermore, the K_m value of AlyA towards polyG (3.2 mM) was lower than that towards sodium alginate (5.6 mM) and polyM (6.7 mM). TLC (Thin Layer Chromatography) and ESI-MS (Electrospray Ionization Mass Spectrometry) were used to study the action mode of the enzyme, showing that it can hydrolyze the substrates in an endolytic manner to release a series of oligosaccharides such as disaccharide, trisaccharide, and tetrasaccharide. This study provided extended insights into the substrate recognition and degrading pattern of the alginate lyases, with a broad substrate specificity.

Keywords: *Isoptericola halotolerans*; bifunctional alginate lyase; oligosaccharides

1. Introduction

Alginate is the major component of the cell wall of brown algae [1]. It is a linear anionic heteropolysaccharide comprising of two uronic acids, α-L-guluronic (G) and β-D-mannuronic acid (M) [2]. The G and M subunits are covalently linked by 1,4-glycosidic linkage in three different types of blocks, homopolymeric α-L-guluronic acid (polyG), homopolymeric β-D-mannuronic acid (polyM), and heteropolymeric α-L-guluronic acid-β-D-mannuronic acid (polyMG) [3]. Because of the high viscosity, gelling properties, and versatile activities, alginate has been widely applied in food, chemical, and pharmaceutical industries [4–6]. However, the applications of this polysaccharide are still limited, and are subjected to the high molecular weight and poor solubility [7]. Thus, alginate oligosaccharide (AOS) has attracted more and more attention, as it retains the physiological functions and activities of alginate, but possessed smaller molecule weights and good bioavailability [8]. Yang et al. found that AOS can enhance the uptake of LDL (Low-Density Lipoprotein) by regulating the expression of LDLR (Low-Density Lipoprotein Receptor) and PCSK9 (Proprotein Convertase Subtilisin/kexin

Type 9) [9]. Iwamoto et al. found that G8 (octaguluronic acid) and M7 (heptamannuronic acid) can induce RAW264.7 cells to produce cytokine furthest [10]. The similar effect of mannuronate oligomers have also been reported by Yamamoto et al. [11].

Alginate lyase can catalyze the alginate by the β-elimination and release unsaturated AOS with double bonds between C4 and C5 [12]. So far, hundreds of alginate lyases have been found in marine and terrestrial bacteria, marine mollusks, and algae, according to the CAZy database [13–19]. According to the substrate specificities, they can be classified into three types, polyM-specific lyases (EC 4.2.2.3), polyG-specific lyases (EC 4.2.2.11), and bifunctional lyases (EC 4.2.2.) [20]. Additionally, based on the sequence similarity, the enzymes are generally grouped into seven PL (Polysaccharide Lyases) families (PL-5, -6, -7, -14, -15, -17, and -18) [21]. Furthermore, alginate lyases can be grouped into endolytic and exolytic alginate lyases in terms of the mode of action [22]. Endolytic enzymes can cleave the glycosidic bonds inside the alginate with unsaturated oligosaccharides as the main products [23], while the exolytic ones degrade alginate into monomers [24]. So far, many endolytic alginate lyases have been widely used to produce AOS in food and nutraceutical industries [25]. Furthermore, they have also been used to elucidate the fine structures of alginate, prepare protoplast of brown algae [26,27], and treat cystic fibrosis, combined with antibiotics [28]. So far, many alginate lyases originating from marine microorganisms have been identified, gene-cloned, purified, and well characterized. However, few have been commercially used because of the poor substrate specificity and low activity [29–31]. In addition, there are few reports about the bifunctional alginate lyase with broad substrate specificity. Thus, it will be of great importance to explore the novel enzymes with excellent characteristics, such as a high activity and broad substrate specificity.

In this work, a new bifunctional alginate lyase with a high activity and broad substrate specificity has been cloned from *Isoptericola halotolerans* NJ-05, followed by being identified and characterized. The enzymatic kinetics were further characterized and the degrading products were also analyzed, which suggested it a good candidate to expand the applications of alginate lyases in food and nutraceutical industries.

2. Results and Discussions

2.1. Screening and Identification of Strain NJ-05

According to the screening results by the plates and activity assay, the strain NJ-05 from the rotten brown algae obtained from the East China Sea showed the maximal activity of alginate lyase and then was identified for further investigation. The 16S rRNA sequence of the strain was then sequenced (GeneBank No. MH390700) and according to the phylogenetic analysis of the 16S rRNA sequence, the strain was assigned to the genus *Isoptericola* and named *Isoptericola halotolerans* NJ-05 (Figure 1).

2.2. Sequence Analysis of Alginate Lyase

The genome of *Isoptericola halotolerans* NJ-05 was firstly sequenced. After the annotation of the genomic information, several putative alginate lyase genes were found in a gene cluster for the alginate metabolism and the AlyA was identified as a putative alginate lyase by multiple sequences alignment. Then, the gene AlyA was cloned from the genome of *Isoptericola halotolerans* NJ-05 and sequenced (GeneBank No. MH390701). It can be observed that the open reading frame (ORF) consists of 771 bp, which encodes a putative alginate lyase with 256 amino acids, including a signal peptide with 32 amino acids (Supplementary Materials). According to the conserved domain analysis, AlyA possesses only a C-terminal catalytic domain consisting of 213 amino acids (Thr43-Glu255).

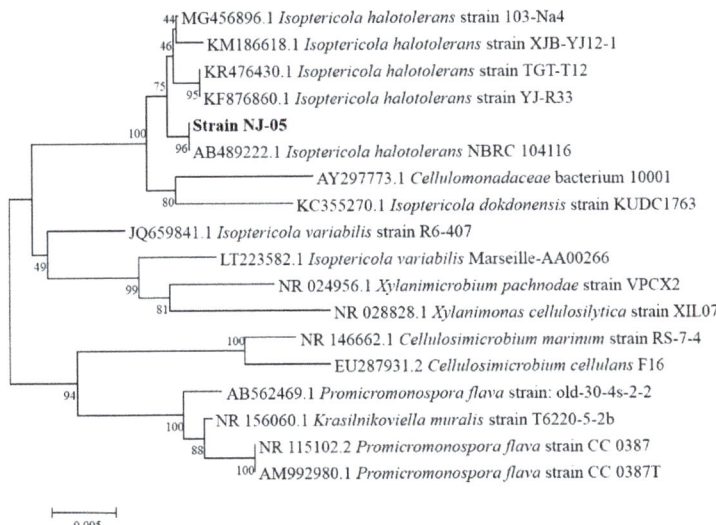

Figure 1. The phylogenetic analysis of strain NJ-05 and other similar strains. The phylogenetic tree was constructed by MEGA 6.0 (https://www.megasoftware.net/), based on the 16S rRNA gene sequences of strain NJ-05 and other known species.

The sequence alignment of AlyA and other alginate lyases of the PL7 family are shown in Figure 2. It can be observed that AlyA shares the highest identity, of 79%, with alginate lyase (ACN56743.1) from *Streptomyces* sp. M3 [32] and exhibited an identity of 67% with alginate lyase (BAA83339.1) from *Corynebacterium* sp. ALY-1 [33]. Additionally, it also contains the conserved regions "PRT/V/SELRE", "YFKA/VGN/VY", and "QIH", which are related to the substrate combination and catalytic activity (Figure 2). Thus, AlyA is a member of the PL7 family. Moreover, as the alginate lyases of PL7 family are further grouped into five families (1–5) according to the amino acid sequence homology, the phylogenetic tree was constructed by comparing the sequence homology of AlyA with the alginate lyases from a different subfamily to determine the subfamily of AlyA (Figure 3). It can be observed that the AlyA clusters represent the enzymes of subfamily 3, indicating that AlyA is the member of the subfamily 3 alginate lyases.

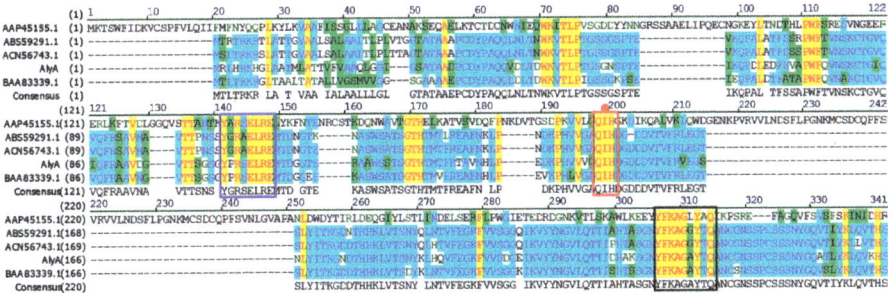

Figure 2. Multiple sequences alignments of alginate lyase (AlyA) and related alginate lyases of the PL7 family. The conserved regions and identical residues were highlighted with bands and red star, respectively.

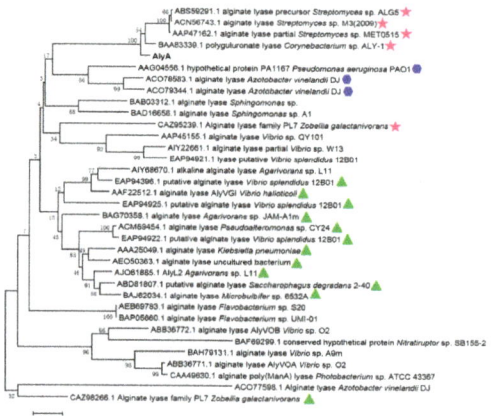

Figure 3. Phylogenetic analysis of AlyA and other alginate lyases of PL7. The phylogenetic tree was generated by the neighbor-joining method using MEGA 6.0 software (https://www.megasoftware.net/). The species names are indicated along with the accession number of the corresponding alginate lyase sequence. Bootstrap values of 1000 trials are presented in the branching points. The scale bar indicating ten nucleotide substitutions per 100 nucleotides is indicated at the bottom. The alginate lyases of subfamily 1, 3, and 5 were marked with a blue hexagon, red pentacles, and green triangles, respectively.

2.3. Expression and Purification of AlyA

For a better characterization, the recombinant protein was expressed by firstly inserting the AlyA gene into the pET-21a (+) vector, and then being purified by Ni-NTA(Ni- Nitrilotriacetic acid) column (The summary of purification was shown in Table S1). The molecular mass of the recombinant protein with (His)$_6$-tag is calculated to be 25.32 kDa, which is smaller than two other bifunctional alginate lyases, the Aly-SJ02 of *Pseudoalteromonas* sp. SM0524 with a bigger molecular mass of 32 kDa [34] and the Aly202 of *Alteromonas* sp. No. 272 with a similar molecular mass of 33.9 kDa [35]. The recombinant AlyA was further purified by Ni-NTA Sepharose affinity chromatography and analyzed by sodium dodecyl sulfate polyacrylamide gel electrophoresis (SDS-PAGE) (Figure 4). A single band of purified AlyA was observed at the gel, and it can be further used for downstream biochemical characterization.

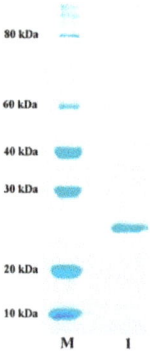

Figure 4. The sodium dodecyl sulfate polyacrylamide gel electrophoresis (SDS-PAGE) analysis of purified alginate lyase, AlyA. Lane M—the protein molecular weight standard; lane 1—the purified AlyA.

2.4. Substrate Specifity and Enzymatic Kinetics of the Enzyme

As shown in Table 1, AlyA possessed a broad substrate specificity, showing a higher activity towards sodium alginate (7984.82 U/mg) and polyG (8643.29 U/mg), and a lower activity towards polyM (7658.63 U/mg). Thus, it can be concluded that AlyA is a bifunctional alginate lyase. Compared with another bifunctional alginate lyase, Aly-SJ02, which displayed almost the same activity towards sodium alginate (4802.7 U/mg) and polyM (4153.8 U/mg), but lower activity towards polyG (3073.7 U/mg) [34], it exhibited a much higher activity to all three of the substrates. Similarly, alginate substrates with various MM (Manuronate-Manuronate) and GG (Guluronate- Guluronate) ratios were used to further confirm the substrate affinity and substrate specificity of FAly and SALy. FAly prefer to degrade polyM, as the initial rates towards polyM are almost double that towards polyG, while SALy performed slightly better on polyG than on polyM [36].

Table 1. The substrate specificity and enzymatic kinetics of alginate lyase (AlyA) towards various substrates.

Parameters	Sodium Alginate	PolyG	PolyM
Specific activity (U/mg)	7984.82	8643.29	7658.63
K_m (mM)	5.6	3.2	6.7
V_{max} (nmol/s)	3.22	1.74	1.89
k_{cat} (s^{-1})	45.92	24.82	26.95
k_{cat}/K_m (s^{-1}·mM^{-1})	8.20	7.56	4.02

The enzymatic kinetics of AlyA towards sodium alginate, polyM, and polyG were also studied according to the hyperbolic regression analysis. As shown in Table 1, the K_m values of AlyA towards sodium alginate, polyM, and polyG were 5.6 mM, 6.7 mM, and 3.2 mM, respectively. Thus, it has higher affinity towards sodium alginate and polyG than that to polyM. The k_{cat}/K_m values of AlyA towards sodium alginate (8.20 mM^{-1}·s^{-1}), polyG (7.56 mM^{-1}·s^{-1}), and polyM (4.02 mM^{-1}·s^{-1}) were also calculated, indicating that the enzyme exhibited a higher catalytic efficiency towards sodium alginate and polyG than to polyM. Therefore, AlyA prefers the G block or MG block to the M block as the fit substrate, which is consistent with the substrate specificity of AlyA. It can be reasonable, as the conserved residue "I" in AlyA can recognize the polyG block or MG blocks [20]. As to the calculation of the kinetics data, the K_m and V_{max} values of some alginate lyases were determined by double-reciprocal plots of Lineweaver and Burk. For instance, Algb from *Vibrio* sp. W13 showed lower K_m values toward alginate (0.67 mg/mL) polyG (1.04 mg/mL) than that to polyG (6.90 mg/mL) [17], and Alg-S5 from *Exiguobacterium* sp. Alg-S5 exhibited a high affinity towards the alginate with a lower K_m value of 0.91 mg/mL [37]. To conveniently calculate the k_{cat} and k_{cat}/K_m values, we referred the calculation method in the literature [38], and described the details of the calculation of the molar concentration and molecular weight in our recent publications [23,30,39–41].

2.5. Biochemical Characterization of AlyA

AlyA was further characterized biochemically. The optimal pH for the enzyme activity was 7.5 and retained more than 60% activity after being incubated at a broad pH range of pH 5.5–9.0 for 24 h (Figure 5A). Additionally, the enzyme was mostly stable at pH 7.0. As previously reported, the alginate lyases of the PL7 family were active at a neutral pH. The Alg7D from *Saccharophagus degradans*, AlgNJ-04 from *Vibrio* sp. NJ-04, FsAlgA from *Flammeovirga* sp. NJ-04, and AlyA1 from *Zobellia galactanivorans* all showed their maximal activity at pH 7.0 [13,39,41,42]. However, they usually show a high activity in a narrow pH range, and exhibit instability under alkaline conditions. For instance, the AlySJ-02 from *Pseudoalteromonas* sp. SM0524 exhibited its maximal activity at pH 8.0 and retained its stability between pH 7.0–9.0 [34]. The AlyIH from *Isoptericola halotolerans* CGMCC5336 showed the highest activity at pH 7.0 and was stable at pH 7.0–8.0 [43]. Similarly, the A9m from *Vibrio* sp. A9mT exhibited its maximal activity at pH 7.5 and could maintain its stability between pH 7.0–9.0 [44].

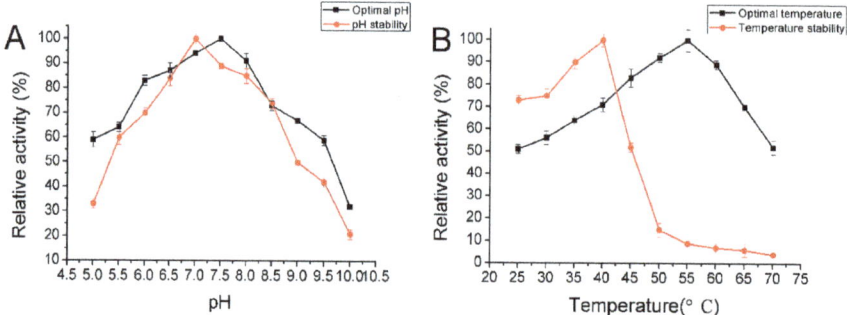

Figure 5. The biochemical characteristics of AlyA. (**A**) The optimal pH and the pH stability of AlyA. (**B**) The optimal temperature and the thermal stability of AlyA. The assay was then incubated at 40 °C for 10 min. Each value represents the mean of three replicates ± standard deviation.

AlyA showed maximum activity at 55 °C and was stable below 40 °C (Figure 5B). This enzyme possessed an approximately 80% activity after incubation at 40 °C for 30 min, and was gradually inactivated as the temperature increased. Similarly, most of the characterized enzymes of the PL7 family showing maximal activity around 30–40 °C (Table 2). For example, the AlgNJ-04 from *Vibrio* sp. NJ-04 [39] and AlgNJU-03 from *Vibrio* sp. NJU-03 [40] both possessed the optimal temperature of 40 °C. AlgC-PL7 from *Cobetia* sp. NAP1, OalY1 from *Halomonas* sp. QY114, and an alginate lyase from an unknown marine bacterium showed their maximal temperature at 45 °C [45–47]. Remarkably, Aly-SJ02 from *Pseudoalteromonas* sp. SM0524, FsAlgA from *Flammeovirga* sp. NJ-04, Alg7D from *Saccharophagus degradans*, AlgMsp from *Microbulbifer* sp. 6532A, and OalY2 from *Halomonas* sp. QY114 all had a higher optimal temperature of 50 °C [34,41,47]. Therefore, the AlyA possessed a great potential for industrial applications, due to the higher optimal temperature.

The thermal stability of AlyA were further investigated by thermal-induction, as shown in Figure 6. The enzyme could retain almost 80% of its maximal activity after been incubated at 35 °C for 60 min. Similarly, the OalY1 and OalY2 from *Halomonas* sp. QY114 retained about 80% of the initial activities after incubation at 30 °C for 1 h [47]. While the AlgC-PL7 maintained approximately 80% of it activity at 70 °C [45]. As with the alginate lyase from the unknown marine bacterium, its activity remained without a noticeable loss up to 70 °C, with a monotonic decrease beyond this temperature [46].

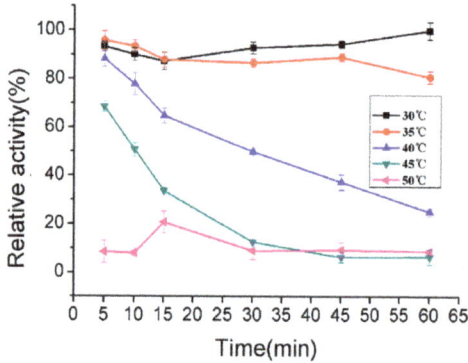

Figure 6. The thermal degeneration curve of AlyA. The maximal activity of the treated enzyme was regarded as 100% and the other relative activity was determined. Each value represents the mean of three replicates ± standard deviation.

Table 2. Comparison of biochemical properties of AlyA and partial enzymes of PL7 family.

Enzyme	Specificity	K_m, V_{max}	Optimal Temperature and pH	Products (Dp)	Reference
AlyA	Bifunctional	5.6 mM, 3.22 nmol/s	55 °C, 7.5	2–5	This study
Algb	polyG > polyM	0.67 mg/mL, 473.93 U/mg	30 °C, 8.0	2–5	[17]
FsAlgA	polyG > polyM	0.48 mM, 0.19 nmol/s	50 °C, 7.0	2–5	[41]
Alg7D	polyM > polyG	3.0 mg/mL, 6.2 U/g	50 °C, 7.0	2–5	[13]
AlgMsp	polyG > polyM	3.4 mM, 57 pmol/s	50 °C, 8.0	2–5	[38]
AlgNJ-04	polyG > polyM	0.49 mM, 72 pmol/s	40 °C, 7.0	2–5	[41]
AlgNJU-03	polyG > polyM	8.5 mM, 1.67 nmol/s	30 °C, 7.0	2–4	[40]
AlgC-PL7	Bifunctional	-	55 °C, 8.0	1	[45]
A9m	polyG > polyM	-	30 °C, 7.5	-	[44]

The effect of metal ions on the activity of AlyA is shown in Table 3. It can be observed that Na^+, Mg^{2+}, and Ca^{2+} could greatly enhance the activity of the enzyme, while some divalent ions, such as Co^{2+}, Cu^{2+}, and Fe^{3+}, inhibited the activity. Similarly, Ca^{2+} can activate AlyA from *Pseudomonas* sp. E03 [48], the AlyA from *Azotobacter chroococcum* 4A1M [49], and ALYII from *Pseudomonas* sp. OS-ALG-9 [50], as it could enhance the substrate-binding ability of the enzyme.

Table 3. The effect of metal ions on activity of AlyA.

Reagent	Relative Activity (%)
Control	100 ± 0.5
Na^+(100 mM)	126 ± 2.2
Na^+(300 mM)	180 ± 3.1
Na^+(500 mM)	203 ± 4.6
Na^+(700 mM)	136 ± 2.9
Na^+(900 mM)	89 ± 7.9
Zn^{2+}	91 ± 2.3
Cu^{2+}	65 ± 3.2
Mn^{2+}	94 ± 2.1
Co^{2+}	75 ± 3.4
Ca^{2+}	174 ± 1.3
Ca^{2+}(10 mM)	135 ± 5.7
Fe^{3+}	88 ± 2.1
Mg^{2+}	168 ± 2.7
Mg^{2+}(10 mM)	119 ± 2.9
Mg^{2+}(50 mM)	101 ± 3.2
Ni^{2+}	87 ± 1.5

The sequence alignment of AlyA and AlyPG from *Corynebacterium* sp. ALY-1 was constructed by CLSTALW (http://www.clustal.org/) (as shown in Figure 7A). The three-dimensional model of the AlyA was constructed based on the homologues structure of the alginate lyase, AlyPG, of *Corynebacterium* sp. ALY-1 (PDB ID: 1UAI), using PHYRE2 (http://www.sbg.bio.ic.ac.uk/phyre2/html/page.cgi?id=index). As shown in Figure 7C, the overall structure of the AlyA was predicted to fold as a β-sandwich jelly roll formed, using two anti-parallel β sheets. The outer convex sheet includes five β-strands, and the inner concave sheet contains seven β-strands forming a groove that harbors the catalytic active site. In order to identify the key residues for substrate recognition, the structural alignment of AlyA and AlyPG was analyzed (Figure 7C). The two enzymes share a similar structure, and the residues Q159, H161, and R119 essential for substrate reorganization form hydrogen bonds with the carboxyl groups in subsites +1, +2, and +3, respectively (Figure 7A,D).

The ESI-MS analysis of the degradation products is shown in Figure 8; disaccharides, trisaccharides, and tetrasaccharides account for a major fraction of the hydrolysates of two kinds of substrates. Thus, AlyA can be a potential tool to produce oligosaccharides with lower Dps by hydrolyzing sodium alginate. So far, most of the characterized alginate lyases of the PL7 family are endolytic enzymes, which can release oligosaccharides with a low Dp of 2–5 as the main products. Interestingly, AlyA5 from *Zobellia galactanivorans* can release disaccharides in an exolytic manner [42].

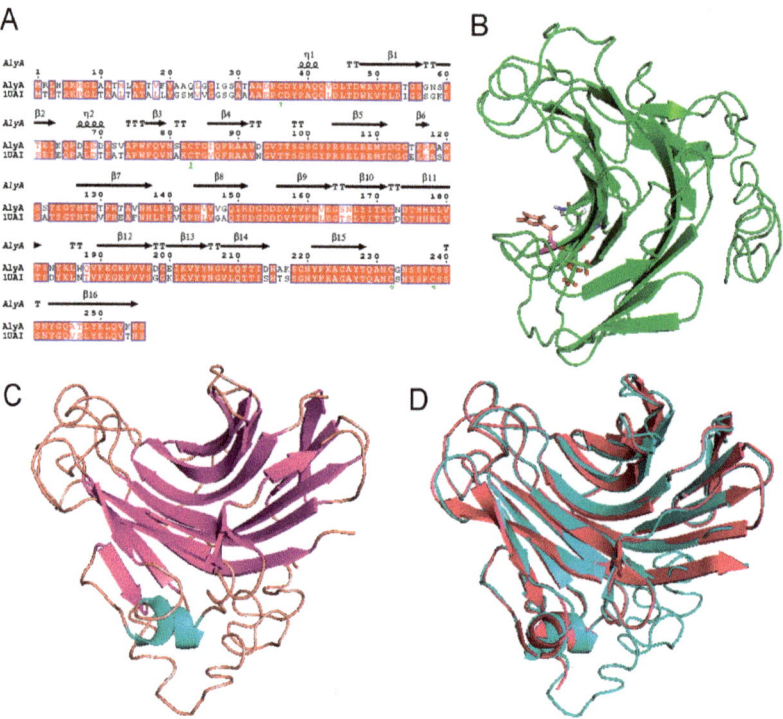

Figure 7. (**A**) The sequence alignment of AlyA and AlyPG from *Corynebacterium* sp. ALY-1, (**B**) the modeling structure of AlyA, (**C**) the structural comparison of AlyA (marked with red) and AlyPG (marked with blue) from *Corynebacterium* sp. ALY-1 (PDB ID: 1UAI), and (**D**) the key residuals for substrate reorganization of AlyA.

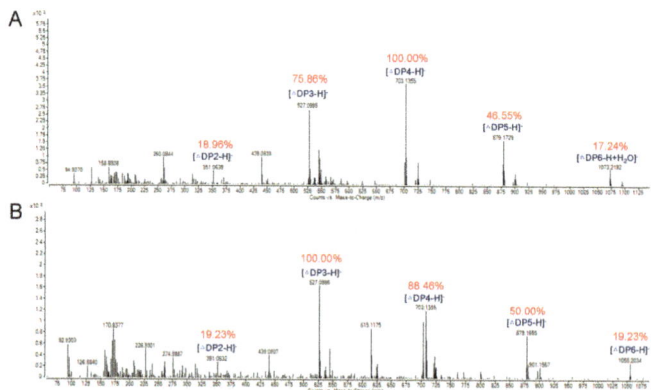

Figure 8. ESI-MS analysis of the degradation products of AlyA with (**A**) polyM and (**B**) polyG as substrate. The reaction mixtures (800 μL) containing 1 μg purified enzyme and 2 mg substrates (polyG and polyM) were incubated at 30 °C for 48 h.

3. Materials and Methods

3.1. Materials

Sodium alginate derived from brown seaweed was purchased from Sigma (St. Louis, MO, USA, M/G ratio: 77/23). PolyM (purity: about 99%) and polyG (purity: about 99%) were purchased from Qingdao BZ Oligo Biotech Co., Ltd. (Qingdao, China). Other chemicals and reagents used in this study were of analytical grade.

3.2. Screening and Identification of Strain NJ-05

The decaying seaweed samples were collected from the coast of the East China Sea (123°11′ E, 25°10′ N), washed by sterilized sea water, and then spread on sodium alginate-agar plates (modified marine broth 2216 medium containing 5 g/L $(NH_4)_2SO_4$, 19.45 g/L NaCl, 12.6 g/L $MgCl_2·6H_2O$, 6.64 g/L $MgSO_4·7H_2O$, 0.55 g/L KCl, 0.16 g/L $NaHCO_3$, 1 g/L ferric citrate, 15 g/L agar, and 10 g/L sodium alginate). The strains with ability to produce alginate lyase were screened according to the procedures previously reported [30]. Furthermore, the activity of alginate lyase was determined by 3, 5-dinitrosalicylic acid (DNS) colorimetry [51]. Among the isolates, the most active strain NJ-05 was selected and identified for further studies by the alignment of the 16S rRNA sequence. A phylogenetic tree was constructed using CLUSTAL X (http://www.clustal.org/) and MEGA 6.0 (https://www.megasoftware.net/) through the neighbor-joining method.

3.3. Cloning, Expression, and Purification of the Alginate Lyase

The strain *Isoptericola halotolerans* NJ-05 was genome-sequenced and its genomic information was analyzed. Alginate utilization loci has been found and there are three putative alginate lyases within the cluster. Therefore, the primers for cloning AlyA were designed on the basis of the sequence of the putative alginate lyase gene sequence (No.chr_1816) within the genome of *Isoptericola halotolerans* NJ-05. The AlyA gene was amplified with primers designed as follows: the forward primer, 5′-ATGCGCCTGCATCGCAAAC-3′, and the reverse primer, 5′-GCTATGTTTCACCTGCAGTT-3′, from the genomic DNA of *Isoptericola halotolerans* NJ-05.

The alginate lyase gene was then subcloned into the pET-21a (+) expression vector for heterologously expression with restriction sites of *NdeI* and *XhoI*. The recombinant *E. coli* BL21 (DE3) harboring the pET-21a (+)/AlyA was cultured in an LB medium (containing 100 µg ampicillin/mL) for 2–3 h with shaking at 200 rpm and 37 °C up to an OD600 of 0.4–0.6. The cells were induced by adding 0.1 mM IPTG and then cultured at 20 °C for 30 h. The AlyA was purified by an NTA-column, as previously described [44]. The active fraction was collected, desalted, and then analyzed by 12% sodium dodecyl sulfate polyacrylamide gel electrophoresis (SDS-PAGE).

3.4. Enzyme Activity Assay

The assay including the purified enzyme (0.1 mL, 1.78 mg/mL) and 0.9 mL Tris-HCl (20 mM, pH 8.0, 1% sodium alginate) was then incubated at 40 °C for 10 min, as previously described [30]. The reaction was stopped by heating in boiling water for 10 min. The enzyme activity was then assayed by measuring the increased absorbance at 235 nm, and the enzymatic activity (one unit) was defined as the amount of enzyme required to increase the absorbance at 235 nm by 0.01 per min [38].

3.5. Substrate Specificity and Kinetic Measurement of Alginate Lyase

The assays of the enzyme activity for sodium alginate, polyM, and polyG were defined as described previously for investigating the substrate specificity. The kinetic parameters of the purified enzyme toward different substrates, including sodium alginate, polyM, and polyG, were determined by measuring the enzyme activity with substrates at different concentrations (0.1–8.0 mg/mL). The concentrations of the different substrates was calculated according to the method previously described [38]. As alginate consists of random combinations of mannuronic acid and guluronic acid residues with the same molecular weight (MW), substrate molarity can be calculated using the MW of 176 g/mol for each monomer of uronic acid in the polymer (i.e., 194 g/mol monomer MW − 18 g/mol for the loss of H_2O during polymerization). The concentrations of the product were determined from the increase in absorbance at 235 nm using the extinction coefficient of 6150 M^{-1} cm^{-1}. The velocity (V) at the tested substrate concentration was calculated as follows: V (mol/s) = (milliAU/min × min/60 s × AU/1000 milli AU × 1 cm)/(6150 M^{-1} cm^{-1}) × (2 × 10^{-4} L). In addition, the K_m and V_{max} values were calculated by hyperbolic regression analysis, as described previously [38]. Additionally, the turnover number (k_{cat}) of the enzyme was calculated by the ration of V_{max} versus the enzyme concentration ([E]).

3.6. Biochemical Characterization of AlyA

The effects of temperatures (25–70 °C) on the purified enzyme were investigated at pH 9.0. The thermal stability of the enzyme was determined at pH 7.5 under the assay conditions described previously after incubating the purified enzyme at 25–70 °C for 30 min. In addition, the thermally-induced denaturation was also investigated by incubating the enzyme at 30–50 °C for 0–60 min at pH 7.5. Moreover, the effects of pH on the enzyme activity were evaluated by incubating the purified enzyme in buffers with different pHs (5.0–10.0) at 40 °C under the assay conditions described previously. The pH stability depended on the residual activity after the enzyme was incubated in buffers with different pHs (5.0–10.0) for 24 h, and then the residual activity was determined at 40 °C under the assay conditions. Meanwhile, the buffers with different pHs were used for phosphate–citrate (pH 5.0), NaH_2PO_4–Na_2HPO_4 (pH 6.0–8.0), Tris–HCl (pH 7.0–9.0), and glycine–NaOH (pH 10.0).

The influence of metal ions on the activity was performed by incubating the purified enzyme with various metal compounds at a concentration of 1 mM at 4 °C for 24 h. Then, the activity was measured under standard test conditions. The reaction mixture without any metal ion was taken as the control.

3.7. Molecular Modeling and Structural Alignment

The three-dimensional structure of AlyA was constructed using Protein Homology/analogY Recognition Engine V 2.0 (http://www.sbg.bio.ic.ac.uk/phyre2/html/page.cgi?id=index), on the basis of the homologues of the known structure (alginate lyase AlyA of *Corynebacterium* sp. ALY-1 (PDB ID: 1UAI) [52]. PyMOL (http://www.pymol.org) was used to visualize and analyze the modeled structure and to construct graphical presentations and illustrative figures.

3.8. ESI-MS Analysis of the Degradation Products of AlyA

For investigating the degradation pattern of AlyA, the reaction mixtures (800 µL) with pH 7.5 containing 1 µg purified enzyme and 2 mg substrates (polyG and polyM, the average Dp of the two substrates are about 40) were incubated at 30 °C for 0–48 h. After incubation, the mixture solutions were boiled for 10 min and then centrifuged at 12,000 rpm for 10 min to remove the unsolved materials. The hydrolysates were loaded onto a carbograph column (Alltech, Grace Davison Discovery Sciences, United Kingdom) to remove the salts after removing the proteins, and then concentrated, dried, and re-dissolved in 1 mL methanol. The supernatant (2 µL) was loop-injected to an LTQ XL linear ion trap mass spectrometer (Thermo Fisher Scientific, Waltham, MA, USA) after centrifugation, to further determine the composition of the products. The oligosaccharides were detected in a negative-ion mode using the following settings, previously described, with the scanning mass range 150–2000 *m/z* [30].

4. Conclusions

In this work, an alginate lyase-producing bacterium was isolated and identified to be *Isoptericola halotolerans* NJ-05. A novel bifunctional alginate lyase, AlyA, with high activity and broad substrate specificity was cloned and characterized. It exhibited the highest activity (7984.82 U/mg) at pH 7.5 and 55 °C. Additionally, it possessed broad substrate specificity, showing high activities towards not only polyM but also polyG. Furthermore, the K_m value of AlyA towards polyG (3.2 mM) was lower than that towards sodium alginate (5.6 mM) and polyM (6.7 mM). The TLC and ESI-MS analyses indicated that it can hydrolyze the substrates in an endolytic manner to release a series of oligosaccharides such as disaccharide, trisaccharide, and tetrasaccharide. This study provided extended insights into the substrate recognition and degrading pattern of alginate lyases with broad substrate specificity.

Supplementary Materials: The following are available online at http://www.mdpi.com/1660-3397/16/8/258/s1, The nucleotide and protein sequence of AlyA and Table S1. Purification of the recombinant AlyA.

Author Contributions: L.N. and Y.J. conceived and designed the experiments; M.C. and Y.H. performed the experiments; L.N. and Y.J. analyzed the data; L.N. wrote the paper; B.Z. revised the paper. All of the authors reviewed the manuscript.

Funding: This research was funded by the National Natural Science Foundation of China (No. 31601410 and 81503463).

Acknowledgments: The authors gratefully acknowledge the financial support of the National Natural Science Foundation of China (No. 31601410 and 81503463).

Conflicts of Interest: The authors declare no conflict of interest.

References

1. Gacesa, P. Enzymic degradation of alginates. *Int. J. Biochem.* **1992**, *24*, 545–552. [CrossRef]
2. Pawar, S.N.; Edgar, K.J. Alginate derivatization: A review of chemistry, properties and applications. *Biomaterials* **2012**, *33*, 3279–3305. [CrossRef] [PubMed]
3. Mabeau, S.; Kloareg, B. Isolation and analysis of the cell walls of brown algae: *Fucus spiralis, F. ceranoides, F. vesiculosus, F. Serratus, Bifurcaria bifurcata* and *Laminaria digitata*. *J. Exp. Bot.* **1987**, *38*, 1573–1580.
4. Fujihara, M.; Nagumo, T. An influence of the structure of alginate on the chemotactic activity of macrophages and the antitumor activity. *Carbohydr. Res.* **1993**, *243*, 211–216. [CrossRef]
5. Otterlei, M.; Ostgaard, K.; Skjak-Braek, G.; Smidsrod, O.; Soon-Shiong, P.; Espevik, T. Induction of cytokine production from human monocytes stimulated with alginate. *J. Immunother.* **1991**, *10*, 286–291. [CrossRef] [PubMed]
6. Bergero, M.F.; Liffourrena, A.S.; Opizzo, B.A.; Fochesatto, A.S.; Lucchesi, G.I. Immobilization of a microbial consortium on Ca-alginate enhances degradation of cationic surfactants in flasks and bioreactor. *Int. Biodeter. Biodegr.* **2017**, *117*, 39–44. [CrossRef]
7. Yang, J.S.; Xie, Y.J.; He, W. Research progress on chemical modification of alginate: A review. *Carbohydr. Polym.* **2011**, *84*, 33–39. [CrossRef]
8. Tai, H.B.; Tang, L.W.; Chen, D.D.; Irbis, C.; Bioconvertion, L.O. Progresses on preparation of alginate oligosaccharide. *Life Sci. Res.* **2015**, *19*, 75–79.
9. Yang, J.H.; Bang, M.A.; Jang, C.H.; Jo, G.H.; Jung, S.K.; Ki, S.H. Alginate oligosaccharide enhances LDL uptake via regulation of LDLR and PCSK9 expression. *J. Nutr. Biochem.* **2015**, *26*, 1393–1400. [CrossRef] [PubMed]
10. Iwamoto, M.; Kurachi, M.; Nakashima, T.; Kim, D.; Yamaguchi, K.; Oda, T.; Iwamoto, Y.; Muramatsu, T. Structure-activity relationship of alginate oligosaccharides in the induction of cytokine production from RAW264.7 cells. *FEBS Lett.* **2005**, *579*, 4423–4429. [PubMed]
11. Yamamoto, Y.; Kurachi, M.; Yamaguchi, K.; Oda, T. Induction of multiple cytokine secretion from RAW264.7 cells by alginate oligosaccharides. *Biosci. Biotechnol. Biochem.* **2007**, *71*, 238–241. [CrossRef] [PubMed]
12. Wong, T.Y.; Preston, L.A.; Schiller, N.L. Alginate lyase: Review of major sources and enzyme characteristics, structure-function analysis, biological roles, and applications. *Annu. Rev. Microbiol.* **2000**, *54*, 289–340. [CrossRef] [PubMed]
13. Kim, H.T.; Ko, H.J.; Kim, N.; Kim, D.; Lee, D.; Choi, I.G.; Woo, H.C.; Kim, M.D.; Kim, K.H. Characterization of a recombinant endo-type alginate lyase (Alg7D) from *saccharophagus degradans*. *Biotechnol. Lett.* **2012**, *34*, 1087–1092. [PubMed]
14. Inoue, A.; Mashino, C.; Uji, T.; Saga, N.; Mikami, K.; Ojima, T. Characterization of an eukaryotic PL-7 alginate lyase in the marine red alga *pyropia yezoensis*. *Curr. Biotechnol.* **2015**, *4*, 240–248. [CrossRef] [PubMed]

15. Inoue, A.; Anraku, M.; Nakagawa, S.; Ojima, T. Discovery of a novel alginate lyase from *Nitratiruptor* sp. SB155-2 thriving at deep-sea hydrothermal vents and identification of the residues responsible for its heat stability. *J. Biol. Chem.* **2016**, *291*, 15551–15563. [CrossRef] [PubMed]
16. Hata, M.; Kumagai, Y.; Rahman, M.M.; Chiba, S.; Tanaka, H.; Inoue, A.; Ojima, T. Comparative study on general properties of alginate lyases from some marine gastropod mollusks. *Fisheries Sci.* **2009**, *75*, 755. [CrossRef]
17. Zhu, B.; Tan, H.; Qin, Y.; Xu, Q.; Du, Y.; Yin, H. Characterization of a new endo-type alginate lyase from *vibrio* sp. W13. *Int. J. Biol. Macromol.* **2015**, *75*, 330–337. [CrossRef] [PubMed]
18. Elyakova, L.A.; Favorov, V.V. Isolation and certain properties of alginate lyase VI from the mollusk *Littorina* sp. *Biochim. Biophys. Acta* **1974**, *358*, 341–354. [CrossRef]
19. Lombard, V.; Bernard, T.; Rancurel, C.; Brumer, H.; Coutinho, P.M.; Henrissat, B. A hierarchical classification of polysaccharide lyases for glycogenomics. *Biochem. J.* **2010**, *432*, 437–444. [CrossRef] [PubMed]
20. Zhu, B.; Yin, H. Alginate lyase: Review of major sources and classification, properties, structure-function analysis and applications. *Bioengineered* **2015**, *6*, 125–131. [CrossRef] [PubMed]
21. Li, F.L.; Lu, M.; Ji, S.Q.; Wang, B. Biochemical and structural characterization of alginate lyases: an update. *Curr. Biotechnol.* **2015**, *4*, 223–239.
22. Xu, F.; Wang, P.; Zhang, Y.Z.; Chen, X.L. Diversity of three-dimensional structures and catalytic mechanisms of alginate lyases. *Appl. Environ. Microbiol.* **2018**, *84*, e02040-17. [PubMed]
23. Zhu, B.; Chen, M.; Yin, H.; Du, Y.; Ning, L. Enzymatic hydrolysis of alginate to produce oligosaccharides by a new purified endo-type alginate lyase. *Mar. Drugs* **2016**, *14*, 108. [CrossRef] [PubMed]
24. Wang, L.; Li, S.; Yu, W.; Gong, Q. Cloning, overexpression and characterization of a new oligoalginate lyase from a marine bacterium, *shewanella* sp. *Biotechnol. Lett.* **2015**, *37*, 665–671. [PubMed]
25. Zhang, Z.; Yu, G.; Guan, H.; Zhao, X.; Du, Y.; Jiang, X. Preparation and structure elucidation of alginate oligosaccharides degraded by alginate lyase from *Vibro* sp. 510. *Carbohydr. Res.* **2004**, *339*, 1475–1481. [CrossRef] [PubMed]
26. Inoue, A.; Kagaya, M.; Ojima, T. Preparation of protoplasts from *Laminaria japonica* using native and recombinant abalone alginate lyases. *J. Appl. Phycol.* **2008**, *20*, 633–640.
27. Inoue, A.; Mashino, C.; Kodama, T.; Ojima, T. Protoplast preparation from *Laminaria japonica* with recombinant alginate lyase and cellulase. *Mar. Biotechnol.* **2011**, *13*, 256–263. [CrossRef] [PubMed]
28. Islan, G.A.; Bosio, V.E.; Castro, G.R. Alginate lyase and ciprofloxacin co-immobilization on biopolymeric microspheres for cystic fibrosis treatment. *Macromol. Biosci.* **2013**, *13*, 1238–1248. [CrossRef] [PubMed]
29. Farrell, E.K.; Tipton, P.A. Functional characterization of AlgL, an alginate lyase from pseudomonas aeruginosa. *Biochemistry* **2012**, *51*, 10259–10266. [CrossRef] [PubMed]
30. Zhu, B.; Hu, F.; Yuan, H.; Sun, Y.; Yao, Z. Biochemical characterization and degradation pattern of a unique pH-stable PolyM-specific alginate lyase from newly isolated *Serratia marcescens* NJ-07. *Mar. Drugs* **2018**, *16*, 129. [CrossRef] [PubMed]
31. Kam, N.; Park, Y.J.; Lee, E.Y.; Kim, H.S. Molecular identification of a polym-specific alginate lyase from *Pseudomonas* sp. Strain ks-408 for degradation of glycosidic linkages between two mannuronates or mannuronate and guluronate in alginate. *Can. J. Microbiol.* **2011**, *57*, 1032–1041. [CrossRef] [PubMed]
32. Kim, H.S. Cloning and expression of alginate lyase from a marine bacterium, *Streptomyces* sp. M3. *J. Life Sci.* **2009**, *19*, 1522–1528.
33. Purification and Characterization of Extracellular Alginate Lyase from *Corynebacterium* sp. Aly-1 Strain. 1995. Available online: http://xueshu.baidu.com/s?wd=paperuri%3A%2841681fcc50d2b5e7270247da75b131af%29&filter=sc_long_sign&tn=SE_xueshusource_2kduw22v&sc_vurl=http%3A%2F%2Fci.nii.ac.jp%2Fnaid%2F110002947783%2Fen&ie=utf-8&sc_us=10692311932531899694 (accessed on 30 July 2018).
34. Li, J.W.; Dong, S.; Song, J.; Li, C.B.; Chen, X.L.; Xie, B.B.; Zhang, Y.Z. Purification and characterization of a bifunctional alginate lyase from *Pseudoalteromonas* sp. SM0524. *Mar. Drugs* **2011**, *9*, 109–123. [CrossRef] [PubMed]
35. Iwamoto, Y.; Araki, R.; Iriyama, K.; Oda, T.; Fukuda, H.; Hayashida, S.; Muramatsu, T. Purification and characterization of bifunctional alginate lyase from *Alteromonas* sp. Strain no. 272 and its action on saturated oligomeric substrates. *Biosci. Biotechnol. Biochem.* **2001**, *65*, 133–142. [CrossRef] [PubMed]
36. Manns, D.; Nyffenegger, C.; Saake, B.; Meyer, A.S. Impact of different alginate lyases on combined cellulase–lyase saccharification of brown seaweed. *Rsc Adv.* **2016**, *6*, 45392–45401. [CrossRef]

37. Mohapatra, B.R. Kinetic and thermodynamic properties of alginate lyase and cellulase co-produced by *Exiguobacterium* species Alg-S5. *Int. J. Biol. Macromol.* **2017**, *98*, 103–110. [CrossRef] [PubMed]
38. Swift, S.M.; Hudgens, J.W.; Heselpoth, R.D.; Bales, P.M.; Nelson, D.C. Characterization of AlgMsp, an alginate lyase from *Microbulbifer* sp. 6532A. *PLoS ONE* **2014**, *9*, e112939.
39. Zhu, B.; Ni, F.; Ning, L.; Sun, Y.; Yao, Z. Cloning and characterization of a new pH-stable alginate lyase with high salt tolerance from marine *Vibrio* sp. Nj-04. *Int. J. Biol. Macromol.* **2018**, *115*, 1063–1070. [CrossRef] [PubMed]
40. Zhu, B.; Sun, Y.; Ni, F.; Ning, L.; Yao, Z. Characterization of a new endo-type alginate lyase from *Vibrio* sp. Nju-03. *Int. J. Biol. Macromol.* **2018**, *108*, 1140–1147. [CrossRef] [PubMed]
41. Zhu, B.; Ni, F.; Sun, Y.; Yao, Z. Expression and characterization of a new heat-stable endo-type alginate lyase from deep-sea bacterium *Flammeovirga* sp. Nj-04. *Extremophiles* **2017**, *21*, 1027–1036. [PubMed]
42. Thomas, F.; Lundqvist, L.C.E.; Jam, M.; Jeudy, A.; Barbeyron, T.; Sandström, C.; Michel, G.; Czjzek, M. Comparative characterization of two marine alginate lyases from *Zobellia galactanivorans* reveals distinct modes of action and exquisite adaptation to their natural substrate. *J. Biol. Chem.* **2013**, *288*, 23021. [CrossRef] [PubMed]
43. Dou, W.; Wei, D.; Li, H.; Li, H.; Rahman, M.M.; Shi, J.; Xu, Z.; Ma, Y. Purification and characterisation of a bifunctional alginate lyase from novel *Isoptericola halotolerans* CGMCC 5336. *Carbohydr. Polym.* **2013**, *98*, 1476–1482. [CrossRef] [PubMed]
44. Uchimura, K.; Miyazaki, M.; Nogi, Y.; Kobayashi, T.; Horikoshi, K. Cloning and sequencing of alginate lyase genes from deep-sea strains of *Vibrio* and *Agarivorans* and characterization of a new *Vibrio* enzyme. *Mar. Biotechnol.* **2010**, *12*, 526–533. [CrossRef] [PubMed]
45. Yagi, H.; Fujise, A.; Itabashi, N.; Ohshiro, T. Purification and characterization of a novel alginate lyase from the marine bacterium *Cobetia* sp. NAP1 isolated from brown algae. *Biosci. Biotechnol. Biochem.* **2016**, *80*, 2338–2346. [PubMed]
46. Takeshita, S.; Sato, N.; Igarashi, M.; Muramatsu, T. A highly denaturant-durable alginate lyase from a marine bacterium: Purification and properties. *Biosci. Biotechnol. Biochem.* **1993**, *57*, 1125. [CrossRef] [PubMed]
47. Yang, X.M.; Li, S.Y.; Wu, Y.; Yu, W.G.; Han, F. Cloning and characterization of two thermo- and salt-tolerant oligoalginate lyases from marine bacterium *Halomonas* sp. *Fems. Microbiol. Lett.* **2016**, *363*, fnw079. [CrossRef] [PubMed]
48. Zhu, B.W.; Huang, L.S.; Tan, H.D.; Qin, Y.Q.; Du, Y.G.; Yin, H. Characterization of a new endo-type polyM-specific alginate lyase from *Pseudomonas* sp. *Biotechnol. Lett.* **2015**, *37*, 409–415. [CrossRef] [PubMed]
49. Haraguchi, K.; Kodama, T. Purification and propertes of poly(β-D-mannuronate) lyase from *Azotobacter chroococcum*. *Appl. Microbiol. Biot.* **1996**, *44*, 576–581. [CrossRef]
50. Maki, H.; Mori, A.; Fujiyama, K.; Kinoshita, S.; Yoshida, T. Cloning, sequence analysis and expression in escherichia coli of a gene encoding an alginate lyase from Pseudomonas sp. OS-ALG-9. *J. Gen. Microbiol.* **1993**, *139*, 987–993. [CrossRef] [PubMed]
51. Miller, G.L. Use of dinitrosalicylic acid reagent for determination of reducing sugar. *Anal. Biochem.* **1959**, *31*, 426–428. [CrossRef]
52. Osawa, T.; Matsubara, Y.; Muramatsu, T.; Kimura, M.; Kakuta, Y. Crystal structure of the alginate (poly α-L-guluronate) lyase from *Corynebacterium* sp. at 1.2 Å resolution. *J. Mol. Biol.* **2005**, *345*, 1111. [CrossRef] [PubMed]

© 2018 by the authors. Licensee MDPI, Basel, Switzerland. This article is an open access article distributed under the terms and conditions of the Creative Commons Attribution (CC BY) license (http://creativecommons.org/licenses/by/4.0/).

Article

Biochemical Characterization and Degradation Pattern of a Unique pH-Stable PolyM-Specific Alginate Lyase from Newly Isolated *Serratia marcescens* NJ-07

Benwei Zhu [†,*], Fu Hu [†], Heng Yuan, Yun Sun and Zhong Yao *

College of Food Science and Light Industry, Nanjing Tech University, Nanjing 211816, China; hufu@njtech.edu.cn (F.H.); yuanheng17@njtech.edu.cn (H.Y.); sunyun_food@njtech.edu.cn (Y.S.)
* Correspondence: zhubenwei@njtech.edu.cn (B.Z.); yaozhong@njtech.edu.cn (Z.Y.); Tel.: +86-25-5813-9419 (B.Z.)
† These authors contributed equally to this work.

Received: 28 March 2018; Accepted: 12 April 2018; Published: 15 April 2018

Abstract: Enzymatic preparation of alginate oligosaccharides with versatile bioactivities by alginate lyases has attracted increasing attention due to its featured characteristics, such as wild condition and specific products. In this study, AlgNJ-07, a novel polyM-specific alginate lyase with high specific activity and pH stability, has been purified from the newly isolated marine bacterium *Serratia marcescens* NJ-07. It has a molecular weight of approximately 25 kDa and exhibits the maximal activity of 2742.5 U/mg towards sodium alginate under 40 °C at pH 9.0. Additionally, AlgNJ-07 could retain more than 95% of its activity at pH range of 8.0–10.0, indicating it possesses excellent pH-stability. Moreover, it shows high activity and affinity towards polyM block and no activity to polyG block, which suggests that it is a strict polyM-specific alginate lyase. The degradation pattern of AlgNJ-07 has also been explored. The activity of AlgNJ-07 could be activated by NaCl with a low concentration (100–300 mM). It can be observed that AlgNJ-07 can recognize the trisaccharide as the minimal substrate and hydrolyze the trisaccharide into monosaccharide and disaccharide. The TLC and ESI-MS analysis indicate that it can hydrolyze substrates in a unique endolytic manner, producing not only oligosaccharides with Dp of 2–5 but also a large fraction of monosaccharide. Therefore, it may be a potent tool to produce alginate oligosaccharides with lower Dps (degree of polymerization).

Keywords: *Serratia marcescens*; polyM-specific; alginate lyase; oligosaccharides

---

1. Introduction

Alginate is the major component of cell wall of brown algae [1]. It is a linear anionic polysaccharide and consists of α-L-guluronate (G) and its C5 epimer β-D-mannuronate (M), which are linked by α-1, 4-glycosidic bonds [2]. The two monomeric units are arranged into three groups: poly-α-L-guluronate (polyG), poly-β-D-mannuronate (polyM), and the heteropolymer (polyMG) [3]. Due to its high viscosity, gelling properties, and versatile activities, alginate has been widely applied in food, chemical, and pharmaceutical industries [4–6]. However, the applications of this polysaccharide are still limited by its high molecular weight and poor solubility [7]. The alginate oligosaccharide, as the degradation product of alginate, retains various specific physiological functions and activities of polysaccharide but possesses good bioavailability [8]. For instance, Pack et al. found that alginate oligosaccharide (AOS) can reduce plasma LDL-cholesterol levels by regulating the expression of LDLR [9]. Iwamoto et al. studied the effect of AOS with different structures on the induction of cytokine production from RAW264.7 cells and found that G8 and M7 showed the most potent activity [10]. Yamamoto et al. reported that mannuronate oligomers (M3–M7) could induce the production and secretion of multiple cytokines, such as tumor necrosis factor-α (TNF-α), granulocyte colony-stimulating factor (GCSF), and monocyte chemoattractant protein-1 (MCP-1) [11].

Alginate lyase, a member of polysaccharide lyase, can catalyze the alginate by the β-elimination, producing unsaturated oligosaccharides with double bonds between C4 and C5 [12]. Until now, a number of alginate lyases have been identified, gene-cloned, purified, and characterized from various sources, such as marine and terrestrial bacteria, marine mollusks, and algae [13–18]. According to the substrate specificities, alginate lyases can be classified into three types: polyM-specific lyases (EC 4.2.2.3), polyG-specific lyases (EC 4.2.2.11), and bifunctional lyases (EC 4.2.2.-) [19]. Additionally, the alginate lyases are generally organized into seven polysaccharide lyase (PL) families according to the sequence similarity, namely PL-5, -6, -7, -14, -15, -17, and -18 families [20]. Moreover, in terms of the mode of action, alginate lyases can be grouped into endolytic and exolytic alginate lyases [21]. Endolytic enzymes can cleave glycosidic bonds inside alginate polymer and release unsaturated oligosaccharides as main products [22], while exolytic ones can further degrade oligosaccharides into monomers [23]. Now alginate lyases, especially endolytic enzymes, have been widely used to produce alginate oligosaccharides for food and nutraceutical industries [24,25]. Moreover, the enzymes can also be used to elucidate the fine structures of alginate and prepare protoplast of brown algae [26–28]. Furthermore, alginate lyases also show great potential in the treatment of cystic fibrosis by degrading the polysaccharide biofilm of pathogen bacterium [29]. So far, many alginate lyases originating from marine microorganisms have been well characterized. However, few of these enzymes have been commercially used in the food and nutraceutical industries due to the poor substrate specificity and low activity [30–36]. Thus, to explore novel enzymes with high activity and high substrate specificity will be of great importance for both research and commercial purposes.

In this work, a new alginate lyase with high substrate specificity and pH stability has been identified and characterized from *Serratia marcescens* NJ-07. To evaluate the enzyme for potential use in the food and nutraceutical industries, the kinetics and analysis of degrading products has also been characterized, which suggests that it would be a potential candidate for expanding applications of alginate lyases.

2. Results and Discussions

2.1. Screening and Identification of Strain NJ-07

The strain was isolated from rotten red algae from the Yellow Sea. The 16S rRNA sequence of the strain was sequenced and submitted to GeneBank (accession number MH119760). According to the phylogenetic analysis of 16S rRNA sequence (Figure 1), the strain was assigned to the genus Serratia and named *Serratia marcescens* NJ-07.

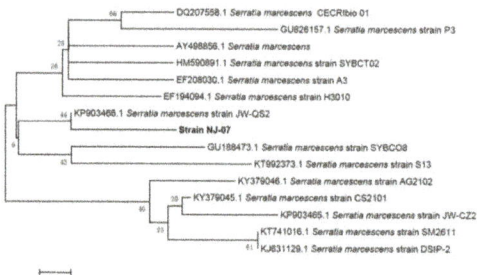

Figure 1. The phylogenetic analysis of strain NJ-07 and other similar strains. The phylogenetic tree was constructed by MEGA 6.0 on the basis of the 16S rRNA gene sequences of strain AlgNJ-07 and other known Serratia species.

2.2. Purification of Alginate Lyase

The strain NJ-07 was cultured in optimized liquid medium for 40 h until alginate lyase reached the highest activity. The supernatant containing alginate lyase was subjected to further purification by anion exchange chromatography with Source 15Q. After purification, the alginate lyase was purified 7.43-fold with a yield of 68.1%. The final specific activity of the purified alginate lyase was 2742.5 U/mg towards sodium alginate. The result of SDS-PAGE showed a single protein band with a molecular weight of 25 kDa (Figure 2), which was designated as AlgNJ-07. The alginate lyases are grouped into three types based on their molecular weights: small alginate lyases (25–30 kDa), medium-sized alginate lyases (around 40 kDa), and large alginate lyases (>60 kDa). As a result, the AlgNJ-07 belongs to the small ones. Similarly, the AlyA from *Azotobacter chroococcum* 4A1M has a small molecular weight of 24 kDa [31]. While the AlyA from *Pseudomonas* sp. E03, AlyA from *Pseudomonas aeruginosa*, and AlyA from *Pseudomonas* sp. strain KS-408 possess the medium-sized molecular weights of 40.4 kDa, 43 kDa, and 44.5 kDa, respectively [33,34,36]. The ALYII from *Pseudomonas* sp. OS-ALG-9 has a large molecular weight of 79 kDa [32].

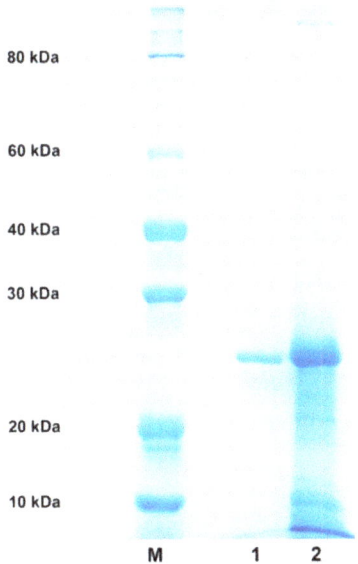

Figure 2. The SDS-PAGE analysis of purified alginate lyase AlgNJ-07. Lane M: the protein molecular weight standard; lane 1: the purified AlgNJ-07; lane 2: the crude enzyme from supernatant.

2.3. Substrate Specifity and Enzymatic Kinetics of the Enzyme

Seven kinds of polysaccharide substrates were used to investigate the substrate specificity of the enzyme (Table 1). The alginate lyase showed higher activity towards sodium alginate and polyM, but no activity towards polyG. Additionally, the AlgNJ-07 displayed no activity towards pullulan, pectin, xylan, and heparin. Therefore, the AlgNJ-07 is a novel polyM-specific alginate lyase. Until now, hundreds of alginate lyases have been identified and characterized. However, only a few enzymes exhibited the polyM-specific activity, such as AlgA from *Pseudomonas* sp. E03 [34], ALYII from *Pseudomonas* sp. OS-ALG-9 [32], the AlyA from *Azotobacter chroococcum* 4A1M [31], AlgL from *Pseudomonas aeruginosa* [30], AlyA from *Pseudomonas aeruginosa* [36], AlyA from *Pseudomonas* sp. strain KS-408, and AlyM from unknown marine bacterium [33,35]. They all displayed preference to polyM substrate and very low activity toward polyG substrate. However, compared with these characterized enzymes, AlgNJ-07 showed no activity toward polyG, indicating it is a novel alginate lyase with strict polyM-specific substrate specificity.

Table 1. The substrate specificity of AlgNJ-07 towards various substrates.

Substrate	Activity (U/mg)
Sodium alginate	2742.5
PolyM	3842.3
PolyG	N.D. *
Pullulan	N.D.
Pectin	N.D.
Xylan	N.D.
Heparin	N.D.

* No activity detected.

The kinetics of AlgNJ-07 towards sodium alginate and polyM were calculated according to the hyperbolic regression analysis. As shown in Table 2, the K_m values of AlgNJ-07 with sodium alginate and polyM as substrates were 0.53 mM and 0.27 mM. The results showed that AlgNJ-07 had a much lower K_m values towards polyM than sodium alginate, indicating that it showed higher affinity towards polyM than that to sodium alginate. The k_{cat}/K_m values of AlgNJ-07 towards polyM (115 mM^{-1}·s^{-1}) was higher than alginate (64 mM^{-1}·s^{-1}), which indicates that the enzyme possesses higher catalytic efficiency towards M block than to MG block. The polyM-specific alginate lyase AlgL from *Pseudomonas aeruginosa* showed different K_m and k_{cat} values towards polyM substrates with various Dps and it exhibited different affinity and catalytic efficiency towards those substrates. The variation in k_{cat}/K_m with substrate length suggests that AlgL operates in a processive manner [30].

Table 2. The kinetics parameters of AlgNJ-07.

Substrate	Sodium Alginate	polyM
K_m (mM)	0.53	0.27
V_{max} (nmol/s)	74	67
k_{cat} (s^{-1})	34	31
k_{cat}/K_m (s^{-1}/mM)	64	115

2.4. Biochemical Characterization of AlgNJ-07

The enzyme showed maximum activity at 40 °C (Figure 3A) and was stable below 40 °C (Figure 3B). It possessed approximately 50% activity after incubation at 40 °C for 30 min and was gradually inactivated as the temperature increased. The thermal degeneration curve of AlgNJ-07 was shown in Figure 4. The enzyme could retain more than 70% of its total activity after being incubated at 40 °C for 60 min, which indicates it possesses better thermal stability. The optimal temperature for polyM-specific alginate lyase from *Pseudomonas* sp. strain KS-408 was 37 °C [33]. The AlgA from *Pseudomonas* sp. E03 and ALYII from *Pseudomonas* sp. OS-ALG-9 both exhibited their maximal activity at 30 °C [33,34]. While the AlyA from *Azotobacter chroococcum* 4A1M showed the highest activity at 60 °C, which shows potential in industrial applications [31].

The optimal pH for the enzyme activity was 9.0 (Figure 3C) and retained more than 80% activity at a broad pH range from pH 8.0 to 10.0 (Figure 3D) after incubation for 24 h. However, this enzyme was mostly stable at pH 9.0 and retained more than 80% activity at a broad pH range from 7.0 to 10.0. Interestingly, it could retain about 40% of its activity at pH 11.0. Thus, AlgNJ-07 was an alkaline-stable lyase and it could retain stability in a broader pH range. While most of the other characterized polyM-specific alginate lyases exhibited their maximal activity around neutral pH. For instance, the AlgA from *Pseudomonas* sp. E03 possessed its optimal pH of 8.0 [34], the AlyA from

Pseudomonas sp. strain KS-408 displayed its maximal activity at pH of 9.0 [34]. While the AlyA from *Azotobacter chroococcum* 4A1M had a lower optimal pH of 6.0 [31].

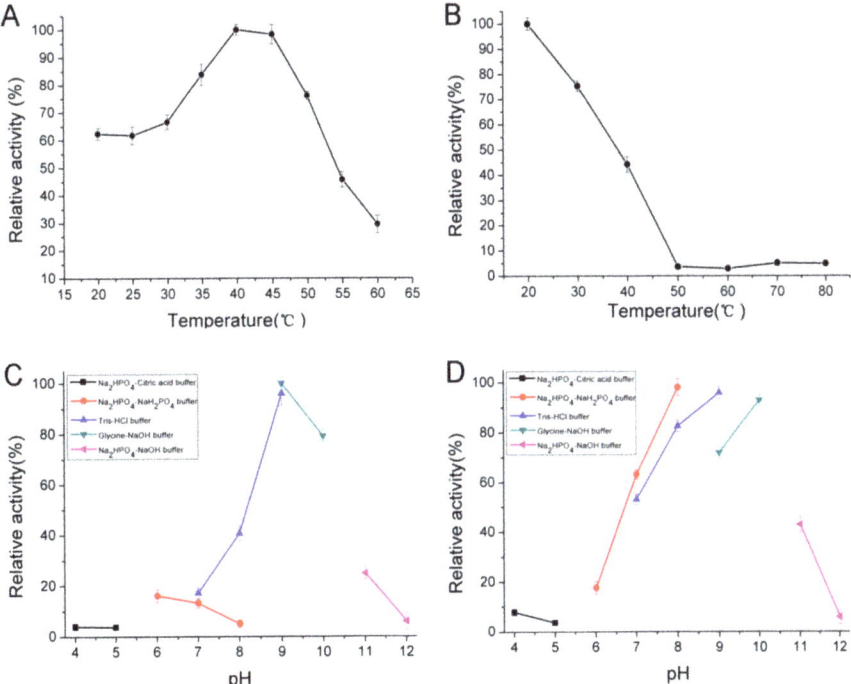

Figure 3. The biochemical characteristics of AlgNJ-07. (**A**) The optimal temperature of AlgNJ-07. (**B**) The thermal stability of AlgNJ-07. (**C**) The optimal pH of AlgNJ-07. (**D**) The pH stability of AlgNJ-07. Each value represents the mean of three replicates ± standard deviation.

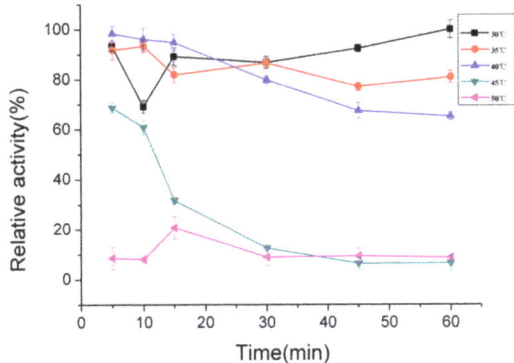

Figure 4. The thermal degeneration curve of AlgNJ-07. The maximal activity of the treated enzyme was regarded as 100% and the other relative activity was determined.

The effects of metal ions on the activity of AlgNJ-07 are shown in Table 3. It was observed that Na$^+$ could enhance the activity of the enzyme, while some divalent ions such as Zn^{2+}, Cu^{2+}, Mn^{2+}, and Co^{2+} inhibited the activity. Interestingly, the reported activators such as Mg^{2+} and Ca^{2+} displayed slight inhibitory effects on activity of AlgNJ-07. While Ca^{2+} can activate the activities of the AlyA from *Pseudomonas* sp. strain KS-408 [33], the AlyA from *Pseudomonas* sp. E03 [34], the AlyA from *Azotobacter chroococcum* 4A1M [31], and ALYII from *Pseudomonas* sp. OS-ALG-9 [32] could enhance the substrate-binding ability of the enzyme.

Table 3. The effect of metal ions on activity of AlgNJ-07.

Reagent	Relative Activity (%)
Control	100 ± 0.5
K^+ (100 mM)	87 ± 0.5
K^+ (300 mM)	94 ± 0.3
K^+ (500 mM)	92 ± 2.2
Na^+ (100 mM)	106 ± 0.6
Na^+ (300 mM)	120 ± 1.1
Na^+ (500 mM)	103 ± 2.6
Zn^{2+}	1 ± 0.3
Cu^{2+}	5 ± 0.5
Mn^{2+}	4 ± 0.1
Co^{2+}	25 ± 0.3
Ca^{2+}	90 ± 0.3
Fe^{3+}	18 ± 0.1
Mg^{2+}	98 ± 0.5
Ni^{2+}	72 ± 1.2

To determine the number of substrate binding subsites in the active tunnel of AlgNJ-07, we compared the degrading capability of AlgNJ-07 to oligosaccharide substrates with different Dps. As shown in Figure 5, purified disaccharide cannot be further degraded by the enzyme even under more focused conditions (high enzyme concentration and prolonged incubation time). The trisaccharide was the shortest chain that can be recognized and cleaved by AlgNJ-07, producing monosaccharide and disaccharide. The result indicated that trisaccharide was the shortest substrate for AlgNJ-07.

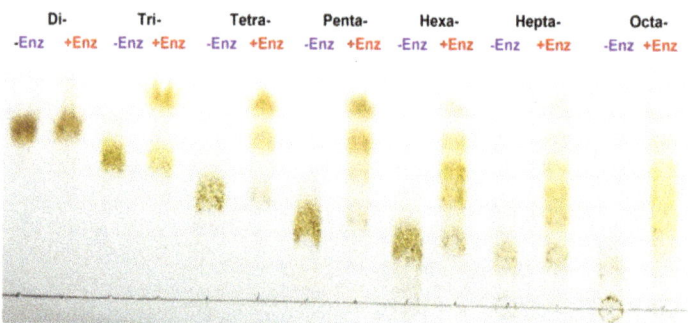

Figure 5. TLC analysis of hydrolysis products of oligosaccharides with Dps (2–8) for determination of substrate binding sites of AlgNJ-07 (−Enz: enzyme free; +Enz: AlgNJ-07 added).

The degradation products of sodium alginate and polyM by AlgNJ-07 were analyzed by TLC plate (Figure 6). As the proceeding of hydrolysis, oligosaccharides with high Dp (6–8) appeared. After incubation for 48 h, dimers, trimers, and tetramers turned out to be the main hydrolysis products for sodium alginate and polyM. Interestingly, the enzyme could release monosaccharide with processing of the hydrolysis. The distributions of the degradation products for the above two kinds of substrates were similar, and the results indicate that AlgNJ-07 can hydrolyze the substrates in a unique endolytic manner.

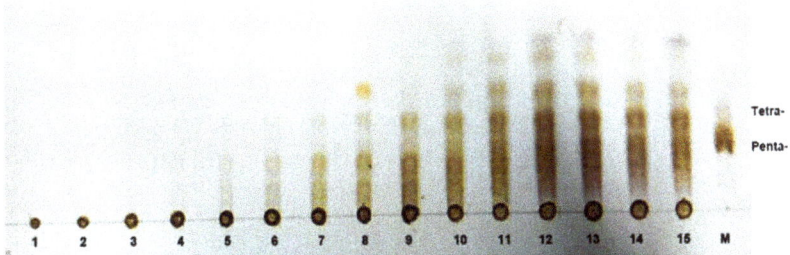

Figure 6. TLC analysis of the AlgNJ-07 hydrolysis products for different times. Lane 1–15, the samples taken by 0 min, 1 min, 3 min, 5 min, 10 min, 15 min, 30 min, 45 min, 60 min, 2 h, 4 h, 12 h, 24 h, 36 h, and 48 h. Lane M, the oligosaccharide standards of tetramer and pentamer.

In order to further determine the composition of the degradation products, the hydrolysates (1 mL) were then loaded onto a carbograph column to remove salts after removing other proteins, followed by being concentrated, dried, and re-dissolved in 1 mL methanol with the final concentration of 1 mg/mL. The degradation products were then analyzed by ESI-MS. As shown in Figure 7, monosaccharides, disaccharides, and trisaccharides account for a major fraction of the hydrolysates of two kinds of substrates. This result indicate that AlgNJ-07 may be a potential tool for the enzymatic hydrolysis of sodium alginate to produce oligosaccharides with lower Dps. The distribution of degradation products of other polyM-specific enzymes is similar, such as AlgA from *Pseudomonas* sp. E03 [34] and AlyA from *Pseudomonas* sp. strain KS-408 [33], which mainly produced oligosaccharides with Dp of 2–5 in an endolytic manner. However, the AlgL from *Pseudomonas aeruginosa* generated dimeric and trimeric products, and the rapid-mixing chemical quench studies indicate that AlgL can operate as an exopolysaccharide lyase [30]. None of those enzymes could produce monosaccharide during the hydrolytic procedure, which indicates that the AlgNJ-07 possesses a unique manner for releasing products.

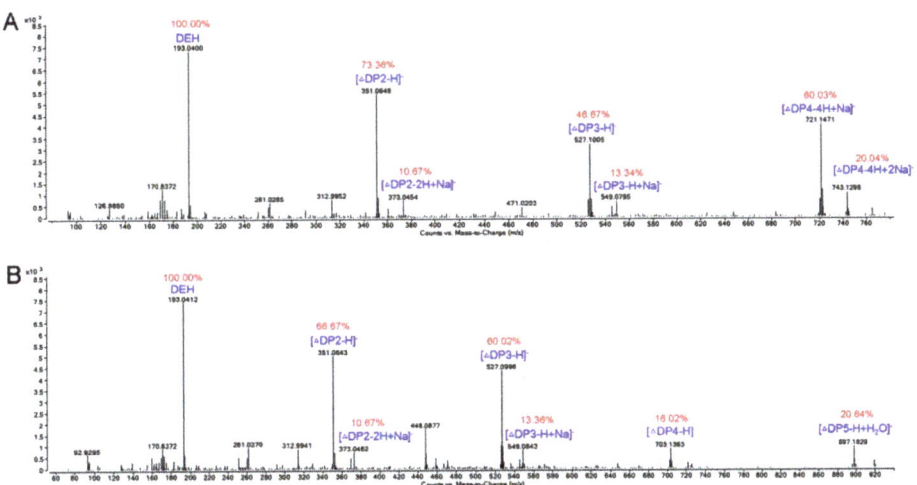

Figure 7. ESI-MS analysis of the degradation products of AlgNJ-07 with (**A**) alginate and (**B**) the polyM as substrate. The data highlighted in red represent the relative abundance of peaks.

3. Materials and Methods

3.1. Materials

Sodium alginate derived from brown seaweed was purchased from Sigma (St. Louis, MO, USA). PolyM (purity: about 99%) and polyG (purity: about 99%) were purchased from Qingdao BZ Oligo Biotech Co., Ltd. (Qingdao, China). The SOURCE™ 15Q 4.6/100 PE column was purchased from GE HealthCare Bio-Sciences (Uppsala, Sweden). Other chemicals and reagents used in this study were of analytical grade.

3.2. Screening and Identification of Strain NJ-07

The samples were collected from the coast of the Yellow Sea, washed by sterilized sea water and then spread on sodium alginate-agar plates. The plates were incubated at 30 °C for 36 h and the positive colonies showing clear zones were picked out from the selection plates. The re-screening process was conducted as follows. Strains with clear hydrolytic zones were selected and incubated aerobically in a fermentation medium (modified marine broth 2216 medium containing 5 g/L $(NH_4)_2SO_4$, 19.45 g/L NaCl, 12.6 g/L $MgCl_2·6H_2O$, 6.64 g/L $MgSO_4·7H_2O$, 0.55 g/L KCl, 0.16 g/L $NaHCO_3$, 1 g/L ferric citrate, and 10 g/L sodium alginate) at 30 °C and 200 rpm. Furthermore, the activity of alginate lyase was determined by 3,5-dinitrosalicylic acid (DNS) colorimetry [37]. Among the isolates, the most active strain NJ-07 was selected for further studies. To identify the NJ-07 strain, the 16S rRNA gene of the strain was amplified through PCR by using universal primers. The purified PCR fragment was sequenced and compared with reported 16S rRNA sequences in GenBank by using BLAST. A phylogenetic tree was constructed using CLUSTAL X and MEGA 6.0 through neighbor-joining method [38].

3.3. Production and Purification of the Alginate Lyase

The strain NJ-07 was propagated in a fermentation medium with shaking for 40 h at 30 °C. The culture medium was centrifuged (10,000× g, 60 min) to completely remove the sludge and the cell-free supernatant was fractionated at 30% and 80% ammonium sulfate saturation. The precipitated protein with 30% ammonium sulfate saturation was discarded, and the precipitated protein with 80% ammonium sulfate saturation was suspended in distilled water and dialyzed in a dialysis bag (MWCO: 8000–14,000 Da) against the distilled water and freeze-dried successively. Protein contents were determined by the Bradford method [39]. The obtained enzyme powder was dissolved in 5 mL Tris-HCl buffer (pH 9.0) with 4% as the final concentration, then the enzyme solution was applied to a SOURCE™ 15Q 4.6/100 PE column equilibrated with a linear gradient of 0–0.5 M NaCl in an equilibrating buffer under a flow rate of 1 mL/min. The eluents were monitored continuously at 280 nm for protein and fractions were assayed for activity against sodium alginate. Fractions were collected and monitored for the presence of alginate lyase. The purity of the fractions was assessed by SDS-PAGE. Pure fractions with activity were stored at −80 °C.

3.4. Enzyme Activity Assay

The purified enzyme (0.1 mL) was mixed with 0.9 mL Tris-HCl (20 mM, pH 8.0, 1% sodium alginate) and incubated at 40 °C for 10 min. The reaction was stopped by heating in boiling water for 10 min. The enzyme activity was then assayed by measuring the increased absorbance at 235 nm due to the formation of double bonds between C4 and C5 at the nonreducing terminus by β-elimination. One unit was defined as the amount of enzyme required to increase the absorbance at 235 nm by 0.01 per min [40].

3.5. Substrate Specificity and Kinetic Measurement of Alginate Lyase

The purified enzyme was reacted with 1% of sodium alginate, polyM, polyG, pectin, xylan, and heparin. The assays of enzyme activity for sodium alginate, polyM, and polyG were defined as described previously, whereas the assays for pectin, xylan, and heparin were determined by using the DNS method. The kinetic parameters of the purified enzyme toward sodium alginate and polyM were determined by measuring the enzyme activity with substrates at different concentrations (0.1–8.0 mg/mL). As sodium alginate is a polymer consisting of random combinations of mannuronic acid and guluronic acid residues. Since they both have the same molecular weight (MW), substrate molarity was calculated using the MW of 176 g/mol for each monomer of uronic acid in the polymer. The concentrations of the product were determined by monitoring the increase in absorbance at 235 nm using the extinction coefficient of 6150 M^{-1} cm^{-1}. Velocity (V) at the tested substrate concentration was calculated as follows: V (mol/s) = (milliAU/min × min/60 s × AU/1000 milliAU × 1 cm)/(6150 M^{-1} cm^{-1}) × (2 × 10^{-4} L). The K_m and V_{max} values were calculated by hyperbolic regression analysis as described previously [41]. Additionally, the turnover number (k_{cat}) of the enzyme was calculated by the ration of V_{max} versus enzyme concentration ([E]).

3.6. Biochemical Characterization of AlgNJ-07

The effects of pH on the enzyme activity were evaluated by incubating the purified enzyme in buffers with different pHs (4.0–12.0) at 40 °C under the assay conditions described previously. The pH stability depended on the residual activity after the enzyme was incubated in buffers with different pH (4.0–12.0) for 24 h and then residual activity was determined at 40 °C under the assay conditions. Meanwhile, the effects of temperatures (20–60 °C) on the purified enzyme were investigated at pH 9.0. The thermal stability of the enzyme was determined at pH 9.0 under the assay conditions described previously after incubating the purified enzyme at 30–50 °C for 30 min. The buffers with different pHs used were phosphate-citrate (pH 4.0–5.0), NaH_2PO_4-Na_2HPO_4 (pH 6.0–8.0), Tris–HCl (pH 7.0–9.0), glycine-NaOH (pH 9.0–10.0), and Na_2HPO_4–NaOH (pH 11.0–12.0). In addition, the thermally-induced denaturation was also investigated by incubating the enzyme at 30–50 °C for 0–60 min.

The influence of metal ions on the activity of the enzyme was performed by incubating the purified enzyme at 4 °C for 24 h in the presence of various metal compounds at a concentration of 1 mM. Then, the activity was measured under standard test conditions. The reaction mixture without any metal ions was used as a control.

3.7. Substrate Binding Subsites of AlgNJ-07

To determine the smallest substrate and the number of substrate binding subsites in its catalytic tunnel of AlgNJ-07, hydrolysis reactions were carried out using oligosaccharides with different Dps (Dp 2–8) at a concentration of 10 mg/mL in 10 µL reaction mixture (pH 9.0). The reaction mixtures were incubated at 40 °C with AlgNJ-07 for 24 h. The hydrolysates were loaded onto a carbograph column (Alltech, Grace Davison Discovery Sciences, Carnforth, UK) to remove salts after removing proteins, and then concentrated, dried, and re-dissolved in 1 mL methanol. The degradation products were analyzed by TLC with the solvent system (1-butanol/formic acid/water 4:6:1) and visualized by heating TLC plate at 130 °C for 5 min after spraying with 10% (v/v) sulfuric acid in ethanol.

3.8. TLC and ESI-MS Analysis of the Degradation Products of AlgNJ-07

To investigate the degradation pattern of AlgNJ-07, the reaction mixtures (800 µL) containing 1 µg purified enzyme and 2 mg substrates (sodium alginate and polyM) were incubated at 30 °C for 0–48 h. The hydrolysis products were analyzed by TLC as above. To further determine the composition of the products, ESI-MS was used. The supernatants (2 µL) were loop-injected to an LTQ XL linear ion trap mass spectrometer (Thermo Fisher Scientific, Waltham, MA, USA) after centrifugation. Samples were introduced by direct infusion into the electrospray ionization source (ESI) and mass spectra (MS) were collected. To help elucidate the structure of the ESI-MS peaks, the MS spectra were collected concurrently by isolating specific m/z anions, and the oligosaccharides were detected in a negative-ion mode using the following settings: ion source voltage, 4.5 kV; capillary temperature, 275–300 °C; tube lens, 250 V; sheath gas, 30 arbitrary units (AU); and scanning the mass range, 150–2000 m/z.

4. Conclusions

An alginate lyase-producing bacterium was isolated and identified as *Serratia marcescens* NJ-07. The alginate lyase AlgNJ-07 was purified by anion-exchange chromatography. It had a molecular weight of approximately 25 kDa and exhibited the maximal activity of 2742.52 U/mg under 40 °C at pH 9.0. Additionally, AlgNJ-07 could retain more than 95% of its activity at pH range of 8.0–10.0, which indicates it possesses excellent pH-stability. It showed high activity and affinity toward polyM block and no activity on polyG block, suggesting it is a strict polyM-specific alginate lyase. TLC and ESI-MS analysis indicated that it can hydrolyze substrates in a unique endolytic manner and produce oligosaccharides with Dp of 2–5 and a large fraction of monosaccharides. Therefore, it may be a potent tool to produce alginate oligosaccharides with lower Dps.

Acknowledgments: The authors gratefully acknowledge the financial support of the National Natural Science Foundation of China (No. 31601410).

Author Contributions: B.Z. and F.H. conceived and designed the experiments; B.Z. and F.H. performed the experiments; Y.S., H.Y. and Z.Y. analyzed the data; B.Z. wrote the paper. All authors reviewed the manuscript.

Conflicts of Interest: The authors declare no conflict of interest.

References

1. Gacesa, P. Enzymic degradation of alginates. *Int. J. Biochem.* **1992**, *24*, 545–552. [CrossRef]
2. Pawar, S.N.; Edgar, K.J. Alginate derivatization: A review of chemistry, properties and applications. *Biomaterials* **2012**, *33*, 3279–3305. [CrossRef] [PubMed]
3. Mabeau, S.; Kloareg, B. Isolation and analysis of the cell walls of brown algae: *Fucus spiralis, F. Ceranoides, F. Vesiculosus, F. Serratus*, bifurcaria bifurcata and laminaria digitata. *J. Exp. Bot.* **1987**, *38*, 1573–1580. [CrossRef]
4. Fujihara, M.; Nagumo, T. An influence of the structure of alginate on the chemotactic activity of macrophages and the antitumor activity. *Carbohydr. Res.* **1993**, *243*, 211–216. [CrossRef]
5. Otterlei, M.; Ostgaard, K.; Skjak-Braek, G.; Smidsrod, O.; Soon-Shiong, P.; Espevik, T. Induction of cytokine production from human monocytes stimulated with alginate. *J. Immunother.* **1991**, *10*, 286–291. [CrossRef] [PubMed]
6. Bergero, M.F.; Liffourrena, A.S.; Opizzo, B.A.; Fochesatto, A.S.; Lucchesi, G.I. Immobilization of a microbial consortium on ca-alginate enhances degradation of cationic surfactants in flasks and bioreactor. *Int. Biodeterior. Biodegrad.* **2017**, *117*, 39–44. [CrossRef]
7. Yang, J.S.; Xie, Y.J.; He, W. Research progress on chemical modification of alginate: A review. *Carbohydr. Polym.* **2011**, *84*, 33–39. [CrossRef]
8. Tai, H.B.; Tang, L.W.; Chen, D.D.; Irbis, C.; Bioconvertion, L.O. Progresses on preparation of alginate oligosaccharide. *Life Sci. Res.* **2015**, *19*, 75–79.
9. Do, J.R.; Back, S.Y.; Kim, H.K.; Lim, S.D.; Jung, S.K. Effects of aginate oligosaccharide on lipid metabolism in mouse fed a high cholesterol diet. *J. Korean Soc. Food Sci. Nutr.* **2014**, *43*, 491–497.
10. Iwamoto, M.; Kurachi, M.; Nakashima, T.; Kim, D.; Yamaguchi, K.; Oda, T.; Iwamoto, Y.; Muramatsu, T. Structure-activity relationship of alginate oligosaccharides in the induction of cytokine production from raw264.7 cells. *FEBS Lett.* **2005**, *579*, 4423–4429. [CrossRef] [PubMed]
11. Yamamoto, Y.; Kurachi, M.; Yamaguchi, K.; Oda, T. Induction of multiple cytokine secretion from raw264.7 cells by alginate oligosaccharides. *J. Agric. Chem. Soc. Jpn.* **2007**, *71*, 238–241. [CrossRef]
12. Wong, T.Y.; And, L.A.P.; Schiller, N.L. Alginate lyase: Review of major sources and enzyme characteristics, structure-function analysis, biological roles, and applications. *Annu. Rev. Microbiol.* **2000**, *54*, 289–340. [CrossRef] [PubMed]

13. Hashimoto, W.; Miyake, O.; Momma, K.; Kawai, S.; Murata, K. Molecular identification of oligoalginate lyase of sphingomonas sp. Strain a1 as one of the enzymes required for complete depolymerization of alginate. *J. Bacteriol.* **2000**, *182*, 4572–4577. [CrossRef] [PubMed]
14. Zhu, B.; Tan, H.; Qin, Y.; Xu, Q.; Du, Y.; Yin, H. Characterization of a new endo-type alginate lyase from *Vibrio* sp. W13. *Int. J. Biol. Macromol.* **2015**, *75*, 330–337. [CrossRef] [PubMed]
15. Rahman, M.M.; Inoue, A.; Tanaka, H.; Ojima, T. Isolation and characterization of two alginate lyase isozymes, akaly28 and akaly33, from the common sea hare aplysia kurodai. *Comp. Biochem. Phys.* **2010**, *157*, 317–325. [CrossRef] [PubMed]
16. Li, J.W.; Dong, S.; Song, J.; Li, C.B.; Chen, X.L.; Xie, B.B.; Zhang, Y.Z. Purification and characterization of a bifunctional alginate lyase from *Pseudoalteromonas* sp. Sm0524. *Mar. Drugs* **2011**, *9*, 109–123. [CrossRef] [PubMed]
17. Zhu, X.; Li, X.; Shi, H.; Zhou, J.; Tan, Z.; Yuan, M.; Yao, P.; Liu, X. Characterization of a novel alginate lyase from marine *Bacteriumvibrio furnissii* H1. *Mar. Drugs* **2018**, *16*, 30. [CrossRef] [PubMed]
18. Chen, P.; Zhu, Y.; Men, Y.; Zeng, Y.; Sun, Y. Purification and characterization of a novel alginate lyase from the marine *Bacterium bacillus* sp. Alg07. *Mar. Drugs* **2018**, *16*, 86. [CrossRef] [PubMed]
19. Zhu, B.; Yin, H. Alginate lyase: Review of major sources and classification, properties, structure-function analysis and applications. *Bioengineered* **2015**, *6*, 125–131. [CrossRef] [PubMed]
20. Henrissat, B. A classification of glycosyl hydrolases based on amino acid sequence similarities. *Biochem. J.* **1991**, *280 Pt 2*, 309–316. [CrossRef] [PubMed]
21. Kim, H.S.; Lee, C.G.; Lee, E.Y. Alginate lyase: Structure, property, and application. *Biotechnol. Bioprocess Eng.* **2011**, *16*, 843. [CrossRef]
22. Kim, H.T.; Ko, H.J.; Kim, N.; Kim, D.; Lee, D.; Choi, I.G.; Woo, H.C.; Kim, M.D.; Kim, K.H. Characterization of a recombinant endo-type alginate lyase (alg7d) from *Saccharophagus degradans*. *Biotechnol. Lett.* **2012**, *34*, 1087–1092. [CrossRef] [PubMed]
23. Heetaek, K.; Jaehyuk, C.; Wang, D.M.; Jieun, L.; Heechul, W.; Ingeol, C.; Kyoungheon, K. Depolymerization of alginate into a monomeric sugar acid using alg17c, an exo-oligoalginate lyase cloned from *Saccharophagus degradans* 2–40. *Appl. Microbiol. Biotechnol.* **2012**, *93*, 2233–2239.
24. Zhang, Z.; Yu, G.; Guan, H.; Zhao, X.; Du, Y.; Jiang, X. Preparation and structure elucidation of alginate oligosaccharides degraded by alginate lyase from vibro sp. 510. *Carbohydr. Res.* **2004**, *339*, 1475–1481. [CrossRef] [PubMed]
25. Zhu, B.; Chen, M.; Yin, H.; Du, Y.; Ning, L. Enzymatic hydrolysis of alginate to produce oligosaccharides by a new purified endo-type alginate lyase. *Mar. Drugs* **2016**, *14*, 108. [CrossRef] [PubMed]
26. Min, K.H.; Sasaki, S.F.; Kashiwabara, Y.; Umekawa, M.; Nisizawa, K. Fine structure of smg alginate fragment in the light of its degradation by alginate lyases of pseudomonas sp. *J. Biochem.* **1977**, *81*, 555–562. [CrossRef] [PubMed]
27. Boyen, C.; Kloareg, B.; Polne-Fuller, M.; Gibor, A. Preparation of alginate lyases from marine molluscs for protoplast isolation in brown algae. *Phycologia.* **1990**, *29*, 173–181. [CrossRef]
28. Inoue, A.; Kagaya, M.; Ojima, T. Preparation of protoplasts from laminaria japonica using native and recombinant abalone alginate lyases. *J. Appl. Phycol.* **2008**, *20*, 633–640. [CrossRef]
29. Alkawash, M.A.; Soothill, J.S.; Schiller, N.L. Alginate lyase enhances antibiotic killing of mucoid *Pseudomonas aeruginosa* in biofilms. *APMIS* **2006**, *114*, 131–138. [CrossRef] [PubMed]
30. Farrell, E.K.; Tipton, P.A. Functional characterization of algl, an alginate lyase from *Pseudomonas aeruginosa*. *Biochemistry* **2012**, *51*, 10259–10266. [CrossRef] [PubMed]
31. Haraguchi, K.; Kodama, T. Purification and propertes of poly(β-D-mannuronate) lyase from azotobacter chroococcum. *Appl. Microbiol. Biotechnol.* **1996**, *44*, 576–581. [CrossRef]
32. Kraiwattanapong, J.; Motomura, K.; Ooi, T.; Kinoshita, S. Characterization of alginate lyase (alyii) from *Pseudomonas* sp. Os-alg-9 expressed in recombinant escherichia coli. *World J. Microb. Biotechnol.* **1999**, *15*, 105–109. [CrossRef]
33. Kam, N.; Park, Y.J.; Lee, E.Y.; Kim, H.S. Molecular identification of a polym-specific alginate lyase from *Pseudomonas* sp. Strain ks-408 for degradation of glycosidic linkages between two mannuronates or mannuronate and guluronate in alginate. *Can. J. Microbiol.* **2011**, *57*, 1032–1041. [CrossRef] [PubMed]
34. Zhu, B.W.; Huang, L.S.X.; Tan, H.D.; Qin, Y.Q.; Du, Y.G.; Yin, H. Characterization of a new endo-type polym-specific alginate lyase from *Pseudomonas* sp. *Biotechnol. Lett.* **2015**, *37*, 409–415. [CrossRef] [PubMed]

35. Romeo, T.; Iii, J.F.P. Purification and structural properties of an extracellular (1–4)-.beta.-D-mannuronan-specific alginate lyase from a marine bacterium. *Biochemistry* **1986**, *25*, 8385–8391. [CrossRef]
36. Eftekhar, F.; Schiller, N.L. Partial purification and characterization of a mannuronan-specific alginate lyase from *Pseudomonas aeruginosa*. *Curr. Microbiol.* **1994**, *29*, 37–42. [CrossRef]
37. Miller, G.L. Use of dinitrosalicylic acid reagent for determination of reducing sugar. *Anal. Biochem.* **1959**, *31*, 426–428. [CrossRef]
38. Saitou, N.; Nei, M. The neighbor-joining method: A new method for reconstructing phylogenetic trees. *Mol. Biol. Evol.* **1987**, *4*, 406–425. [PubMed]
39. Kruger, N.J. The bradford method for protein quantitation. *Methods Mol. Biol.* **1988**, *32*, 9–15.
40. Zhu, B.; Ni, F.; Sun, Y.; Yao, Z. Expression and characterization of a new heat-stable endo-type alginate lyase from deep-sea bacterium *Flammeovirga* sp. Nj-04. *Extremophiles* **2017**, *21*, 1027–1036. [CrossRef] [PubMed]
41. Swift, S.M.; Hudgens, J.W.; Heselpoth, R.D.; Bales, P.M.; Nelson, D.C. Characterization of algmsp, an alginate lyase from *Microbulbifer* sp. 6532a. *PLoS ONE* **2014**, *9*, e112939. [CrossRef] [PubMed]

© 2018 by the authors. Licensee MDPI, Basel, Switzerland. This article is an open access article distributed under the terms and conditions of the Creative Commons Attribution (CC BY) license (http://creativecommons.org/licenses/by/4.0/).

Article

AlgM4: A New Salt-Activated Alginate Lyase of the PL7 Family with Endolytic Activity

Guiyuan Huang [1], Qiaozhen Wang [1], Mingqian Lu [1], Chao Xu [1], Fei Li [1], Rongcan Zhang [1], Wei Liao [1,2] and Shushi Huang [1,*]

[1] Guangxi Key Laboratory of Marine Natural Products and Combinatorial Biosynethesis Chemistry, Guangxi Academy of Sciences, Nanning 530007, China; guiyuan1105@163.com (G.H.); wqzh-333@163.com (Q.W.); lumq520@126.com (M.L.); 18174662721@163.com (C.X.); lifei@gxas.cn (F.L.); zhangrongcan@126.com (R.Z.); laoweimail@163.com (W.L.)
[2] The Food and Biotechnology, Guangxi Vocational and Technical College, Nanning 530226, China
* Correspondence: hshushi@gxas.cn; Tel.: +86-771-250-3990

Received: 6 March 2018; Accepted: 3 April 2018; Published: 6 April 2018

Abstract: Alginate lyases are a group of enzymes that catalyze the depolymerization of alginates into oligosaccharides or monosaccharides. These enzymes have been widely used for a variety of purposes, such as producing bioactive oligosaccharides, controlling the rheological properties of polysaccharides, and performing structural analyses of polysaccharides. The *algM4* gene of the marine bacterium *Vibrio weizhoudaoensis* M0101 encodes an alginate lyase that belongs to the polysaccharide lyase family 7 (PL7). In this study, the kinetic constants V_{max} (maximum reaction rate) and K_m (Michaelis constant) of AlgM4 activity were determined as 2.75 nmol/s and 2.72 mg/mL, respectively. The optimum temperature for AlgM4 activity was 30 °C, and at 70 °C, AlgM4 activity dropped to 11% of the maximum observed activity. The optimum pH for AlgM4 activity was 8.5, and AlgM4 was completely inactive at pH 11. The addition of 1 mol/L NaCl resulted in a more than sevenfold increase in the relative activity of AlgM4. The secondary structure of AlgM4 was altered in the presence of NaCl, which caused the α-helical content to decrease from 12.4 to 10.8% and the β-sheet content to decrease by 1.7%. In addition, NaCl enhanced the thermal stability of AlgM4 and increased the midpoint of thermal denaturation (Tm) by 4.9 °C. AlgM4 exhibited an ability to degrade sodium alginate, poly-mannuronic acid (polyM), and poly-guluronic acid (polyG), resulting in the production of oligosaccharides with a degree of polymerization (DP) of 2–9. AlgM4 possessed broader substrate, indicating that it is a bifunctional alginate lyase. Thus, AlgM4 is a novel salt-activated and bifunctional alginate lyase of the PL7 family with endolytic activity.

Keywords: *Vibrio weizhoudaoensis*; alginate lyase; PL7 family; salt-activated enzyme

1. Introduction

Alginic acid has the ability to form viscous solutions and gels in aqueous media and is nontoxic to living organisms. Therefore, it has been widely used in the pharmaceutical, cosmetic, food, and biotech industries [1]. In addition, the degradation products of alginic acid—alginate oligosaccharides—have a wide range of biological activities, such as the promotion of growth and the alleviation of abiotic stress in plants; antitumour, antibacterial, anti-inflammatory, anticoagulant, antioxidative, and immunomodulatory activities; and the reduction of free radicals and blood glucose and lipids. Alginate oligosaccharides have broad application prospects in the green agriculture, medical, food, and household chemical industries, among others [2].

Alginic acid, also known as algin or alginate, is a straight-chain polysaccharide composed of β-D-mannuronic acid (M) and its C5 stereoisomer α-L-guluronic acid (G), which are randomly linked via α-1,4-glycosidic bonds. Alginate molecules can have the sugar monomers M and G arranged in three ways: M monomers can be linked in succession, forming poly-mannuronic acid (polyM); G monomers can be linked in succession, forming poly-guluronic acid (polyG); and M and G monomers can be randomly and alternately linked, forming poly-guluronic acid -mannuronic acid (polyMG) [3]. The relative proportions of M and G vary among the alginic acids derived from different organisms [4,5].

Alginate lyases catalyze the cleavage of the 1,4-glycosidic bonds between the uronic acid monomers in alginate, resulting in the production of oligouronic acids or uronic acid monomers. Alginate lyases are present in a wide range of organisms and can be isolated from marine algae, molluscs, and microorganisms, as well as soil microorganisms. Currently, the primary sources of alginate lyases are marine bacteria, including members of the genera *Pseudomonas* and *Vibrio* [6].

The catalytic mechanisms of alginate lyases are well understood. Alginate lyases catalyze the cleavage of 1,4-glycosidic bonds in alginic acid via a β-elimination reaction. In addition, a double bond is formed between C4 and C5 of the saccharide ring containing the 4-*O* glycosidic bond, generating an oligomer with 4-deoxy-L-erythro-hex-4-enepyranosyluronate at the nonreducing end [7]. Based on their catalytic characteristics, alginate lyases are divided into endolytic and exolytic alginate lyases [8,9]. According to the substrate specificity, endolytic alginate lyases can be divided into mannuronate lyases (polyM lyase, EC 4.2.2.3) and guluronate lyases (polyG lyase, EC 4.2.2.11). Bifunctional enzymes that exhibit activities towards both polyG and polyM have also been identified [10]. While numerous alginate lyases have been discovered, few studies have focused on the enzymatic properties of alginate lyases. In recent years, a number of alginate lyases have been discovered and reported, including the cold-adapted alginate lyases [11–13], thermostable alginate lyases [14], high-alkaline alginate lyases [15], and salt-activated alginate lyases [12,15–17]. According to the evolution and homology of amino acid sequences, most alginate lyases belong to seven families of polysaccharide lyases (PL-5, PL-7, PL-14, PL-15, PL-17, and PL-18) [18]. Most of the reported alginate lyases have endolytic activity [15,19–21] and hydrolyze sodium alginate (SA) to produce oligosaccharides. Alginate lyase A1-IV, produced by the bacterial strain *Sphingomonas* sp. A1 [22], and alginate lyase Atu3025, produced by *Agrobacterium tumefaciens* C58 [23], possess exolytic activity and hydrolyze SA into mannuronate or guluronate. Alginate lyases that degrade alginate into monosaccharides are part of the PL15 family [22,23], whereas alginate lyases with endolytic activity belong to the PL-5, 6, 7, 14, 16, 17, and 18 families [24].

The marine bacterium *Vibrio weizhoudaoensis* M0101 harbors the *algM4* gene, which encodes an alginate lyase belonging to polysaccharide lyase family 7 (PL7). In this study, we purified exogenously expressed AlgM4 and observed it to exhibit high salt tolerance, as AlgM4 activity increased more than sevenfold in the presence of 1 mol/L NaCl. The result is different from those for other salt-activated alginate lyases for which enzyme activity is decreased at 1 mol/L NaCl [15,25]. In the depolymerization of a high content of sodium alginate, Alg2A generated equal total molar amounts of oligosaccharides, but the amounts of oligosaccharides with DP of 5–10 were higher than those for both of the two commercial enzymes [26]. AlgM4 showed activities toward both polyM and polyG, which may degrade alginate more effectively. Moreover, AlgM4 catalyzing polyM released oligosaccharides with DP 7–9 from the polyM, which was different from the previously reported endolytic alginate lyases, despite their diverse substrate specificities [27,28]. Therefore, the unique endolytic reaction mode of AlgM4 gives it a distinct advantage in facilitating uronic acid oligosaccharides with high DPs. AlgM4 could be a good tool for the preparation of alginate oligosaccharides. AlgM4 not only functions as a key enzyme in the preparation and functional study of oligosaccharides but also plays an important role in utilization of alginate for ethanol fermentation.

2. Results and Discussion

2.1. Analysis of the AlgM4 Sequence

The alginate lyase gene *algM4* is 1563 bp in length and encodes a 520-amino-acid protein. A signal peptide analysis of AlgM4 using SignalP 4.0 (http://www.cbs.dtu.dk/services/SignalP/) predicted that the N-terminus of AlgM4 contains a 24-amino-acid signal peptide (Figure 1). A sequence alignment using the National Center for Biotechnology Information (NCBI) BLAST (https://blast.ncbi.nlm.nih.gov/) engine revealed that AlgM4 is a dual-domain protease, containing an N-terminal F5/8 type C domain and a C-terminal alginate-lyase 2 domain. The alginate lyases of the PL7 family contain three highly conserved domains: SA3 (RXEXR), SA4 (YXKAGXYXQ), and SA5 (QXH). The BLAST sequence analysis showed that AlgM4 contains the conserved amino acid sequences of the SA3 domain (RTEMR), the SA4 domain (YFKAGVYNQ), and the SA5 catalytic domain (QIH) (Figure 1). Therefore, AlgM4 belongs to the PL7 family of alginate lyases. The amino acid sequence of AlgM4 was compared with the sequences of other alginate lyases of the PL7 family selected in the CAZy database (http://www.cazy.org/), and a phylogenetic tree was constructed using the neighbor joining method (Figure 2). AlgM4 was most closely related to an alginate lyase from *Vibrio litoralis* BZM-2 (ALP75562.1), with an amino acid sequence similarity of 74% observed between AlgM4 and ALP75562.1, without annotation by genome analysis. These results suggest that AlgM4 is a new alginate lyase of the PL7 family.

MKHKFLKTLLASSVLFAVGCTSNGNDTSQLHPQSDTGAPVLTPVAIEASSHDGNGPDRLFDQDINTR
    Signal peptide
WSASGDGEWAVLDYGSVHEFDAIRAAFSKGNERQSIFDILVSTDGETWTPVLENVQSSGGVIGYERFE
FAPVQARYVKYVGHGNTANGWNSVTELAAIKCGVNACPSNHIITPEVIAAEQGLIAKQKAEEAARK
EARKDLRKGNFGAPAVYPCQTTVKCGKTALPVPTGLPATPKAGNAPSENFDLTSWYLSQPFDHNND
NRPDDVSEWDLANGYSHPDVFYTADDGGLVFKSFVKGVRTSPNTKYARTEMREMLRRGDTSIPTKG
                                                                   SA3
VNKNNWVFSSAPVADQKAAGGVDGVMEATLKIDHTTTTGEAGEVGRFIIGQIHDQDDEPIRLYYRKL
                                                                                   SA5
PNQDKGTVYFAHENTLKGTDQYFDLVGGMTGEIGDDGIALGEKFSYRIEVKGNTLTVTVMREGKPDV
KQVVDMSESGYDVGGKYMYFKAGVYNQNISGEMDDYVQATFYKLEKSHGSYKG-
       SA4

Figure 1. The deduced amino acid sequences of AlgM4. The signal peptide is underlined; SA3, SA5, and SA4 are indicated with blue symbols.

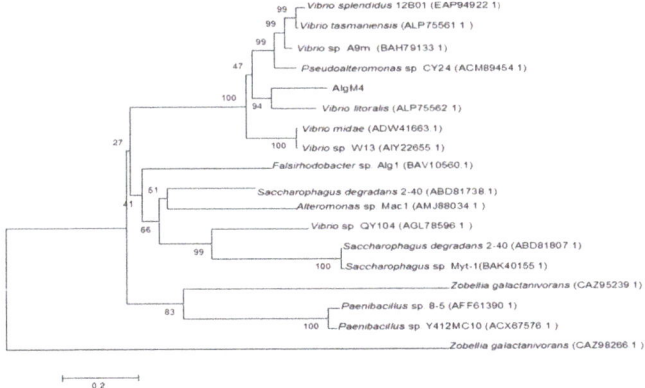

Figure 2. Neighbor-joining phylogenetic tree of *V. weizhoudaoensis* strain M0101 based on putative AlgM4 protein sequences.

2.2. Purification and Enzymatic Activity of AlgM4

Using *V. weizhoudaoensis* M0101 genomic DNA as a template, the *algM4* gene was PCR amplified without the N-terminal signal peptide sequence or the stop codon and ligated into the pET30a(+) plasmid. The plasmid was then transformed into *Escherichia coli* BL21 (DE3) cells for AlgM4 expression. Purified AlgM4 protein with a C-terminal 6×histidine (His) tag was obtained by Ni^{2+} affinity chromatography. Sodium dodecyl sulfate polyacrylamide gel electrophoresis (SDS-PAGE) showed a single protein band with a molecular weight of approximately 55 kDa (Figure 3). The specific activity of the purified AlgM4 was 4638 U/mg (Table 1), exhibiting a 36.7% increase in specific activity compared with the crude enzyme extract.

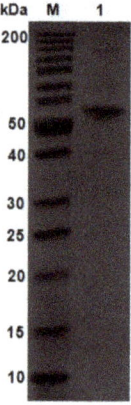

Figure 3. SDS-PAGE analysis of purified AlgM4. Lane M, molecular weight markers; Lane 1, purified AlgM4.

Table 1. Summary of AlgM4 purification.

Steps	Total Protein (mg)	Total Activity (U)	Specific Activity (U/mg)	Recover (%)	Purification (Fold)
Crude enzyme AlgM4	58.352	73878	1266	100	1
Purified AlgM4	11.55	53570	4638	72.5	3.7

2.3. V_{max} (Maximum Reaction Rate) and K_m (Michaelis Constant) of AlgM4

AlgM4 activity was measured when incubated with various concentrations of substrate (SA) in a water bath at 30 °C for 10 min, and the resulting values were used to calculate the kinetic constants of AlgM4 activity. The V_{max} and K_{cat} values of AlgM4 were 2.75 nmol/s and 30.25 S^{-1}, respectively, and the K_m value was 2.72 mg/mL. Two other NaCl-activated enzymes—AlyPM from *Pseudoalteromonas* sp. SM0524 and AlySY08 from *Nitratiruptor* sp. SB155-2—were previously shown to have K_m values of 74.39 [17] and 0.36 mg/mL [14], respectively. The oligoalginate lyase Alg17C derived from *Saccharophagus degradans* 2-40 had an observed K_m value of 35.2 mg/mL [18]. *Vibrio splendidus* 12B01 expresses three oligoalginate lyases—OalA, OalB, and OalC—which had observed K_m values of 3.25, 0.76, and 0.53 mg/mL, respectively [29].

2.4. Enzymatic Characteristics of AlgM4

AlgM4 exhibited high enzymatic activity at 20–40 °C, which decreased to 11% of the maximal observed activity when incubated at 70 °C. Similar to other NaCl-activated alginate lyases in the PL7 family (such as A1m [15], rA9mT [16], and AlyPM [17]), AlgM4 had highest enzymatic activity at 30 °C (Figure 4A).

Many alginate lyases have optimal pH values between pH 7 and 8, in contrast to the optimal high-alkaline pH values of pectate lyases [20]. The optimum pH for AlgM4 activity was assessed, and the results are shown in Figure 4B. The optimum reaction pH was 8.5, indicating that AlgM4 was weakly basophilic. In slightly alkaline environments (pH 7.5–9.0), AlgM4 retained 80% activity, whereas AlgM4 was completely inactive at pH 11.

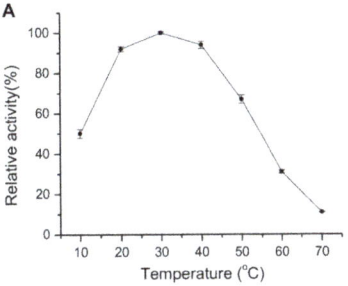
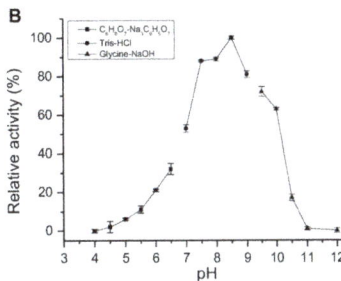

Figure 4. Effect of temperature and pH on AlgM4 activity. (**A**) The optimal temperature for AlgM4 activity. The activity of AlgM4 at 30 °C was completely retained; (**B**) The optimal pH for AlgM4 activity. The activity of AlgM4 at pH 8.5 was completely retained.

The activity of AlgM4 remained stable for 30 min at 25 °C, regardless of the presence or absence of 1 mol/L NaCl (Figure 5A). After a 30 min incubation at 30 or 35 °C in the presence of 1 mol/L NaCl, AlgM4 retained 92% of its initial activity, which decreased rapidly at temperatures exceeding 40 °C and decreased by 63% at 45 °C. In the absence of NaCl, the enzymatic activity of AlgM4 was reduced by 94% at 45 °C. An examination of temperature tolerance showed that NaCl improved the thermal stability of AlgM4.

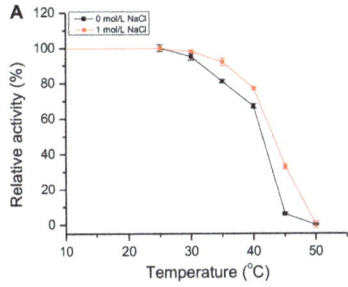
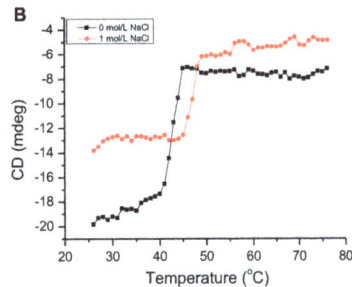

Figure 5. The thermal stability and melting temperature (Tm) of AlgM4. (**A**) The thermal stability of AlgM4. The residual activity of AlgM4 at 25 °C was completely retained; (**B**) Circular dichroism signals at 218 nm were used for analysis of the Tm value.

The effects of metal ions and surfactants on AlgM4 activity are shown in Table 2. Ca^{2+} did not influence enzyme activity, unlike for other alginate lyases [12,17,20]. While Mg^{2+} promoted the activity of AlgM4, Cu^{2+}, Mn^{2+}, and Zn^{2+} inhibited AlgM4 activity. The most significant inhibitory effect on AlgM4 was observed by Zn^{2+}, which caused an 82% reduction in AlgM4 activity. Ethylene diamine tetraacetic acid (EDTA) and SDS suppressed AlgM4 activity to varying degrees. The anionic surfactant SDS strongly inhibited the enzymatic activity of AlgM4, causing a 97% reduction in AlgM4 activity, while EDTA reduced AlgM4 activity by 35%.

Table 2. Effect of chemical reagents on AlgM4 activity.

Reagent	Concentration (mmol/L)	Relative Activity (%)
None	-	100 ± 0.3
$CaCl_2$	1	97 ± 0.1
$MgCl_2$	1	114 ± 1.1
KCl	1	91 ± 0.9
$CuCl_2$	1	75 ± 1.7
$MnCl_2$	1	77 ± 2.4
$ZnCl_2$	1	18 ± 4.5
EDTA	1	65 ± 6.2
SDS	1	3 ± 3.1

The data are expressed as the means ± SD, $n = 3$. The activity of AlgM4 in the absence of chemical reagents was completely retained.

2.5. Effect of NaCl on AlgM4 Activity

Different concentrations of NaCl were added to 0.1 mg/mL purified AlgM4, and the enzymatic activity of AlgM4 was then examined. AlgM4 was promoted at NaCl concentrations of 0.1–1.4 mol/L and was greatest in the presence of 1.0 mol/L NaCl, exhibiting more than 7 times the activity observed in the absence of NaCl (Table 3). The activity of AlyPM in the presence of 0.5–1.2 mol/L NaCl was 6 times that observed in the absence of NaCl [17]. Under optimum reaction conditions, the activity of A1m in the presence of 0.6–0.8 mol/L NaCl was 20 times that observed in the absence of NaCl [15]. The activity of rA9mT in the presence of 0.4 mol/L NaCl was 24 times that observed in the absence of NaCl [16].

Table 3. Effect of NaCl on AlgM4 activity.

NaCl Concentration (mol/L)	Relative Activity (%)
0	100 ± 0.2
0.1	229 ± 5.6
0.2	286 ± 6.4
0.3	356 ± 3.8
0.4	447 ± 1.4
0.5	481 ± 0.8
0.6	528 ± 3.1
0.7	572 ± 0.4
0.8	628 ± 3.4
0.9	664 ± 1.9
1.0	741 ± 3.2
1.2	462 ± 5.8
1.4	236 ± 4.5

The data are expressed as the means ± SD, $n = 3$. The activity of AlgM4 in the absence of NaCl was completely retained.

2.6. Determination of the Secondary Structure and Thermal Denaturation Temperature of AlgM4

The thermal denaturation temperature of AlgM4 was determined by circular dichroism (CD) spectroscopy at 25–75 °C. In addition, the secondary structure of AlgM4 was determined by ultraviolet–visible (UV-Vis) spectroscopy and CD spectroscopy. The CD absorption values of AlgM4 protein at 218 nm in different temperatures are shown in Figure 5B, and the CD values were used for analysis of the Tm value. The CD values of AlgM4 relatively remained stable at 25–35 °C in the absence of NaCl; subsequently, CD values slowly increased with the increase of temperature and the denaturation of AlgM4, then increased dramatically at 40–45 °C and remained stable at temperatures exceeding 45 °C. The AlgM4 protein was denatured completely when the temperature was higher than 45 °C. However, in the presence of 1 mol/L NaCl, the CD values increased rapidly at 45–50 °C and remained stable at temperatures exceeding 50 °C. The results showed that NaCl enhanced the ability of AlgM4 to resist thermal denaturation.

As shown in Figure 6A, the UV absorbance of AlgM4 at 220–240 nm was significantly reduced in the presence of 1 mol/L NaCl. CD spectroscopy was used to determine the secondary structure of the purified AlgM4, and the results are shown in Figure 6B. The absorption spectrum of AlgM4 was characterized by positive and negative peaks at approximately 198 and 218 nm, respectively, and the intensity of these peaks was reduced by the addition of 1 mol/L NaCl. The results showed that the α-helix and β-sheet contents in the secondary structure of AlgM4 were decreased after addition of 1 mol/L NaCl. Specifically, the α-helix and β-sheet contents were reduced from 12.4% and 38.2% to 10.8% and 36.5%, respectively. The changes in the contents of secondary structural elements are summarized in Table 4. The secondary structure of AlgM4 was altered in the presence of 1 mol/L NaCl, which may enhance the affinity of the enzyme for its substrates and facilitate enzymolysis. The alginate lyase AlyPM, derived from *Pseudoalteromonas* sp. SM0524, differs from AlgM4 in that the presence of NaCl did not alter the secondary structure of AlyPM. However, NaCl enhanced the affinity of AlyPM for its substrates, thereby promoting enzymolysis [17]. In addition, AlyPM is a cold-adapted enzyme, and its thermal denaturation temperature is relatively low, with a Tm of 37 °C [17]. At 1 mol/L, NaCl not only altered the secondary structure of AlgM4 but also enhanced the ability of AlgM4 to resist thermal denaturation. As a result, the midpoint of thermal denaturation (Tm) was increased from 43.3 to 48.2 °C (Figure 5B).

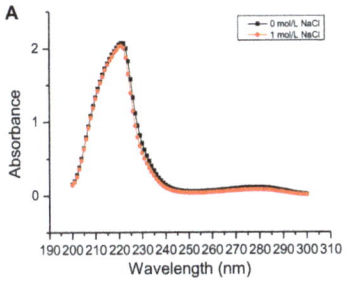 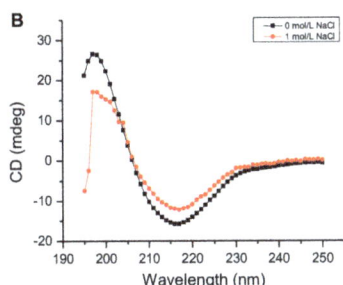

Figure 6. Determination of the secondary structure of the AlgM4. (**A**) Determination of the secondary structure by UV–Vis absorption spectra; (**B**) Determination of the secondary structure by circular dichroism.

Table 4. Secondary structure of AlgM4 as estimated by CD.

Enzyme	α-Helix (%)	β-Sheet (%)	β-Turn (%)	Random Coil (%)
AlgM4	12.4	38.2	21.3	28
AlgM4 + 1 mol/L NaCl	10.8	36.5	22.9	29.6

2.7. Analysis of the Products of AlgM4-Mediated Enzymolysis of Alginate by Ultra-Performance Liquid Chromatography (UPLC)–Quadrupole Time-of-Flight (QTOF)–Mass Spectrometry (MS)/MS

Alginate lyases with endolytic characteristics generally act on glycosidic bonds within the linear polysaccharide chain of alginate, generating unsaturated oligosaccharides that are dominated by disaccharides, trisaccharides, and tetrasaccharides [27]. Exolyases further depolymerize these oligosaccharides into mannuronic acids [18,29,30]. AlgM4 degrades both SA (Figure 7A) and polyG (Figure 7B) to produce oligosaccharides with degree of polymerization DP 2–6 [31,32]. The content of each oligosaccharide can be determined only after quantitative determination. Unlike SA and polyG, degradation of polyM produces oligosaccharides DP7, DP8, and DP9 (Figure 9). Furthermore, AlgM4 may be useful in the preparation of oligosaccharides, especially with high DP 7–9, and the study of their biological functions.

In recent years, the biological activities of oligosaccharides and the application of oligosaccharides in the medical and biotechnology fields has attracted the attention of researchers [33]. Oligosaccharides have higher degrees of polymerization and possibly better bioactivity [26]. An et al. observed that oligosaccharides (DP 6–8) derived from SA stimulated the accumulation of phytoalexin and induced phenylalanine ammonia lyase in soybean cotyledons, resulting in their acquired resistance to *Pseudomonas aeruginosa* [34]. In addition, trisaccharides, tetrasaccharides, pentasaccharides, and hexasaccharides obtained by enzymatic degradation of SA promoted the growth of lettuce seedlings [35].

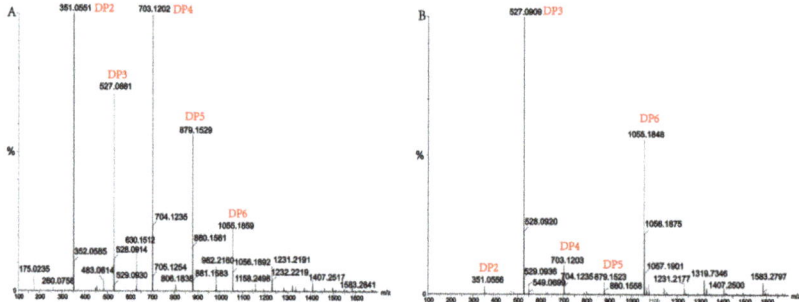

Figure 7. (**A**) Ultra-Performance Liquid Chromatography (UPLC)–Quadrupole Time-of-Flight (QTOF)–MS/MS analysis of hydrolysates of AlgM4 with sodium alginate as the substrate; (**B**) UPLC–QTOF–MS/MS analysis of hydrolysates of AlgM4 with polyG as the substrate. DP indicates the degree of polymerization of oligosaccharides from the alginate lyase hydrolysates.

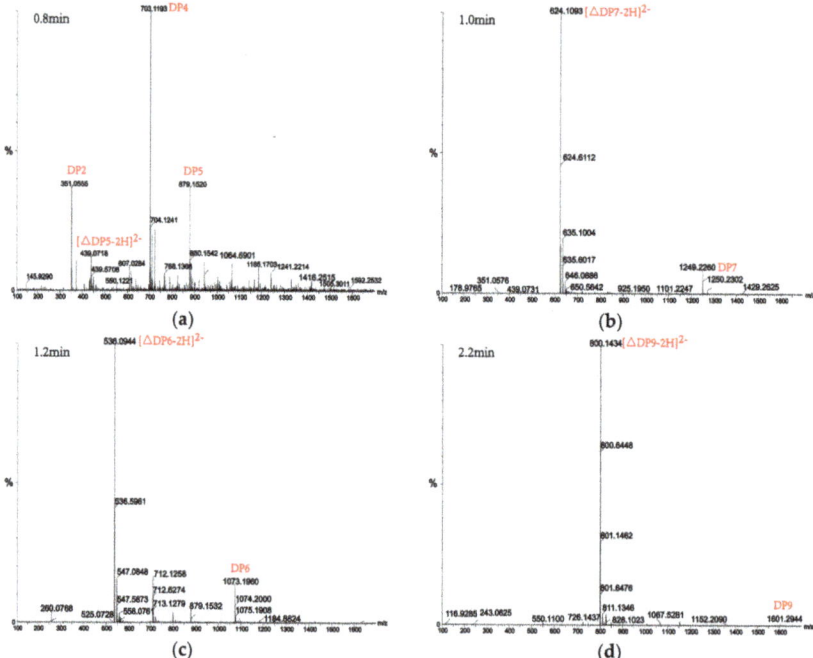

Figure 8. *Cont.*

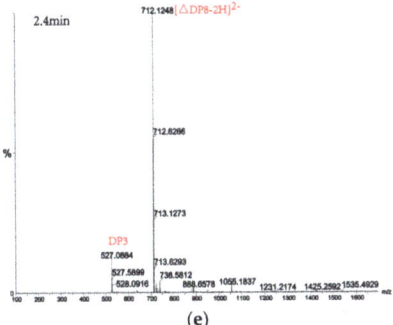

(e)

Figure 9. UPLC–QTOF–MS/MS analysis of hydrolysates of AlgM4 with polyM as the substrate. DP indicates the degree of polymerization of oligosaccharides from the alginate lyase hydrolysates. The reaction products of disaccharide, tetrasaccharide and pentasaccharide were detected at 0.8 min (**a**). The major peaks of DP7, DP6, DP9 and DP8 were detected at 1.0 min (**b**), 1.2 min (**c**), 2.2 min (**d**) and 2.4 min (**e**), respectively.

3. Materials and Methods

3.1. The Bacterium

The marine bacterium *V. weizhoudaoensis* M0101 was isolated from rotten Sargassum collected from Weizhou Island, Beihai, Guangxi Province, China.

3.2. Cloning and Expression of the algM4 Gene

The primers were designed according to the nucleic acid sequence of the *algM4* gene. The following primers were used to amplify the *algM4* gene: upstream primer, 5′-GGA ATTCCATATGCTTGCATCTTCTGTG-3′ (the *Nde*I restriction site is underlined); downstream primer, 5′-CCGCTCGAGACCTTTATAAGAACCGTG-3′ (the *Xho*I restriction site is underlined). The parameters of the polymerase chain reaction (PCR) were as follows: 94 °C for 2 min; followed by 30 cycles of 94 °C for 30 s, 58 °C for 30 s, and 72 °C for 1 min 40 s; with a final incubation at 72 °C for 10 min. The PCR product was double digested with the *Nde*I and *Xho*I restriction enzymes and then ligated into the pET30a(+) vector to construct the recombinant plasmid pET30a-*algM4*. To induce the expression of AlgM4, the recombinant plasmid was transformed into the *E. coli* strain BL21 (DE3). The positive transformants were picked, inoculated into 10 mL of LB (Luria-Bertani) medium containing 50 µg/mL kanamycin (Kan) and cultured at 37 °C with shaking (200 rpm) until the OD600 of the culture reached 0.5. The culture was then inoculated into LB medium containing 50 µg/mL Kan (inoculum volume: 1% (*v*/*v*)) and cultivated until the OD600 value reached 0.5. Subsequently, isopropyl β-D-1-thiogalactopyranoside (IPTG) was added at a final concentration of 0.5 mmol/L. After the addition of IPTG, the bacteria were cultured for another 12 h at 25 °C with shaking (200 rpm).

3.3. Purification of AlgM4

Protein isolation and purification was carried out at 0–4 °C. The IPTG-induced bacteria were harvested by centrifugation at 6000 rpm for 10 min. Subsequently, the bacteria were resuspended in buffer (10 mmol/L imidazole, 300 mmol/L NaCl, and 20 mmol/L Tris-HCl; pH 7.0) and lysed by ultrasonication. The lysates were centrifuged at 12,000 rpm for 30 min, and the resulting crude enzyme solution was collected. After washing off the protein impurities with buffer (50 mmol/L imidazole, 300 mmol/L NaCl, and 20 mmol/L Tris-HCl; pH 7.0), AlgM4 bound to the Ni^{2+} column was eluted with elution buffer containing 150 mmol/L imidazole, 300 mmol/L NaCl, and 20 mmol/L

Tris-HCl (pH 7.0). The purified protein was analyzed by SDS-PAGE, and the protein concentration was determined using the Bradford method.

3.4. Determination of the Enzymatic Properties of AlgM4

Enzymatic properties were determined using 0.1 mg/mL AlgM4. Nine hundred microliters of a 1.0% (w/v) SA solution was mixed with 100 µL of purified AlgM4 and incubated in a water bath at 30 °C for 10 min. The mixture was then boiled in a water bath for 5 min to terminate the reaction. After the reaction system was cooled to room temperature, the absorbance was measured at 235 nm. The unit of enzymatic activity (U) was defined as an increase in absorbance of 0.01 per minute. To determine the optimum reaction temperature, the enzymatic activity of AlgM4 was measured at 10–70 °C in 100 mmol/L Tris-HCl (pH 7.0) buffer using 0.2% (w/v) SA as a substrate. To determine the optimum reaction pH, the activity of AlgM4 was evaluated in 100 mmol/L citric acid–sodium citrate buffer (pH value: 4.5–6.5), 100 mmol/L Tris-HCl buffer (pH value: 7.0–9.0), or 100 mmol/L glycine-sodium hydroxide buffer (pH value: 9.5–12.0) at the optimum reaction temperature. To test the thermal stability of the enzyme, AlgM4 was first incubated at 25–50 °C for 30 min, and the residual AlgM4 activity was then measured under the optimum reaction conditions. To examine the effects of metal ions and surfactants on the enzymatic activity of AlgM4, metal ions (Ca^{2+}, K^+, Mg^{2+}, Mn^{2+}, Zn^{2+}, and Cu^{2+}), EDTA, or SDS were added to 50 µL of enzyme solution at a final concentration of 1 mmol/L. The activity of AlgM4 was then measured under the optimum reaction conditions using a 0.2% (w/v) SA solution as a substrate. The effect of NaCl on AlgM4 activity was also examined. NaCl was added to the enzyme solution at final concentrations of 0.1–1.4 mol/L, and the activity of AlgM4 was then compared to that observed under optimum reaction conditions.

3.5. Determination of the Kinetic Constants for AlgM4 Activity

The enzymatic activity of AlgM4 (0.1 mg/mL) was measured under the optimum reaction conditions using 3,5-dinitrosalicylic acid (DNS) assay. The concentrations of the substrate, SA, assayed were 0.5–10 mg/mL. Lineweaver-Burk (double-reciprocal) plots were generated using $1/[S]$ as the abscissa and $1/V$ as the ordinate. The V_{max} and K_m values were calculated using the Michaelis-Menten equation.

3.6. The Effects of NaCl on the Secondary Structure and Thermal Stability of AlgM4

In one group, NaCl was added to a 300 µL reaction containing 0.28 mg/mL of purified AlgM4 (final concentration of NaCl: 1 mol/L), while another group was not exposed to NaCl. The secondary structure of the purified AlgM4 protein was determined at 25 °C using a TU-1901 Dual Beam Ultraviolet Spectrophotometer (PERSEE, Beijing, China) (spectral range: 200–350 nm) and a Chirascan Circular Dichroism Spectrometer (Applied Photophysics Ltd., Surrey, UK) (spectral range: 195–250 nm; optical path length: 10 mm; bandwidth: 0.5 nm). The composition of the secondary structural elements was analyzed using CDpro software (http://sites.bmb.colostate.edu/sreeram/CDPro/CDPro.htm). The thermal denaturation temperature of AlgM4 was measured under the following conditions: spectral range, 200–260 nm; bandwidth, 0.7 nm; 25–75 °C. The Tm value was calculated using Global 3 software (Applied Photophysics Ltd., Surrey, UK).

3.7. Analysis of the Products of AlgM4-Mediated Enzymolysis Using UPLC–QTOF–MS/MS

The purified AlgM4 (0.25 mg/mL) enzyme was mixed with an equal volume of 1.0% (w/v) SA, polyM, or polyG; incubated in a water bath at 30 °C for 6 h; and then concentrated in vacuo. After high-speed centrifugation, the supernatants were collected and filtered through filter membranes with a 0.22 µm pore size. The filtrates were analyzed using liquid chromatography (LC)-MS. The equipment used in the analysis was a UPLC-QTOF-MS/MS system (Waters Corporation, Milford, MA, USA), which consisted of a UPLC I-Class instrument (Waters Corporation, Singapore) and a XEVO G2-S mass spectrometer (Waters Corporation, Milford, MA, USA). The operating conditions of LC were as follows: ACQUITY UPLC HSS T3 C18 column (2.1 mm × 100 mm, 1.8 µm, Waters Corporation, Milford, MA, USA); gradient elution with 0.1% formic acid-water (A) and formic acid-acetonitrile (B); flow rate, 0.5 mL/min; column temperature, 35 °C; analytic time, 10 min; injection volume, 1.0 µL. The MS conditions were as follows: ion scan mode, negative-ion ESI mode; scan range, 100–1700 Da; ion source temperature, 100 °C; desolvation gas temperature, 400 °C; desolvation gas flow rate, 1000 L/h; capillary voltage, 2.5 kV; cone voltage, 40 V; low collision energy, 6 V; high collision energy, 35–50 V; data acquisition software, MassLynx 4.1 SCN 884 (Waters Corporation, Milford, MA, USA); data acquisition mode, MSE.

4. Conclusions

AlgM4 is a new salt-activated and bifunctional alginate lyase of the PL7 family with endolytic activity derived from the marine bacterium *V. weizhoudaoensis* M0101. Compared with the observed AlgM4 activity in the absence of NaCl, the enzymatic activity of AlgM4 increased in the presence of various concentrations of NaCl (0.1–1.4 mol/L). The addition of 1 mol/L NaCl resulted in a more than sevenfold increase in AlgM4 activity. Therefore, AlgM4 tolerates high-salinity environments. NaCl not only altered the composition of the secondary structural elements in AlgM4 but also enhanced its thermal stability. AlgM4 hydrolyzed SA, polyM, and polyG via its endolytic activity, producing oligosaccharides (DP 2–9). The alginate lyase AlgM4 has an important application value in the preparation of bioactive oligosaccharides and in the processes of alginate saccharification and ethanol fermentation.

Acknowledgments: This study was supported by grants from National Natural Science Foundation of China (No. 31560017), Key Program of National Natural Science Foundation of Guangxi (No. 2014GXNSFDA118012), Key Research and Development Program of Guangxi (No. AB16380071), The Special Project for the Base of Guangxi Science and Technology and Talents (No. AD17129019), the Fundamental Research Funds for Guangxi Academy of Sciences (2017YJJ23020), high-level innovation teams of Guangxi colleges and universities and academic excellence program (Gui-Jiao-Ren, No. 2016/42).

Author Contributions: Shushi Huang conceived and designed the experiments; Shushi Huang and Guiyuan Huang took charge of the preparation of the manuscript. Guiyuan Huang, Qiaozhen Wang, Mingqian Lu, Chao Xu and Fei Li performed genome analysis of strain *V. weizhoudaoensis* M0101, expression and purification of recombinant enzyme, and determination of activity and enzymatic kinetics of AlgM4 lyases. Guiyuan Huang, Rongcan Zhang and Wei liao analyzed the data of UPLC-QTOF-MS/MS from alginate using AlgM4 lyases. Guiyuan Huang and Chao Xu performed the determination of the secondary structure of AlgM4 with UV-Vis and circular dichroism under NaCl stress.

Conflicts of Interest: The authors declare no conflict of interest.

References

1. Hernandez-carmona, G.; Mchugh, D.J.; Iopez-gutierrez, F. Pilot plant scale extraction of alginates from *Macrocystis pyrifera*. *Appl. Phycol.* **2000**, *11*, 493–502. [CrossRef]
2. Liu, H.; Yin, H.; Zhang, Y.H.; Zhao, X.M. Research progress on biological activities of alginate oligosaccharides. *Nat. Prod. Res.* **2012**, *S1*, 201–204.
3. Gacesa, P. Alginate. *Carbohydr. Polym.* **1988**, *8*, 161–182. [CrossRef]
4. Tako, M. Chemical characterization of acetyl fucoidan and alginate from commercially cultured cladosiphon okamuranus. *Bot. Mar.* **2000**, *43*, 393–398. [CrossRef]
5. Tako, M.; Kiyuna, S.; Uechi, S.; Hongo, F. Isolation and characterization of alginic acid from commercially cultured Nemacystus decipiens (Itomozuku). *Biosci. Biotechnol. Biochem.* **2001**, *65*, 654–657. [CrossRef] [PubMed]
6. Qian, L.; Tang, L.W.; Huang, S.S.; Chagan, I. Research progress of bioethanol from alginate fermentation. *China Biotechnol.* **2013**, *33*, 122–127.
7. Schaumann, K.; Weide, G. Enzymatic degradation of alginate by marine fungi. *Hydrobiologia* **1990**, *204*, 589–596. [CrossRef]

8. Lee, S.I.; Choi, S.H.; Lee, E.Y.; Kim, H.S. Molecular cloning, purification, and characterization of a novel polyMG-specific alginate lyase responsible for alginate MG block degradation in *Stenotrophomas maltophilia* KJ-2. *Appl. Microbiol. Biotechnol.* **2012**, *95*, 1643–1653. [CrossRef] [PubMed]
9. Rahman, M.M.; Inoue, A.; Tanaka, H.; Ojima, T. cDNA cloning of an alginate lyase from a marine gastropod *Aplysia kurodai* and assessment of catalytically important residues of this enzyme. *Biochimie* **2011**, *93*, 1720–1730. [CrossRef] [PubMed]
10. Rahman, M.M.; Wang, L.; Inoue, A.; Ojima, T. cDNA cloning and bacterial expression of a PL-14 alginate lyase from a herbivorous marine snail *Littorina brevicula*. *Carbohydr. Res.* **2012**, *360*, 69–77. [CrossRef] [PubMed]
11. Sheng, D.; Jie, Y.; Zhang, X.Y.; Mei, S.; Song, X.Y.; Chen, X.L.; Zhang, Y.Z. Cultivable Alginate Lyase-Excreting Bacteria Associated with the Arctic Brown Alga *Laminaria*. *Mar. Drugs* **2012**, *10*, 2481–2491.
12. Xiao, L.; Han, F.; Yang, Z.; Lu, X.Z.; Yu, W.G. A Novel Alginate Lyase with High Activity on Acetylated Alginate of *Pseudomonas aeruginosa*, FRD1 from *Pseudomonas* sp. QD03. *World J. Microbio. Biotechnol.* **2006**, *22*, 81–88. [CrossRef]
13. Li, S.; Yang, X.; Zhang, L.; Yu, W.; Han, F. Cloning, Expression, and Characterization of a Cold-Adapted and Surfactant-Stable Alginate Lyase from Marine Bacterium *Agarivorans* sp. L11. *J. Microbiol. Biotechnol.* **2015**, *25*, 681–686. [CrossRef] [PubMed]
14. Inoue, A.; Anraku, M.; Nakagawa, S.; Ojima, T. Discovery of a Novel Alginate Lyase from *Nitratiruptor* sp. SB155-2 Thriving at Deep-sea Hydrothermal Vents and Identification of the Residues Responsible for Its Heat Stability. *J. Biol. Chem.* **2016**, *291*, 15551–15563. [CrossRef] [PubMed]
15. Kobayashi, T.; Uchimura, K.; Miyazaki, M.; Nogi, Y.; Horikoshi, K. A new high-alkaline alginate lyase from a deep-sea bacterium *Agarivorans* sp. *Extremophiles* **2009**, *13*, 121–129. [CrossRef] [PubMed]
16. Uchimura, K.; Miyazaki, M.; Nogi, Y.; Kobayashi, T.; Horikoshi, K. Cloning and sequencing of alginate lyase genes from deep-sea strains of *Vibrio* and *Agarivorans* and characterization of a new *Vibrio* enzyme. *Mar. Biotechnol.* **2010**, *12*, 526–533. [CrossRef] [PubMed]
17. Chen, X.L.; Sheng, D.; Fei, X.; Fang, D.; Li, P.Y.; Zhang, X.Y.; Zhou, B.C.; Zhang, Y.Z.; Xie, B.B. Characterization of a New Cold-Adapted and Salt-Activated Polysaccharide Lyase Family 7 Alginate Lyase from *Pseudoalteromonas* sp. SM0524. *Front. Microbiol.* **2016**, *7*, e30105. [CrossRef] [PubMed]
18. Kim, H.T.; Chung, J.H.; Wang, D.; Lee, J.; Woo, H.C.; Choi, I.G.; Kim, K.H. Depolymerization of alginate into a monomeric sugar acid using Alg17C, an exo-oligoalginate lyase cloned from *Saccharophagus degradans* 2-40. *Appl. Microbiol. Biotechnol.* **2012**, *93*, 2233–2239. [CrossRef] [PubMed]
19. Matsubara, Y.; Iwasaki, K.; Muramatsu, T. Action of poly (alpha-L-guluronate)lyase from *Corynebacterium* sp. ALY-1 strain on saturated oligoguluronates. *J. Agric. Chem. Soc. Jpn.* **1998**, *62*, 1055–1060.
20. Thiangyian, W.; Preston, L.A.; Schiller, N.L. Alginate lyase: Review of major sources and enzyme characteristics, structure-function analysis, biological roles, and applications. *Annu. Rev. Microbiol.* **2000**, *54*, 289–340.
21. Sawabe, T.; Takahashi, H.; Ezura, Y.; Gacesa, P. Cloning, sequence analysis and expression of *Pseudoalteromonas elyakovii* IAM 14594 gene (alyPEEC) encoding the extracellular alginate lyase. *Carbohydr. Res.* **2001**, *335*, 11–21. [CrossRef]
22. Miyake, O.; Hashimoto, W.; Murata, K. An exotype alginate lyase in *Sphingomonas* sp. A1: Overexpression in *Escherichia coli*, purification, and characterization of alginate lyase IV (A1-IV). *Protein Expr. Purif.* **2003**, *29*, 33–41. [CrossRef]
23. Ochiai, A.; Yamasaki, M.; Mikami, B.; Hashimoto, W.; Murata, K. Crystal structure of exotype alginate lyase Atu3025 from *Agrobacterium tumefaciens*. *J. Biol. Chem.* **2010**, *285*, 24519–24528. [CrossRef] [PubMed]
24. Garron, M.L.; Cygler, M. Structural and mechanistic classification of uronic acid-containing polysaccharide lyases. *Glycobiology* **2010**, *20*, 1547–1573. [CrossRef] [PubMed]
25. Zhu, Y.; Wu, L.; Chen, Y.; Ni, H.; Xiao, A.; Cai, H. Characterization of an extracellular biofunctional alginate lyase from marine *microbulbifer* sp. ALW1 and antioxidant activity of enzymatic hydrolysates. *Microbiol. Res.* **2016**, *182*, 49–58. [CrossRef] [PubMed]
26. Huang, L.; Zhou, J.; Li, X.; Peng, Q.; Lu, H.; Du, Y. Characterization of a new alginate lyase from newly isolated *Flavobacterium* sp. S20. *J. Ind. Microbiol. Biotechnol.* **2013**, *40*, 113–122. [CrossRef] [PubMed]

27. Kim, H.T.; Ko, H.J.; Kim, N.; Kim, D.; Lee, D.; Choi, I.G.; Woo, H.C.; Kim, M.D.; Kim, K.H. Characterization of a recombinant endo-type alginate lyase (Alg7D) from *Saccharophagus degradans*. *Biotechnol. Lett.* **2012**, *34*, 1087–1092. [CrossRef] [PubMed]
28. Zhu, B.; Chen, M.; Yin, H.; Du, Y.; Ning, L. Enzymatic hydrolysis of alginate to produce oligosaccharides by a new purified endo-type alginate lyase. *Mar. Drugs* **2016**, *14*, 108. [CrossRef] [PubMed]
29. Jagtap, S.S.; Hehemann, J.H.; Polz, M.F.; Lee, J.K.; Zhao, H. Comparative biochemical characterization of three exolytic oligoalginate lyases from *Vibrio splendidus* reveals complementary substrate scope, temperature, and pH adaptations. *Appl. Environ. Microbio.* **2014**, *80*, 4207–4214. [CrossRef] [PubMed]
30. Park, H.H.; Kam, N.; Lee, E.Y.; Kim, H.S. Cloning and characterization of a novel oligoalginate lyase from a newly isolated bacterium *Sphingomonas* sp. MJ-3. *Mar. Biotechnol.* **2012**, *14*, 189–202. [CrossRef] [PubMed]
31. Li, S.; Yang, X.; Bao, M.; Wu, Y.; Yu, W.; Han, F. Family 13 carbohydrate-binding module of alginate lyase from *Agarivorans* sp. L11 enhances its catalytic efficiency and thermostability, and alters its substrate preference and product distribution. *FEMS Microbiol. Lett.* **2015**, *362*. [CrossRef] [PubMed]
32. Kurakake, M.; Kitagawa, Y.; Okazaki, A.; Shimizu, K. Enzymatic Properties of Alginate Lyase from *Paenibacillus* sp. S29. *Appl. Biochem. Biotechnol.* **2017**, 1–10. [CrossRef] [PubMed]
33. Tusi, S.K.; Khalaj, L.; Ashabi, G.; Kiaei, M.; Khodagholi, F. Alginate oligosaccharide protects against endoplasmic reticulum- and mitochondrial-mediated apoptotic cell death and oxidative stress. *Biomaterials* **2011**, *32*, 5438–5458. [CrossRef] [PubMed]
34. An, Q.D.; Zhang, G.L.; Wu, H.T.; Zhang, Z.C.; Zheng, G.S.; Luan, L.; Murata, Y.; Li, X. Alginate-deriving oligosaccharide production by alginase from newly isolated *Flavobacterium* sp. LXA and its potential application in protection against pathogens. *J. Appl. Microbiol.* **2009**, *106*, 161–170. [CrossRef] [PubMed]
35. Iwasaki, K.I.; Matsubara, Y. Purification of Alginate Oligosaccharides with Root Growth-promoting Activity toward Lettuce. *Biosci. Biotechnol. Biochem.* **2000**, *64*, 1067–1070. [CrossRef] [PubMed]

© 2018 by the authors. Licensee MDPI, Basel, Switzerland. This article is an open access article distributed under the terms and conditions of the Creative Commons Attribution (CC BY) license (http://creativecommons.org/licenses/by/4.0/).

Article

Purification and Characterization of a Novel Alginate Lyase from the Marine Bacterium *Bacillus* sp. Alg07

Peng Chen [1,2], Yueming Zhu [1], Yan Men [1], Yan Zeng [1] and Yuanxia Sun [1,*]

[1] National Engineering Laboratory for Industrial Enzymes, Tianjin Institute of Industrial Biotechnology, Chinese Academy of Sciences, Tianjin 300308, China; chen_p@tib.cas.cn (P.C.); zhu_ym@tib.cas.cn (Y.Z.); men_y@tib.cas.cn (Y.M.); zeng_y@tib.cas.cn (Y.Z.)
[2] University of Chinese Academy of Sciences, Beijing 100049, China
* Correspondence: Sun_yx@tib.cas.cn; Tel.: +86-022-8486-1960

Received: 9 February 2018; Accepted: 5 March 2018; Published: 9 March 2018

Abstract: Alginate oligosaccharides with different bioactivities can be prepared through the specific degradation of alginate by alginate lyases. Therefore, alginate lyases that can be used to degrade alginate under mild conditions have recently attracted public attention. Although various types of alginate lyases have been discovered and characterized, few can be used in industrial production. In this study, AlgA, a novel alginate lyase with high specific activity, was purified from the marine bacterium *Bacillus* sp. Alg07. AlgA had a molecular weight of approximately 60 kDa, an optimal temperature of 40 °C, and an optimal pH of 7.5. The activity of AlgA was dependent on sodium chloride and could be considerably enhanced by Mg^{2+} or Ca^{2+}. Under optimal conditions, the activity of AlgA reached up to 8306.7 U/mg, which is the highest activity recorded for alginate lyases. Moreover, the enzyme was stable over a broad pH range (5.0–10.0), and its activity negligibly changed after 24 h of incubation at 40 °C. AlgA exhibited high activity and affinity toward poly-β-D-mannuronate (polyM). These characteristics suggested that AlgA is an endolytic polyM-specific alginate lyase (EC 4.2.2.3). The products of alginate and polyM degradation by AlgA were purified and identified through fast protein liquid chromatography and electrospray ionization mass spectrometry, which revealed that AlgA mainly produced disaccharides, trisaccharides, and tetrasaccharide from alginate and disaccharides and trisaccharides from polyM. Therefore, the novel lysate AlgA has potential applications in the production of mannuronic oligosaccharides and poly-α-L-guluronate blocks from alginate.

Keywords: alginate lyase; marine bacterium; *Bacillus* sp. Alg07; purification; alginate oligosaccharides

1. Introduction

Alginate is a linear copolymer that is composed of homopolymeric blocks of (1–4)-linked α-L-guluronic acid (G) and its C5 epimer β-D-mannuronic acid (M), which forms three types of blocks: poly-α-L-guluronate (polyG), poly-β-D-mannuronate (polyM), and random heteropolymeric sequences (polyMG) [1]. Alginate is the most abundant carbohydrate in brown algae, and it accounts for up to 10–45% of the dry weight of brown algae [2]. Some bacteria that belong to the genera *Azotobacter* [3] and *Pseudomonas* [4] produce alginate as an extracellular polysaccharide. In contrast to algal alginate, bacterial alginate is acetylated. Commercial alginate manufactured from brown algae has been used as a thickening agent or gelling agent in the food and pharmaceutical industries [5]. Alginate can be degraded into alginate oligosaccharides (AOS) through a chemical process or by alginate lyase. Given that AOS can stimulate the growth of endothelial cells [6] and the production of multiple cytokines [7], they may be applied as growth-promoting agents in some plants [8] and bifidobacteria [9]. Furthermore, AOS demonstrate excellent antioxidant activity [10] and havepotential uses in protection against pathogens [11].

In alginate degradation, alginate lyases cleave the (1–4)-linked glucosidic bond of alginate via a β-elimination mechanism and generate unsaturated oligosaccharides with 4-deoxy-alpha-L-erythrohex-4-enopyranuronosyl uronate as the nonreducing terminal residue [12]. Numerous alginate lyases have been isolated from various organisms, such as marine algae [13], marine mollusks [14], marine and terrestrial bacteria [15,16], marine fungi [17], and viruses [18]. Alginate lyases can be categorized into polyM-specific, polyG-specific, and polyMG-specific lyases on the basis of their substrate preferences [19] or into endo- or exo-alginate lyases on the basis of their cleavage mode [19]. In the carbohydrate-active enzyme database, alginate lyases belong to the polysaccharide lyase family [20]. The structures of some alginate lyases have been elucidated.

Alginate lyases are widely used in many fields. For example, alginate lyases have been employed to explain the fine structures of alginate [21] and to prepare red and brown algal protoplasts [22]. These enzymes may be utilized in the treatment of cystic fibrosis [23] and have been used as catalysts for AOS production [24]. The application of alginate lyases in alginate degradation under mild conditions has recently attracted public attention given the high efficiency and specifi of these enzymes. Nevertheless, present studies on alginate lyases remain in development, and the low catalytic efficiency and poor thermostability of alginate lyases limit their utility in AOS production. Therefore, high-efficiency and thermostable alginate lyases should be identified for use in AOS production.

In our work, we isolated and identified *Bacillus* sp. Alg07, a novel marine bacterium. AlgA, the alginate lyase secreted by this strain, showed extremely high activity. Hence, we purified and characterized AlgA to confirm that it has potential applications in AOS production.

2. Results and Discussion

2.1. Screening and Identification of Strain Alg07

Twenty-one strains with alginate lyase activity were isolated using alginate as the sole carbon source. Alg07 secreted the alginate lyase with the highest activity in the fermentation culture.

The 16S rRNA gene of Alg07 was cloned, sequenced, and submitted to GeneBank (accession number KM040772) for strain identification. The alignment of 16S rRNA gene sequences from different *Bacillus* species showed that strain Alg07 is closely related to *Bacillus litoralis* S20409 (97%) and *Bacillus simplex* J2S3 (97%). However, the low similarity shared by the 16s rRNA gene sequence of Alg07 with that of other known *Bacillus* species indicated that Alg07 may be a novel *Bacillus* species. In accordance with the neighbor-joining phylogenetic tree, the strain was assigned to the genus *Bacillus* and designated as *Bacillus* sp. Alg07 (Figure 1).

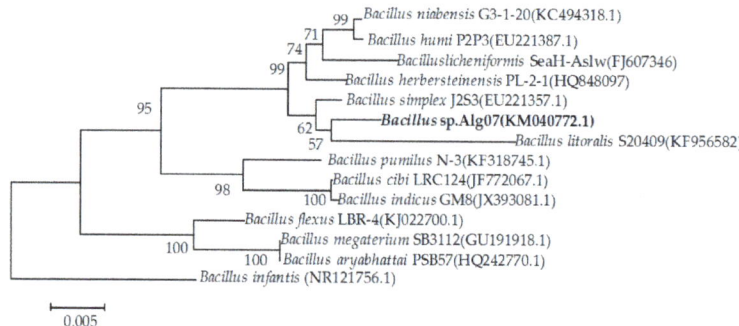

Figure 1. Neighbor-joining phylogenetic tree generated on the basis of the 16S rRNA gene sequences of strain Alg07 and other known *Bacillus* species.

2.2. Purification of Alginate Lyase from Bacillus sp. Alg07

Bacillus sp. Alg07 was cultured in optimized liquid medium for 24 h until its alginate lyase reached the highest activity. The supernatant containing 510 U/mL of alginate lyase was subjected to further purification through the two simple steps of tangential flow filtration concentration and anion exchange chromatography with Source 15Q (Figure S1). After purification, the alginate lyase was purified 8.34-fold with a yield of 62.4% (Table 1). The final specific activity of the purified alginate lyase was 8306.7 U/mg. The purified alginate lyase from Bacillus sp. Alg07 was designated as AlgA.

Table 1. Summary of the purification of AlgA.

Step	Total Protein (mg)	Total Activity (U)	Specific Activity (U/mg)	Folds	Yield (%)
Culture broth	23.82	23,724.72	996.01	1.00	100
Vivaflow50	10.30	22,672.14	2201.18	2.21	95.4
Source 15Q	1.78	14,785.9	8306.71	8.34	62.4

The activities of AlgA and those of other well-studied strains are shown in Table 2, which shows that AlgA has the highest activity among all reported alginate lyases. The simple purification, high recovery, and high specific enzyme activity of AlgA indicated that it may be produced industrially on a large scale.

Table 2. Comparison of the properties of AlgA with those of alginate lyases from different microorganisms.

Enzyme	Source	Specific Activity (U/mg)	Molecular Mass (kDa)	Optimal Temperature (°C)	Optimal pH	Cation Activators	Substrate Specificity
AlgA	This study	8306.71	60	40	7.5	Na^+, Mg^{2+}, Ca^{2+}, Mn^{2+}	PM
Oal17	Shewanella sp.Kz7 [25]	32	82	50	6.2	Na^+	PM
Cel32	Cellulophaga sp. NJ-1 [15]	2417.8	32	50	8.0	Ca^{2+}, Mg^{2+}, K^+	PM, PG
Alg7D	Saccharophagus degradans [26]	4.6	63.2	50	7.0	Na^+	PM, PG
AlySY08	Vibrio sp. Aly08 [27]	1070.2	33	40	7.6	Na^+, K^+, Ca^{2+}, Mg^{2+}	PM, PG
AlyV5	Vibrio sp. QY105 [28]	2152	37	38	7.0	Na^+, Mg^{2+}, Ca^{2+}, Mn^{2+}	PM, PG
Alm	Agarivorans sp. JAM-Alm [29]	108.5	31	30	10.0	Na^+, K^+	PG, PMG
FlAlyA	Flavabacterium sp. UMI-01 [16]	2347.8	30	55	7.7	Na^+, K^+, Ca^{2+}, Mg^{2+}	PM

Figure 2 shows that the purified AlgA exhibited a clear and unique band on sodium dodecyl sulfate polyacrylamide gel electrophoresis (SDS-PAGE). This result suggested that the two-step purification process is successful. The molecular weight of AlgA was approximately 60 kDa, which is similar to molecular weights of alginate lyases from Vibrio sp.YKW-34 [30] and Saccharophagus degradans [26]. AlgA belongs to the class of large alginate lyases on the basis of its molecular weight [31].

The N-terminal amino acid sequence of the purified AlgA was analyzed. The sequence was identified as Glutamic acid–Glutamic acid–Glutamic acid–Glutamic acid–Aspartic acid–Valine–Threonine–Tyrosine (Figure S2). The results from homology search using BLASTp and CLUSTAL X indicated that the N-terminal amino acid sequence of AlgA is absent from the sequences of the previously reported alginate lyases. Furthermore, a protein with four Glu residues at its N-terminal is unusual. Therefore, AlgA may be a novel alginate lyase.

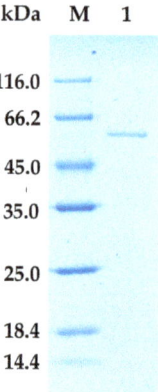

Figure 2. Sodium dodecyl sulfate polyacrylamide gel electrophoresis (SDS-PAGE) result for AlgA. Lane M, protein ladder; Lane 1, purified AlgA.

2.3. Biochemical Characterization of AlgA

The optimal temperature of AlgA is 40 °C, which is similar to alginate lyases derived from *Vibrio* sp.YKW-34 [30], *Flavobacterium* sp. LXA [11], *Pseudomonas aeruginosa* PA1167 [32], and *Stenotrophomnas maltophilia* KJ-2 [33]. The activity of AlgA significantly decreased under temperatures exceeding 50 °C (Figure 3A). Thermostability analysis indicated that the activity of AlgA remained relatively unchanged after 5 h of incubation at 40 °C and did not decrease even after 24 h of incubation at 40 °C (data not shown). However, the enzyme was less stable under high temperatures, and the half-life of AlgA is approximately 3 h under 45 °C and 0.75 h under 50 °C (Figure 3B). This result indicated that at its optimal temperature, AlgA might be the most stable enzyme among all reported alginate lyases. Alginate lyases from *Favobacterium* sp. LXA [11], *Vibrio* sp. W13 [34], and *Zobellia galactanivorans* [35] are stable only under temperatures less than 40 °C. The activities of many alginate lyases considerably decrease after incubation at 40 °C. AlyL2 from *Agarivorans* sp. L11 has a half-life of 125 min at 40 °C [36], and OalS17 from *Shewanella* sp. Kz7 retains 88% of its activity after 1 h of incubation at 40 °C [25]. Furthermore, alginate lyases from *Flammeovirga* sp. MY04 [37] and *Vibrio* sp. SY08 [27] retain approximately 80 and 75%, respectively, of their activities after 2 h of incubation at 40 °C. Thermostable enzymes are more advantageous than thermolabile counterparts due totheir long half-lives and low production costs. Thus, AlgA has potential industrial applications given its excellent thermostability.

The optimal pH of AlgA was 7.5 in 20 mM Tris-HCl buffer. AlgA exhibited more than 90% of its maximal activity in pH 8.0 buffer. These results suggested that AlgA is basophilic (Figure 3C). Similarly, the optimal activities of most alginate lyases from marine bacteria are observed at pH 7.5–8.0 (Table 2). The results of the pH stability assay showed that AlgA presented the highest stability for an extended period over a pH range of 5.0–10.0 (Figure 3D). The good stability of AlgA over a wide pH range indicated its suitability for industrial application. Furthermore, the use of AlgA may decrease production costs given that pH adjustment will be unnecessary even when various alginates from different sources are employed as substrates.

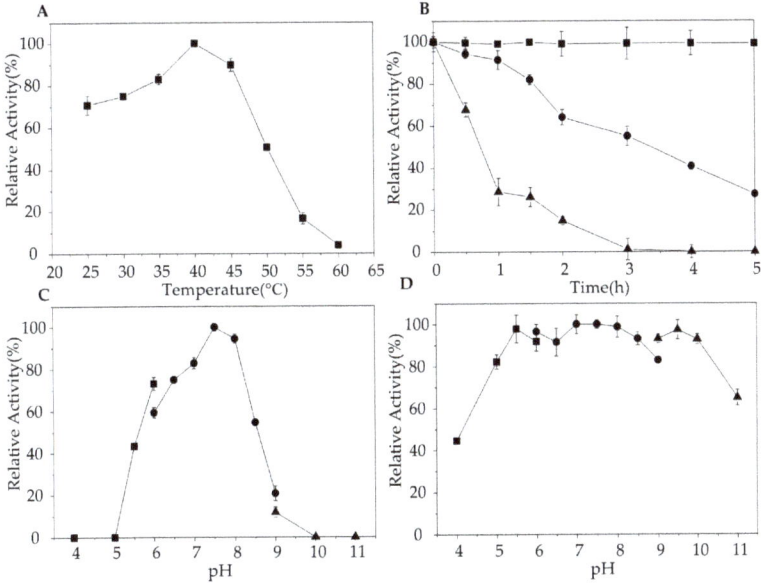

Figure 3. Effects of temperature and pH on the relative activity of AlgA. (**A**) Optimal temperature of AlgA. (**B**) Thermostability of AlgA at 40 °C (filled square), 45 °C (filled circle), and 50 °C (filled triangle). (**C**) Optimal pH for the relative activity of AlgA was determined in 20 mM CH_3COOH-CH_3COONa buffer (filled square), 20 mM Tris-HCl buffer (filled circle), or 20 mM Glycine-NaOH buffer (filled triangle). (**D**) pH stability of AlgA in 20 mM CH_3COOH-CH_3COONa buffer (filled square), 20 mM Tris-HCl buffer (filled circle), and 20 mM Glycine-NaOH (filled triangle).

To determine the effect of NaCl on AlgA, the activity of AlgA in the presence of various NaCl concentrations was measured. As shown in Figure 4A, the optimal NaCl concentration for AlgA activity was 200 mM, and no activity was detected in the absence of NaCl. These results indicated that the activity of AlgA is dependent on NaCl. Thus, NaCl concentration is crucial for the activity of AlgA, which is a salt-activated alginate lyase. However, high NaCl concentrations decreased the activity of AlgA. Similarly, the activities of marine bacterial alginate lyases, such as AlyV5 from *Vibrio* sp. QY105 [28] and AlyYKW-34 from *Vibrio* sp. YKW-34 [30], are dependent on NaCl.

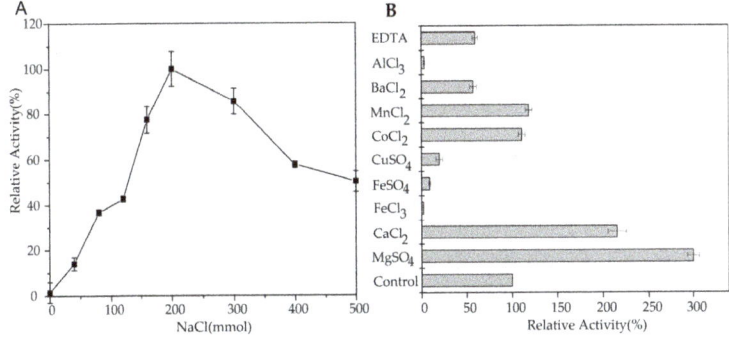

Figure 4. Effect of NaCl (**A**) and metal ions (**B**) on the activity of AlgA.

The effects of metal ions and EDTA on the activity of AlgA were determined in the presence of 200 mM NaCl. Mg^{2+} or Ca^{2+} significantly enhanced enzymatic activity by 300% or 215%, and Co^{2+} and Mn^{2+} slightly increased enzymatic activity (Figure 4B). By contrast, Hg^{2+}, Fe^{3+}, Fe^{2+}, and Cu^{2+} completely inhibited lyase activity. Ba^{2+} and EDTA partially inhibited lyase activity. Ca^{2+} and Mg^{2+} increase the activity of many alginate lyases, such as aly-SJ02 from *Pseudoalteromonas* sp. SM0524 [38] and AlyV5 from *Vibrio* sp. QY105 [28]. However, Mg^{2+} and Ca^{2+} decrease the activity of Alg7D from *S. degradans* [26].

2.4. Substrate Specificity and Kinetic Parameters of AlgA

AlgA exhibited activity toward alginate but not toward pectin, hyaluronan, chitin, or agar. This behavior suggested that AlgA is indeed an alginate lyase. As shown in Figure 5, the relative activities of AlgA toward alginate, polyM, and polyG blocks are 100 ± 4.3, 87.2 ± 6.9, and $10.5 \pm 1.7\%$, respectively. The slight activity toward polyG block might result from the presence of a few M residues in polyG substrates. These results indicated that the polyM block of substrates is the preferred substrate of AlgA. Thus, AlgA is a mannuronate lyase. Alginate lyases from *Flavobacterium* sp. UMI-01 [16], *Vibrio* sp. JAM-A9m [39], and *Pseudomonas* sp. QD03 [40] also belong to the mannuronate lyase class of enzymes.

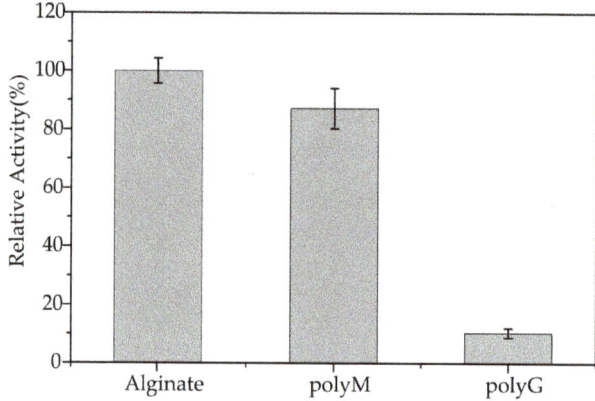

Figure 5. Relative activities of AlgA toward alginate, polyM, and polyG.

The kinetic parameters of AlgA were determined through nonlinear regression analysis (Figure S3) and are shown in Table 3. AlgA has a lower Km value for polyM than for sodium alginate. This result suggested that AlgA has high affinity for polyM blocks and further confirmed that AlgA is a mannuronate lyase. However, the $kcat$ values of AlgA for sodium alginate were higher than those for polyM. Therefore, AlgA has equivalent catalytic efficiency for sodium alginate and polyM.

Table 3. Kinetic parameters of the activity of AlgA toward sodium alginate and polyM blocks.

Parameter	Sodium Alginate	PolyM
$Vmax$ (U mg of protein^{-1})	1052.0 ± 214.6	547.6 ± 22.4
Km (mg mL^{-1})	9.0 ± 3.3	3.6 ± 0.4
$kcat$ (s^{-1})	911.7 ± 185.9	474.6 ± 19.4
$kcat/Km$ (mg^{-1} mL s^{-1})	101.3 ± 20.7	132.0 ± 5.4

2.5. Fast Protein Liquid Chromatography and Electrospray Ionization Mass Spectrometry Analysis of the Degradation Products of AlgA

To investigate the action patterns of AlgA, degradation of alginate by this enzyme was performed. The degradation products at different time intervals were analyzed through gel chromatography. Fast protein liquid chromatography (FPLC) analysis indicated that AOS with different degrees of polymerization (DP) gradually accumulated (Figure 6). Therefore, AlgA is an endo-type alginate lyase.

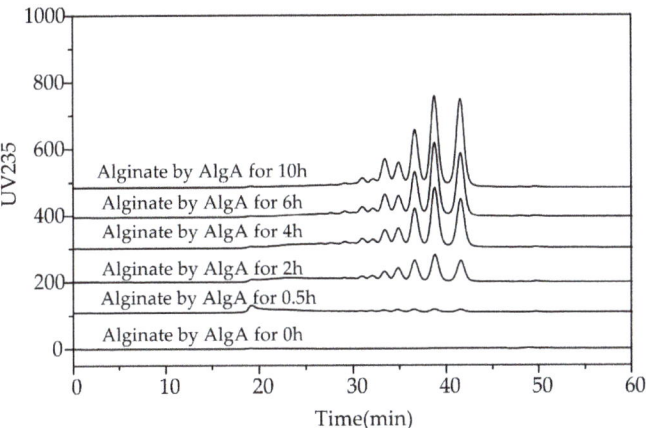

Figure 6. Patterns of the polysaccharide degradation products of AlgA. Enzymatic degradation products collected at 0.5, 2, 4, 6, and 10 h were subjected to gel filtration with a Superdex peptide 10/300 GL column. The absorbances of the products were monitored at 235 nm.

In addition, the alginate was completely digested with an excess of AlgA at 40 °C for 24 h. The products were separated through gel chromatography. The elution profiles (Figure 7) of the degradation products presented three major fractions (peaks 1, 2, and 3).

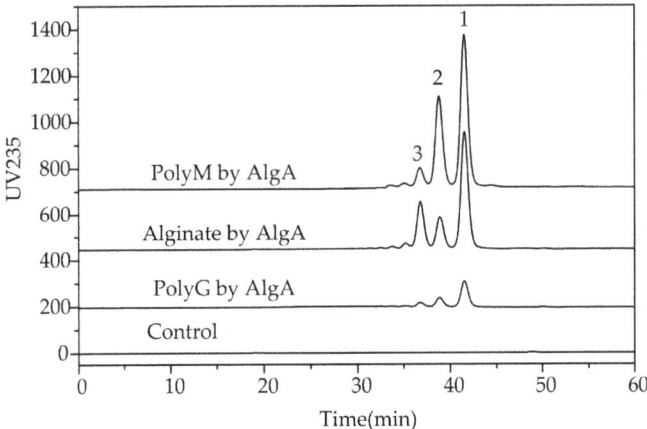

Figure 7. Final products of alginate, polyM, and polyG after degradation by AlgA. Oligosaccharide products were gel-filtered through a Superdex peptide 10/300 GL column and monitored at a wavelength of 235 nm.

To identify the final oligosaccharide products of AlgA degradation and to determine their DP, three major fractions were subjected to electrospray ionization mass spectrometry (ESI-MS). The molecular masses of oligosaccharides in peaks 1, 2, and 3 were determined to be 351.06, 527.09, and 703.12, respectively (Figure 8). These results indicated that the main degradation products are di-, tri- and tetra-saccharides. The relative contents of di-, tri- and tetra-saccharides in alginate were 61.19%, 15.59%, and 23.22%, respectively, whereas those of di-, tri- and tetra-saccharides in the polyM substrate were 58.30%, 34.26%, and 7.43%, respectively. Meanwhile, AlgA has limited activity toward polyG. Therefore, AlgA may be used in the production of mannuronic oligosaccharides from polyM blocks and the preparation of polyG blocks from sodium alginate via the degradation of polyM blocks. The products of mannuronic oligosaccharides and polyG blocks possess special biological activity and have potential applications in many fields. For example, mannuronate oligosaccharides can promote the secretion of multiple cytokines [7], and polyG demonstrates higher macrophage-stimulation activity than polyM [41].

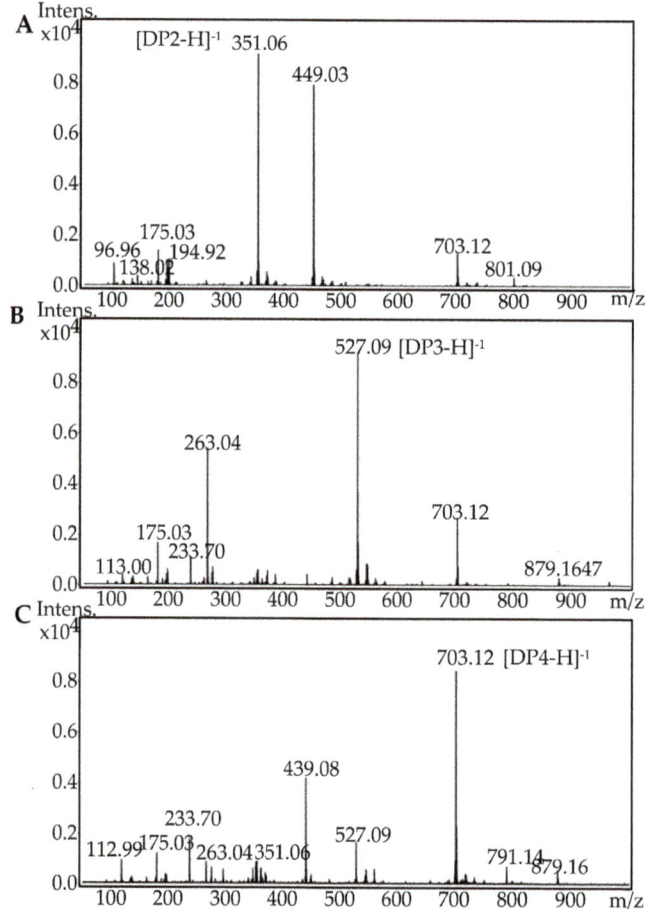

Figure 8. Electrospray ionization mass spectrometry (ESI-MS) analysis of the final oligosaccharide products. (**A**) Fraction peak 1 separated through fast protein liquid chromatography (FPLC), (**B**) fraction peak 2 separated through FPLC, and (**C**) fraction peak 3 separated through FPLC.

3. Materials and Methods

3.1. Materials

Sodium alginate derived from brown seaweed was purchased from Sigma (St. Louis, MO, USA). SOURCE™ 15Q 4.6/100 PE and Superdex peptide 10/300 gel filtration columns were purchased from GE HealthCare Bio-Sciences (Uppsala, Sweden). DNA polymerase, protein molecular weight markers, and polyacrylamide were purchased from New England Biolabs (Ipswich, MA, USA). Other chemicals and reagents used in this study were of analytical grade.

3.2. Screening and Identification of Strain Alg07

Sea mud and rotten kelp samples were collected from a seaweed farm in Weihai, China. Five grams of the samples were added to 45 mL of modified marine broth 2216 medium containing 5 g/L $(NH_4)_2SO_4$, 19.45 g/L NaCl, 12.6 g/L $MgCl_2 \cdot 6H_2O$, 6.64 g/L $MgSO_4 \cdot 7H_2O$, 0.55 g/L KCl, 0.16 g/L $NaHCO_3$, 1 g/L ferric citrate, and 10 g/L sodium alginate. After 48 h of enrichment at 30 °C, the culture was serially diluted with deionized water and spread on modified marine broth 2216 agar containing 10 g/L sodium alginate. The plates were incubated at 30 °C for two days until colonies appeared. Single colonies were inoculated into marine broth 2216 medium containing 10 g/L sodium alginate and incubated for 48 h at 30 °C. Then, the activities of alginate lyase in the supernatants were determined. The alginate lyase from one of the isolates, strain Alg07, showed the highest activity among all lyases.

To identify the Alg07 strain, the 16S rRNA gene of the strain was amplified through PCR by using universal primers. The purified PCR fragment was sequenced and compared with reported 16s rRNA sequences in Genbank by using BLAST. A phylogenetic tree was constructed using CLUSTAL X and MEGA 6.0 [42] through neighbor-joining method [43].

3.3. Production and Purification of AlgA

Strain Alg07 was cultured for 24 h at 30 °C and 180 rpm in the optimized liquid medium, which contained 1g/L peptone, 3 g/L yeast extract, 9 g/L sodium alginate, 5 g/L NaCl, 1 g/L $MgSO_4 \cdot 7H_2O$, 5 g/L KCl, and 4 g/L $CaCl_2$ (pH 6.5). The supernatant was collected after 30 min of centrifugation at 10,000 rpm and 4 °C and then concentrated and desalted using a tangential flow filtration system (Vivaflow 50, Sartorius, Goettingen, Germany). The concentrated solution was subjected to AKTA FPLC (GE Healthcare Life Science, Marlborough, MA, USA) equipped with a SOURCE™ (Barrie, ON, Canada) 15Q 4.6/100 PE column that had been equilibrated with 20 mM Tris–HCl buffer (pH 7.0). Adsorbed proteins were eluted with a linear gradient of 0–0.5 M NaCl in equilibrating buffer under a flow rate of 1 mL/min. Fractions possessing the highest specific activity among all fractions were pooled and dialyzed against 20 mM Tris-HCl buffer (pH 7.0) for further enzyme characterization. All purification procedures were performed at 4 °C. Protein concentration was determined with a protein quantitative analysis kit (Solarbio, Beijing, China) using bovine serum albumin as the calibration standard. The purity of the isolated alginate lyase was analyzed through 12.5% SDS-PAGE in accordance with the method of Laemmli (1970) [44].

3.4. Enzyme Activity Assay

To determine the activity of alginate lyase, 100 µL of appropriately diluted enzyme was added to 1900 µL of substrate solution containing 10 g/L sodium alginate, 20 mM Tris-HCl, and 200 mM NaCl (pH 7.5). The reaction was allowed to proceed for 20 min at 40 °C and terminated by the addition of 20 µL of 10 M NaOH. Absorbance at 235 nm was recorded. One unit (U) was defined as the amount of enzyme required to increase the absorbance at 235 nm by 0.1 per min. For kinetic parameter analysis, the 3,5-dinitrosalicylic acid method was performed to determine alginate lyase activity based on the release of reducing sugars from substrates [45]. One unit (U) was defined as the amount of enzyme required to release 1 µmol of reducing sugar per min. All enzyme reactions were performed in triplicate, and reaction parameters were expressed as mean ± standard deviation.

3.5. Characterization of AlgA

Enzyme reactions were carried out at different temperatures (25 °C–60 °C) to determine the effect of temperature on AlgA. To evaluate the thermal stability of AlgA, the enzyme was incubated for different intervals at 40 °C, 45 °C, and 50 °C. Then, the residual activities of the enzyme were tested. The activity of the enzyme stored at 4 °C was used to represent 100% enzyme activity. The effect of different pH values on AlgA was determined by calculating the residual activities of AlgA after 20 h of incubation at 4 °C in 20 mM sodium acetate (pH 3.0–6.0), Tris-HCl (pH 6.0–9.0), or glycine-NaOH (pH 9.0–11.0) buffers. The initial activities in different pH buffers represented 100% enzyme activity.

Enzyme reactions were performed in the presence of different concentrations of NaCl (0–500 mM) to evaluate the effect of NaCl on AlgA. To determine the effects of metal ions and EDTA on the activity of AlgA, the highest enzyme activity was considered as 100% enzyme activity. AlgA was subjected to an activity assay after 12 h of incubation at 4 °C in the presence of 2 mM of different metal ions and EDTA.

3.6. Analysis of the N-Terminal Amino Acid Sequence of AlgA

After SDS-PAGE, the protein band of AlgA was electro-transferred onto a polyvinylidene difluoride membrane (Imobulon; Millipore, Darmstadt, Germany). Amino acid sequences were determined with a PPSQ-31A protein sequencer (Shimadzu Corporation; Kyoto, Japan).

3.7. Substrate Specificity and Kinetic Parameters of AlgA

A standard enzymatic assay was performed to test the substrate preference of AlgA by using pectin, hyaluronan, chitin, agar, sodium alginate, polyM, and polyG as substrates. PolyM and polyG were prepared in accordance with the method of Haug et al. [46].

The kinetic parameters of AlgA toward alginate and polyM were determined by measuring the initial velocities of enzyme activity under various substrate concentrations and were calculated on the basis of the nonlinear regression fitting of the Michaelis–Menten equation using Prism 6 (GraphPad Software, Inc., La Jolla, CA, USA).

3.8. FPLC and ESI-MS Analysis of the Degradation Products of AlgA

To elucidate the mode of action of AlgA toward alginate, alginate degradation was performed at 40 °C with 10 g/L sodium alginate as a substrate. The reaction was initiated by the addition of 2 µg of purified AlgA in a 10 mL reaction volume. Reaction solutions were withdrawn at appropriate time intervals, and AlgA was inactivated by 5 min of boiling. The samples were analyzed through FPLC with a Superdex peptide 10/300 gel filtration column (GE Health, Marlborough, MA, USA) with 0.2 M ammonium bicarbonate as the mobile phase at a flow rate of 0.4 mL/min [34]. The reaction was monitored at 235 nm.

To determine the oligosaccharide compositions of the final digests, an excess of AlgA was used to completely degrade 10 g/L sodium alginate, polyM, or polyG. The reaction was carried out in a 10 mL reaction volume at 40 °C for 24 h and then terminated by boiling for 5 min. Oligosaccharides were separated through gel filtration as described above. Peak fractions containing unsaturated oligosaccharide products were collected and repeatedly freeze-dried to remove NH_4HCO_3 for ESI-MS analysis. The molecular weight of each oligosaccharide fraction was determined using the ESI-MS method on microTOF-Q II equipment (Bruker, Billerica, MA, USA) with the following conditions: capillary voltage of 4 kV, dry temperature of 180 °C, gas flow rate of 4.0 L/min, and scan range of 50–1000 m/z.

4. Conclusions

An alginate lyase–producing marine bacterium was isolated and identified as *Bacillus* sp. Alg07. AlgA, the alginate lyase derived from *Bacillus* sp. Alg07, was purified through the two simple steps of tangential flow filtration concentration and anion-exchange chromatography. The results of SDS-PAGE indicated that the molecular mass of AlgA is approximately 60 kDa. The optimal temperature and pH for the activity of AlgA is 40 °C and 7.5, respectively. The activity of AlgA is dependent on NaCl and is promoted by the addition of Mg^{2+} and Ca^{2+}. Under optimal conditions, the specific activity of AlgA reaches up to 8306.7 U/mg, which is the highest activity recorded for all reported alginate lyases. Moreover, AlgA is stable over a broad pH range (5.0–10.0) and under its optimal temperature (40 °C). AlgA is an endolytic polyM-specific alginate lyase and mainly produces disaccharides, trisaccharides, and tetrasaccharides from alginate and disaccharides and trisaccharides from polyM. The highly efficient and thermostable AlgA can have potential applications in the production of mannuronic oligosaccharides and polyG blocks from alginate.

Supplementary Materials: The following are available online at www.mdpi.com/1660-3397/16/3/86/s1, Figure S1: Purification of alginate lyase by anion exchange chromatograph (Source 15Q); Figure S2. The N-terminal amino acid sequence of the purified AlgA; Figure S3. Effects of substrate concentration on the activity of AlgA.

Acknowledgments: This work was supported by the National key R&D Program of China (2017YFD200900), the Science Technology Planning Project of Tianjin (16YFXTNC00160), and Youth Innovation Promotion Association, Chinese Academy of Sciences (to Yueming Zhu).

Author Contributions: Peng Chen and Yueming Zhu conceived and designed the experiments; Peng Chen performed the experiments; Peng Chen and Yan Zeng analyzed the data; Yan Men contributed reagents/materials/analysis tools; Peng Chen wrote the paper. Yuanxia Sun edited the paper.

Conflicts of Interest: The authors declare no conflict of interest.

References

1. Gacesa, P. Enzymic degradation of alginates. *Int. J. Biochem.* **1992**, *24*, 545–552. [CrossRef]
2. Mabeau, S.; Kloareg, B. Isolation and Analysis of the Cell Walls of Brown Algae: *Fucus spiralis, F. ceranoides, F. vesiculosus, F. serratus, Bifurcaria bifurcata* and *Laminaria digitata*. *J. Exp. Bot.* **1987**, *38*, 1573–1580. [CrossRef]
3. Gorin, P.A.J.; Spencer, J.F.T. Exocellular alginic acid from *Azotobacter vinelandii*. *Can. J. Chem.* **1966**, *44*, 993–998. [CrossRef]
4. Evans, L.R.; Linker, A. Production and Characterization of the Slime Polysaccharide of *Pseudomonas aeruginosa*. *J. Bacteriol.* **1973**, *116*, 915–924. [PubMed]
5. Scott, C.D. Immobilized cells: A review of recent literature. *Enzym. Microb. Technol.* **1987**, *9*, 66–72. [CrossRef]
6. Kawada, A.; Hiura, N.; Tajima, S.; Takahara, H. Alginate oligosaccharides stimulate VEGF-mediated growth and migration of human endothelial cells. *Arch. Dermatol. Res.* **1999**, *291*, 542–547. [CrossRef] [PubMed]
7. Yamamoto, Y.; Kurachi, M.; Yamaguchi, K.; Oda, T. Stimulation of multiple cytokine production in mice by alginate oligosaccharides following intraperitoneal administration. *Carbohydr. Res.* **2007**, *342*, 1133–1137. [CrossRef] [PubMed]
8. Iwasaki, K.I.; Matsubara, Y. Purification of Alginate Oligosaccharides with Root Growth-promoting Activity toward Lettuce. *Biosci. Biotechnol. Biochem.* **2000**, *64*, 1067–1070. [CrossRef] [PubMed]
9. Wang, Y.; Han, F.; Hu, B.; Li, J.; Yu, W. In vivo prebiotic properties of alginate oligosaccharides prepared through enzymatic hydrolysis of alginate. *Nutr. Res.* **2006**, *26*, 597–603. [CrossRef]
10. Falkeborg, M.; Cheong, L.Z.; Gianfico, C.; Sztukiel, K.M.; Kristensen, K.; Glasius, M.; Xu, X.; Guo, Z. Alginate oligosaccharides: enzymatic preparation and antioxidant property evaluation. *Food Chem.* **2014**, *164*, 185–194. [CrossRef] [PubMed]
11. An, Q.D.; Zhang, G.L.; Wu, H.T.; Zhang, Z.C.; Zheng, G.S.; Luan, L.; Murata, Y.; Li, X. Alginate-deriving oligosaccharide production by alginase from newly isolated *Flavobacterium* sp. LXA and its potential application in protection against pathogens. *J. Appl. Microbiol.* **2009**, *106*, 161–170. [CrossRef] [PubMed]
12. Kim, H.S.; Lee, C.G.; Lee, E.Y. Alginate lyase: Structure, property, and application. *Biotechnol. Bioprocess Eng.* **2011**, *16*, 843–851. [CrossRef]
13. Inoue, A.; Mashino, C.; Uji, T.; Saga, N.; Mikami, K.; Ojima, T. Characterization of an Eukaryotic PL-7 Alginate Lyase in the Marine Red Alga*Pyropia yezoensis*. *Curr. Biotechnol.* **2015**, *4*, 240–258. [CrossRef] [PubMed]

14. Suzuki, H.; Suzuki, K.; Inoue, A.; Ojima, T. A novel oligoalginate lyase from abalone, *Haliotis discus hannai*, that releases disaccharide from alginate polymer in an exolytic manner. *Carbohydr. Res.* **2006**, *341*, 1809–1819. [CrossRef] [PubMed]
15. Zhu, B.; Chen, M.; Yin, H.; Du, Y.; Ning, L. Enzymatic Hydrolysis of Alginate to Produce Oligosaccharides by a New Purified Endo-Type Alginate Lyase. *Mar. Drugs.* **2016**, *14*, 108. [CrossRef] [PubMed]
16. Inoue, A.; Takadono, K.; Nishiyama, R.; Tajima, K.; Kobayashi, T.; Ojima, T. Characterization of an Alginate Lyase, FlAlyA, from *Flavobacterium* sp. Strain UMI-01 and Its Expression in *Escherichia coli*. *Mar. Drugs* **2014**, *12*, 4693–4712. [CrossRef] [PubMed]
17. Singh, R.P.; Gupta, V.; Kumari, P.; Kumar, M.; Reddy, C.R.K.; Prasad, K.; Jha, B. Purification and partial characterization of an extracellular alginate lyase from *Aspergillus oryzae* isolated from brown seaweed. *J. Appl. Phycol.* **2011**, *23*, 755–762. [CrossRef]
18. Suda, K.; Tanji, Y.; Hori, K.; Unno, H. Evidence for a novel *Chlorella* virus-encoded alginate lyase. *FEMS Microbiol. Lett.* **1999**, *180*, 45–53. [CrossRef] [PubMed]
19. Wong, T.Y.; Preston, L.A.; Schiller, N.L. ALGINATE LYASE: Review of major sources and enzyme characteristics, structure-function analysis, biological roles, and applications. *Annu. Rev. Microbiol.* **2000**, *54*, 289–340. [CrossRef] [PubMed]
20. Cantarel, B.L.; Coutinho, P.M.; Rancurel, C.; Bernard, T.; Lombard, V.; Henrissat, B. The Carbohydrate-Active EnZymes database (CAZy): an expert resource for Glycogenomics. *Nucleic Acids Res.* **2009**, *37*, 233–238. [CrossRef] [PubMed]
21. Aarstad, O.A.; Tøndervik, A.; Sletta, H.; Skjåkbræk, G. Alginate Sequencing: An Analysis of Block Distribution in Alginates Using Specific Alginate Degrading Enzymes. *Biomacromolecules* **2012**, *13*, 106–116. [CrossRef] [PubMed]
22. Inoue, A.; Mashino, C.; Kodama, T.; Ojima, T. Protoplast preparation from *Laminaria japonica* with recombinant alginate lyase and cellulase. *Mar. Biotechnol.* **2011**, *13*, 256–263. [CrossRef] [PubMed]
23. Islan, G.A.; Martinez, Y.N.; Illanes, A.; Castro, G.R. Development of novel alginate lyase cross-linked aggregates for the oral treatment of cystic fibrosis. *RSC Adv.* **2014**, *4*, 11758–11765. [CrossRef]
24. Li, L.; Jiang, X.; Guan, H.; Wang, P. Preparation, purification and characterization of alginate oligosaccharides degraded by alginate lyase from *Pseudomonas* sp. HZJ 216. *Carbohydr. Res.* **2011**, *346*, 794–800. [CrossRef] [PubMed]
25. Wang, L.; Li, S.; Yu, W.; Gong, Q. Cloning, overexpression and characterization of a new oligoalginate lyase from a marine bacterium, *Shewanella* sp. *Biotechnol. Lett.* **2015**, *37*, 665–671. [CrossRef] [PubMed]
26. Kim, H.T.; Ko, H.J.; Kim, N.; Kim, D.; Lee, D.; Choi, I.G.; Woo, H.C.; Kim, M.D.; Kim, K.H. Characterization of a recombinant endo-type alginate lyase (Alg7D) from *Saccharophagus degradans*. *Biotechnol. Lett.* **2012**, *34*, 1087–1092. [CrossRef] [PubMed]
27. Li, S.; Wang, L.; Hao, J.; Xing, M.; Sun, J.; Sun, M. Purification and Characterization of a New Alginate Lyase from Marine Bacterium *Vibrio* sp. SY08. *Mar. Drugs* **2017**, *15*, 1. [CrossRef] [PubMed]
28. Wang, Y.; Guo, E.W.; Yu, W.G.; Han, F. Purification and characterization of a new alginate lyase from a marine bacterium *Vibrio* sp. *Biotechnol. Lett.* **2013**, *35*, 703–708. [CrossRef] [PubMed]
29. Kobayashi, T.; Uchimura, K.; Miyazaki, M.; Nogi, Y.; Horikoshi, K. A new high-alkaline alginate lyase from a deep-sea bacterium *Agarivorans* sp. *Extremophiles* **2009**, *13*, 121–129. [CrossRef] [PubMed]
30. Xiao, T.F.; Hong, L.; Sang, M.K. Purification and characterization of a Na^+/K^+ dependent alginate lyase from turban shell gut *Vibrio* sp. YKW-34. *Enzym. Microb. Technol.* **2007**, *41*, 828–834.
31. Zhu, B.; Yin, H. Alginate lyase: Review of major sources and classification, properties, structure-function analysis and applications. *Bioengineered* **2015**, *6*, 125–131. [CrossRef] [PubMed]
32. Yamasaki, M.; Moriwaki, S.; Miyake, O.; Hashimoto, W.; Murata, K.; Mikami, B. Structure and function of a hypothetical *Pseudomonas aeruginosa* protein PA1167 classified into family PL-7. *J. Biol. Chem.* **2004**, *279*, 31863–31872. [CrossRef] [PubMed]
33. Lee, S.I.; Choi, S.H.; Lee, E.Y.; Kim, H.S. Molecular cloning, purification, and characterization of a novel polyMG-specific alginate lyase responsible for alginate MG block degradation in *Stenotrophomas maltophilia* KJ-2. *Appl. Microbiol. Biotechnol.* **2012**, *95*, 1643–1653. [CrossRef] [PubMed]
34. Zhu, B.; Tan, H.; Qin, Y.; Xu, Q.; Du, Y.; Yin, H. Characterization of a new endo-type alginate lyase from *Vibrio* sp. W13. *Int. J. Biol. Macromol.* **2015**, *75*, 330–337. [CrossRef] [PubMed]

35. Thomas, F.; Lundqvist, L.C.E.; Jam, M.; Jeudy, A.; Barbeyron, T.; Sandström, C.; Michel, G.; Czjzek, M. Comparative Characterization of Two Marine Alginate Lyases from *Zobellia galactanivorans* Reveals Distinct Modes of Action and Exquisite Adaptation to Their Natural Substrate. *J. Biol. Chem.* **2013**, *288*, 23021–23037. [CrossRef] [PubMed]
36. Li, S.; Yang, X.; Bao, M.; Wu, Y.; Yu, W.; Han, F. Family 13 carbohydrate-binding module of alginate lyase from *Agarivorans* sp. L11 enhances its catalytic efficiency and thermostability, and alters its substrate preference and product distribution. *FEMS Microbiol. Lett.* **2015**, *362*. [CrossRef] [PubMed]
37. Han, W.; Gu, J.; Cheng, Y.; Liu, H.; Li, Y.; Li, F. Novel Alginate Lyase (Aly5) from a Polysaccharide-Degrading Marine Bacterium, *Flammeovirga* sp. Strain MY04: Effects of Module Truncation on Biochemical Characteristics, Alginate Degradation Patterns, and Oligosaccharide-Yielding Properties. *Appl. Environ. Microbiol.* **2015**, *82*, 364–374. [CrossRef] [PubMed]
38. Li, J.W.; Dong, S.; Song, J.; Li, C.B.; Chen, X.L.; Xie, B.B.; Zhang, Y.Z. Purification and Characterization of a Bifunctional Alginate Lyase from *Pseudoalteromonas* sp. SM0524. *Mar. Drugs* **2011**, *9*, 109–123. [CrossRef] [PubMed]
39. Uchimura, K.; Miyazaki, M.; Nogi, Y.; Kobayashi, T.; Horikoshi, K. Cloning and sequencing of alginate lyase genes from deep-sea strains of *Vibrio* and *Agarivorans* and characterization of a new *Vibrio* enzyme. *Mar. Biotechnol.* **2010**, *12*, 526–533. [CrossRef] [PubMed]
40. Xiao, L.; Han, F.; Yang, Z.; Lu, X.Z.; Yu, W.G. A Novel Alginate Lyase with High Activity on Acetylated Alginate of *Pseudomonas aeruginosa* FRD1 from *Pseudomonas* sp. QD03. *World J. Microbiol. Biotechnol.* **2006**, *22*, 81–88. [CrossRef]
41. Ueno, M.; Hiroki, T.; Takeshita, S.; Jiang, Z.; Kim, D.; Yamaguchi, K.; Oda, T. Comparative study on antioxidative and macrophage-stimulating activities of polyguluronic acid (PG) and polymannuronic acid (PM) prepared from alginate. *Carbohydr. Res.* **2012**, *352*, 88–93. [CrossRef] [PubMed]
42. Tamura, K.; Stecher, G.; Peterson, D.; Filipski, A.; Kumar, S. MEGA6: Molecular Evolutionary Genetics Analysis Version 6.0. *Mol. Biol. Evol.* **2013**, *30*, 2725–2729. [CrossRef] [PubMed]
43. Saitou, N.; Nei, M. The neighbor-joining method: A new method for reconstructing phylogenetic trees. *Mol. Biol. Evol.* **1987**, *4*, 406–425. [PubMed]
44. Laemmli, U.K. Cleavage of Structural Proteins during the Assembly of the Head of Bacteriophage T4. *Nature* **1970**, *227*, 680–685. [CrossRef] [PubMed]
45. Miller, G.L. Use of Dinitrosalicylic Acid Reagent for Determination of Reducing Sugar. *Anal. Biochem.* **1959**, *31*, 426–428. [CrossRef]
46. Haug, A.; Larsen, B. A study on the constitution of alginic acid by partial acid hydrolysis. *Acta Chem. Scand.* **1966**, *20*, 183–190. [CrossRef]

© 2018 by the authors. Licensee MDPI, Basel, Switzerland. This article is an open access article distributed under the terms and conditions of the Creative Commons Attribution (CC BY) license (http://creativecommons.org/licenses/by/4.0/).

Article

Characterization of Properties and Transglycosylation Abilities of Recombinant α-Galactosidase from Cold-Adapted Marine Bacterium *Pseudoalteromonas* KMM 701 and Its C494N and D451A Mutants

Irina Bakunina [1,*], Lubov Slepchenko [1,2], Stanislav Anastyuk [1], Vladimir Isakov [1], Galina Likhatskaya [1], Natalya Kim [1], Liudmila Tekutyeva [2], Oksana Son [2] and Larissa Balabanova [1,2]

[1] Laboratory of Enzyme Chemistry, Laboratory of Marine Biochemistry, Laboratory of Bioassays and Mechanism of action of Biologically Active Substances, Laboratory of Instrumental and Radioisotope Testing Methods, Group of NMR-Spectroscopy of G.B. Elyakov Pacific Institute of Bioorganic Chemistry, Far Eastern Branch, Russian Academy of Sciences, Vladivostok 690022, Russia; lubov99d@mail.ru (L.S.); Sanastyuk@piboc.dvo.ru (S.A.); ivv43@mail.ru (V.I.); galin56@mail.ru (G.L.); natalya_kim@mail.ru (N.K.); lbalabanova@mail.ru (L.B.)

[2] School of Economics and Management, School of Natural Sciences of Far Eastern Federal University, Russky Island, Vladivostok 690022, Russia; tekuteva.la@dvfu.ru (L.T.); oksana_son@bk.ru (O.S.)

* Correspondence: bakun@list.ru; Tel.: +7-432-231-0705-3; Fax: +7-432-231-0705-7

Received: 16 August 2018; Accepted: 21 September 2018; Published: 24 September 2018

Abstract: A novel wild-type recombinant cold-active α-D-galactosidase (α-PsGal) from the cold-adapted marine bacterium *Pseudoalteromonas* sp. KMM 701, and its mutants D451A and C494N, were studied in terms of their structural, physicochemical, and catalytic properties. Homology models of the three-dimensional α-PsGal structure, its active center, and complexes with D-galactose were constructed for identification of functionally important amino acid residues in the active site of the enzyme, using the crystal structure of the α-galactosidase from *Lactobacillus acidophilus* as a template. The circular dichroism spectra of the wild α-PsGal and mutant C494N were approximately identical. The C494N mutation decreased the efficiency of retaining the affinity of the enzyme to standard p-nitrophenyl-α-galactopiranoside (pNP-α-Gal). Thin-layer chromatography, matrix-assisted laser desorption/ionization mass spectrometry, and nuclear magnetic resonance spectroscopy methods were used to identify transglycosylation products in reaction mixtures. α-PsGal possessed a narrow acceptor specificity. Fructose, xylose, fucose, and glucose were inactive as acceptors in the transglycosylation reaction. α-PsGal synthesized -α(1→6)- and -α(1→4)-linked galactobiosides from melibiose as well as -α(1→6)- and -α(1→3)-linked p-nitrophenyl-digalactosides (Gal$_2$-pNP) from pNP-α-Gal. The D451A mutation in the active center completely inactivated the enzyme. However, the substitution of C494N discontinued the Gal-α(1→3)-Gal-pNP synthesis and increased the Gal-α(1→4)-Gal yield compared to Gal-α(1→6)-Gal-pNP.

Keywords: α-D-galactosidase; homology model; GH 36 family; mutation; transglycosylation; marine bacteria; *Pseudoalteromonas* sp. KMM 701

1. Introduction

α-D-Galactosidases (EC 3.2.1.22) catalyze the hydrolysis of the nonreducing terminal α-D-galactose (Gal) from α-D-galactosides, galactooligosaccharides, and polysaccharides, such as galactomannans, galactolipids, and glycoproteins. According to the classification of carbohydrate-active enzymes (CAZy) [1], α-D-galactosidases mostly belong to 27, 36, and 110 families of glycoside hydrolases

(GH). They are found also in the GH 4, GH 57, and GH 97 families. The GH 27 and GH 36 enzymes, with a common mechanism of catalysis, have the protein structural $(\beta/\alpha)_8$-barrel fold in the catalytic domain and similar topology of their active centers, typical for a clan GH D [2]. The GH 27 and GH 36 family members are classical retaining glycoside hydrolyses in accordance with Koshland's classification [3]. These enzymes catalyze the hydrolysis of O-glycosidic bonds by a double displacement mechanism through the galactosyl-enzyme covalent intermediate, as well as the transglycosylation reaction under the specific conditions [4].

α-D-Galactosidases are widespread among terrestrial plants, animals, and microorganisms. These enzymes have found many practical uses in different fields from biomedicine to enzymatic synthesis [4]. The enzymes occur frequently in marine bacteria, especially in γ-Proteobacteria and Bacteroidetes [5–9]. Currently, the genes encoding these enzymes can be found in the genomes of marine bacteria, annotated in the National Center for Biotechnology Information NCBI database. For the first time, α-PsGal was isolated from a cold-adapted marine bacterium *Pseudoalteromonas* sp. strain KMM 701 inhabiting in the cold water in the Sea of Okhotsk. The enzyme attracted our attention due to its ability to reduce the serological activity of B red blood cells. The marine bacterium's α-PsGal was more efficient in the model of B-erythrocyte antigen, than a well-known α-D-galactosidase from green coffee beans, which was usually used in experiments on transformation of donor blood erythrocytes for intravenous injection [10]. The enzyme also interrupted the adhesion of *Corynebacterium diphtheria* to buccal epithelium cells at neutral pH values [11], as well as stimulated the growth of biofilms of some bacteria [12]. These properties of the enzyme determined the possible directions for its practical application in biomedicine. According to the structural CAZy classification, α-PsGal belongs to the GH 36 family [11]. The enzyme is a retaining glycoside hydrolase [13], cleaving the terminal Gal from melibiose Gal-α(1→6)-Glc, raffinose Gal-α(1→6)-Glc-β(1→4)-(Fru), digalactoside Gal-α(1→3)-Gal, and B-trisaccharide Gal-α(1→3)-(Fuc-α(1→2)-Gal) [10]. However, the most important glycosynthase properties of the enzyme have yet to be studied.

The present article aimed to compare the properties of recombinant α-D-galactosidase from marine bacterium *Pseudoalteromonas* sp. KMM 701 (α-PsGal) and its mutants, where the predicted functionally important residues D451 and C494 of the active center were replaced by the less reactive alanine (A) and asparagine (N) residues, respectively. Major attention was focused on the regioselectivity of the transglycosylation reaction.

2. Results

2.1. Bioinformatics Analysis and Homology Modeling of α-PsGal Protein 3D-Structure

Bioinformatics analysis and homology modeling of the protein structure was completed to elucidate the amino acid residues roles of α-PsGal for catalysis. The results of homology modeling of the α-PsGal protein three-dimensional (3D) structure are shown in Figure 1.

The homology model of the α-PsGal 3D-structure was constructed by the package Molecular Operating Environment version 2018.01 (MOE) [14] using the crystal structure of α-galactosidase from *Lactobacillus acidophilus* of the GH 36 family [15] as a template (Figure 1a). The amino acid sequence of α-galactosidase from the marine bacterium has 28.7% identity and 44% similarity with the sequence of the prototype. The superposition (root mean squared difference (RMSD) of the C_α-atoms = 0.8 Å) of the α-PsGal homology model with the template active site revealed that the D451 and C494 residues superimposed on the nucleophile/base D482 and the substrate binding C530 residues in the template, respectively (Figure 1b) [15]. Thus, the predicted structure of α-PsGal is applicable for in silico mutagenesis and molecular docking studies. The evidence from Figure 1c suggests that C494 takes part in forming a network of hydrogen bonds between the catalytic residues D451, D516, and the hydrolysis product D-galactose (D-Gal). To eliminate the roles of D451 and C494 residues, they were substituted by A451 and N494, respectively.

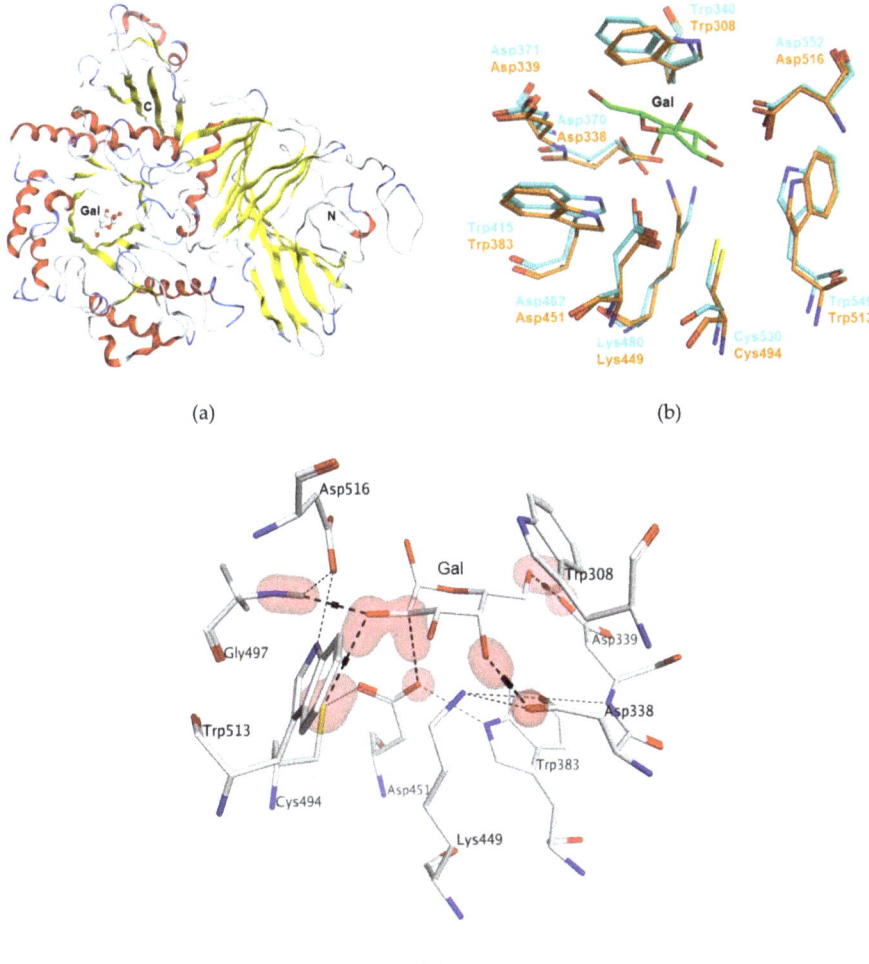

Figure 1. Homology model of α-PsGal three-dimensional (3D) structure generated using X-ray structure of the α-galactosidase of *Lactobacillus acidophilus* (PDB ID 2XN2) as a template: (**a**) 3D-model of α-PsGal structure in a ribbon diagram representation: α-helixes (red), β-strands (yellow), coils (white), and turns (blue); (**b**) superimposition of the α-PsGal homology model (orange) with template active sites (turquoise); D-galactose is shown by sticks (green); and (**c**) the binding site of D-galactose in the active center of α-PsGal homology model. Hydrogen-bond contacts were determined using the Protein Contacts module of Molecular Operating Environment version 2018.01 (MOE) program (Chemical Computing Group ULC: 1010 Sherbrooke St. West, Suite #910, Montreal, QC, Canada, H3A 2R7, 2018) and are shown with a dashed line.

2.2. Enzyme Production and Purification

The recombinant wild α-PsGal and its mutants D451A and C494N were expressed and purified successfully as soluble proteins with 97% purity according to the sodium dodecyl sulfate polyacrylamide gel electrophoresis (SDS-PAGE) data. The gel electorophoregrams of the enzyme preparations were obtained in different experiments and are summarized in Figure 2.

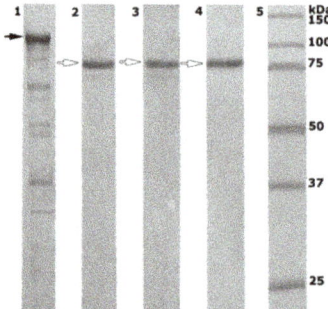

Figure 2. SDS-PAGE (12.5%) of the recombinant wild α-PsGal (lanes 1 and 2 before and after final purification stage, respectively) and its D451A and C494N mutants after the final stages of purification (lanes 3 and 4, respectively); molecular weight markers are shown in lane 5.

The molecular weight of the protein fraction of the cellular extract *Escherichia coli* Rosetta (DE3)/40Gal (Figure 2, lane 1) corresponds to the chimeric recombinant α-PsGal fused with the plasmid pET-40 b (+) chaperone protein DsbC overhang (80 kDa (α-PsGal) + 32.5 kDa (DsbC) = 112.5 kDa). After the final stage of purification and treatment with enterokinase, the molecular weight of mature recombinant proteins α-PsGal, D451A, and C494N were ca. 80 kDa (Figure 2, lanes 2–4, respectively).

2.3. Properties of Recombinant Wild-Type α-PsGal and Mutant C494N

The mutant D451A did not exhibit any hydrolytic activity against either melibiose or pNP-α-Gal, indicating the extreme importance of this residue for the functioning of this enzyme. The specific activities of the recombinant wild α-PsGal and mutant C494N, with the use of pNP-α-Gal as a substrate, were 90.0 and 0.87 U/mg, respectively. Further comparative studies showed some similarities and differences in the enzymatic properties of the wild recombinant α-PsGal and its mutant C494N.

2.3.1. Circular Dichroism Spectra of Wild-Type α-PsGal and Mutant C494N

To identify the similarity of the secondary structure of the wild α-PsGal and C494N mutant; circular dichroism (CD) spectroscopy was used (Figure 3).

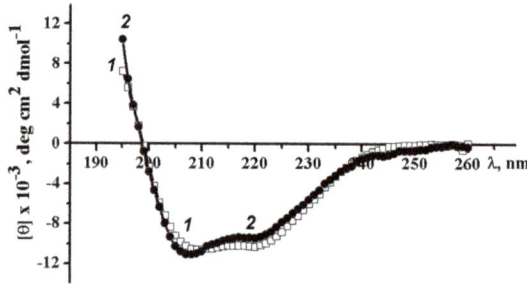

Figure 3. Circular dichroism spectra of wild α-PsGal (1) and C494N mutant (2) with 0.1 M sodium phosphate buffer (pH 7.0), 25 °C, and 0.1 cm cell.

The CD spectra of the wild α-PsGal and C494N mutant were approximately identical binshape and amplitude of the bands (Figure 3, spectrum 1 and 2, respectively). Calculation of the secondary protein structure elements (see p. 4.4.1.) indicated the presence of 27.9% and 27.9% α-helices, 19.9% and 21.4% β-structures, and 28.9% and 28.9% disordered structure, including 23.3% and 21.8% β-turns for the wild α-PsGal and C494N mutant, respectively. The determination of the tertiary structure class using the same software package established that the wild α-PsGal and C494N mutant belong to α + β tertiary structure class of proteins. Thus, the C494N point mutation in the active site does not significantly affect the secondary structure of the enzyme as a whole.

2.3.2. Effect of pH on Wild-Type α-PsGal and C494N Mutant Activities

It is evident from Figure 4 that the wild α-PsGal (Figure 4a) retains activity in a wide pH range (6.5–8.0).

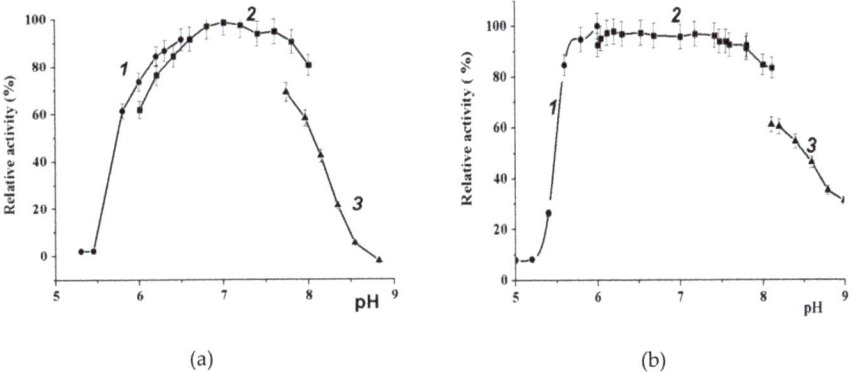

Figure 4. Effect of pH on the activity of enzymes: (**a**) wild-type α-PsGal and (**b**) mutant C494N. Fragments of curves correspond to 0.1 M sodium citrate buffer (1), 0.1 M sodium phosphate buffer (2), and 0.1 M Tris HCl buffer (3).

Citrate and phosphate anions were preferable for enzyme activity. The tris ion was an effective inhibitor of the activity of wild α-PsGal and mutant C494N. The replacement of cysteine 494 with asparagine residue did not significantly change the pH effect on the enzyme activity (Figure 4).

2.3.3. Effect of Temperature on Wild-Type α-PsGal and Mutant C494N Activities

It is evident that wild α-PsGal is a cold-active enzyme due to retaining about 30% of its activity at 5 °C (Figure 5a). The activity of the mutant C494N reached a maximum at higher temperature values than wild α-PsGal (Figure 5b).

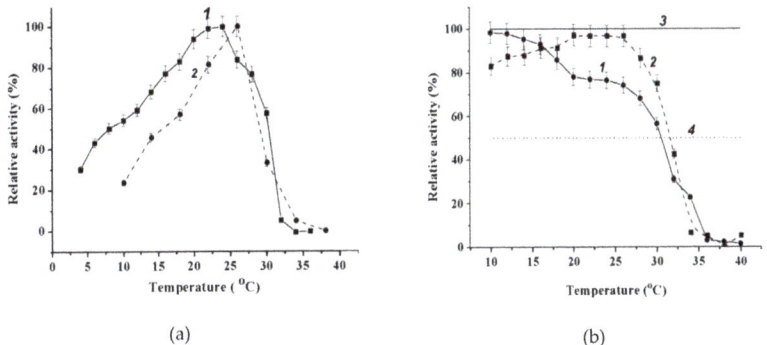

Figure 5. Effect of temperature on activity of enzymes: (**a**) the dependence of relative activity on temperature of wild α-PsGal (1) and C494N mutant (2) and (**b**) thermal stability of wild α-PsGal (1) and C494N mutant (2). The solid line (3) indicates 100% activity and the dashed line (4) indicates 50% activity.

Although the middle of the temperature transition lay at the same temperature of ~32 °C, the temperature inactivation in the wild α-PsGal started at lower temperatures than in the C494N mutant. Both the recombinant wild α-PsGal and the C494N mutant proved to be more thermostable enzymes than the natural α-galactosidase from the marine bacterium *Pseudoalteromonas* sp. KMM 701 [10].

2.3.4. Kinetic Parameters of Catalytic Reactions for Wild-Type α-PsGal and Mutant C494N

Michaelis-Menten constant (K_m) and maximal rate (V_{max}) of pNP-α-Gal hydrolysis were defined from Lineweaver-Burk graphs.

Catalytic parameters of the reaction are summarized in Table 1.

Table 1. Catalytic properties of wild α-PsGal and mutant C494N.

Enzyme	K_m (mM)	V_{max} (μmol min^{-1} ml^{-1})	k_{cat} (s^{-1})	k_{cat}/K_m (mM^{-1} s^{-1})
Wild-type α-PsGal	0.40 ± 0.030	0.32 ± 0.003	324.4 ± 3.5	820
C494N mutant	0.30 ± 0.005	0.024 ± 0.0003	2.84 ± 0.02	3.86

Conditions of reactions: 0.05 M sodium phosphate, pH 7.0, 20 °C; concentration of α-PsGal: 34 μ/mL, mutant C494N: 22 μ/mL.

It is evident from Table 1 that the replacement of cysteine 494 in the active site of α-PsGal with asparagine residue leads to an approximately 200-fold decrease in the efficiency (k_{cat}/K_m) of the enzyme retaining the identical affinity (K_m) of both enzymes to the standard substrate pNP-α-Gal. This indicates the extreme importance of C494 in the manifestation of α-PsGal activity (Table 1).

2.3.5. Theoretical Model of the D-Gal Complexes with Wild α-PsGal and Mutant C494N

Figure 6 shows two-dimensional (2D) diagrams of the D-Gal complexes with the active center of the wild α-PsGal (Figure 6a) and mutant C494N (Figure 6b) built by molecular docking in the MOE program.

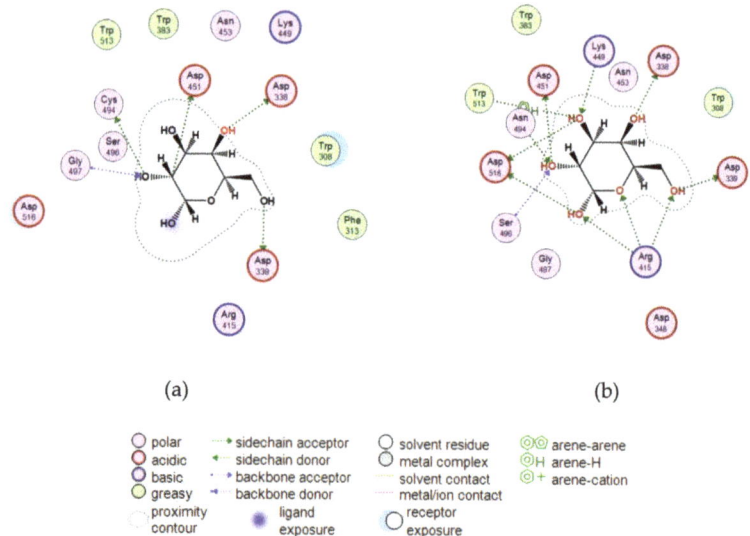

Figure 6. Two-dimensional (2D) diagrams of the D-Gal binding sites in (a) wild α-PsGal and (b) mutant C494N.

In silico analysis of the wild α-PsGal-D-Gal and mutant C494N-D-Gal complexes (Figure 6) showed that substitution C494N led to the emergence of new hydrogen bonds (Figure 6b) and to the increase in binding energy of the reaction product D-Gal in the binding site of the enzyme (Table S1). This probably reflected a decrease in the values of the k_{cat} and k_{cat}/K_m constants of the enzyme.

2.4. Acceptor Specificity of Recombinant α-PsGal

The results of the preliminary determination of the acceptor specificity in the transglycosylation reaction of the wild α-PsGal showed that the enzyme possesses narrow acceptor specificity, as new sugars fructose, xylose, fucose, and glucose were not found in the experimental conditions (Table 2).

Table 2. Acceptor specificity of recombinant wild-type α-PsGal.

Substrate	Acceptor	New Products
pNP-α-Gal	D-Glucose	−
pNP-α-Gal	D-Fructose	−
pNP-α-Gal	D-Xylose	−
pNP-α-Gal	L-Fucose	−
pNP-α-Gal	D-Galactose	+
Gal-α(1→6)-Glcα,β	D-Xylose	−
Gal-α(1→6)-Glcα,β	D-Fructose	−
Gal-α(1→6)-Glcα,β	L-Fucose	−

(−) not detected; (+) detected.

2.5. Production of Transglysolylation Reactions

The transglycosylation properties of the purified wild α-PsGal, mutant D451A, and C494N were tested using melibiose (1.10 mmol/mL) and pNP-α-Gal (0.25 mmol/mL) as substrates at pH 7.0 and 8 °C and 20 °C. The concentration of the substrates significantly exceeded the enzyme concentration. The composition of incubation mixtures and units of the enzymes are shown in Table 3.

Table 3. Composition of reaction mixtures for monitoring of transglycosylation with various forms of the recombinant α-galactosidase from the marine bacterium *Pseudoalteromonas* sp. KMM 701.

Enzyme	Substrate	Acceptor	Substrate (mol)	Acceptor (mol)	V (mL)	U or mg	t (°C)	τ (h)
Wild	Gal-α(1→6)-Glcα,β	-	0.11	0	0.1	2.2	20	48
Wild	Gal-α(1→6)-Glcα,β	-	0.31	0	0.3	4.5	8	7 days
Wild	pNP-α-Gal	-	0.05	0	0.2	1.0	20	48
Wild	pNP-α-Gal	-	0.07	0	0.2	0.015	8	7 days
Wild	Gal-α(1→6)-Glcα,β	pNP-α-Gal	0.06	0.13	0.3	2.6	20	48
C494N	pNP-α-Gal	-	0.07	0	0.3	0.012	8	7 days
D451A	Gal-α(1→6)-Glcα,β	NaN₃	0.06	0.3	0.1	(0.008)	20	75
D451A	pNP-α-Gal	NaN₃	0.07	0.3	0.3	(0.024)	20	75
D451A	pNP-α-Gal	-	0.07	0	0.2	(0.016)	20	75

Since the activity of mutant C494N was much lower compared with the wild type (Table 3), the reaction mixtures containing weakly active C494N (0.012 U) and wild-type α-PsGal with an activity of 0.015 U were incubated for seven days. In order to avoid the thermal inactivation of the enzymes, the reactions were carried out in a refrigerator at 8 °C. Therefore, the products from the reactions at low (8 °C) and moderate (20 °C) temperatures were compared. The conditions of rescue experiments and the investigation of glycosynthase properties of the mutant D451A in the presence of sodium azide, as an external nucleophile, are also shown in Table 3.

2.5.1. Thin Layer Chromatography (TLC)

The results of TLC analyses of the reaction mixtures after the action of recombinant α-PsGal, as well as D451A and C494N mutants, on melibiose (Gal-α(1→6)-Glc), and pNP-α-Gal were obtained under different conditions (Figure 7).

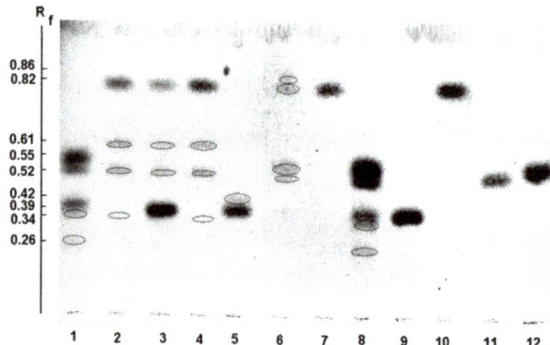

Figure 7. Thin layer chromatography (TLC) profiles of the hydrolysis and transglycosylation products produced by recombinant α-PsGal. Lanes 1 and 8—α-PsGal with Gal-α(1→6)-Glcα,β at 20 °C and 8 °C, respectively. Lanes 2 and 4—α-PsGal with pNP-α-Gal at 20 °C and 8 °C, respectively. Lane 3—α-PsGal with mixture of Gal-α(1→6)-Glcα,β/pNP-α-Gal at 20 °C. Lane 5—D451A mutant with Gal-α(1→6)-Glcα,β/NaN$_3$. Lane 6—D451A mutant with pNP-α-Gal/NaN$_3$. Lane 7—D451A mutant with pNP-α-Gal. Blank mixtures: lane 9—Gal-α(1→6)-Glcα,β, lane 10—pNP-α-Gal, lane 11—Gal, and lane 12—Glc.

Lanes 1–8 correspond to each experiment. The melibiose, pNP-α-Gal, galactose (Gal), and glucose (Glc) standards (Figure 7, lanes 9–12, respectively) spots on the chromatogram had retardation factors (R_f) of 0.39, 0.83, 0.52, and 0.56, respectively. The conditions of reactions are listed in Table 3.

After separation of the reaction mixtures with the melibiose processed by the recombinant α-PsGal at 20 °C and 8 °C (Figure 7; lanes 1 and 8; respectively), two new spots with R_f 0.35 and 0.26 occurred in the chromatogram; in addition to the hydrolysis products Gal and Glc (R_f 0.52 and 0.55; respectively). These new sugars could be interpreted as transglycosylation products. In the mixture of the reaction products of pNP-α-Gal treated with the recombinant α-PsGal at 20 °C and 8 °C; the spots with R_f 0.61 and 0.34, corresponding to new sugars along with the spots of pNP-α-Gal and Gal (R_f = 0.82 and 0.52; respectively), were observed (Figure 7; lanes 2 and 4; respectively). It is interesting to note that the compound with the R_f value of 0.34 was not formed by the action of the recombinant wild α-PsGal on the mixture of melibiose (donor)/pNP-α-Gal (acceptor) (Figure 7; lane 3). The D451A mutant showed no activity toward pNP-α-Gal (Figure 7; lane 7). Unfortunately, reactivation experiments with external nucleophile sodium azide failed to restore the activity of the nucleophile mutant enzyme (data not shown). However, in the presence of sodium azide (Figure 7; lanes 5 and 6), the traces of unidentified compounds (R_f value of 0.42 and 0.86; respectively) were formed under an action of D451A on melibiose and pNP-α-Gal

2.5.2. MALDI Mass Spectrometry

Matrix-assisted laser desorption/ionization (MALDI) mass spectra were recorded for five samples (S1): α-PsGal with a mixture of melibiose and pNP-α-Gal, α-PsGal with pNP-α-Gal, D451A with pNP-α-Gal and NaN$_3$, C494N with pNP-α-Gal, as well as standard mixtures: melibiose, pNP-α-Gal, Gal, and Glc. The molecular weights of sugars (both products of hydrolysis and substrates) were registered as sodium adducts $[M + Na]^+$ in positive-ion mode, where M represents the neutral molecule: $[Hex + Na]^+$ at m/z 203.06, $[Hex_2 + Na]^+$ at m/z 365, $[pNP-Hex_2 + Na]^+$ at m/z 486, and $[Hex_3 + Na]^+$ at m/z 527. pNP-α-Gal was found as $[pNP-α-Gal + Na]^+$ ion at m/z 324. MALDI mass spectra showed only the semiqualitative composition of the transglycosylation products (Table 4).

Table 4. TLC and matrix-assisted laser desorption/ionization (MALDI)-mass spectrometry (MS) characteristics of reaction products of wild α-PsGal and its mutant C494N.

Enzyme	Substrate	Product Number	TLC (R_f)	MALDI-MS (m/z)
Wild α-PsGal	Gal-α(1→6)-Glcα,β		0.39	365
		1	0.52	203
		2	0.55	203
		3	0.35	365
		4	0.35	365
		5	0.26	527
	pNP-α-Gal		0.83	324
		1	0.52	203
		6	0.61	486
		4	0.34	365
Mutant C494N	pNP-α-Gal		–	324
		1	–	203
		3	–	365
		6	–	486

According to the mass spectral data, the signal of the [Hex_2 + Na]$^+$ ion at m/z 365 of disaccharide (Hex_2) was major in the MALDI MS of the reaction mixture of α-PsGal with melibiose/pNP-α-Gal. In addition, there were signals of the new [pNP-Hex_2 + Na]$^+$ and [Hex + Na]$^+$ ions in the spectrum (Figure S1a) at m/z 486 and 203, respectively. Along with signals of the hydrolysis product [Hex + Na]$^+$ ion at m/z 203 and the remainder of the substrate [pNP-Hex + Na]$^+$ ion at m/z 324, new signals of [Hex_2 + Na]$^+$ and [pNP-Hex_2 + Na]$^+$ ions were observed in the matrix-assisted laser desorption ionization-mass spectroscopy (MALDI-MS) of the reaction mixture of the wild α-PsGal with pNP-α-Gal at 8 °C (Figure S1b). The signals of new carbon compounds were not detected in the MALDI-MS of the product mixture obtained under the action of mutant D451A on pNP-α-Gal in the presence of NaN$_3$ (Figure S1c). The main signal of the [Hex_2 + Na]$^+$ ion at m/z 365 was identified in the spectrum of the reaction mixture of the C494N mutant with pNP-α-Gal. The spectrum also contained two minor signals of [pNP-Hex_2 + Na]$^+$ at m/z 486 and [Hex + Na]$^+$ ion at m/z 203 (Figure S1d).

To identify the regioselectivity of the transglycosylation, we used tandem electrospray ionization mass spectrometry (EISMS/MS) with collisional induced dissociation (CID) in positive ion mode. The EISMS profiles of the reaction products are illustrated in Figures S2, S3, and S4. The linkage identifications were based on the fragmentation rules described earlier [16] for negative ion mode, and further supported by positive ion mode [17]. In brief, the absence of fragment ions from cross-ring cleavages in a disaccharide suggests a 1,3-type linkage, the $^{0,2}A_2$-type fragment ion suggests 1,4-type linkages, and both $^{0,2}A_2$ and $^{0,3}A_2$-type ions suggest 1,6-type linkages in disaccharides. The nomenclature for the mass spectrometric fragmentation of glycoconjugates was suggested by Domon and Costello [18].

Ion signals of [Hex_n + Na]$^+$, n = 1–3, at m/z 203, 365, and 527 were the major components found by EISMS among the reaction products of melibiose and α-PsGal (Figure S2a). The CID ESIMS/MS fragmentation pattern of a trisaccharide ion suggested Hex-(1→4)-Hex-(1→6)-Hex structure (Figure S2b). Fragment ion $^{0,2}A_2$ at m/z 305 and $^{0,3}A_2$ at m/z 275 indicated the presence of both 1→4- and 1→6-O-glycosidic links between two hexoses in disaccharide (Figure S2c). The question concerning the presence of the (1→3)-linked hexoses remained unclear, since the disaccharide Galα-(1→3)-Gal could be identified only by the absence of fragment ions from cross-ring cleavages [16]. Ion signals of [Hex_n + Na]$^+$, n = 1,2, at m/z 203 and 365 were the major components found by EISMS among the reaction products of melibiose and mutant C494N (Figure S3a). The CID ESIMS/MS fragmentation pattern of a disaccharide ion suggested Hex-(1→4)-Hex only (Figure S3b), so the fragment ion $^{0,2}A_2$ at m/z 305 was observed. The type of O-glycosidic bond in the pNP-glycosides could not be established because the mobile proton at the glycosyl hydroxyl was blocked. In this case, fragmentation was not observed (Figure S3c–e and Figure S4b) [19].

We used heavy ^{18}O-water for the transglycosylation experiment; we deemed it the most interesting substrate. The use of buffered heavy-oxygen water and liquid chromatography (LC), coupled with ESI-MS/MS (LC-ESIMS/MS) allowed simultaneous observing of the products of hydrolysis and transglycosylation. Since the transfer of heavy ^{18}OH-group produces a +2 mass shift, it was possible to distinguish between fragment ions, retaining the positive charge on the reducing end from the charge on the nonreducing end. Figure 8 shows the kinetics of the consumption and accumulation of the hydrolysis and synthesis products with the use of heavy-oxygen water.

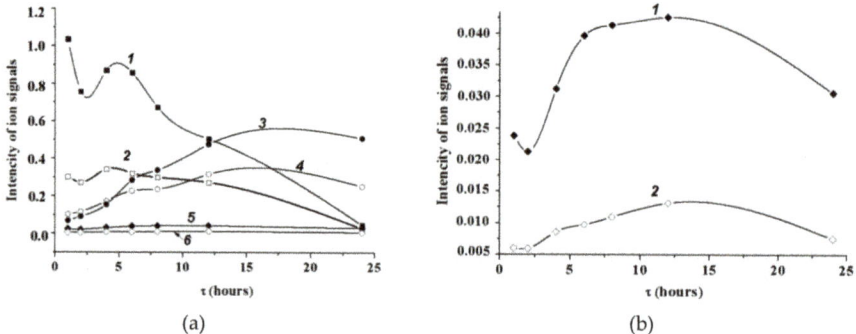

Figure 8. Mass spectrometry monitoring of the reaction of melibiose hydrolysis and transglycosylation catalyzed by wild α-PsGal in the buffered heavy-oxygen water at 20 °C. (**a**) Experimental time-dependent changes in integral intensity of matrix-assisted laser desorption ionization-mass spectroscopy (MALDI-MS) signals of the Hex$_2$ ions at 365 m/z (1) and 367 m/z (2), Hex at 203 m/z (3) and 205 m/z (4), and Hex$_3$ at 527 m/z (5) and 529 m/z; (**b**) time dependences of the MALDI-MS signals intensity of Hex$_3$ ions at 527 m/z (1) and 529 m/z (2) in an expanded scale.

According to the results (Figure 8), the consumption of Hex$_2$ (Figure 8a, curves 1 and 2) was accompanied by the appearance of Hex (Figure 8a, curves 3 and 4), new Hex$_2$ (Figure 8a, curves 1 and 2), and Hex$_3$ (Figure 8b).

2.5.3. NMR Spectroscopy

Nuclear magnetic resonance (NMR) spectroscopy was used to elucidate the transglycosylation regioselectivity and product structures. To avoid loss of information about minor transglycosylation products, the proton (^{1}H) and carbon (^{13}C) NMR spectra were analyzed without separating the reaction mixtures into individual compounds. The ^{1}H and ^{13}C signals of the anomeric atoms of the substrates and the products were assigned according to the respective reference data [20–22], as well as by H,H-Correlation Spectroscopy (COSY) and Heteronuclear Single-Quantum Correlation (HSQC) experiments. The identified signals are shown in Table 5.

Table 5. Anomer signals of nuclear magnetic resonance (NMR) spectra of substrates and products of hydrolysis and transglycosylation catalyzed by wild α-PsGal and its mutant C494.

Sugars	Product Number	δ_H (J in Hz)			δ_C		
		A	B		A	B	
		αH1	αH1	βH1	αC1	αC1	βC1
Gal-α(1→6)-Glcα,β A B		5.00 (3.7)	5.24	4.67 (7.8)	98.2	92.2	96.1
Galα,β	1		5.26 (3.7)	4.64 (7.9)		97.7	93.6
Glcα,β	2		5.23 (3.7)	4.58 (7.8)		97.6	93.6
Gal-α(1→6)-Galα,β A B	3	4.98 (4.8)	5.27	4.59 (7.9)	97.7		99.8
Gal-α(1→4)-Glcα,β A B	4	5.22 (3.2)	N/D*		102.2	N/D	N/D
Gal-(1→4)-Gal-(1→6)-Glc	5		undetected				
pNP-α-Gal			5.85 (3.74)			98.1	
Gal-α(1→6)-Gal-α-pNP A B	6	4.82 (3.85)	5.91 (3.82)		98.8	97.6	
Gal-α(1→3)-Gal-α-pNP A B	7	5.87 (3.6)	N/D		N/D	N/D	

* N/D = not defined.

The composition and yields of transglycosylation products in each case are shown in Table 6.

Table 6. The yield and structure of the products of transglycosylation reactions catalyzed by the recombinant wild α-PsGal and its mutant C494N based on NMR data.

Enzyme	Substrate	T (°C)	Substrate Conversion (%)	Structure of the Hydrolysis and Transglycosylation Products		Yield of Products (%)
Wild	Gal-α(1→6)-Glcα,β	20	88.5	Gal Glc Gal-α(1→6)-Galα,β Gal-α(1→4)-Galα,β	1 2 3 4	45.6.27.3.5.5.0.6
Wild	Gal-α(1→6)-Glcα,β	8	90.2	Gal Glc Gal-α(1→6)-Galα,β Gal-α(1→4)-Galα,β	1 2 3 4	46.0.35.0.7.8.1.4
Wild	pNP-α-Gal	20	67.2	Gal Gal-α(1→6)-Gal-α-pNP Gal-α(1→6)-Galα,β Gal-α(1→4)-Galα,β Gal-α(1→3)-Gal-α-pNP	1 6 3 4 7	21.5.8.0.3.0.1.2
Wild	pNP-α-Gal	8	15.9	Gal Gal-α(1→6)-Gal-α-pNP Gal-α(1→6)-Galα,β Gal-α(1→4)-Galα,β Gal-α(1→3)-Gal-α-pNP	1 6 3 4 7	9.8.4.11.20.78<1
Wild	Gal-α(1→6)-Glcα,β + pNP-α-Gal	20	32.0	Gal Gal-α(1→6)-Gal-α-pNP Gal-α(1→4)-Galα,β Gal-α(1→6)-Galα,β	1 6 4 3	30.0.9.0.1.6.0.8
C494N	pNP-α-Gal	8	19.0	Gal Gal-α(1→6)-Galα,β Gal-α(1→6)-Gal-α-pNP Gal-α(1→4)-Galα,β	1 3 6 4	5.0.2.8.2.0.1.2

In the ^{1}H NMR spectrum of the reaction mixture obtained after the action of recombinant α-PsGal on melibiose, the α,β1H signals of galactose (Gal) (**1**) and glucose (Glc) (**2**), which are the hydrolysis products, along with the major α,β1H signal of transglycosylation product Gal-α(1→6)-Galα,β (**3**) and a minor signal of Gal-α(1→4)-Galα,β (**4**), were observed (Tables 5 and 6). In the ^{1}H NMR spectrum of the products obtained after the action of recombinant α-PsGal on pNP-α-Gal, the signals of the α1H atoms were registered and characterized for the major autocondensation product Gal-α(1→6)-Gal-α-pNP (**6**) and a minor autocondensation product Gal-α(1→3)-Gal-α-pNP (**7**), as well as for the transglycosylation products Gal-α(1→6)-Galα,β (**3**) and Gal-α(1→4)-Galα,β (**4**) (Tables 5 and 6). The substituted bigalactoside Gal-α(1→6)-Gal-α-pNP (**6**) was a major product of transglycosylation obtained under the action of the recombinant α-PsGal on the mixture of melibiose (donor) and pNP-α-Gal (acceptor), whereas unsubstituted bigalactosides, Gal-α(1→6)-Galα,β (**3**) and Gal-α(1→4)-Galα,β (**4**), were synthesized in small amounts. Signals of α1H of Gal-α(1→6)-Gal-α-pNP (**6**) were found in the ^{1}H NMR spectrum of reaction products observed in the reaction mixture with the C494N mutant with pNP-α-Gal as the substrate. In the last case, there were no α1H signals found for compound Gal-α(1→3)-Gal-α-pNP (**7**). Analysis of the reaction mixture after the action of the D451A mutant on pNP-α-Gal did not reveal new signals in the NMR spectra except for the signals corresponding to pNP-α-Gal.

3. Discussion

The catalytic properties and structure-function relationships for the marine bacterial α-galactosidase from the GH 36 family, whose genes frequently occur in the genomes of marine bacteria, were characterized for the first time for recombinant α-galactosidase from the marine bacterium *Pseudoalteromonas* sp. KMM 701 (α-PsGal). As a result of our bioinformatic analysis of the amino acid sequence of the enzyme and homologous modeling of the 3D structure, presumably catalytic (D451 and D516) and substrate-binding (C494) residues—extremely important for the functioning of the enzyme—were identified. The predicted nucleophilic residue D451 and substrate-binding residue C494 were replaced with A451 and N494, respectively. Properties of the mutant D451A and C494N were investigated with comparison to wild α-PsGal.

We showed that α-PsGal and its mutant C494N are cold-active enzymes characterized by their neutral pH-optima (6.5–8.0) and low thermostability of 20 to 30 °C among the known α-galactosidases. The wild enzyme exhibited about 30% activity at 5 °C. No data on temperature and pH effects on the activity were available in the literature for the prototype α-galactosidases from the mesophiles *L. acidophilus*. The α-galactosidases from different mesophilic lactobacilli showed an acidic optimum activity, in the pH range from 5.2 to 5.9, and maximum activity between higher temperatures of 38 to 42 °C [23] compared with α-PsGal. The optimal temperature for the activity of the AgaA enzyme from psychrophilic lactic acid bacterium *Carnobacterium piscicola* was 32 to 37 °C [24]. The optimum temperature of the enzyme from *Lactobacillus fermentum* was found to be 45 °C. The enzyme was inactivated at temperatures higher than 55 °C and stable in wide ranges of temperatures and pH [25,26]. As for thermophilic enzymes from bacteria-thermophiles and hyperthermophiles *Bifidobacterium adolescentis* DSM 20083, *B. stearothermophilus*, *Thermus brockianus* [27], *Thermus* sp. T2 [28], *Thermoanaerobacterium polysaccharolyticum* [29], and *Thermotoga maritime*, their temperature optimums were 75 to 100 °C. The last enzyme was inactive at 30 °C [30].

Capability of catalyzing a transglycosylation reaction is an inherent property of all members of the retaining α-D-galactosidases of the GH 27 and GH 36 families [4]. The inverting α-D-galactosidases of the GH 110 family [31], as well as NAD^+- and Mn^{2+}-dependent α-D-galactosidases found in the family GH 4 [32], have lost their transglycosylation properties. To date, there is no information about the transglycosylation ability of α-D-galactosidases from the GH 97 and GH 57 families.

It is known that the first step in the catalytic reaction is cleavage of the glycosidic bond of the melibiose or pNP-α-Gal molecules, as well as the formation of the covalent galactosyl-enzyme intermediate. The molecules of Glc and pNP are leaving groups. In the second step, water or some carbohydrate molecules attack the covalent galactosyl-enzyme intermediate, and then hydrolysis or transglycosylation, respectively, can be observed. In the case where the substrate is an attacking molecule, we can observe an autocondensation reaction (Figure S5).

α-PsGal catalyzed synthesis with a total yield of transglycosylation products ranging from 6.0% to 12% (Table 6), similar to the known retaining bacterial galactosidases of the GH 36 family. It was difficult to identify the structures of the transglycosylation products without appropriate standards. However, the use of three methods (TLC, MALDI MS in conjunction with ESIMS/MS, and NMR) provided a suggestion of the relationships between the sugars' molecular weights and the type of O-glycoside bonds in the synthesized oligosaccharides.

TLC is commonly used to analyze low-molecular-weight sugars and their derivatives that differ in the number of carbon atoms, configurations, and molecule sizes. If two carbohydrates have one of these three different characteristics, they can be separated [33]. Based on the results, we assumed that the spot with R_f of 0.35 corresponds to bihexoses, distinguishable from melibiose by the configuration of the stereocenter, but the spot with R_f of 0.26 corresponds to sugars distinct from melibiose by the degree of polymerization (Figure 7, lane 1). The ^{1}H NMR signals of the anomeric atoms of the trisaccharides were not detected. However, the signal of [Hex$_3$ + Na]$^+$ at m/z 527 was observed in the MALDI MS (Figure S2a). The structure of the trisaccharide Hex-(1→4)-Hex-(1→6)-Hex was established by electrospray ionization tandem mass spectrometry as Gal-(1→4)-Gal-(1→6)-Glc (Figure S2b). The kinetic of accumulation and consumption of Gal-(1→4)-Gal-(1→6)-Glc was registered with the use of heavy-oxygen water (Figure 8b). In the NMR spectrum of the reaction mixture, we found the 1H signals of the anomer atoms (1→6)-α- and (1→4)-α-linked bigalactosides only. In this connection, we think that the spot with an R_f of 0.35 corresponds to two poorly shared (1→6)-α- and (1→4)-α-linked bigalactosides Gal-α(1→6)-Galα,β (3) and Gal-α(1→4)-Galα,β (4), respectively. This assumption was confirmed by EISMS-MS (Figure S2c).

Thus, when melibiose was used as the substrate, the enzyme synthesized the (1→6)-α-linked bigalactosides (Figure S2c), similar to all known melibiases of the GH 36 family [16,34–42] and to their closely-related GH 27 representatives [43–52]. Furthermore, α-PsGal formed the (1→4)-α-linked bigalactosides as described for mesophilic terrestrial α-D-galactosidases from *Bifidobacterium breve* 203 [35], *Lactobacillus reuteri* [16], and the acidic GH 27 family α-D-galactosidases (AgaBf3S) from the bacterium *Bacteroides fragilis*. The latter was able to transfer galactosyl residues from pNP-α-Gal in lactose Gal-β(1→4)-Glc with the efficiency and strict (1→4)-α-regioselectivity [52], whereas α-PsGal synthesized both (1→6)-α- and (1→4)-α-O-glycoside bonds in the bigalactosides from melibiose in the ratio of 9:1 at 20 °C and 5:1 at 8 °C. It is interesting to note that glucose, which is released from melibiose, did not participate in the transglycosylation reaction as an acceptor because its content in the mixtures was almost half of all products without any change in the course of the reaction (Table 3).

Similarly, we established the structure of the autocondensation products in the mixtures of α-PsGal and pNP-α-Gal (Figure 7, lanes 2 and 4, respectively). α-PsGal was able to produce novel compounds by catalyzing the autocondensation reaction of pNP-α-Gal. Both the substituted Gal$_2$-pNP with (1→6)-α- and (1→3)-α-O-glycoside bonds and unsubstituted Gal$_2$ with (1→6)-α- and (1→4)-α-O-glycoside bonds were found in the reaction mixture. The ratio of (1→6)-α-:(1→3)-α-linked Gal$_2$-pNP was 7:1, but the ratio for unsubstituted (1→6)-α-:(1→4)-α-linked bigalactosides was 3:1 at 20 °C. The ratio of (1→6)-α-:(1→4)-α-linked bigalactosides reached up to 2:1 at 8 °C (Table 6).

The transglycosylation properties are well-studied for the highly thermoresistant GH 36 α-D-galactosidase from the hyperthermophilic bacterium *Thermotoga maritima* (TmGal36A). This enzyme catalyzes an autocondensation reaction with pNP-α-Gal as a substrate, forming substituted (1→2)-α-, (1→3)-α- and (1→6)-α-linked Gal$_2$-pNP [22]. In total, the wild TmGal36A can produce up to 5.5% transglycosylation products. The mechanism of the hydrolysis and synthesis in TmGal36A is not favorable for the formation and breaking of the (1→4)-α-O-glycosidic linkage [22], unlike α-D-galactosidases from human intestine [34–37] and α-PsGal from marine bacterium.

The replacement of the predictive nucleophilic residue D451 to A451 in the active center led to complete loss of the ability of α-PGal to catalyze the hydrolysis. For unknown reasons, the rescue strategy, with an addition of the external nucleophilic sodium azide, proved to be ineffective in this case. Molar concentrations of sodium azide or sodium formate were unable to restore or increase the activity of the mutant D425G of α-D-galactosidase from archaeon *Sulfolobus solfataricus* [53]. Sodium azide did not inhibit any activity of the wild enzyme α-PsGal [10], but it did not restore the activity in its mutant D451A. Galactosyl-β-azide was not found both either of the reaction products of mutant D451A and pNP-α-Gal substrate, as occurred in the experiment with TmGal36A [54]. The D327G mutant of TmGal36A lost hydrolytic properties but retained glycosynthase properties and became an effective α-galactosynthase, which could produce various galactosylated disaccharides from galactosyl-β-azide as a donor and pNP-α(β)-galactosides as acceptors [55].

The mutation C494N changed the specificity for α-PsGal in the synthesis of O-glycoside bonds. Under the action of the C494N mutant on pNP-α-Gal, the yield of pNP-Gal-α-(1→6)-Gal (6) decreased, whereas pNP-Gal-(1→3)-α-Gal was not observed. In addition, the content of Gal-(1→4)-α-Gal (4) increased two-fold (Table 6). In the literature, it has been reported that the substitution of some bulk residues in the active site of α-D-galactosidase Aga A from *Bacillus stearothermophilus* KVE39 resulted in a 4.5-fold increase in the yield of substituted (1→3)-α-linked compared with pNP-Gal-α-(1→6)-Gal [37]. A number of single and double substitutions of protruded residues in the active site of α-D-galactosidase from *Bifidobacterium adolescentis* DSM 20083 led to an increase in the yield of the total transglycosylation products but they did not change the regioselectivity of the reaction [22,38].

4. Materials and Methods

4.1. Materials

The 4-nitrophenyl-α-D-galactopyranoside (pNP-α-Gal), melibiose (Gal-α-(1→6)-Glc), galactose (Gal), glucose (Glc), Bovine serum albumin (BSA), NaN_3, and 2,5-dihydroxybenzoic acid were purchased from Sigma Chemical Company (St. Louis, MO, USA). Encyclo DNA-polymerase and enterokinase were purchased from Evrogen (Moscow, Russian Federation). Sodium phosphates, one- and two-substituted, were purchased from PanReac AppliChem GmbH (Darmstadt, Germany). IMAC Ni^{2+} Sepharose, Q-Sepharose, Mono-Q, and Superdex-200 PG were purchased from GE Healthcare (Uppsala, Sweden). Heavy-oxygen water was purchased from Component Reactive (Moscow, Russia).

4.2. Homology Model of α-PsGal 3D Structure

The target-template alignment customization of the modeling process and 3D model building of α-PsGalA (GenBank: ABF72189.2) were carried out using the Molecular Operating Environment version 2018.01 [14] package (Chemical Computing Group ULC: 1010 Sherbrooke St. West, Suite #910, Montreal, QC, Canada, H3A 2R7, 2018) using the forcefield Amber12: EHT. The α-D-galactosidase from *Lactobacillus acidophilus* NCFM (PDB code: 2XN2) with a high-resolution crystal structure was used as a template. The evaluation of structural parameters, contact structure analysis, physicochemical properties, molecular docking, and visualization of the results were carried out with the Ligand interaction and Dock modules in the MOE 2018.01 program (Chemical Computing Group ULC: 1010 Sherbrooke St. West, Suite #910, Montreal, QC, Canada, H3A 2R7, 2018). The results were obtained using the equipment of Shared Resource Center Far Eastern Computing Resource of Institute of Automation and Control Processes Far Eastern Branch of the Russian Academy of Sciences (IACP FEB RAS) [56]

4.3. Production of Recombinant Enzymes

The recombinant wild α-D-galactosidase α-PsGal was produced as described earlier [13]. The D451A and C494N mutants were produced by polymerase chain reaction (PCR)-mediated site-directed mutagenesis using the full-length wild gene of α-PsGal. The mutations were inserted in the sequences of synthetic oligonucleotides for each DNA chain of the wild gene:

(1) D451A dir 5′-TTAAGTACATTAAATGGG**C**TATGAACCGCGA-3′
    D451A rev 5′-GTTAATATCGCGGTTCATA**G**CCCATTTAATG-3′
(2) C494N dir 5′-AGGGCTTGAAATAGAAAGC**AA**TTCGTCAGGTGG-3′
    C494N rev 5′-ACGTGCACCACCTGACGAA**TT**GCTTTCTATTTC-3′

The plasmid DNA pET40 containing an insertion of the α-D-galactosidase gene of the marine bacterium *Pseudoalteromonas* sp. KMM 701 (α-PsGal) or its D451A and C494N mutants were transformed in the *E. coli* strain Rosetta (DE3). Heterological expression was carried out at optimal conditions into *E. coli*, as described previously [57]. Purification of the recombinant α-PaGal and its D451A and C494N mutant forms were performed according to the procedures described previously [13].

4.4. Enzyme and Protein Essays

To determine the activity, 0.02 mL of an enzyme solution were mixed with 0.38 mL of the pNP-α-Gal solution (1 mg/mL in 0.05 M sodium phosphate buffer, pH 7.0). The reaction mixture was incubated at 20 °C for 10 min. The reaction was stopped by addition of 0.6 mL of 1 M Na_2CO_3. One unit of activity (U) was determined as the amount of enzyme that releases 1 µmol of pNP per 1 min at 20 °C in 0.05 M sodium phosphate buffer at pH 7.0. The amount of released pNP was determined spectrophotometrically (ε_{400} = 18300 M^{-1} cm^{-1}). The specific activity was calculated as U/mg of protein. Protein concentration was determined by the Bradford method and calibrated with BSA as a standard [58].

4.4.1. Circular Dichroism Spectra

The CD spectra were recorded in the ultraviolet (UV) region of 190 to 250 nm with Chirascan plus CD spectrometers (Applied Photophysics Ltd., Leatherhead, UK), equipped with an optional Peltier temperature controller for rapid and precise temperature control of the sample cell (Quantum North West, 22910 E Appleway Avenue, Suite 4 Liberty Lake, WA, USA), in 0.01 M sodium phosphate buffer (pH 7.0) and 20 °C. The average molecular weight of the amino acid residue for calculation of molar ellipticity [Θ] (degree cm^2 $dmol^{-1}$) was assumed to be 112 Da. The secondary structure elements were calculated by the Provencher–Glöcker method CONTIN/LL modified by Sreerama N. of the CDPro software package, 2000 (Colorado State University, Fort Collins, Colorado, USA, http://lamar.colostate.edu/sreeram/CDPro) [59,60]

4.4.2. UV Absorption Spectra

Absorption spectra of proteins were recorded with a UV-Visible spectrophotometer UV-1601 PC (Shimadzu, Kyoto, Japan) in quartz cells with an optical path length of 1 cm, 0.1 cm, and 0.01 cm in the range of 190 to 400 nm. The molar extinction coefficient of enzyme ε_{280} = 100,770 M^{-1} cm^{-1} was calculated from the content of aromatic amino acids using the ExPASy server [61].

4.5. Effect of pH and Temperature

The pH optimums of purified enzymes were determined with pNP-α-Gal as the substrate in the pH range of 5.2 to 6.5 using 0.1 M sodium citrate buffer, in the pH range of 6.2 to 8.0 using 0.1 M sodium phosphate buffer, and in the pH range of 7.8 to 9.0 with 0.1 M Tris-HCl buffer. The temperature optimums for the purified enzymes were determined at pH 7.0 in the temperature range of 5 to 40 °C. The temperature stabilities of the enzymes were investigated after incubation for 60 min at 10 to 40 °C.

4.6. Determination of Kinetic Parameters

All kinetic studies were performed in 0.1 M sodium phosphate buffer, pH 7.0, at 20 °C. The Michaelis–Menthen constants, K_m and V_{max}, were determined from the coefficients of linear regression of the Lineweaver–Burk plot. The substrate concentrations (mM) were 3.24, 2.59, 1.62, 1.29, 0.81, 0.65, 0.40, and 0.32 for wild α-PsGal and 2.50, 2.0, 1.25, 1.0, 0.62, 0.50, 0.31, and 0.25 for the C494N mutant.

4.7. Transglycosylation

4.7.1. Acceptor Specificity of Transglycosylation

For preliminary determination of acceptor specificity of transglycosylation, the synthesis reactions were performed at 20 °C for 24 h in a mixture (10 µL) containing 0.01 U of an enzyme, 10 mM of pNP-α-Gal or melibiose as the substrate, and 20 mM of glucose, galactose, fructose, fucose, or xylose as acceptor in 0.05 M sodium phosphate buffer (pH 7.0). The reaction was stopped by heating at 100 °C for 5 min and the reaction mixture was centrifuged at 14,000 rpm. Sugars were analyzed by TLC, mass spectrometry, and NMR spectroscopy methods.

4.7.2. Transglycosylation Using Heavy-Oxygen Water ($H_2^{18}O$)

An experiment using mass spectrometry and heavy-oxygen water was prepared similarly as described in Section 4.7.1. above; but the concentrations were significantly lowered. Seven identical reaction mixtures were created. Each mixture contained 1.4 mg melibiose, 10 µL enzyme (1 U), and 70 µL H_2O^{18} (0.02 M sodium phosphate buffer, 0.05 M NaCl, pH 7.0). The mixtures were incubated for 1, 2, 4, 6, 8, 12, and 24 h. Each reaction mixture was dissolved in 1 mL methanol and introduced into the mass spectrometer. For ESIMS and ESIMS-MS experiments, direct injection was performed using a syringe pump (KD Scientific, Hollison City, MA, USA) at a flow rate 5 µL/min. For LC-ESIMS experiments, samples were further diluted 10 times in methanol.

4.7.3. Identification of Transglycosylation Products

The recombinant α-PsGal or D451A and C494N mutants, in an aqueous solution of 0.05 M sodium phosphate buffer (pH 7.0), were added to the preweighed dry samples of substrates melibiose or pNP-α-Gal or their mixture and were incubated for a certain time (Table 3) at 20 °C or 8 °C. The standard units of activity (U) or milligrams enzyme added (mg), the incubation time (τ) and reaction temperature are shown in Table 2. The reaction was stopped by heating at 100 °C for 5 min. The samples were centrifuged to remove the denatured protein and dried on Refrigerated CentriVap Concentrater (Labconco, Kansas City, MO, USA). The qualitative composition of the hydrolysis and transglycosylation products were analyzed by TLC and MALDI-MS without their separation from the reaction mixtures. Identification of oligosaccharides in the mixture and their output was performed via NMR spectroscopy.

4.7.4. Thin-Layer Chromatography Analysis

Mono- and oligosaccharide composition of the hydrolysis and transglycosylation products were analyzed on silica gel TLC plates on aluminum foil (Sigma-Aldrich, St. Louis, MO, USA) with a 254 nm fluorescent indicator. The pore diameter was 60 A. R_f was calculated for every stain.

Weighed freeze-dried reaction product mixtures were placed in 0.5 mL Eppendorf, dissolved in distilled water to a concentration of 5 mg/mL, and centrifuged at 10,000 rpm to remove the denatured protein. A small spot of the analyzed mixture and standard compounds were applied at the start line of the TLC plate and chromatographed over 30 minutes in a sealed chamber Latch-Lid ChromatoTank (General Glass Blowing Co. Inc., Richmond, CA, USA), containing 100 mL of the mobile phase butanol/acetic acid/water (3:1:1; $v/v/v$). For visualization of stains, the plate was treated three times with 5% sulfuric acid solution, drying by warm air after each spraying.

4.7.5. Mass Spectrometry Analysis

The molecular weights of the oligosaccharide ions were recorded as sodium adducts [M + Na]$^+$ using MALDI time-of-flight mass spectrometer, Ultra Flex III (Bruker BioSpin GmbH, Rheinstetten/Karlsruhe, Germany) equipped with a smartbeam laser (355 nm, Bruker Daltonik GmbH, Bremen, Germany) in reflector mode at an accelerating voltage of 21 kV, using the saturated solution (acetonitrile-water, 1:1) of 2,5-dihydroxybenzoynoic acid as a matrix.

The composition of the oligosaccharide mixture after enzymatic transformation was performed using an Ultimate 3000 rapid separation liquid chromatography (RSLC) nano system (Dionex, Thermo Fisher Scientific, Waltham City, MA, USA) connected to a Bruker Impact II quadrupole time-of-flight (Q-TOF) mass spectrometer (Bruker Daltonics, Bremen, Germany). An Acclaim (Thermo Fisher Scientific, Waltham City, MA, USA) PepMap RSLC column (75 μm × 150 mm, C18, 2 μm, 100 A) was used for chromatographic separation. The mobile phases were 0.1% formic acid in H$_2$O (eluent A) and 0.1% formic acid in acetonitrile (eluent B). The gradient program was: isocratic at 1% of eluent B from start to 5 min, from 1% to 10% eluent B from 5 to 10 min, from 10% to 95% eluent B from 10 to 11 min, and isocratic at 95% of eluent B to 15 min. After returning to the initial conditions, equilibration was achieved after 10 min. Chromatographic separation was performed at a 0.4 μL/min flow rate at 40 °C. Injection volume was 0.2 μL. The mass spectrometry detection was performed using CaptiveSpray (Bruker Daltonics, Bremen, Germany) ionization source at a capillary voltage of 1.3 kV. Collision induced dissociation (CID)-produced ion mass spectra were recorded in auto-MS/MS mode with collision energy 43 eV. The precursor ions were isolated with an isolation width of 1 mass unit.

The mass spectrometer was calibrated using the ESI-L Low Concentration Tuning Mix (Agilent Technologies, Santa Clara, CA, USA). The instrument was operated using the OTOFControl software (version 4.0, Bruker Daltonics, Bremen, Germany) and data were analyzed using Data Analysis software (version 4.3, Bruker Daltonics, Bremen, Germany).

4.7.6. NMR Spectroscopy Analysis

The structure of disaccharides was characterized by NMR spectroscopy. Signals in the NMR spectra of sugars were assigned by two-dimensional correlated spectroscopy (H,H-COSY) and two-dimensional heteronuclear multiple bond correlation spectroscopy (HSQC) experiments. One-dimensional ^{1}H-NMR and ^{13}C-NMR, and two-dimensional H,H-COSY and HSQC spectra were recorded with a Bruker Avance III 500 HD (Bruker BioSpin GmbH, Rheinstetten/Karlsruhe, Germany) spectrometer in D$_2$O at 50 °C with acetone as internal standard (δ = 31.45 and 2.20 ppm for ^{13}C NMR and ^{1}H NMR spectra, respectively). ^{1}H NMR anomer signals of α-pNP-galactopyranose, β-galactopyranose, melibiose, and bigalactosides, as well as proton signals of the free 4-nitrophenol ring were analyzed and integrated by the standard software TopSpin 3.2.

The depth of the pNP-α-Gal conversion ($H_{pNP\alpha Gal}$, %) was calculated by

$$H_{pNP\alpha Gal} = \{I_{Gal}/(I_{pNP\alpha Gal} + I_{Gal2pNP} + I_{Gal2})\} \times 100 \quad (1)$$

where I_{Gal} is integrated intensities of all 1H signals of liberated Galα,β, $I_{pNP\alpha Gal}$ is the integrated intensity of 1H proton signals in substrate pNP-α-Gal, and $I_{Gal2pNP}$ and I_{Gal2} are the integrated intensity of 1H proton signals in transglycosylation products.

The depth of melibiose conversion (H_{Mel}, %) was calculated by

$$H_{Mel} = (I_{Gal}/I_{Mel}) \times 100 \quad (2)$$

where I_{Mel} is the integrated intensities of signals 1H of initial mixture of melibiose.

Yield of oligosaccharides in the total reaction mixture (Y, %) was calculated by

$$Y = \{I_{product} / (I_{Gal} + I_{Glc} + I_{Gal2} + I_{Gal2pNP})\} \times 100 \qquad (3)$$

where $I_{product}$ is the integrated intensity of all 1H signals of a particular oligosaccharide.

5. Conclusions

The recombinant and mutated α-PsGal were shown to be expressed as soluble proteins with the use of pET-40 b (+)-based constructions, and were purified successfully with the His-tag approach. The wild α-PsGal from the cold-adapted marine bacterium *Pseudoalteromonas* sp. KMM 701 has the traits of a cold-active enzyme that catalyzes the hydrolysis and weak transglycosylation reactions. The combination of TLC, MALDI-MS, and NMR spectroscopy methods to analyze the reaction mixture sugars allowed us to define the regioselectivity of the transglycosylation reaction, to qualitatively and quantitatively identify the reaction products obtained under the action of the recombinant analogues of α-PsGal, and to evaluate the yield of these products. The yield of transglycosylation products ranged from 6 to 12%. α-PsGal has a narrow acceptor specificity but rather wide regioselectivity. D-glucose, D-fructose, L-fucose, and D-xylose are not acceptors in the transglycosylation reaction. Together with the major α(1→6)-links under experimental conditions, the enzyme produced minor (traces) α(1→4)- or -α(1→3)-links in bigalactosides at the saturating concentrations of melibiose and pNP-α-Gal, respectively. The point mutation D451A resulted in the completely loss of α-PsGal activity, indicating crucial significance of the residue A451 in the performance of the α-PsGal-mediated hydrolysis as well as transglycosylation. The C494N mutation slightly changed the structure, properties, and substrate specificity of the enzyme. Thus, *Pseudoalteromonas* KMM 701 α-D-galactosidase of the GH 36 family, which is important in biomedical technology, demonstrates weak glycosynthase properties in vitro.

Supplementary Materials: The following are available online at http://www.mdpi.com/1660-3397/16/10/349/s1, Figure S1: MALDI-MS of mixtures of reaction products with: (**a**) melibiose and pNP-α-Gal as donor and acceptor, respectively; (**b**) wild α-PsGal with pNP-α-Gal as substrate; (**c**) D451A mutant with pNP-α-Gal; (**d**) C494N mutant with pNP-α-Gal; and (**e**) MALDI-MS of cellobiose and melibiose as standards, Figure S2: Electrospray ionization mass (ESIMS-MS) spectra of the transglycosylation products catalyzed by wild α-PsGal with melibiose as substrate: (**a**) ESIMS-MS spectra; (**b**) fragmentation of *m/z* 527 ion; (**c**) fragmentation of *m/z* 365 ion, Figure S3: ESIMS-MS spectra of the transglycosylation products from pNP-α-Gal catalyzed by wild α-PsGal: (**a**) fragmentation of *m/z* 365 ion; (**b**) fragmentation of *m/z* 486 ion, Figure S4: ESIMS-MS spectra of the transglycosylation products catalyzed by mutant C494N: (**a**) melibiose as substrate; (**b**) fragmentation of *m/z* 365 ion; (**c**) pNP-α-Gal as substrate; (**d**) fragmentation of *m/z* 486 ion; (**e**) and fragmentation of *m/z* 347 ion. Figure S5: Schemes for presumable mechanism of hydrolysis and transglycosylation of substrates melibiose (**a**) and pNP-α-Gal (**b**) with recombinant α-PaGal, Table S1: Ligand Interactions Report.

Author Contributions: Conceptualization, enzyme investigation, writing-original draft preparation: I.B.; construction, expression, and purification of recombinant and mutant enzymes: L.S. and L.B.; mass-spectra registration and interpretation: S.A.; NMR experiments: V.I.; CD spectra registration and analysis: N.K.; bioinformatics analysis and computer modeling of protein structure: G.L; formal analysis: O.S.; resources: L.T.

Funding: Financial support was provided by Ministry of Education and Science of Russia (Agreement 02.G25.31.0172, 01.12.2015); Mass spectrometric experiments using heavy-oxygen water were funded by Russian Science Foundation (Grand No. 16-13-10185).

Conflicts of Interest: The authors declare no conflicts of interest.

References

1. Cantarel, B.L.; Coutinho, P.M.; Rancurel, C.; Bernard, T.; Lombard, V.; Henrissat, B. The Carbohydrate-Active EnZymes database (CAZy): An expert resource for glycogenomics. *Nucleic Acids Res.* **2009**, *37*, D233–D238. [CrossRef] [PubMed]
2. Henrissat, B.; Callebaut, I.; Fabrega, S.; Lehn, P.; Mornon, J.P.; Davies, G. Conserved catalytic machinery and the prediction of a common fold for several families of glycosyl hydrolases. *Proc. Natl. Acad. Sci. USA* **1995**, *92*, 7090–7094. [CrossRef] [PubMed]

3. Koshland, D.E., Jr. Stereochemistry and the mechanism of enzymatic reactions. *Biol. Rev. Camb. Phil. Soc.* **1953**, *28*, 416–436. [CrossRef]
4. Bakunina, I.Y.; Balabanova, L.A.; Pennacchio, A.; Trincone, A. Hooked on α-D-galactosidases: from biomedicine to enzymatic synthesis. *Crit. Rev. Biotechnol.* **2016**, *36*, 233–245. [CrossRef] [PubMed]
5. Ivanova, E.P.; Bakunina, I.Y.; Nedashkovskaya, O.I.; Gorshkova, N.M.; Mikhailov, V.V.; Elyakova, L.A. Incidence of marine microorganisms producing β-N-acetylglucosaminidases, α-D-galactosidases and α-N-acetylgalactosaminidases. *Rus. J. Mar. Biol.* **1998**, *24*, 365–372.
6. Bakunina, I.Y.; Nedashkovskaya, O.I.; Alekseeva, S.A.; Ivanova, E.P.; Romanenko, L.A.; Gorshkova, N.M.; Isakov, V.V.; Zvyagintseva, T.N.; Mikhailov, V.V. Degradation of fucoidan by the marine proteobacterium *Pseudoalteromonas citrea*. *Microbiology* **2002**, *71*, 41–47. [CrossRef]
7. Bakunina, I.Y.; Ivanova, E.P.; Nedashkovskaya, O.I.; Gorshkova, N.M.; Elyakova, L.A.; Mikhailov, V.V. Search for α-D-galactosidase producers among marine bacteria of the genus *Alteromonas*. *Prikl. Biokh. Mikrobiol.* **1996**, *32*, 624–628.
8. Bakunina, I.Y.; Nedashkovskaya, O.I.; Kim, S.B.; Zvyagintseva, T.N.; Mihailov, V.V. Diversity of glycosidase activities in the bacteria of the phylum *Bacteroidetes* isolated from marine algae. *Microbiology* **2012**, *81*, 688–695. [CrossRef]
9. Bakunina, I.Y.; Nedashkovskaya, O.I.; Balabanova, L.A.; Zvyagintseva, T.N.; Rasskasov, V.V.; Mikhailov, V.V. Comparative analysis of glycoside hydrolases activities from phylogenetically diverse marine bacteria of the genus *Arenibacter*. *Mar. Drugs* **2013**, *11*, 1977–1998. [CrossRef] [PubMed]
10. Bakunina, I.Y.; Sova, V.V.; Nedashkovskaya, O.I.; Kuhlmann, R.A.; Likhosherstov, L.M.; Martynova, M.D.; Mihailov, V.V.; Elyakova, L.A. α-D-galactosidase of the marine bacterium *Pseudoalteromonas* sp. KMM 701. *Biochemisrty* **1998**, *63*, 1209–1215.
11. Balabanova, L.A.; Bakunina, I.Y.; Nedashkovskaya, O.I.; Makarenkova, I.D.; Zaporozhets, T.S.; Besednova, N.N.; Zvyagintseva, T.N.; Rasskazov, V.A. Molecular characterization and therapeutic potential of a marine bacterium *Pseudoalteromonas* sp. KMM 701 α-D-galactosidase. *Mar. Biotechnol.* **2010**, *12*, 111–120. [CrossRef] [PubMed]
12. Terentieva, N.A.; Timchenko, N.F.; Balabanova, L.A.; Golotin, V.A.; Belik, A.A.; Bakunina, I.Y.; Didenko, L.V.; Rasskazov, V.A. The influence of enzymes on the formation of bacterial biofilms. *Health Med. Ecol. Sci.* **2015**, *60*, 86–93.
13. Bakunina, I.Y.; Balabanova, L.A.; Golotin, V.A.; Slepchenko, L.V.; Isakov, V.V.; Rasskazov, V.A. Stereochemical course of hydrolytic reaction catalyzed by alpha-galactosidase from cold adapTable marine bacterium of genus *Pseudoalteromonas*. *Front. Chem.* **2014**, *2*, 89. [CrossRef] [PubMed]
14. *Molecular Operating Environment (MOE)*; 2018.01; Chemical Computing Group ULC: Montreal, QC, Canada, 2018.
15. Fredslund, F.; Hachem, M.A.; Larsen, R.J.; Sørensen, P.G.; Coutinho, P.M.; Lo Leggio, L.; Svensson, B. Crystal structure of α-galactosidase from *Lactobacillus acidophilus* NCFM: Insight into tetramer formation and substrate binding. *J. Mol. Biol.* **2011**, *412*, 466–480. [CrossRef] [PubMed]
16. Wang, Y.; Black, B.A.; Curtis, J.M.; Gänzle, M.G. Characterization of α-galacto-oligosaccharides formed via heterologous expression of α-D-galactosidases from *Lactobacillus reuteri* in *Lactococcus lactis*. *Appl. Microbiol. Biotechnol.* **2014**, *98*, 2507–2517. [CrossRef] [PubMed]
17. Menshova, R.V.; Ermakova, S.P.; Anastyuk, S.D.; Isakov, V.V.; Dubrovskaya, Y.V.; Kusaykin, M.I.; Um, B.H.; Zvyagintseva, T.N. Structure, enzymatic transformation and anticancer activity of branched high molecular weight laminaran from brown alga *Eisenia bicyclis*. *Carbohydr. Polym.* **2014**, *99*, 101–109. [CrossRef] [PubMed]
18. Domon, B.; Costello, C.E. A Systematic nomenclature for carbohydrate fragmentations in FAB-MS/MS spectra of glycoconjugates. *Glycoconj. J.* **1988**, *5*, 397–409. [CrossRef]
19. Zaia, J.; Miller, M.J.C.; Seymour, J.L.; Costello, C.E. The role of mobile protons in negative ion CID of oligosaccharides. *J. Am. Soc. Mass Spectrom.* **2007**, *18*, 952–960. [CrossRef] [PubMed]
20. Weignerova, L.; Hunkova, Z.; Kuzma, M.; Kren, V. Enzymatic synthesis of three pNP-α-galactobiopyranosides: application of the library of fungal α-D-galactosidases. *J. Mol. Catal. B: Enzym.* **2001**, *11*, 219–224. [CrossRef]
21. Borisova, A.S.; Ivanen, D.R.; Bobrov, K.S.; Eneyskaya, E.V.; Rychkov, G.N.; Sandgren, M.; Kulminskaya, A.A.; Sinnott, M.L.; Shabalin, K.A. α-Galactobiosyl units: Thermodynamics and kinetics of their formation by

transglycosylations catalysed by the GH36 α-D-galactosidase from *Thermotoga maritima*. *Carbohydr. Res.* **2015**, *401*, 115–121. [CrossRef] [PubMed]
22. Bobrov, K.S.; Borisova, A.S.; Eneyskaya, E.V.; Ivanen, D.R.; Shabalin, K.A.; Kulminskaya, A.A.; Rychkov, G.N. Improvement of efficiency of transglycosylation catalyzed by α-D-galactosidase from *Thermotoga maritima* by protein engineering. *Biochemistry* **2013**, *78*, 1112–1123. [CrossRef] [PubMed]
23. Mitfal, B.K.; Shallenberger, R.S.; Stainkraus, K.H. α-Galactosidase activity of lactobacilli. *Appl. Microbiol.* **1973**, *26*, 783–788.
24. Coombs, J.; Brenchley, J.E. Characterization of two new glycosyl hydrolases from the lactic acid bacterium *Carnobacterium piscicola* Strain BA. *Appl. Environ. Microbiol.* **2001**, *67*, 5094–5099. [CrossRef] [PubMed]
25. Garro, M.S.; Degiori, G.S.; Devaldez, G.F.; Oliver, G. Characterization of α-galactosidase from *Lactobacillus fermentum*. *J. Appl. Bacteriol.* **1993**, *75*, 485–488. [CrossRef]
26. Carrera-Silva, E.A.; Silvestroni, A.; LeBlanc, J.G.; Piard, J.C.; de Giori, G.S.; Sesma, F. A thermostable α-galactosidase from *Lactobacillus fermentum* CRL722: Genetic characterization and main properties. *Curr. Microbiol.* **2006**, *53*, 374–378. [CrossRef] [PubMed]
27. Fridjonsson, O.; Watzlawick, H.; Gehweiler, A.; Rohrhirsch, T.; Mattes, R. Cloning of the gene encoding a novel thermostable α-galactosidase from *Thermus brockianus* ITI360. *Appl. Environ. Microbiol.* **1999**, *65*, 3955–3963. [PubMed]
28. Ishiguro, M.; Kaneko, S.; Kuno, A.; Koyama, Y.; Yoshida, S.; Park, G.G.; Sakakibara, Y.; Kusakabe, I.; Kobayashi, H. Purification and characterization of the recombinant *Thermus* sp. strain T2 α-galactosidase expressed in *Escherichia coli*. *Appl. Environ. Microbiol.* **2001**, *67*, 1601–1606. [PubMed]
29. King, M.R.; White, B.A.; Blaschek, H.P.; Chassy, B.M.; Mackie, R.I.; Cann, I.K. Purification and characterization of a thermostable α-galactosidase from *Thermoanaerobacterium polysaccharolyticum*. *J. Agric. Food Chem.* **2002**, *50*, 5676–5682. [CrossRef] [PubMed]
30. Liebl, W.; Wagner, B.; Schellhase, J. Properties of an α-galactosidase, and structure of its gene galA, within an α- and β-galactoside utilization gene cluster of the hyperthermophilic bacterium *Thermotoga maritime*. *Syst. Appl. Microbiol.* **1998**, *21*, 1–11. [CrossRef]
31. Liu, Q.P.; Yuan, H.; Bennett, E.P.; Levery, S.B.; Nudelman, E.; Spence, J.; Pietz, G.; Saunders, K.; White, T.; Olsson, M.L.; et al. Identification of a GH110 subfamily of α-1,3-galactosidases, novel enzymes for removal of the a3Gal xenotransplantation antigen. *J. Biol. Chem.* **2008**, *283*, 8545–8554. [CrossRef] [PubMed]
32. Liu, Q.P.; Sulzenbacher, G.; Yuan, H.; Bennett, E.P.; Pietz, G.; Saunders, K.; Spence, J.; Nudelman, E.; Levery, S.B.; White, T.; et al. Bacterial glycosidases for the production of universal red blood cells. *Nat. Biotechnol.* **2007**, *25*, 454–464. [CrossRef] [PubMed]
33. Robyt, J.F. Carbohydrates/Thin-Layer (Planar) Chromatography. In *Encyclopedia of separation science*; Wilson, I.D., Cooke, M., Poole, C.F., Eds.; Academic New Press: Pittsburgh, PA, USA, 2000; pp. 2235–2244.
34. Van Laere, K.M.J.; Hartemink, R.; Beldman, G.; Pitson, S.; Dijkema, C.; Schols, H.A.; Voragen, A.G.J. Transglycosidase activity of *Bifidobacterium adolescentis* DSM 20083 α-D-galactosidase. *Appl. Microbiol. Biotechnol.* **1999**, *52*, 681–688. [CrossRef] [PubMed]
35. Zhao, H.; Lu, L.; Xiao, M.; Wang, Q.; Lu, Y.; Liu, C.; Wang, P.; Kumagai, H.; Yamamoto, K. Cloning and characterization of a novel α-D-galactosidase from *Bifidobacterium breve* 203 capable of synthesizing Gal-α-1,4 linkage. *FEMS Microbiol. Lett.* **2008**, *285*, 278–283. [CrossRef] [PubMed]
36. Cervera-Tison, M.; Tailford, L.E.; Fuell, C.; Bruel, L.; Sulzenbacher, G.; Henrissat, B.; Berrin, J.G.; Fons, M.; Giardina, T.; Jugea, N. Functional analysis of family GH-36 α-D-galactosidases from *Ruminococcus gnavus* E1: Insights into the metabolism of a plant oligosaccharide by a human gut symbiont. *Appl. Environ. Microbiol.* **2012**, *78*, 7720–7732. [CrossRef] [PubMed]
37. Dion, M.; Nisole, A.; Spangenberg, P.; André, C.; Glottin-Fleury, A.; Mattes, R.; Tellier, C.; Rabiller, C. Modulation of the regioselectivity of a *Bacillus* α-D-galactosidase by directed evolution. *Glycoconj. J.* **2001**, *18*, 215–223. [CrossRef] [PubMed]
38. Hinz, S.W.A.; Doeswijk-Voragen, C.H.L.; Schipperus, R.; Broek, L.A.M.; Vincken, J.P.; Voragen, A.G.J. Increasing the transglycosylation activity of α-D-galactosidase from *Bifidobacterium adolescentis* DSM 20083 by site-directed mutagenesis. *Biotechnol. Bioeng.* **2006**, *93*, 122–131. [CrossRef] [PubMed]
39. Nakai, H.; Baumann, M.J.; Petersen, B.O.; Westphal, Y.; Hachem, M.A.; Dilokpimol, A.; Duus, J.Ø.; Schols, H.A.; Svensson, B. *Aspergillus nidulans* α-D-galactosidase of glycoside hydrolase family 36 catalyses

the formation of α-galacto-oligosaccharides by transglycosylation. *FEBS J.* **2010**, *277*, 3538–3551. [CrossRef] [PubMed]

40. Spangenberg, P.; Andre, C.; Dion, M.; Rabiller, C.; Mattes, R. Potential of new sources of α-D-galactosidases for the synthesis of disaccharides. *Carbohydr. Res.* **2000**, *329*, 65–73. [CrossRef]
41. Schroder, S.; Kroger, L.; Mattes, R.; Thiem, J. Transglycosylations employing recombinant α- and β-galactosidases and novel donor substrates. *Carbohydr. Res.* **2015**, *403*, 157–166. [CrossRef] [PubMed]
42. Zhou, J.; Lu, Q.; Zhang, R.; Wang, Y.; Wu, Q.; Li, J.; Tang, X.; Xu, B.; Ding, J.; Huang, Z. Characterization of two glycoside hydrolase family 36 α-D-galactosidases: Novel transglycosylation activity, lead–zinc tolerance, alkaline and multiple pH optima, and low-temperature activity. *Food Chem.* **2016**, *194*, 156–166. [CrossRef] [PubMed]
43. Koizumi, K.; Tanimoto, T.; Okada, Y.; Hara, K.; Fujita, K.; Hashimoto, H.; Kitahata, S. Isolation and characterization of novel heterogeneous branched cyclomalto-oligosaccharides (cyclodextrins) produced by transgalactosylation with α-D-galactosidase from coffee bean. *Carbohydr. Res.* **1995**, *278*, 129–142. [CrossRef]
44. Spangenberg, P.; Andre, C.; Langlois, V.; Dion, M.; Rabiller, C. α-Galactosyl fluoride in transfer reactions mediated by the green coffee beans α-D-galactosidase in ice. *Carbohydr. Res.* **2002**, *337*, 221–228. [CrossRef]
45. Savel'ev, A.N.; Ibatyllin, F.M.; Eneyskaya, E.V.; Kachurin, A.M.; Neustroev, K.N. Enzymatic properties of α-D-galactosidase from *Trichoderma reesei*. *Carbohydr. Res.* **1996**, *296*, 261–273. [CrossRef]
46. Eneyskaya, E.V.; Golubev, A.M.; Kachurin, A.M.; Savel'ev, A.N.; Neustroev, K.N. Transglycosylation activity of α-D-galactosidase from *Trichoderma reesei*. An investigation of the active site. *Carbohydr. Res.* **1998**, *305*, 83–91. [CrossRef]
47. Shabalin, K.A.; Kulminskaya, A.A.; Savel'ev, A.N.; Shishlyannikov, S.M.; Neustroev, K.N. Enzymatic properties of α-D-galactosidase from *Trichoderma reesei* in the hydrolysis of galactooligosaccharides. *Enzyme Microb. Technol.* **2002**, *30*, 231–239. [CrossRef]
48. Shivam, K.; Mishra, S.K. Purification and characterization of a thermostable α-D-galactosidase with transglycosylation activity from *Aspergillus parasiticus* MTCC-2796. *Process Biochem.* **2010**, *45*, 1088–1093. [CrossRef]
49. Wang, C.; Wang, H.; Ma, R.; Shi, P.; Niu, C.; Luo, H.; Yang, P.; Yao, B. Biochemical characterization of a novel thermophilic α-D-galactosidase from *Talaromyces leycettanus* JCM12802 with significant transglycosylation activity. *J. Biosci. Bioeng.* **2016**, *121*, 7–12. [CrossRef] [PubMed]
50. Kurakake, M.; Moriyama, Y.; Sunouchi, R.; Nakatani, S. Enzymatic properties and transglycosylation of α-D-galactosidase from *Penicillium oxalicum* SO. *Food Chem.* **2011**, *126*, 177–182. [CrossRef]
51. Vic, G.; Scigelova, M.; Hastings, J.J.; Howarth, O.W.; Crout, D.H.G. Glycosydase catalysed synthesis of oligosaccharides: trisaccharides with the α-D-Gal-(1→3)-D-Gal terminus responsible for the hyper acute rejection response in cross-species transplant rejection from pigs to man. *Chem. Commun.* **1996**, *12*, 1473–1474. [CrossRef]
52. Gong, W.; Xu, L.; Gu, G.; Lu, L.; Xiao, M. Efficient and regioselective synthesis of globotriose by a novel α-D-galactosidase from *Bacteroides fragilis*. *Appl. Microbiol. Biotechnol.* **2016**, *100*, 6693–6702. [CrossRef] [PubMed]
53. Brouns, J.J.S.; Smits, N.; Wu, H.; Snijders, A.P.L.; Wright, P.C.; de Vos, W.M.; van der Oost, J. Identification of a novel alpha-galactosidase from the hyperthermophilic archaeon *Sulfolobus solfataricus*. *J. Bacteriol.* **2006**, *188*, 2392–2399. [CrossRef] [PubMed]
54. Comfort, D.A.; Bobrov, K.S.; Ivanen, D.R.; Shabalin, K.A.; Harris, J.M.; Kulminskaya, A.A.; Brumer, H.; Kelly, R.M. Biochemical analysis of *Thermotoga maritima* GH36 α-D-galactosidase (*Tm*GalA) confirms the mechanistic commonality of clan GH-D glycoside hydrolases. *Biochemistry* **2007**, *46*, 3319–3330. [CrossRef] [PubMed]
55. Cobucci-Ponzano, B.; Zorzetti, C.; Strazzulli, A.; Carillo, S.; Bedini, E.; Corsaro, M.M.; Comfort, D.A.; Kelly, R.M.; Rossi, M.; Moracci, M. A novel α-D-galactosynthase from *Thermotoga maritima* converts β-D-galactopyranosyl azide to α-galacto-oligosaccharides. *Glycobiology* **2011**, *21*, 448–456. [CrossRef] [PubMed]
56. Shared Resource Center Far Eastern Computing Resource of Institute of Automation and Control Processes Far Eastern Branch of the Russian Academy of Sciences (IACP FEB RAS). Available online: https://cc.dvo.ru (accessed on 12 January 2018).

57. Golotin, V.A.; Balabanova, L.A.; Noskova, Y.A.; Slepchenko, L.V.; Bakunina, I.Y.; Vorobieva, N.S.; Terentieva, N.A.; Rasskazov, V.A. Optimization of cold-adapted α-D-galactosidase expression in *Escherichia coli*. *Protein Expr. Purif.* **2016**, *123*, 14–18. [CrossRef] [PubMed]
58. Bradford, M.M. A rapid and sensitive method for the quantitation of microgram quantities of protein utilizing the principle of protein-dye binding. *Anal. Biochem.* **1976**, *72*, 248–254. [CrossRef]
59. Provencher, S.W.; Glockner, J. Estimation of globular protein secondary structure from circular dichroism. *Biochemistry* **1981**, *20*, 33–37. [CrossRef] [PubMed]
60. Sreerama, N.; Woody, R.W. Estimation of protein secondary structure from circular dichroism spectra: comparison of CONTIN, SELCON, and CDSSTR methods with an expanded reference set. *Anal. Biochem.* **2000**, *287*, 252–260. [CrossRef] [PubMed]
61. ExPASy (Expert Protein Analysis System) Proteomics Server. Available online: http://cn.expasy.org/tools/protparam.html (accessed on 15 September 2015).

© 2018 by the authors. Licensee MDPI, Basel, Switzerland. This article is an open access article distributed under the terms and conditions of the Creative Commons Attribution (CC BY) license (http://creativecommons.org/licenses/by/4.0/).

Article

Cloning, Expression and Characterization of a Novel Cold-Adapted β-galactosidase from the Deep-sea Bacterium *Alteromonas* sp. ML52

Jingjing Sun [1,2,*], Congyu Yao [1,3], Wei Wang [1,2], Zhiwei Zhuang [4], Junzhong Liu [1,2], Fangqun Dai [1,2] and Jianhua Hao [1,2,5,*]

1. Key Laboratory of Sustainable Development of Polar Fishery, Ministry of Agriculture and Rural Affairs, Yellow Sea Fisheries Research Institute, Chinese Academy of Fishery Sciences, Qingdao 266071, China; yaocongyv@foxmail.com (C.Y.); weiwang@ysfri.ac.cn (W.W.); qdjz99@163.com (J.L.); dai@ysfri.ac.cn (F.D.)
2. Laboratory for Marine Drugs and Bioproducts, Laboratory for Marine Fisheries Science and Food Production Processes, Qingdao National Laboratory for Marine Science and Technology, Qingdao 266071, China
3. College of Food Science and Technology, Shanghai Ocean University, Shanghai 201306, China
4. New Hope Liuhe Co. Ltd., Qingdao 266071, China; zzw19680@163.com
5. Jiangsu Collaborative Innovation Center for Exploitation and Utilization of Marine Biological Resource, Lianyungang 222005, China
* Correspondence: sunjj@ysfri.ac.cn (J.S.); haojh@ysfri.ac.cn (J.H.); Tel.: +86-532-8584-1193 (J.S.); +86-532-8581-9525 (J.H.)

Received: 11 October 2018; Accepted: 23 November 2018; Published: 6 August 2020

Abstract: The bacterium *Alteromonas* sp. ML52, isolated from deep-sea water, was found to synthesize an intracellular cold-adapted β-galactosidase. A novel β-galactosidase gene from strain ML52, encoding 1058 amino acids residues, was cloned and expressed in *Escherichia coli*. The enzyme belongs to glycoside hydrolase family 2 and is active as a homotetrameric protein. The recombinant enzyme had maximum activity at 35 °C and pH 8 with a low thermal stability over 30 °C. The enzyme also exhibited a K_m of 0.14 mM, a V_{max} of 464.7 U/mg and a k_{cat} of 3688.1 S^{-1} at 35 °C with 2-nitrophenyl-β-D-galactopyranoside as a substrate. Hydrolysis of lactose assay, performed using milk, indicated that over 90% lactose in milk was hydrolyzed after incubation for 5 h at 25 °C or 24 h at 4 °C and 10 °C, respectively. These properties suggest that recombinant *Alteromonas* sp. ML52 β-galactosidase is a potential biocatalyst for the lactose-reduced dairy industry.

Keywords: *Alteromonas*; deep sea; cold-adapted enzyme; β-galactosidase; lactose-free milk

1. Introduction

Beta-galactosidase (EC 3.2.1.23), a glycoside hydrolase enzyme, catalyzes the hydrolysis of terminal non-reducing β-D-galactose residues into β-D-galactosides and also catalyzes transgalactosylation reactions [1–3]. Beta-galactosidases exist naturally in many organisms, including microorganisms, plants and animals [4,5]. Most industrial β-galactosidases are obtained from microorganisms. For example, the enzymes isolated from bacteria [6] and yeast [7], with neutral optimum pH, were used in milk products, and fungal [8] enzymes with an acid optimum pH were used in acid whey products. The main application of β-galactosidase is to hydrolyze lactose in milk in the dairy industry to provide lactose-free milk for lactose-intolerant consumers [9]. Another application of β-galactosidase is to transfer lactose and monosaccharide to a series of galacto-oligosaccharides (GOS) which are functional galactosylated products [10–12]. However, β-galactosidase catalyzed at moderate temperatures may cause some issues, e.g., increased production costs, wasted energy and producing undesirable microbial contamination [13]. Cold-adapted β-galactosidases, with low optimum temperatures, could catalyze hydrolysis or transgalactosylation reactions at refrigerating temperatures (4–10 °C), thus potentially overcoming these shortcomings. It may be especially beneficial to the dairy industry which could improve the hydrolysis of lactose at low temperatures.

While a minority of β-galactosidases from fungus are secreted to the extracellular medium, e.g., an acid β-galactosidase from *Aspergillus* spp. [14], β-galactosidases are generally intracellular enzymes in yeast and bacteria. Most reported β-galactosidases are recombinant enzymes derived from heterologous expression than from a natural source. In recent years, the number of cold-adapted β-galactosidases were isolated from psychrophilic and psychrotrophic microorganisms obtained from isothermal cold environments such as polar [15–17], deep-sea [18] and high mountainous regions [19]. The main source of enzymes has been obtained from bacterial strains such as *Arthrobacter psychrolactophilus* strain F2 [20], *Arthrobacter* sp. 32c [21], *Halomonas* sp. S62 [22], *Paracoccus* sp. 32d [23], *Pseudoalteromonas haloplanktis* [24] and *Rahnella* sp. R3 [19]. Only a few cold-adapted β-galactosidases have been discovered from other sources, including psychrophilic-basidiomycetous yeast *Guehomyces pullulan* [25] and Antarctic haloarchaeon *Halorubrum lacusprofundi* [26]. The β-galactosidase from *Arthrobacter psychrolactophilus* strain F2 showed the lowest optimum temperature at 10 °C with an optimum pH of 8.

Based on the specific features of sequence, structure, substrate specificity and reaction mechanism, β-galactosidases have been classified into GH1, GH2, GH35 and GH42 families [27]. Most reported microorganism β-galactosidases belong to the GH2 [15,20,23,24,28] and GH42 [19,21,29] families. A typical GH2 β-galactosidase from *E. coli* is made up of five sequential domains and forms a functional tetramer [30]. Most of the characterized cold-adapted β-galactosidases from the GH2 family are tetrameric enzymes, except for a dimeric enzyme from *Paracoccus* sp. 32d and a hexameric enzyme from *Arthrobacter* sp. C2-2 [31]. Hitherto, three crystal structures of cold-adapted β-galactosidases have been obtained: GH42 β-galactosidase from *Planococcus* sp. L4 [32] and two GH2 β-galactosidases from Antarctic bacteria *Arthrobacter* sp. C2-2 [31] and *Paracoccus* sp. 32d [33].

In this study, we report on a gene of β-galactosidase from the marine bacterium *Alteromonas* sp. ML52, isolated from a deep-sea sample. This novel cold-adapted β-galactosidase belongs to the GH2 family and was overexpressed and characterized.

2. Results

2.1. Characterization and Identification of Strain ML52

Strain ML52 was isolated from deep-sea water in the Mariana Trench at a depth of 4000 m and found to produce intracellular β-galactosidase at 4 °C. Database searches showed that strain ML52 is related to the genus *Alteromonas*. As shown in the neighbor-joining tree (Figure 1) [34,35], strain ML52 formed a monophyletic cluster with *Alteromonas addita* R10SW13^T (99.9% identity), *Alteromonas stellipolaris* LMG 21861^T (99.8% identity) and *Alteromonas naphthalenivorans* SN2^T (99.3% identity). Strain ML52 was able to produce β-galactosidase on 2216E X-Gal agar in the presence of either lactose or glucose (Figure 2), while the expression of β-galactosidase in *E. coli* containing a lac operon was repressed by glucose [36].

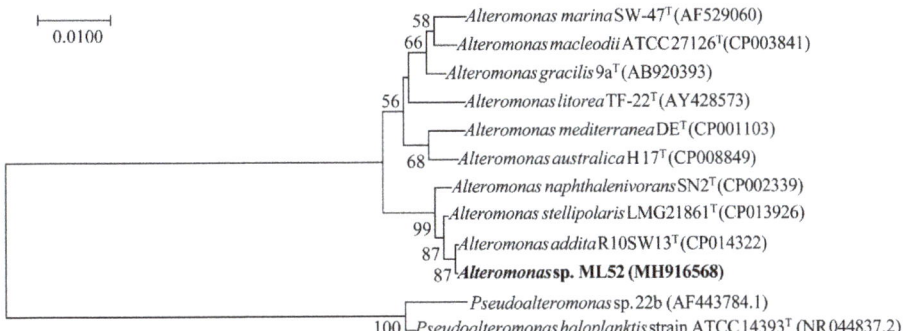

Figure 1. Neighbor-joining tree based on 16S rRNA gene sequences showing the phylogenetic position of strain ML52 and closely related *Alteromonas* and *Pseudoalteromonas* species. Bootstrap values (>50%) were calculated for 1000 replicates. The reference 16S rDNA sequences were collected from EzTaxon-e server (www.bacterio.net/) and the National Center for Biotechnology Information (NCBI) Database and aligned using the ClustalX 2.1 program (Conway Institute UCD Dublin, Dublin, Ireland). The phylogenetic tree was obtained using MEGA 7.0 software (Institute for Genomics and Evolutionary Medicine, Temple University, Tempe, AZ, USA).

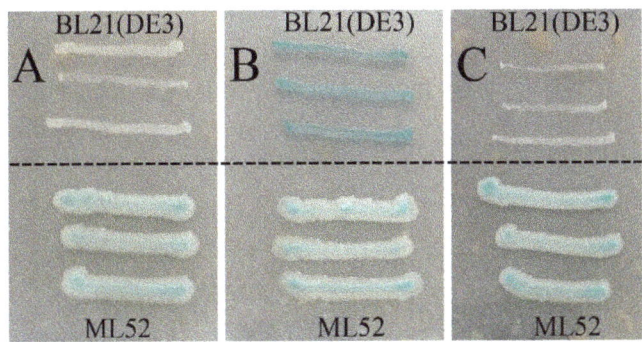

Figure 2. Effects of glucose and lactose on the expression of β-galactosidase of strain ML52. *E. coli* strain BL21(DE3) was used as a control. A, 2216E X-Gal agar with 2% glucose; B, 2216E X-Gal agar with 2% lactose; C, 2216E X-Gal agar with 2% glucose and lactose.

2.2. Molecular Cloning and Sequence Analysis

Deoxyribonucleic acid sequencing showed that *gal* consisted of an open reading frame of 3174 bp, encoding 1058 amino acid residues. The theoretical Mw and pI of the enzyme was 118,543 Da and 4.96, respectively. According to the BLAST results, *gal* had a highest identity of 99% to a putative β-galactosidase from *Alteromonas addita* (WP_062085674.1). At the time of writing, two characterized β-galactosidase genes from *Pseudoalteromonas* sp. 22b (AAR92204.1) and *Pseudoalteromonas haloplanktis* (CAA10470.1) exhibited the highest sequence identity (65%) to *gal*. Based on sequence comparisons, Gal was classified into glycoside hydrolase family 2. Amino acid sequence comparison of Gal and other characterized GH2 β-galactosidases are shown in Figure 3.

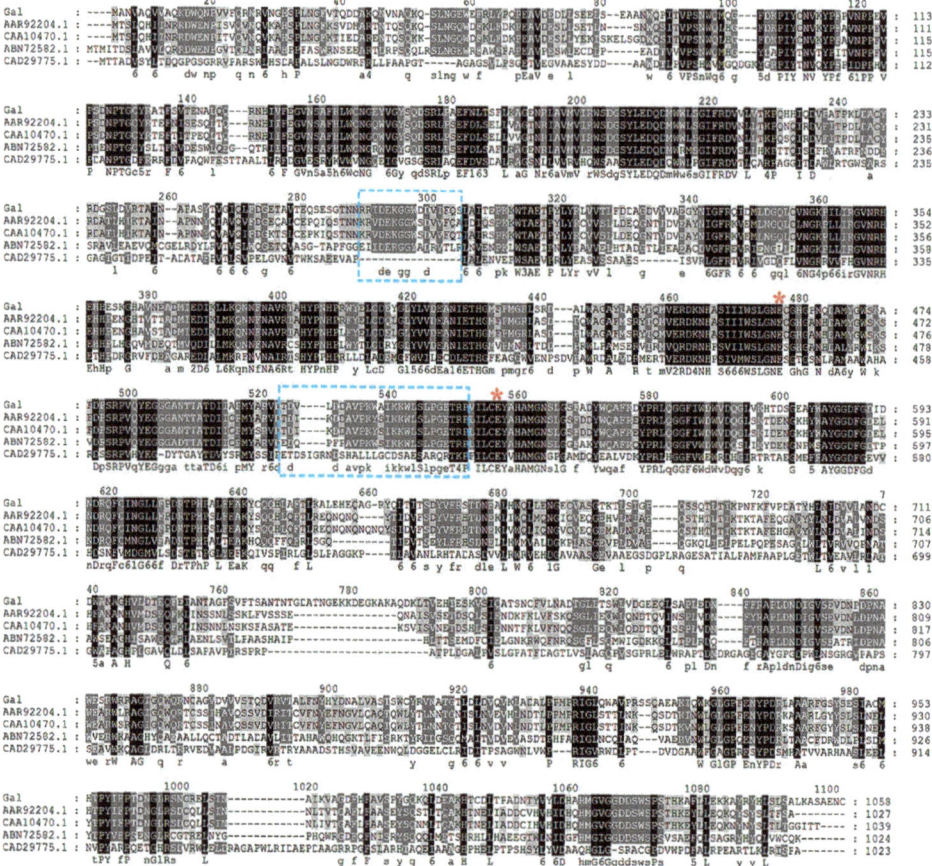

Figure 3. Sequence alignment of Gal and other GH2 β-galactosidase from different microorganisms. Identical residues are shaded in black and conserved residues are shaded in gray. The putative nucleophilic and catalytic amino acids are marked by a red asterisk and the regions relative to the formation of a tetramer are marked by blue boxes. Protein accession numbers and species are as follows: AAR92204.1, *Pseudoalteromonas* sp. 22b; CAA10470.1, *Pseudoalteromonas haloplanktis*; ABN72582.1, *E. coli* K-12; CAD29775.1, *Arthrobacter* sp. C2-2.

2.3. Expression and Purification of Recombinant Gal

The expression vector pET-24a was used for the expression of the *gal* gene from strain ML52 in *E. coli* BL21 (DE3). The recombinant enzyme with a 6-histidine tag was induced by IPTG and purified by Ni-NTA chromatography. After purification, approximately 55.2 mg of pure enzyme was obtained from 1L of induced culture. The apparent molecular weight of recombinant Gal was 126 kDa which was determined by SDS-PAGE (Figure 4), and which corresponded to the theoretical size. The relative molecular weight of recombinant Gal, which was determined by gel-filtration chromatography, was 485 kDa (results not shown). Hence recombinant Gal is probably a tetrameric protein, like *E. coli* β-galactosidase.

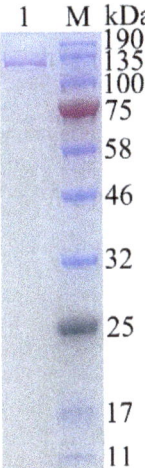

Figure 4. Twelve percent SDS-PAGE analysis of the purified β-galactosidase. Lane 1, purified β-galactosidase; lane M, protein marker. The gel was stained with 0.025% Coomassie blue R250.

2.4. Properties of Recombinant Gal

The optimum temperature for recombinant Gal was close to 35 °C and the enzyme had 19–30% activity at 4–10 °C (Figure 5A). Recombinant Gal maintained high activity over a pH range of 7 to 8.5 with maximum activity at pH 8. The enzyme was stable from a pH range between 6.5 and 9 (Figure 5B). Recombinant Gal was stable at temperatures between 4 and 20 °C, but its activity reduced rapidly at 30 °C and most activity was lost at 50 °C following half hour incubation (Figure 5C).

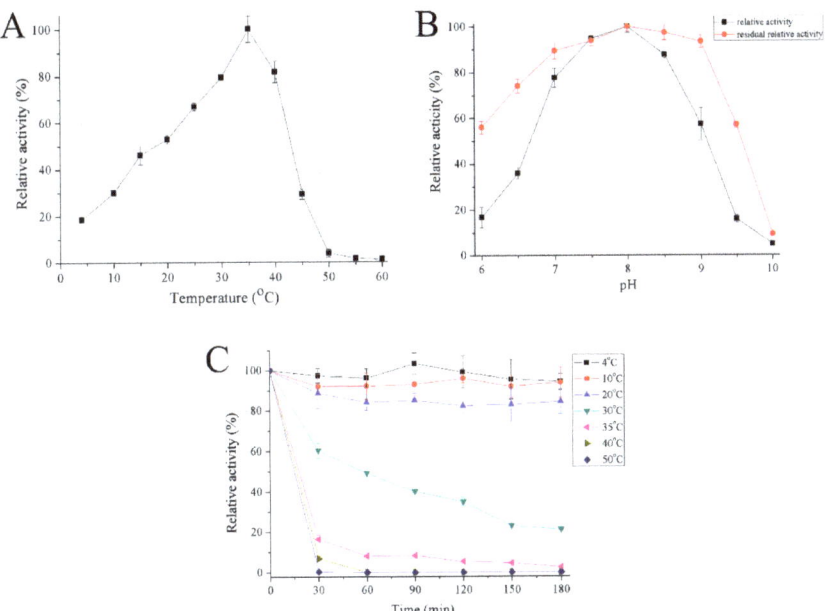

Figure 5. Effect of temperature on activity (**A**) and stability (**C**) of recombinant Gal and effect of pH (**B**) on activity of recombinant Gal. The optimum activity was set as 100% with specific activities of 434.3 U/mg, or 286 U/mg for the effect of temperature or pH, respectively.

Results outlining the effects of metal ions and chemicals on recombinant Gal are shown in Table 1. Reducing agents (DTT and 2-mercaptoethanol) and K^+ could stimulated the enzymatic activity of recombinant Gal. The addition of Ca^{2+} and urea slightly decreased enzyme activity whereas Mg^{2+}, Mn^{2+} and ionic detergent (SDS) had strong inhibitory effects. The presence of chelating agent (EDTA) and the ions Zn^{2+}, Ni^{2+}, and Cu^{2+} completely inhibited enzyme activity. Salt tolerance of the recombinant Gal was also investigated (Figure 6). Recombinant Gal maintained 53% residue activity in the presence of 1.5 M NaCl and 26% residue activity when NaCl concentration increased to 4 M.

Table 1. Effect of metal ions and chemicals on activity of recombinant Gal.

Ion/Reagent	Relative Activity (%)
None [1]	100 ± 1.3
K^+	123 ± 0.6
Ca^{2+}	74.9 ± 5.4
Mg^{2+}	32.4 ± 1.4
Mn^{2+}	21.3 ± 0.9
Ba^{2+}	48.2 ± 2.1
Zn^{2+}	0.1 ± 0.04
Ni^{2+}	0.2 ± 0.03
Cu^{2+}	0
EDTA	0.2 ± 0.02
Urea	89.9 ± 0.3
DTT	116.9 ± 7
2-Mercaptoethanol	108.9 ± 4
SDS	43.5 ± 1

[1] The activity of control (no additions) was set as 100% with a specific activity of 386.7 U/mg.

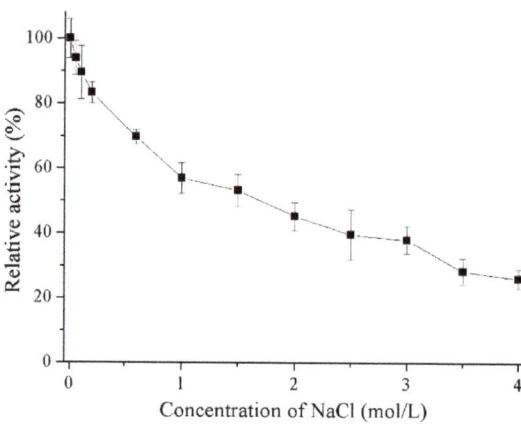

Figure 6. Effects of NaCl on the activity of recombinant Gal. Activity with 0 M NaCl was set as 100% with a specific activity of 445.3 U/mg.

The substrate specificity of recombinant Gal was investigated using seven chromogenic substrates (Table 2). The enzyme was specific to two β-D-galactopyranoside substrates, and showed no activity to other tested substrates. Kinetic parameters of recombinant Gal were determined at optimum temperature with ONPG (2-Nitrophenyl-β-D-galactopyranoside) and lactose as substrates, as shown in Table 3.

Table 2. Substrate specificity of recombinant Gal.

Substrate	Relative Activity (%)
2-Nitrophenyl-β-D-galactopyranoside (ONPG) [1]	100 ± 4.2
4-Nitrophenyl-β-D-galactopyranoside (PNPG)	12.8 ± 0.5
4-Nitrophenyl-α-D-galactopyranoside	0
2-Nitrophenyl-β-D-glucopyranoside	0
4-Nitrophenyl-β-D-glucopyranoside	0
4-Nitrophenyl-α-D-glucopyranoside	0
4-Nitrophenyl-β-D-xylopyranoside	0

[1] The activity of ONPG was set as 100% with a specific activity of 467.2 U/mg.

Table 3. Substrate specificity of recombinant Gal.

Substrate	V_{max} (U/mg)	K_m (mM)	k_{cat} (S^{-1})	k_{cat}/K_m (S^{-1} mM^{-1})
ONPG	464.7	0.14	3688.1	26343.6
Lactose	18.5	7.2	146.5	20.3

2.5. Hydrolysis of Lactose in Milk

The hydrolysis of milk lactose by recombinant Gal was determined at 4 °C, 10 °C and 25 °C (Figure 7 and Figure S1). The conversion rate of lactose in milk reached about 42% during the initial hour at 25 °C and 94% after 5 h incubation. At refrigerating temperatures of 4 °C and 10 °C, 58% and 64% of lactose was hydrolyzed after 8 h incubation, respectively. A lactose conversion rate of greater than 90% was reached after 24 h incubation and almost 100% after 48 h.

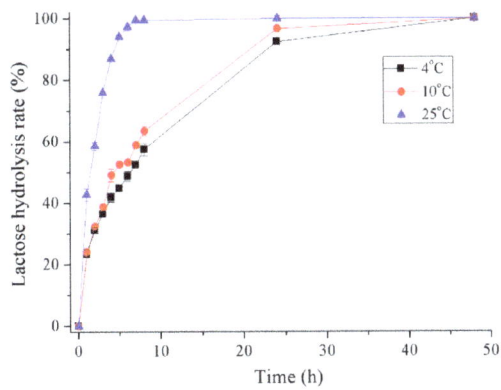

Figure 7. Hydrolysis of lactose in milk of recombinant Gal.

3. Discussion

A deep-sea bacterium *Alteromonas* sp. ML52 with β-galactosidase activity was obtained through plate screening. The β-galactosidase from strain ML52 was considered to be a constitutive enzyme because its expression was normal without lactose and not repressed by glucose. A novel GH2 β-galactosidase gene *gal* was cloned from strain ML52. Sequence alignment of Gal and other GH2 β-galactosidases (Figure 2) showed that Gal shares a conserved acid-base activity site and nucleophilic site typically found in this family of enzyme. Compared with a *E. coli* β-galactosidase, a mesophilic enzyme belonging to the GH2 family, the amino acid sequence composition of Gal showed significant decreases in Arg residues (4.5% versus 6.4%), Arg/Arg+Lys ratio (0.51 versus 0.77) and Pro residues (4.9% versus 6.1%) which are structural features of psychrophilic enzymes [30]. Skálová et al. [31]

compared two crystal structures of GH2 β-galactosidases, a tetrameric enzyme from *E. coli* and a hexameric enzyme from *Arthrobacter* sp. C2-2, and suggested that the outstanding loop region of domain 2, which participates in forming the contacts between subunits of the tetramer along with domain 3 in *E. coli* β-galactosidase; however, this kind of complementation does not occur in *Arthrobacter* sp. C2-2. The sequences relative to the formation of the tetramer of Gal in domain 2 and domain 3 were similar to that of the *E. coli* β-galactosidase, which are labeled in Figure 3. This theoretical prediction was tested by molecular weight determination. Thus, Gal is considered to be a homotetrameric protein.

Recombinant Gal has an optimum temperature of 35 °C and a low thermal stability over 30 °C. Other reported GH2 cold-adapted β-galactosidases possessed optimum temperatures between 10 and 40 °C (Table 4). Thus Gal, which was isolated from a deep-sea bacterium, is considered to be a cold-adapted enzyme. The optimum pH range for enzyme activity and stability was consistent with bacterial β-galactosidases which are normally active within a neutral pH range of 6 to 8. This feature indicates that Gal may be a suitable enzyme for lactose hydrolysis in milk (pH 6.5–6.8). The activating effect of K^+ and the inhibitory action of ions Zn^{2+}, Ni^{2+}, Cu^{2+} to the enzyme activity of recombinant Gal also occurred in a cold-adapted *Pseudoalteromonas* sp. 22b β-galactosidase [28], with the highest sequence identity to Gal. The chelating agent caused a complete inhibition of enzyme activity, indicating that the catalytic reaction of recombinant Gal may rely on the presence of metal ions. It was also noted that recombinant Gal preferred substrate ONPG to substrate PNPG. This phenomenon was also observed in cold-adapted β-galactosidases from *Paracoccus* sp. 32d [23] and *Arthrobacter psychrolactophilus* strain F2 [20], but many cold-adapted β-galactosidases, isolated from *Pseudoalteromonas* sp. 22b [28], *Arthrobacter* sp. 20B [37] and *Arthrobacter* sp. 32c [21], displayed higher levels of activity to PNPG. Remarkably, recombinant Gal showed significantly high affinity and reaction rate to the chromogenic substrate ONPG at optimum temperature compared to other GH2 cold-adapted β-galactosidases from *Arthrobacter psychrolactophilus* strain F2, *Arthrobacter* sp. and *Paracoccus* sp. 32d (Table 4). This property is similar to the *Pseudoalteromonas* sp. 22b β-galactosidase. For lactose, the natural substrate of β-galactosidase, recombinant Gal showed intermediate substrate affinity and catalytic efficiency at optimum temperature. The major industry β-galactosidases were obtained from *Aspergillus* spp. and *Kluyveromyces* spp. [1,4,8]. The β-galactosidases from *Aspergillus* spp. have an optimum pH at acidic range (2.5–5.4) and are not suitable for hydrolysis of lactose in milk. The β-galactosidase from *Kluyveromyces lactis* is one of the most widely used commercial enzymes. The optimum pH (6.6–7) of *K. lactis* β-galactosidase is close to milk because the dairy environment is a natural habitat for this kind of yeast [4]. However, *K. lactis* β-galactosidase showed higher optimum temperature (40 °C) [38] and lower substrate affinity to both ONPG and lactose (K_m, 1.7 mM for ONPG and 17.3 mM for lactose) [8] compared with recombinant Gal.

Table 4. Biochemical characteristics of reported GH2 cold-adapted β-galactosidases.

Strain	Optimum		ONPG		Lactose		References
	Temperature (°C)	pH	K_m (mM)	k_{cat} (S^{-1})	K_m (mM)	k_{cat} (S^{-1})	
Alteromonas sp. ML52	35	8	0.14	3688.1	7.2	146.5	This work
Arthrobacter psychrolactophilus strain F2	10	8	2.7	12.7	42.1	3.02	[20]
Arthrobacter sp. 20B	25	6–8	-	-	-	-	[37]
Arthrobacter sp. C2-2	40	7.5	-	-	53.1	1106	[39]
Arthrobacter sp.	18	7	11.5	5.2	-	-	[17]
Paracoccus sp. 32d	40	7.5	1.17	71.81	2.94	43.23	[23]
Pseudoalteromonas haloplanktis TAE 79	-	8.5	-	203	2.4	33	[24]
Pseudoalteromonas sp. 22b	40	6–8	0.28	312	3.3	157	[28]

The results on lactose digestion in milk indicate that recombinant Gal could hydrolyze 100% of lactose in milk when incubated for 7 h at 25 °C and over 90% lactose was hydrolyzed after incubation for 24 h at 4 °C and 10 °C. No transglycosylation product was observed by HPLC during the hydrolysis process. For the *Pseudoalteromonas* sp. 22b cold-adapted β-galactosidase [28], 90% of lactose was hydrolyzed after 6 h at 30 °C and after 28 h at 15 °C. Another cold-adapted β-galactosidase from the *Arthrobacter psychrolactophilus* strain F2 [20] was able to hydrolyze approximately 70% of the lactose in milk at 10 °C after 24 h and displayed transglycosylation activity by forming a trisaccharide product during the reaction.

4. Materials and Methods

4.1. Isolation and Identification of Bacteria

Deep-sea water samples were collected from the Mariana Trench at a depth of 4000 m in September 2016. They were diluted and spread on marine agar 2216E (MA; BD Difco, Franklin Lakes, NJ, USA) containing 40 µg/mL X-Gal (5-bromo-4-chloro-3-indolyl-β-D-galactopyranoside, Sigma, St. Louis, MO, USA) and 2% lactose. After incubation at 4 °C for 2 weeks, the detectable blue colonies were repeatedly streaked on MA to obtain pure cultures. The 16S rRNA gene of strain ML52 was amplified by PCR from genomic DNA. To test the effects of glucose and lactose on the expression of β-galactosidase, strains were grown on MA containing 40 µg/mL X-gal and 2% glucose or lactose at 25 °C for 2 days.

4.2. Molecular Cloning and Sequence Analysis

Based on the DNA sequence of the predicted β-galactosidase gene (Accession number AMJ93096.1) from *Alteromonas addita* strain R10SW13 (Accession number CP014322.1) reported in the NCBI Database, two primers, forward primer-CG*GGATCC*ATGGCAAATGTTGCTCAA and reverse primer-CC*GCTCGAG*GCAATTTTCAGCACT (*Bam*HI and *Xho*I restriction sites are in italics) were designed. The PCR product was cloned into pMD20-T and sequenced by Tsingke, China. The plasmid pET-24a and amplified *gal* gene were then cleaved with the restriction endonucleases *Bam*HI and *Xho*I and the fragments were ligated with T4 DNA ligase (NEB), resulting in the plasmid pET-*gal*. The nucleotide sequence of *gal* was translated to an amino acid sequence using the ExPASY-Translate tool. A multiple alignment between Gal and other GH2 β-galactosidases from bacteria was constructed using ClustalW 2 (Conway Institute UCD Dublin, Dublin, Ireland) and GeneDoc 2.7 (Karl B. Nicholas et al., San Francisco, CA, USA). The theoretical molecular weight (Mw) and isoelectric point (pI) were calculated using the Compute pI/Mw tool (http://web.expasy.org/compute_pi/).

4.3. Expression and Purification

For expression of recombinant Gal, the plasmid pET-*gal* was transformed into *E. coli* BL21 (DE3) and cultivated at 37 °C in lysogeny broth (LB) containing 30 µg/mL kanamycin until the OD_{600} of the culture reached 0.6–0.8. Following this step, isopropyl-β-D-thiogalactopyranoside (IPTG) was added to the culture with a final concentration of 0.2 mM and the culture was further grown at 16 °C for 12 h. The cells were harvested by centrifugation at 8000 rpm for 10 min at 4 °C and suspended in a lysis solution (50 mM sodium phosphate buffer, pH 8). The mixture was then disrupted by sonication and the cell debris removed by centrifugation at 13,000 rpm for 15 min. The filtered supernatant was applied to a Ni^{2+}-chelating affinity column (His-Trap™ HP, GE, Madison, WI, USA) which was previously equilibrated using an equilibration buffer (50 mM sodium phosphate buffer, 20 mM imidazole, 300 mM NaCl, pH 8) and subsequently eluted using a linear gradient of 20–250 mM imidazole in equilibration buffer. Enzyme purity was analyzed by SDS-PAGE (GenScript, Nanjing, China). Purified recombinant Gal was buffer-exchanged to 50 mM sodium phosphate buffer (pH 8) via centrifugal ultrafiltration (MW cut off 10 kDa, Millipore, Burlington, MA, USA). The protein concentration was determined using the BCA protein assay kit (Solarbio, Beijing, China) with bovine serum albumin as a standard.

4.4. Estimation of Molecular Weight

The purified enzyme was applied to a gel-filtration column (Superdex 200 HR 10/30, GE Healthcare, Madison, WI, USA) and eluted using a buffer containing 50 mM sodium phosphate (pH 7) and 150 mM NaCl. The standard proteins used were thyroglobumin (669 kDa), apoferritin (443 kDa), β-amylase (200 kDa), alcohol dehydrogenase (150 kDa), bovine serum albumin (66 kDa), and carbonic anhydrase (29 kDa) purchased from Sigma (St. Louis, MO, USA).

4.5. Beta-Galactosidase Activity Assay

Beta-galactosidase activity was determined in a 100 µL reaction mixture containing a final concentration of 0.4 ng/µL (0.83 nM) of purified recombinant Gal, 50 mM sodium phosphate buffer (pH 8) and 5 mM 2-nitrophenyl-β-D-galactopyranoside (ONPG). Each reaction was incubated for 5 min at 35 °C and quenched by the addition of 200 µL of 1 M Na_2CO_3. The absorbance of the released o-nitrophenol (ONP) was measured at 420 nm and quantified using an ONP standard curve. One unit (U) of enzyme activity was defined as the amount of enzyme required for the liberation of 1 µmol ONP per minute under the assay conditions. All assays were carried out in triplicate.

4.6. Effect of Temperature and pH

The optimum temperature of recombinant Gal was evaluated by incubating the reaction mixtures at different temperatures ranging from 4 °C to 60 °C. The optimum pH was determined by incubation in 10 mM Britton–Robinson buffers ranging from pH 6 to 10 (in increments of 0.5 pH units) at optimum temperature. The thermostability of recombinant Gal was determined by incubating enzymes at temperatures ranging from 4 °C to 50 °C for 180 min and the residual enzyme activity was determined every 30 min under the optimum conditions. The pH stability was determined after incubating the enzyme in Britton–Robinson buffers ranging from pH 6 to 10 (in increments of 0.5 pH units) at 4 °C for 3 h.

4.7. Effect of Metal Ions and Chemicals

The effect of metal ions and chemicals on recombinant Gal activity was determined after incubating the enzyme in water with 5 mM of KCl, $CaCl_2$, $MgCl_2$, $MnCl_2$, $BaCl_2$, $ZnCl_2$, $NiSO_4$, $CuCl_2$, EDTA, urea,

DTT, 2-mercaptoethanol or SDS at 4 °C for 1 h. The activity of the enzyme when incubated without any additional reagents was considered to be 100%. For salt stability, 0–4 M NaCl (final concentration) was added to the reaction mixtures and the activity determined using optimum conditions.

4.8. Determination of Substrate Specificity and Kinetic Parameters

The substrate specificity of the recombinant Gal was estimated using 5 mM (final concentration) ONPG, 4-Nitrophenyl-β-D-galactopyranoside (PNPG), 4-Nitrophenyl-α-D-galactopyranoside, 2-Nitrophenyl-β-D-glucopyranoside, 4-Nitrophenyl-β-D-glucopyranoside, 4-Nitrophenyl-α-D-glucopyranoside or 4-Nitrophenyl-β-D-xylopyranoside in 50 mM sodium phosphate buffer (pH 8) under the optimum conditions. For the activity of 4-nitrophenol-derived substrates, the absorbance of the released 4-nitrophenol (PNP) was measured at 405 nm and quantified using a PNP standard curve. Kinetic parameters of the recombinant Gal were determined at 35 °C and the reaction rate with ONPG (0.1–5 mM) and lactose (0.2–40 mM) as substrates was determined. The released glucose in the lactose hydrolysis reaction was measured using a commercial glucose oxidase-peroxidase assay kit (Shanghai Rongsheng Biotech Co., Ltd., Shanghai, China). One unit (U) of enzyme activity was defined as 1 µmol of glucose released per minute. The values of the kinetic constants were calculated using the Lineweaver-Burk method.

4.9. Hydrolysis of Lactose in Milk

The hydrolysis of lactose in milk was determined by incubating 100 µg purified Gal in 1 mL of commercial skim milk (Inner Mongolia Yili Industrial Group Co. Ltd., Hohhot, China) at 4 °C, 10 °C, or 25 °C for 48 h. The reaction was terminated by incubating the sample at 60 °C for 5 min and then adding an equal volume of 5% trichloroacetic acid (TCA) to the reaction mixture. The pH of the reaction mixture was adjusted to neutral pH using 1 M NaOH and centrifuged at 10,000 rpm for 10 min. The supernatant was analyzed by HPLC using the ZORBAX Carbohydrate Analysis Column (Agilent, Palo Alto, CA, USA), with 75% acetonitrile used as a mobile phase at a flow rate of 1 mL/min and a Shimadzu Refractive Index Detector (Kyoto, Japan).

4.10. Nucleotide Sequence Accession Numbers

The 16S rRNA gene sequence and β-galactosidase gene (*gal*) of strain ML52 were deposited in GenBank under the accession numbers MH916568 and MH925304, respectively.

5. Conclusions

In this study, a new GH2 β-galactosidase (Gal) was successfully cloned, purified and characterized from the deep-sea bacterium *Alteromonas* sp. ML52. Recombinant Gal is a cold-adapted enzyme and was able to efficiently hydrolyze lactose in milk at refrigerating temperature. These characteristics suggest that Gal could be a potential cold-active biocatalyst and usefully applied to the production of lactose-free milk in the dairy industry.

Supplementary Materials: The following are available online at http://www.mdpi.com/1660-3397/16/12/469/s1, Figure S1: HPLC-RID profiles of hydrolysis of lactose in milk at 4 °C (A), 10 °C (B) and 25 °C (C). (1) Glucose; (2) Galactose; (3) Lactose.

Author Contributions: Data curation, J.S., C.Y. and Z.Z.; Funding acquisition, J.S., W.W. and J.H.; Methodology, J.S., W.W. and J.H.; Project administration, J.S., C.Y., J.L. and F.D.; Writing-original draft, J.S.; Writing-review and editing, J.S. and J.H.

Funding: This research was funded by Central Public-interest Scientific Institution Basal Research Fund, YSFRI, CAFS (No. 20603022018006); the Financial Fund of the Ministry of Agriculture and Rural Affairs, P. R. of China (No. NFZX2013); Central Public-interest Scientific Institution Basal Research Fund (No. 2017HY-ZD1003); the Scientific and Technological Innovation Project Financially Supported by Qingdao National Laboratory for Marine Science and Technology (No. 2016ASKJ14).

Conflicts of Interest: The authors declare no conflict of interest.

References

1. Oliveira, C.; Guimarães, P.M.; Domingues, L. Recombinant microbial systems for improved β-galactosidase production and biotechnological applications. *Biotechnol. Adv.* **2011**, *29*, 600–609. [CrossRef] [PubMed]
2. Ansari, S.A.; Satar, R. Recombinant β-galactosidases—Past, present and future: A mini review. *J. Mol. Catal. B-Enzym.* **2012**, *81*, 1–6. [CrossRef]
3. Park, A.R.; Oh, D.K. Galacto-oligosaccharide production using microbial β-galactosidase: Current state and perspectives. *Appl. Microbiol. Biotechnol.* **2010**, *85*, 1279–1286. [CrossRef] [PubMed]
4. Husain, Q. Beta galactosidases and their potential applications: A review. *Crit. Rev. Biotechnol.* **2010**, *30*, 41–62. [CrossRef] [PubMed]
5. Chandrasekar, B.; van der Hoorn, R.A. Beta galactosidases in Arabidopsis and tomato—A mini review. *Biochem. Soc. Trans.* **2016**, *44*, 150–158. [CrossRef] [PubMed]
6. Vasiljevic, T.; Jelen, P. Lactose hydrolysis in milk as affected by neutralizers used for the preparation of crude β-galactosidase extracts from *Lactobacillus bulgaricus* 11842. *Innov. Food Sci. Emerg. Technol.* **2002**, *3*, 175–184. [CrossRef]
7. Kim, C.S.; Ji, E.S.; Oh, D.K. A new kinetic model of recombinant beta-galactosidase from *Kluyveromyces lactis* for both hydrolysis and transgalactosylation reactions. *Biochem. Biophys. Res. Commun.* **2004**, *316*, 738–743. [CrossRef] [PubMed]
8. Panesar, P.S.; Panesar, R.; Singh, R.S.; Kennedy, J.F.; Kumar, H. Microbial production, immobilization and applications of beta-D-galactosidase. *J. Chem. Technol. Biotechnol.* **2006**, *81*, 530–543. [CrossRef]
9. Horner, T.W.; Dunn, M.L.; Eggett, D.L.; Ogden, L.V. β-Galactosidase activity of commercial lactase samples in raw and pasteurized milk at refrigerated temperatures. *J. Dairy Sci.* **2011**, *94*, 3242–3249. [CrossRef] [PubMed]
10. Rodriguez-Colinas, B.; Fernandez-Arrojo, L.; Santos-Moriano, P.; Ballesteros, A.; Plou, F. Continuous Packed Bed Reactor with Immobilized β-Galactosidase for Production of Galactooligosaccharides (GOS). *Catalysts* **2016**, *6*, 189. [CrossRef]
11. Zhang, J.; Lu, L.; Lu, L.; Zhao, Y.; Kang, L.; Pang, X.; Liu, J.; Jiang, T.; Xiao, M.; Ma, B. Galactosylation of steroidal saponins by β-galactosidase from *Lactobacillus bulgaricus* L3. *Glycoconj. J.* **2016**, *33*, 53–62. [CrossRef] [PubMed]
12. Maischberger, T.; Leitner, E.; Nitisinprasert, S.; Juajun, O.; Yamabhai, M.; Nguyen, T.H.; Haltrich, D. Beta-galactosidase from *Lactobacillus pentosus*: Purification, characterization and formation of galacto-oligosaccharides. *Biotechnol. J.* **2010**, *5*, 838–847. [CrossRef] [PubMed]
13. Feller, G.; Gerday, C. Psychrophilic enzymes: Hot topics in cold adaptation. *Nat. Rev. Microbiol.* **2003**, *1*, 200–208. [CrossRef] [PubMed]
14. Pakula, T.M.; Salonen, K.; Uusitalo, J.; Penttilä, M. The effect of specific growth rate on protein synthesis and secretion in the filamentous fungus *Trichoderma reesei*. *Microbiology* **2005**, *151*, 135–143. [CrossRef] [PubMed]
15. Schmidt, M.; Stougaard, P. Identification, cloning and expression of a cold-active beta-galactosidase from a novel Arctic bacterium, *Alkalilactibacillus ikkense*. *Environ. Technol.* **2010**, *31*, 1107–1114. [CrossRef] [PubMed]
16. Alikkunju, A.P.; Sainjan, N.; Silvester, R.; Joseph, A.; Rahiman, M.; Antony, A.C.; Kumaran, R.C.; Hatha, M. Screening and Characterization of Cold-Active β-Galactosidase Producing Psychrotrophic *Enterobacter ludwigii* from the Sediments of Arctic Fjord. *Appl. Biochem. Biotechnol.* **2016**, *180*, 477–490. [CrossRef] [PubMed]
17. Coker, J.A.; Sheridan, P.P.; Loveland-Curtze, J.; Gutshall, K.R.; Auman, A.J.; Brenchley, J.E. Biochemical characterization of a beta-galactosidase with a low temperature optimum obtained from an Antarctic *Arthrobacter* isolate. *J. Bacteriol.* **2003**, *185*, 5473–5482. [CrossRef] [PubMed]
18. Ghosh, M.; Pulicherla, K.K.; Rekha, V.P.; Raja, P.K.; Sambasiva Rao, K.R. Cold active β-galactosidase from *Thalassospira* sp. 3SC-21 to use in milk lactose hydrolysis: A novel source from deep waters of Bay-of-Bengal. *World J. Microbiol. Biotechnol.* **2012**, *28*, 2859–2869. [CrossRef] [PubMed]
19. Fan, Y.; Hua, X.; Zhang, Y.; Feng, Y.; Shen, Q.; Dong, J.; Zhao, W.; Zhang, W.; Jin, Z.; Yang, R. Cloning, expression and structural stability of a cold-adapted β-galactosidase from *Rahnella* sp. R3. *Protein Expr. Purif.* **2015**, *115*, 158–164. [CrossRef] [PubMed]

20. Nakagawa, T.; Ikehata, R.; Myoda, T.; Miyaji, T.; Tomizuka, N. Overexpression and functional analysis of cold-active beta-galactosidase from *Arthrobacter psychrolactophilus* strain F2. *Protein Expr. Purif.* **2007**, *54*, 295–299. [CrossRef] [PubMed]
21. Hildebrandt, P.; Wanarska, M.; Kur, J. A new cold-adapted beta-D-galactosidase from the Antarctic *Arthrobacter* sp. 32c-gene cloning, overexpression, purification and properties. *BMC Microbiol.* **2009**, *9*, 151. [CrossRef] [PubMed]
22. Wang, G.X.; Gao, Y.; Hu, B.; Lu, X.L.; Liu, X.Y.; Jiao, B.H. A novel cold-adapted β-galactosidase isolated from *Halomonas* sp. S62: Gene cloning, purification and enzymatic characterization. *World J. Microbiol. Biotechnol.* **2013**, *29*, 1473–1480. [CrossRef] [PubMed]
23. Wierzbicka-Woś, A.; Cieśliński, H.; Wanarska, M.; Kozłowska-Tylingo, K.; Hildebrandt, P.; Kur, J. A novel cold-active β-D-galactosidase from the *Paracoccus* sp. 32d—Gene cloning, purification and characterization. *Microb. Cell Fact.* **2011**, *10*, 108. [CrossRef]
24. Hoyoux, A.; Jennes, I.; Dubois, P.; Genicot, S.; Dubail, F.; François, J.M.; Baise, E.; Feller, G.; Gerday, C. Cold-adapted beta-galactosidase from the Antarctic psychrophile *Pseudoalteromonas haloplanktis*. *Appl. Environ. Microbiol.* **2001**, *67*, 1529–1535. [CrossRef] [PubMed]
25. Nakagawa, T.; Ikehata, R.; Uchino, M.; Miyaji, T.; Takano, K.; Tomizuka, N. Cold-active acid beta-galactosidase activity of isolated psychrophilic-basidiomycetous yeast *Guehomyces pullulans*. *Microbiol. Res.* **2006**, *161*, 75–79. [CrossRef] [PubMed]
26. Karan, R.; Capes, M.D.; DasSarma, P.; DasSarma, S. Cloning, overexpression, purification, and characterization of a polyextremophilic β-galactosidase from the Antarctic haloarchaeon *Halorubrum lacusprofundi*. *BMC Biotechnol.* **2013**, *13*, 3. [CrossRef] [PubMed]
27. Henrissat, B.; Davies, G. Structural and sequence-based classification of glycoside hydrolases. *Curr. Opin. Struct. Biol.* **1997**, *7*, 637–644. [CrossRef]
28. Cieśliński, H.; Kur, J.; Białkowska, A.; Baran, I.; Makowski, K.; Turkiewicz, M. Cloning, expression, and purification of a recombinant cold-adapted beta-galactosidase from antarctic bacterium *Pseudoalteromonas* sp. 22b. *Protein Expr. Purif.* **2005**, *39*, 27–34. [CrossRef] [PubMed]
29. Hu, J.M.; Li, H.; Cao, L.X.; Wu, P.C.; Zhang, C.T.; Sang, S.L.; Zhang, X.Y.; Chen, M.J.; Lu, J.Q.; Liu, Y.H. Molecular cloning and characterization of the gene encoding cold-active beta-galactosidase from a psychrotrophic and halotolerant *Planococcus* sp. L4. *J. Agric. Food Chem.* **2007**, *55*, 2217–2224. [CrossRef] [PubMed]
30. Uchil, P.D.; Nagarajan, A.; Kumar, P. β-Galactosidase. *Cold Spring Harb. Protoc.* **2017**, 774–779. [CrossRef] [PubMed]
31. Skálová, T.; Dohnálek, J.; Spiwok, V.; Lipovová, P.; Vondráčková, E.; Petroková, H.; Dusková, J.; Strnad, H.; Králová, B.; Hasek, J. Cold-active beta-galactosidase from *Arthrobacter* sp. C2-2 forms compact 660 kDa hexamers: Crystal structure at 1.9A resolution. *J. Mol. Biol.* **2005**, *353*, 282–294. [CrossRef] [PubMed]
32. Zhang, L.; Wang, K.; Mo, Z.; Liu, Y.; Hu, X. Crystallization and preliminary X-ray analysis of a cold-active β-galactosidase from the psychrotrophic and halotolerant *Planococcus* sp. L4. *Acta Crystallogr. Sect. F Struct. Biol. Cryst. Commun.* **2011**, *67*, 911–913. [CrossRef] [PubMed]
33. Rutkiewicz-Krotewicz, M.; Pietryk-Brzezinska, A.J.; Sekula, B.; Cieśliński, H.; Wierzbicka-Woś, A.; Kur, J.; Bujacz, A. Structural studies of a cold-adapted dimeric β-D-galactosidase from *Paracoccus* sp. 32d. *Acta Crystallogr. D Struct. Biol.* **2016**, *72*, 1049–1061. [CrossRef] [PubMed]
34. Saitou, N.; Nei, M. The neighbor-joining method: A new method for reconstructing phylogenetic trees. *Mol. Biol. Evol.* **1987**, *4*, 406–425. [CrossRef] [PubMed]
35. Kumar, S.; Stecher, G.; Tamura, K. MEGA7: Molecular Evolutionary Genetics Analysis Version 7.0 for Bigger Datasets. *Mol. Biol. Evol.* **2016**, *33*, 1870–1874. [CrossRef] [PubMed]
36. Arukha, A.P.; Mukhopadhyay, B.C.; Mitra, S.; Biswas, S.R. A constitutive unregulated expression of β-galactosidase in *Lactobacillus fermentum* M1. *Curr. Microbiol.* **2015**, *70*, 253–259. [CrossRef] [PubMed]
37. Białkowska, A.M.; Cieśliński, H.; Nowakowska, K.M.; Kur, J.; Turkiewicz, M. A new beta-galactosidase with a low temperature optimum isolated from the Antarctic *Arthrobacter* sp. 20B: Gene cloning, purification and characterization. *Arch. Microbiol.* **2009**, *191*, 825–835. [CrossRef] [PubMed]
38. Zhou, Q.Z.K.; Chen, X.D. Effects of temperature and pH on the catalytic activity of the immobilized β-galactosidase from *Kluyveromyces lactis*. *Biochem. Eng. J.* **2001**, *9*, 33–40. [CrossRef]

39. Karasová-Lipovová, P.; Strnad, H.; Spiwok, V.; Maláa, Š.; Králová, B.; Russell, N.J. The cloning, purification and characterisation of a cold-active β-galactosidase from the psychrotolerant Antarctic bacterium *Arthrobacter* sp. C2-2. *Enzym. Microb. Technol.* **2003**, *33*, 836–844. [CrossRef]

© 2018 by the authors. Licensee MDPI, Basel, Switzerland. This article is an open access article distributed under the terms and conditions of the Creative Commons Attribution (CC BY) license (http://creativecommons.org/licenses/by/4.0/).

Article

Purification and Characterization of a Biofilm-Degradable Dextranase from a Marine Bacterium

Wei Ren [1,2,3], Ruanhong Cai [2,4], Wanli Yan [1,3,5], Mingsheng Lyu [1,3,5], Yaowei Fang [1,3,5] and Shujun Wang [1,3,5,*]

1. Jiangsu Marine Resources Development Research Institute, Huaihai Institute of Technology, Lianyungang 222005, China; renwei1004570447@163.com (W.R.); yanwanli2016@163.com (W.Y.); mslu@hhit.edu.cn (M.L.); foroei@163.com (Y.F.)
2. Key Laboratory of Marine Biology, Nanjing Agricultural University, Nanjing 210000, China; crh1987@163.com
3. Co-Innovation Center of Jiangsu Marine Bio-industry Technology, Huaihai Institute of Technology, Lianyungang 222005, China
4. State Key Laboratory of Marine Environmental Science, College of Ocean and Earth Sciences, Xiamen University, Xiamen 361005, China
5. College of Marine Life and Fisheries, Huahai Institute of Technology, Lianyungang 222005, China
* Correspondence: shjwang@hhit.edu.cn; Tel.: +86-0518-8589-5905

Received: 18 November 2017; Accepted: 31 January 2018; Published: 7 February 2018

Abstract: This study evaluated the ability of a dextranase from a marine bacterium *Catenovulum* sp. (Cadex) to impede formation of *Streptococcus mutans* biofilms, a primary pathogen of dental caries, one of the most common human infectious diseases. Cadex was purified 29.6-fold and had a specific activity of 2309 U/mg protein and molecular weight of 75 kDa. Cadex showed maximum activity at pH 8.0 and 40 °C and was stable at temperatures under 30 °C and at pH ranging from 5.0 to 11.0. A metal ion and chemical dependency study showed that Mn^{2+} and Sr^{2+} exerted positive effects on Cadex, whereas Cu^{2+}, Fe^{3+}, Zn^{2+}, Cd^{2+}, Ni^{2+}, and Co^{2+} functioned as inhibitors. Several teeth rinsing product reagents, including carboxybenzene, ethanol, sodium fluoride, and xylitol were found to have no effects on Cadex activity. A substrate specificity study showed that Cadex specifically cleaved the α-1,6 glycosidic bond. Thin layer chromatogram and high-performance liquid chromatography indicated that the main hydrolysis products were isomaltooligosaccharides. Crystal violet staining and scanning electron microscopy showed that Cadex impeded the formation of *S. mutans* biofilm to some extent. In conclusion, Cadex from a marine bacterium was shown to be an alkaline and cold-adapted endo-type dextranase suitable for development of a novel marine agent for the treatment of dental caries.

Keywords: marine agent; *Catenovulum*; alkaline and cold-adapted dextranase; isomaltooligosaccharides; biofilm; dental caries

---

1. Introduction

Dextranases (α-1,6-D-glucan-6-glucanohydrolase; EC 3.2.1.11) hydrolyze dextran to oligosaccharides at the α-1,6 glucosidic bond, and are widely used in medical, dental, and sugar industries. In clinical applications, specific clinical dextran produced by dextranase can be used as a blood substitute in emergencies [1–3]. In the sugar industry, dextranase has been used to resolve the poor clarification and throughput that dextran can cause in sugarcane juice [4–7]. It is worth noting that this enzyme can be used with commercial dextran to directly synthetize isomaltose and isomaltooligosaccharides which exhibit prebiotic effects [8–10]. A previous report proposed that dextranase may be capable of treating dental plaques [2,3,11]. For this reason, the use of dextranase to treat dental caries has attracted a great deal of attention, particularly with respect to the degradation of dextran in dental plaques. Bacterial cells form biofilms as a protective barrier from external conditions, serving as a mechanism for improving survival and dispersion [12,13]. *Streptococcus mutans* is the main cause of dental decay in human teeth and key modulator of the development of cariogenic biofilms [14,15]. Accumulation of this cariogenic bacterium within the biofilm may lead to the onset of periodontal inflammation. Thus, dislodging the biofilm is the main therapy for periodontal inflammation.

An alkaline dextranase may be suitable for the treatment of dental caries because alkaline tooth-rinse products are expected to be more amenable to enamel than acidic products [16]. In addition, dextranase works efficiently at temperatures of about 37 °C and may also contribute to the degradation of human dental plaques. However, the most common source of dextranase, fungi, produce many acidic and megathermal dextranases, which catalyze at pH values ranging from pH 5.0 to 6.5 and temperatures above 50 °C, and are unstable under alkaline conditions. Dextranases are seldom capable of catalysis under both alkaline and moderate-temperature conditions [1]; however, enzymes from marine microorganisms may be an exception [17–19].

In this study, a dextranase produced by a marine bacterium *Catenovulum* sp. DP03 (CGMCC No. 7386) was sequenced and selected for extracellular experiments [20]. This study demonstrated a method for purifying, characterizing and hydrolyzing products of dextranase from the marine strain DP03 and analyzed its effects on *S. mutans* biofilm. The results provide insights into additional applications for this enzyme.

2. Results

2.1. Purification of Dextranase

A summary of the efficient extraction protocol used to purify *Catenovulum* sp. DP03 dextranase (hereafter called Cadex), which included ultrafiltration, ethanol precipitation, ammonium sulfate precipitation, and ion exchange chromatography (Table 1). The activity of crude dextranase was 77.9 U/mg, and the activity of Cadex sequentially purified by ultrafiltration, ethanol precipitation, and ammonium sulfate precipitation was 163.5 U/mg, 223.3 U/mg and 341.6 U/mg, respectively. When ion exchange chromatography was used for purification, Cadex was purified 29.6-fold with a specific activity of 2309 U/mg protein and a yield of 16.9%. About 5.4% of enzymatic activity was lost and specific enzyme activity was increased by 48.7% after ultrafiltration. Ethanol precipitation and ammonium sulfate precipitation were used to remove polysaccharides and proteins. Ion exchange chromatography was performed to optimize the purification. Ten fractions containing proteins were eluted from Q-sepharose column, among which only one fraction showed dextranase activity. A Q-sepharose column with gradient elution of 45% 800 mM NaCl is an efficient method for Cadex purification. Finally, a homogenetic dextranase was purified. As shown in Figure 1, Cadex appeared as a single band of an estimated molecular weight of 75 kDa. The isoelectric point (pI) was 5.0.

Table 1. Purification of Cadex.

Purification Step	Total Protein (mg)	Total Activity (U)	Specific Activity (U/mg)	Purification (-Fold)	Yield (%)
Culture broth	80.6	6314.3	77.9	1	100
30 kDa ultrafiltration	36.5	5973	163.5	2.1	94.6
Alcohol precipitation	20.9	4664.6	223.3	2.9	73.9
Ammonium sulfate precipitation	6.74	2303.9	341.6	4.4	36.5
Ion exchange chromatography	0.46	1069.5	2309	29.6	16.9

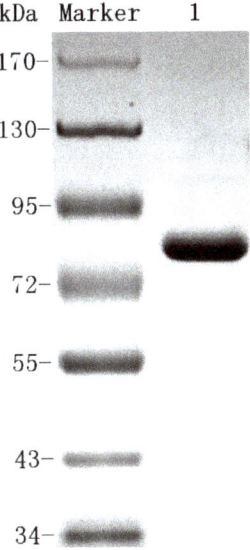

Figure 1. Sodium dodecyl sulfate-polyacrylamide gel electrophoresis (SDS-PAGE) of Cadex. Lane Maker: protein marker. Lane 1: purified Cadex.

2.2. Characterization of Dextranase

2.2.1. Effects of pH and Temperature on Dextranase Activity

The optimum pH for maximum Cadex activity was 8.0 within the range of pH 7.0–9.0 (Figure 2a). The effects of temperature on dextranase activity are shown in Figure 2b. The optimal temperature for dextranase activity was 40 °C within the range of 25–50 °C, showing peak relative activity no less than 87%. However, dextranase activity decreased up to 60 °C. Cadex also showed 33.8% relative activity at 0 °C.

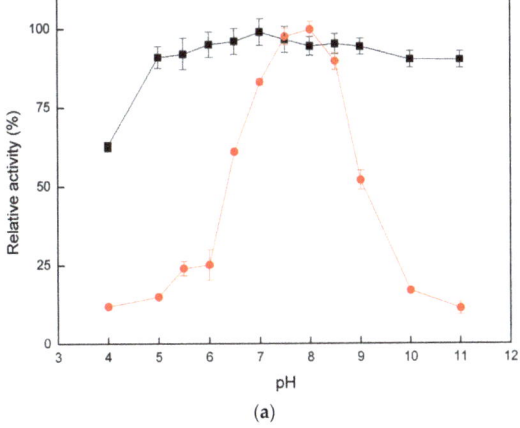

(a)

Figure 2. Cont.

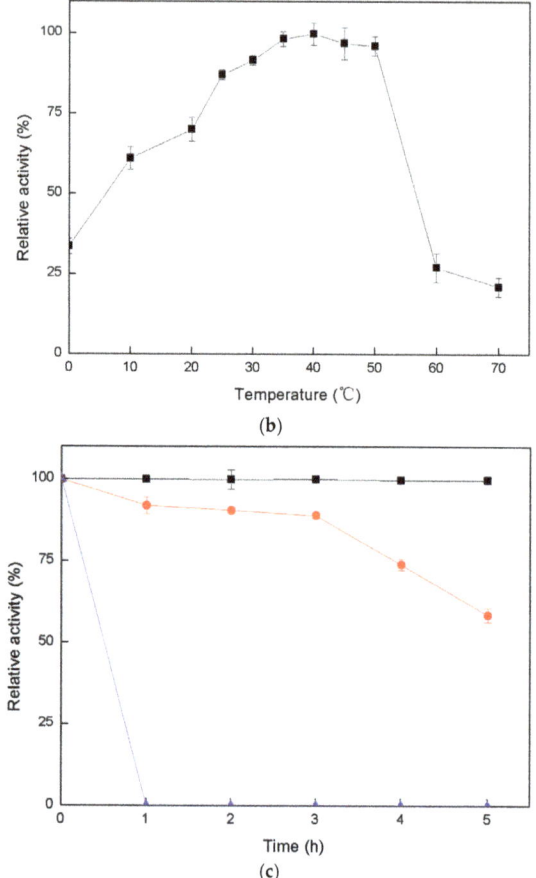

Figure 2. (a) Optimum pH and stability curves of Cadex. For each pH, activity was assayed at 40 °C and considered relative activity (●). The pH stability curve (■) represents residual activity after pre-incubation for 1 h at 25 °C. (b) Effects of temperature on the activity. (c) Effects of temperature on thermal stability of Cadex. Thermal stability: ■ 30 °C, ● 40 °C, ▲ 50 °C.

2.2.2. Enzymatic Stability

Figure 2a shows that Cadex retained more than 90% activity within a pH range of 5–11 (100% activity at pH 7.0). Purified Cadex showed thermal stability at 30 °C (pH 7.5) for 5 h. Under these conditions, almost 100% activity was retained. At 40 °C, 40% of activity was lost after 5 h of exposure. Above 50 °C, activity rapidly declined to undetectable levels at 1 h (Figure 2c).

2.2.3. Effects of Metal Ions and Reagents on Dextranase Activity

Alkaline proteases from microorganism require a divalent cation such as Ca^{2+}, Mg^{2+}, and Mn^{2+} or a combination of these cations, for maximum activity [19]. The effects of various metal ions on Cadex were investigated and the results are shown in Table 2. Enzyme activity was slightly increased upon exposure to 1 mM and 5 mM $MgCl_2$ or $SrCl_2$, respectively. This is consistent with previous observations [21,22]. However, enzyme activity was inhibited by exposure to 1 mM $CuCl_2$, $FeCl_3$, $ZnCl_2$, $CdCl_2$, $NiCl_2$, or $CoCl_2$, and even more so at concentrations of 5 mM, consistent with the results from another report [23], although these data conflict with the results reported by Birol et al. [24]. No apparent effects on dextranase activity were observed after exposure to BaCl$_2$, NH$_4$Cl, CaCl$_2$, KCl, or LiCl. However, the cations Co^{2+} and Ca^{2+} had the opposite effects on dextranase from *Chaetomium erraticum* and *Arthrobacter oxydans* KQ11 [3,10]. The effects of several types of chemical treatments for dental caries on the activity of Cadex are shown in Table 3. The data indicate that 5% ethanol, 0.1% carboxybenzene, sodium fluoride and xylitol had no apparent effects on the Cadex activity.

Table 2. Effects of metal ions on Cadex activity.

Reagents	Relative Activity (%) (1 mM)	Relative Activity (%) (5 mM)
Control	100 ± 1.45	100 ± 0.71
Ba^{2+}	99.82 ± 0.51	97.73 ± 0.95
NH$_4^+$	99.91 ± 1.23	77.8 ± 1.8
Ca^{2+}	97.46 ± 1.1	76.58 ± 2.96
Mg^{2+}	102.77 ± 2.27	104.47 ± 2.91
K$^+$	100.62 ± 3.69	91.87 ± 3.47
Cu^{2+}	1.97 ± 0.85	2.00 ± 1.11
Fe^{3+}	21.96 ± 1.67	0
Zn^{2+}	50.10 ± 1.92	2.47
Li$^+$	99.98 ± 0.4	90.53 ± 0.67
Cd^{2+}	41.14 ± 1.29	14.55 ± 1.65
Ni^{2+}	51.39 ± 1.35	21.35 ± 0.7
Co^{2+}	60.99 ± 2.32	29.00 ± 1.15
Sr^{2+}	103.75 ± 1.77	106.60 ± 1.89

Table 3. Effects of chemical treatment of dental caries on Cadex activity.

Reagents (w/v)	Relative Activity (%)
Control	100 ± 1.15
0.5% sodium lauryl sulfate	7.31 ± 1.39
0.1% sodium fluoride	93.8 ± 1.40
0.1% carboxybenzene	102.2 ± 0.32
0.1% xylitol	100.3 ± 0.50
5% ethanol	105.5 ± 0.70

2.2.4. Substrate Specificity and the Hydrolysis Products of Dextranase

Dextranase activity in the catalyzed hydrolysis of carbohydrates with different glucosidic linkages was determined and used as an indicator of the substrate specificity of Cadex (Table 4). Dextranase showed high specificity towards dextrans containing the α-1,6 glucosidic bond. The poor hydrolysis of soluble starch might indicate that Cadex can only cleave α-1,6 glycosidic bond. Soluble starch was formed using only a few α-1,6 glucosidic and α-1,4 glucosidic linkages [25]. No activity was detected in pullulan, which was mainly formed by α-1,4 glucosidic linkages. The same results were observed with chitin and cellulose, which are formed by β-1,4 glucosidic linkages. First, isomalto-triose (G3), isomalto-tetraose (G4), isomalto–pentaose (G5) and higher molecular weight maltooligosaccharides were found to be the products after a 3 h reaction. The hydrolysis products after a 24 h reaction were glucose (G1), G3, G4, and G5, and G6 was an extra sugar product when the reaction time was extended to 72 h, as shown by thin-layer chromatography (TLC) (Figure 3).

Second, high-performance liquid chromatography (HPLC) indicated that isomaltooligomers were the main products released by Cadex regardless of reaction time (Figure 4). Each hydrolyzed product was quantified by Empower GPC software (Table 5). G1, G2, G3, G4, G5, G6, and G7 were products of glucose, maltose, isomalto-triose, isomalto-tetraose, isomalto-pentaose, isomalto-hexaose, and isomalto-heptaose, respectively. The results indicated that a small amount of glucose was detected, which is similar to most fungal dextranases. They maintained a dynamic equilibrium, which did not increase with longer reaction times. Moreover, the amounts of maltose, isomalto-triose and isomalto-tetraose slightly declined, and an increase in time from 15 min to 5 h did not increase the yield. However, isomalto-hexaose and isomalto-heptaose accumulated more quickly than the other sugars as the time increased from 15 min to 5 h. At the same time, the yield of these intermediate high molecular weight oligomers (15–20%) was significantly higher than that of glucose (about 2%). This was similar to other reports in which macromolecule isomaltooligomers accumulated in the first period (6 h), and the other sugars were hydrolyzed to produce other smaller sugars [26]. The isolation of small amounts of glucose and some intermediate high molecular weight oligomers seemed to be random rather than through the stepwise hydrolysis of polysaccharide, making Cadex an endo-type dextranase.

Table 4. Substrate specificity of Cadex.

Substrate	Main Linkages	Relative Activity (%)
Dextran T20	α-1,6	90.42 ± 0.25
Dextran T40	α-1,6	91.29 ± 0.67
Dextran T70	α-1,6	95.65 ± 1.55
Dextran T500	α-1,6	100 ± 1
Soluble starch	α-1,4, α-1,6	4.93 ± 1.36
Microcrystalline cellulose	β-1,4	0
Chitin	β-1,4	0
Pullulan	α-1,4	0

Table 5. Content of sugar (%) in hydrolysates after enzymatic hydrolysis of dextran by Cadex.

Time of Hydrolysis	Hydrolysis Productions						
	G1	G2	G3	G4	G5	G6	G7
15 min	2.02	18.59	12.31	19.22	21.02	12.49	14.36
30 min	1.95	17.75	12.03	19.13	21.01	13.1	15.03
1 h	1.9	17.44	11.69	18.86	20.77	13.71	15.63
3 h	1.94	16.74	11.25	18.57	21.05	14.53	15.92
5 h	1.96	16.24	9.62	17.7	19.9	16.29	18.3

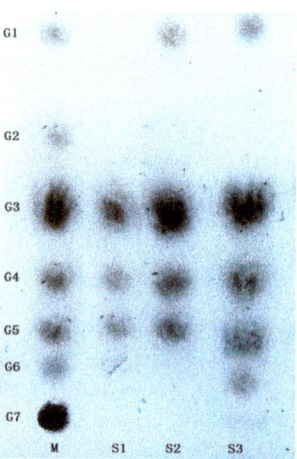

Figure 3. Thin-layer chromatogram of the products from Cadex. Symbols: G1 to G7 a series of authentic sugar standards of glucose, maltose, isomalto-triose, isomalto-tetraose, isomalto-pentaose, isomalto-hexaose, and isomalto-heptaose, respectively. M is the standard marker, S1 to S3 show the 3 h, 24 h, and 72 h reaction times, respectively.

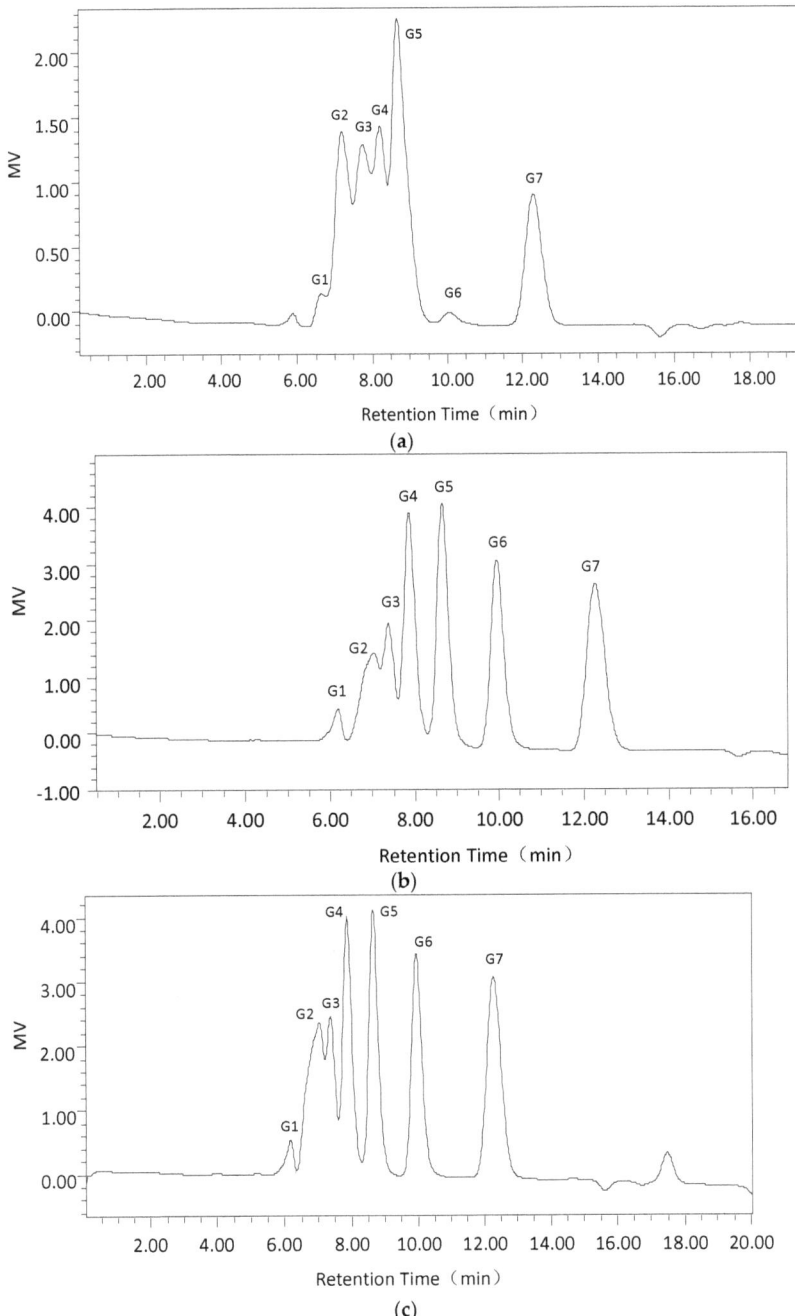

Figure 4. Cont.

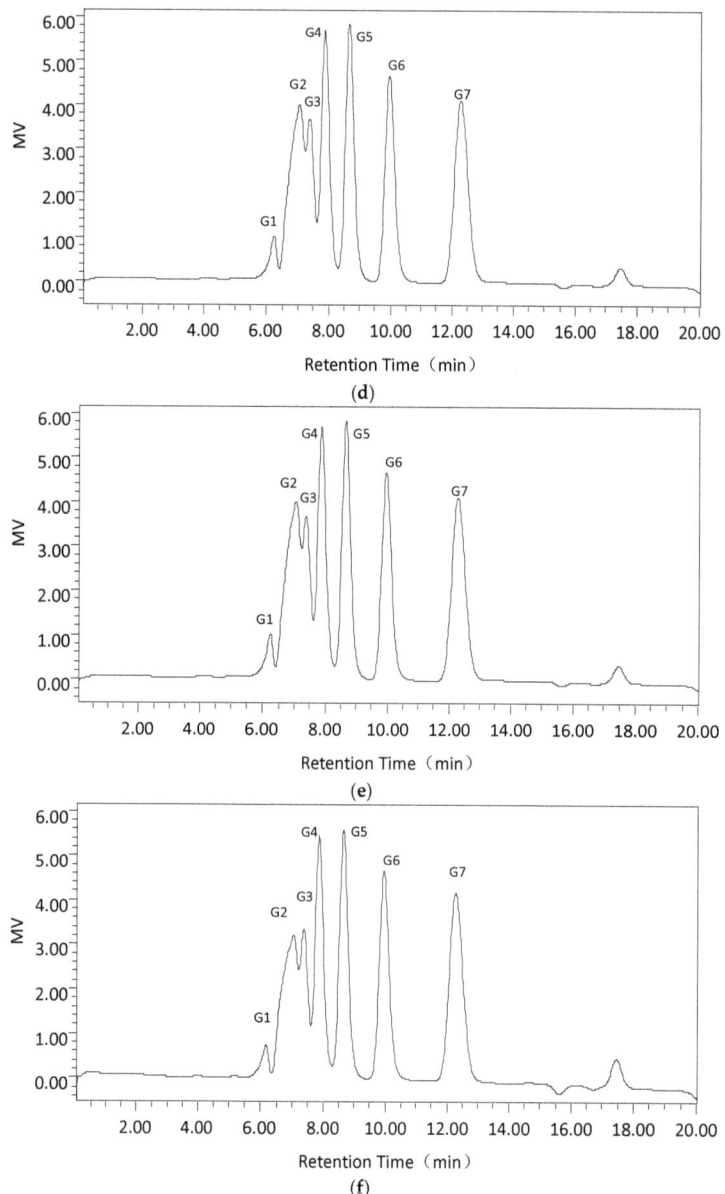

Figure 4. The 3% dextran T70 was treated at 40 °C and pH 8.0 for different periods with the products measured by HPLC: (**a**) the results for standards (G1 to G7 a series of authentic sugar standards of glucose, maltose, isomalto-triose, isomalto-tetraose, isomalto-pentaose, isomalto-hexaose, and isomalto-heptaose, respectively); and (**b–f**) the results for 3% dextran T70 treated for 15 min, 30 min, 1 h, 3 h, and 5 h with Cadex.

Sugars were identified and measured by HPLC. The hydrolysis condition yielded 3% dextran T70 at 40 °C and pH 8.0.

2.3. Effects of Cadex on Biofilm

Table 6 shows the comparison of biofilm inhibitory rates between *Penicillium* dextranase and Cadex under various concentrations. It can be clearly seen that both *Penicillium* dextranase and Cadex impeded the biofilm formation. The minimum biofilm inhibitory concentration for 90% inhibition (MBIC$_{90}$) was calculated when 40 U/mL *Penicillium* dextranase and 30 U/mL Cadex were added to the media after which the effects of the dextranases on *S. mutans* biofilm formation were analyzed. Cadex was more efficient than *Penicillium* dextranase in inhibiting *S. mutans* biofilm formation. Figure 5 shows the bacterial morphology and biofilm, as observed by scanning electron microscopy (SEM.). In the blank control group, *S. mutans* grew well and the biofilm developed smoothly with prolonged time. At 18 h, a thick biofilm with dense cells were seen, with no obvious structural breakdown. In contrast, biofilm did not form easily when *S. mutans* was cultured in brain heart infusion (BHI) medium by adding the MBIC$_{90}$ of Cadex or *Penicillium* dextranase. Dextranase impeded biofilm formation and reduced the number of *S. mutans* cells that adhered to the glass coverslips. Cadex had inhibitory effects on *S. mutans* biofilm formation. A previous report proposed that crude *Catenovulum* dextranase can prevent *S. mutans* from forming biofilms, however, it used crude dextranase and was a preliminary assessment [20].

Table 6. Biofilm inhibitory rates with different concentrations of dextranase.

Concentration of *Penicillium* Dextranase (U/mL)	Biofilm Inhibitory Rate [a] (%)	Concentration of Cadex (U/mL)	Biofilm Inhibitory Rate [a] (%)
0	0	0	0
5	27.46 ± 1.28	5	33.31 ± 0.99
10	39.44 ± 1.33	10	52.39 ± 1.21
15	50.2 ± 1.42	15	62.2 ± 0.92
20	63.13 ± 0.89	20	71.3 ± 0.69
25	73.42 ± 1.13	25	85.45 ± 0.70
30	82.16 ± 0.92	30	91.11 ± 0.83
40	89.34 ± 0.93	35	94.21 ± 1.13

[a] The biofilm inhibitory rate was calculated at an absorbance of 595 (A$_{595}$) of the crystal violet stained biofilm without dextranase subtracted from A$_{595}$ of biofilm with dextranase, and divided by A$_{595}$ of biofilm without dextranase multiplied by 100%.

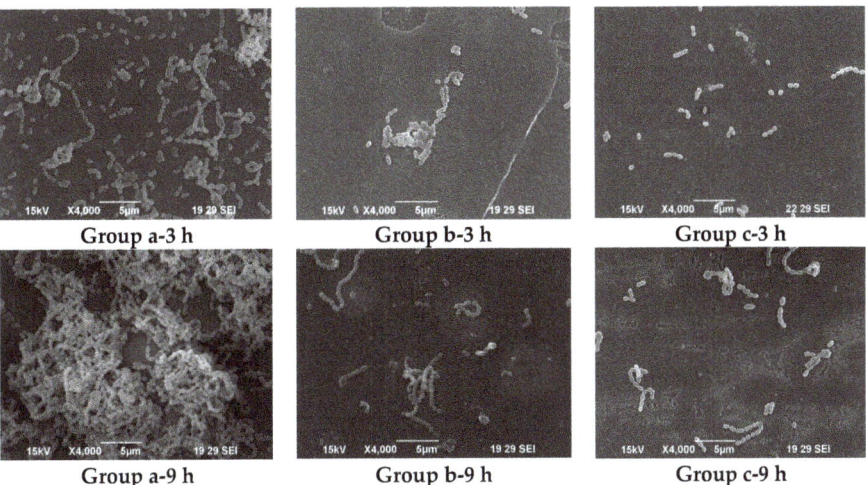

Figure 5. *Cont.*

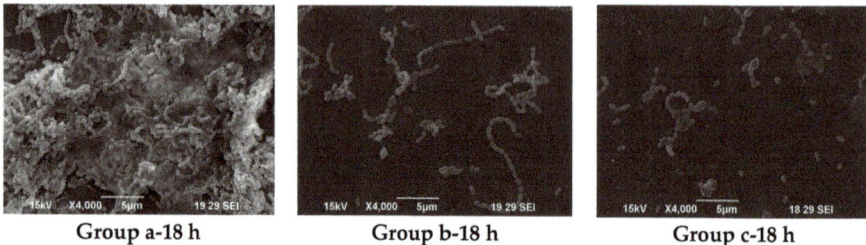

| Group a-18 h | Group b-18 h | Group c-18 h |

Figure 5. Electron microscopy of *S. mutans* biofilm formed on glass coverslips in the presence and absence of dextranase at different periods: (**Group a**) blank control, note the equal volume of cell-free pure water was added to replace dextranase; (**Group b**) biofilm subjected to 40 U/mL *Penicillium* dextranase; and (**Group c**) biofilm subjected to 35 U/mL Cadex.

3. Discussion

A psychrotolerant dextranase-producing bacterium named *Catenovulum* sp. DP03 was previously studied [20]. However, to the best of our knowledge, this is the first report of the purification and characterization of dextranase from *Catenovulum*. Purification of crude dextranase by ammonium sulfate fractionation and Sepharose 6B chromatography, which resulted in a 6.69-fold increase in specific activity and an 11.27% recovery, was previously reported [23]. This system of the aforementioned procedure may be used to produce homogenetic dextranase. The process can easily be scaled up and is cost-effective. The molecular weight of Cadex was about 75 kDa, which resembled that of dextranase from *Sporotrix schencki* (79 kDa) [27]. Bacteria producing dextranases generally have molecular weights ranging from 60 to 114 kDa [28–30]. The smallest dextranase (23 kDa) is from *Lipomyces starkeyi* [24], and the largest (175 kDa) is from *Streptococcus sobrinus* [31].

Endo-type Cadex showed high specificity towards dextrans containing α-1,6 glucosidic bonds. Moreover, the main hydrolysis products of Cadex were isomaltooligomers [30,32,33]. Dextranase from *Chaetomium* [34], *Aspergillus* [35], *Penicillium* [36], and *Fusarium* [37] synthesize comparatively low amounts of glucose and higher amounts of isomaltooligosaccharides. Isomaltooligosaccharides can promote the growth and proliferation of *Bifidobacteria* and *Lactobacillus* [1,38]. Numerous isomaltooligosaccharides are prebiotics, which are produced endodextranases and have garnered much commercial interest [33].

The optimum pH for Cadex activity tends to be alkaline, and recently reported alkalophilic cases were *Streptomyces* sp. NK458 and *Bacillus subtilis* NRC-B233b, which had maximum activities at pH 9.0 and pH 9.2 [7,39]. Evidence is accumulating that alkali generation plays a major role in pH homeostasis which may modulate the initiation and progression of dental caries [40]. Therefore, alkalophilic Cadex may be suitable for the development of novel marine agents for the treatment of this condition [16]. Cadex had catalytic efficiency at 0 °C, similar characteristics to other cold-adapted enzymes: for example, a cold-adapted ι-carrageenase showed 36.5% relative activity at 10 °C [41] and a cold-adapted β-glucosidase retained more than 60% of its activity at temperatures ranging from 15 °C to 35 °C [42]. Cold-adapted enzymes have optimal catalyst temperatures near 30 °C and remain efficient at 0 °C. Cadex can be classified as a cold-adapted enzyme according to the system developed by Margesin and Schinner [43]. The excellent pH stability of Cadex distinguishes it from other dextranases, which are generally unstable across a broad pH range [1,7,22,27]. It would be easier to hydrolyze dextran than dextranases in acid/alkaline catalysis conditions. We speculate that Cadex may be suitable for widespread use. We have classified Cadex as a cold-adapted dextranase, which may explain its lower thermo-stability than terrestrial dextranases [1]. Nevertheless, in our early studies, crude Cadex showed greater thermostability than purified dextranase, as it was stable at 45 °C, and its half-life was 10 h (data not shown).

The present study proposes that purified *Catenovulum* dextranase, namely Cadex, is an alkalophilic and cold-adapted dextranase that is considered to be a novel marine dextranase of dealing with biofilms. The failure of biofilm formation is attributable to the failure of extracellular polysaccharides to form efficiently, possibly due to cleavage of the α-1,6 glucosidic linkages in the biofilm that occurs in the presence of Cadex. The oral *Streptococcus* biofilm is formed by α-(1,3)-glucan and α-(1,6)-glucan, in which the α-1,6 glucosidic linkages are degradable by dextranase while the α-1,3 glucosidic linkages can be cleaved by mutanase [44]. *Penicillium* dextranase is often used as the standard dextranase in studies of this enzyme such as studies showing dextran removal during sugar manufacturing [7]. Cadex was a favorable biofilm inhibitor that surpassed the inhibitory ability of *Penicillium* dextranase at the same concentration. In addition, common teeth rinsing products, such as carboxybenzene, ethanol, sodium fluoride, and xylitol, had no negative effects on Cadex activity. Marine organisms are regarded as a prolific resource of novel bioactive metabolites, including a vast array of macrolides, cyclic peptides, pigments, polyketides, terpenes, steroids and alkaloids [45]. At the same time, marine enzymes are important bioactive metabolites which characterized by high salt tolerance, hyperthermostability, and low ideal temperature tolerance. These beneficial properties make Cadex an attractive candidate for development as a novel reagent for dental plaque treatment [46,47].

4. Materials and Methods

4.1. Chemicals

Q-Sepharose FF, dextran (T20, T40, T70, and T500), and PhastGel IEF 3-9 were obtained from GE Healthcare (Uppsala, Sweden). A prestained protein PAGE ruler was obtained from Fermentas (Waltham, MA, USA). An oligosaccharide kit, an IEF protein mix 3.6–9.3, bovine serum albumin, crystal violet, (CV), and Coomassie brilliant blue R250 and G250 were purchased from Sigma-Aldrich (St. Louis, MO, USA). All other reagents were purchased from Sinopharm Chemical Reagent Corporation (Shanghai, China) and were of the highest analytical grade.

4.2. Crude Dextranase Production

Extracellular dextranase production was performed in medium containing 5 g/L yeast extract, 5 g/L peptone, 10 g/L dextran T20, and 5 g/L NaCl. The pH was adjusted to about 8.0 before autoclaving. Then, the production medium was inoculated with *Catenovulum* sp. DP03. After fermenting at 30 °C for 28 h, a cell-free culture broth was obtained by centrifugation for 20 min at $12,000 \times g$ and 4 °C.

4.3. Purification of Dextranase

First, excess water and other matter were removed from the crude dextranase product by ultrafiltration (Watson Marlow, Cornwall, UK) using a 30,000 NMWC hollow fiber cartridge (GE Healthcare) at room temperature. Deionized water was added three times. The crude enzyme was finally concentrated to one tenth of the original volume of the culture broth. Second, pre-cooled ethanol (−40 °C) was added to the crude enzyme solution slowly and agitated gently for 10 min. An ethanol: enzyme ratio of 0.6:1.2 (*v:v*) was found to be optimal. After centrifugation for 15 min at $12,000 \times g$ and 4 °C, the precipitate was dissolved in 10 mM Tris-HCl buffer (pH 7.5). Third, ammonium sulfate precipitation was performed using a magnetic stirrer. The supernatant from ethanol precipitation was placed in a beaker within an ice tray. Then 25% saturated ammonium sulfate was added to the solution, which then was allowed to stand at 4 °C for 1 h, after which it was centrifuged at $12,000 \times g$ at 4 °C for 20 min. The supernatant was collected, and 60% saturated ammonium sulfate was added. The mixture was allowed to stand for 1 h, and precipitated protein was collected by centrifugation 20 min at $12,000 \times g$ and 4 °C. Then the pellet was dissolved and dialyzed using 10 mM Tris-HCl buffer (pH 7.5). Finally, the enzyme sample was loaded onto a Q-Sepharose column (1.6 cm × 10 cm; GE Healthcare). Chromatographic analysis was conducted by a fast protein liquid chromatography (FPLC;

Bio-Rad, Hercules, CA, USA) at room temperature. Proteins were eluted with several NaCl gradients in 10 mM Tris-HCl (pH 7.5) at a flow rate of 0.8 mL/min. Protein content and enzyme activity for each fraction were monitored. The enzyme-containing fractions were concentrated using ultracel-10k centrifugal filters (Millipore, Burlington, MA, USA).

4.4. SDS-PAGE and Isoelectric Focusing

SDS-PAGE analysis of dextranase was performed according to Laemmli [48] with minor modifications. After electrophoresis, the gel was stained with Coomassie brilliant blue R-250. Isoelectric focusing (IEF) was performed using a Pharmacia PhastGel System (GE Healthcare) on PhastGel IEF 3–9 according to the manufacturer's instructions. An IEF protein marker (mix 3.6–9.3) was used as the pI standard.

4.5. Enzyme Assay and Protein Measurement

Enzyme activity was measured using dextran T70 as the substrate (3%, m/v) in 0.1 M Tris-HCl buffer (pH 8.0) using the 3,5-dinitrosalicylic acid method. Maltose served as the standard. One unit of dextranase activity was defined as the amount of enzyme capable of hydrolyzing dextran to 1 µM of reducing sugar in1 min [49]. Protein concentration was measured by the Bradford protein assay [50] using bovine serum albumin as the standard.

4.6. Enzyme Properties

4.6.1. Effects of pH on Activity and Stability of Cadex

Dextranase activity was measured in buffers of different pH values and containing 3% dextran T70. To determine the effects of pH on dextranase stability, the enzyme solution was incubated at 25 °C for 1 h, and residual activity was measured at the optimum pH. Solutions (50 mM) with different pH values were used as follows: citrate buffer (pH 4.0–6.0), sodium phosphate buffer (pH 6.0–7.5), Tris-HCl (pH 7.5–9.0), and $NaHCO_3$-Na_2CO_3 (pH 9.0–11.0).

4.6.2. Effects of Temperature on Activity and Stability of Cadex

The temperature for the enzymatic reaction was optimized by experimentation at different temperatures (0–70 °C). To assess thermal stability, the enzyme solution was pre-heated at temperatures of 30 °C, 40 °C, and 50 °C for 0–5 h. Residual enzyme activity was measured at each interval.

4.6.3. Effects of Metal Ions and Chemicals on Cadex Activity

The effects of different solutions containing chloride metal ion salts (final concentrations of 1 mM and 5 mM) and chemicals on purified Cadex activities were determined. The relative enzyme activity in the presence of metal ions and chemicals was calculated based on activity in the absence of reagent.

4.6.4. Substrate Specificity

To determine the substrate specificity of Cadex, dextranase activity in the enzymatic hydrolysis of carbohydrates with various glycosidic linkages was determined using the method described by Wu. et al. [22]. Dextranase was incubated in 50 mM Tris-HCl (pH 8.0) with various carbohydrates at 40 °C for 20 min. Relative activity was expressed in percentage values of the highest activity, which was set as 100%.

4.6.5. Products of Cadex Hydrolysis

First, Cadex hydrolysis took place in dextran T70 solution, and the products were analyzed by TLC using a silica gel GF254 plate developed in a chloroform: acetic: acid: water ratio of = 5:7:1 ($v:v:v$). An oligosaccharide kit was used as the standard.

Second, using the optimum dextranase temperature and pH, 3% dextran T70 samples were digested for different periods (15 min, 30 min, 1 h, 3 h, and 5 h). The products were identified and analyzed with the Waters 600 and Waters Sugar-Pak1 (300 mm × 7.8 mm; Waters, Milford, MA, USA) HPLC with a differential refraction detector. The mobile phase was water at 0.4 mL/min, the column temperature was 85 °C and the injection volume was 20 µL. The standard sugars were glucose, maltose, maltotriose, isomaltotriose, isomaltotetraose, isomaltopentose, and isomaltohexose. For quantification, the peak areas were determined. Data acquisition and processing were conducted using Empower GPC software (Waters, Milford, MA, USA).

4.7. Effects of Cadex on Biofilm

4.7.1. Biofilm Mass Assay

The biofilm mass was assayed by CV staining according to the protocol of Cardoso et al. [51] with some modifications. Briefly, biofilms were grown on a flat bottom sterile 96-well plate (Greiner, Frickhausen, Germany) in which the cultured medium was removed. To each well, 0.2 mL of 0.2 M phosphate buffer was added three times to clean the unattached biofilms, which were left to air dry and fixed for 60 min. Then 0.2 mL of 1% CV was added to each well for 5 min. Following the staining step, the CV solution was removed and the biofilms were cleaned and dried, after which 0.2 mL of 95% ethyl alcohol was added to re-solubilize the dyed biofilms. The CV solutions were obtained and transferred to a new 96-well plate and the optical density of the content was measured using a microtiter plate spectrophotometer (Bio-Rad) at 595 nm.

4.7.2. Effects of Cadex on Biofilm Formation

Base on the biofilm mass assay, MBIC of *Streptococcus mutans* ATCC 25175 (American Type Culture Collection (ATCC), Manassas, VA, USA) was measured. The effects of Cadex and a homogenetic purity dextranase from *Penicillium* (SA D8144; Sigma) on *S. mutans* biofilm formation were investigated using SEM at $MBIC_{90}$. *S. mutans* was pre-inoculated in BHI medium without sucrose at 37 °C for 15 h. Then, 1 mL of this precultured solution was inoculated into fresh BHI medium with 1% sucrose (20 mL in 100 mL Erlenmeyer-type flask). Sterile glass coverslips were placed in the BHI medium. The media were co-cultured with *S. mutans* and incubated with Cadex at 37 °C for 3 h, 9 h, and 18 h. An identical assay with an equal volume of cell-free deionized water served as the blank control. All coverslips were collected for fixation, and were dehydrated and dried according to the procedure described by Tao et al. [52]. The coverslips were sputter-coated with gold (JFC-1600, JEOL, Tokyo, Japan) and viewed by SEM (JSM-6390LA; JEOL).

5. Conclusions

Cadex from the marine bacterium *Catenovulum* sp. was purified to 29.6-fold homogeneity. It showed a specific activity of 2309 U/mg protein and a molecular weight of 75 kDa. Its optimum pH and temperature were 8.0 and 40 °C, respectively. The enzyme was stable at temperatures below 30 °C and pH values within 5–11. Some mental ions and chemicals might activate Cadex, but others might inhibit it or leave it unaffected. Cadex was identified as a typical endo-dextranase. The main hydrolysis products were isomaltooligosaccharides which may be included as a prebiotic supplement to promote the growth and proliferation of intestinal flora. Cadex inhibited biofilm formation by *S. mutans*. Thus, this alkaline- and cold-adapted dextranase from the marine bacterium *Catenovulum* sp. appears to be efficacious under both mesophilic and alkalophilic conditions, thus is a potential candidate for development into a novel marine oral biofilm removal drug.

Acknowledgments: This work was supported by the National Natural Science Foundation of China (Grant No: 31471719), the Key Research and Development Program of Jiangsu [Social Development] (Grant No: BE2016702), and the Priority Academic Program Development of Jiangsu Higher Education Institutions.

Author Contributions: Wei Ren, Ruanhong Cai, Yaowei Fang, Mingsheng Lyu, and Shujun Wang designed the experiments. Wei Ren, Ruanhong Cai, and Wanli Yan performed the experiments. Wei Ren, Wanli Yan, Ruanhong Cai, Yaowei Fang, and Mingsheng Lyu analyzed the data. Shujun Wang supervised the study and reviewed the manuscript. Wei Ren and Ruanhong Cai wrote the manuscript. All authors have read and approved the final manuscript.

Conflicts of Interest: The authors have no conflicts of interest to declare.

References

1. Khalikova, E.; Susi, P.; Korpela, T. Microbial dextran-hydrolyzing enzymes: Fundamentals and applications. *Microbiol. Mol. Biol. Rev.* **2005**, *69*, 306–325. [CrossRef] [PubMed]
2. Ren, W.; Wang, S.; Lu, M.; Wang, X.; Fang, Y.; Jiao, Y.; Hu, J. Optimization of four types of antimicrobial agents to increase the inhibitory ability of marine *Arthrobacter oxydans* KQ11 dextranase mouthwash. *Chin. J. Oceanol. Limnol.* **2016**, *34*, 354–366. [CrossRef]
3. Wang, D.; Lu, M.; Wang, S.; Jiao, Y.; Li, W.; Zhu, Q.; Liu, Z. Purification and characterization of a novel marine *Arthrobacter oxydans* KQ11 dextranase. *Carbohydr. Polym.* **2014**, *106*, 71–76. [CrossRef] [PubMed]
4. Eggleston, G.; Monge, A. Optimization of sugarcane factory application of commercial dextranases. *Process Biochem.* **2005**, *40*, 1881–1894. [CrossRef]
5. Bowler, G.; Wones, S. Application of dextranase in UK sugar beet factories. *Zuckerindustrie* **2011**, *136*, 780–783.
6. Park, T.-S.; Jeong, H.J.; Ko, J.-A.; Ryu, Y.B.; Park, S.-J.; Kim, D.; Kim, Y.-M.; Lee, W.S. Biochemical characterization of thermophilic dextranase from a thermophilic bacterium, *Thermoanaerobacter pseudethanolicus*. *J. Microbiol. Biotechnol.* **2012**, *22*, 637–641. [CrossRef] [PubMed]
7. Purushe, S.; Prakash, D.; Nawani, N.N.; Dhakephalkar, P.; Kapadnis, B. Biocatalytic potential of an alkalophilic and thermophilic dextranase as a remedial measure for dextran removal during sugar manufacture. *Bioresour. Technol.* **2012**, *115*, 2–7. [CrossRef] [PubMed]
8. Goulas, A.K.; Cooper, J.M.; Grandison, A.S.; Rastall, R.A. Synthesis of isomaltooligosaccharides and oligodextrans in a recycle membrane bioreactor by the combined use of dextransucrase and dextranase. *Biotechnol. Bioeng.* **2004**, *88*, 778–787. [CrossRef] [PubMed]
9. Bertrand, E.; Pierre, G.; Delattre, C.; Gardarin, C.; Bridiau, N.; Maugard, T.; Strancar, A.; Michaud, P. Dextranase immobilization on epoxy CIM (R) disk for the production of isomaltooligosaccharides from dextran. *Carbohydr. Polym.* **2014**, *111*, 707–713. [CrossRef] [PubMed]
10. Virgen-Ortíz, J.J.; Ibarra-Junquera, V.; Escalante-Minakata, P.; Ornelas-Paz, J.d.J.; Osuna-Castro, J.A.; González-Potes, A. Kinetics and thermodynamic of the purified dextranase from *Chaetomium erraticum*. *J. Mol. Catal. B Enzym.* **2015**, *122*, 80–86. [CrossRef]
11. Keyes, P.H.; Hicks, M.A.; Goldman, M.; McCabe, R.M.; Fitzgerald, R.J. 3. Dispersion of dextranous bacterial plaques on human teeth with dextranase. *J. Am. Dent. Assoc.* **1971**, *82*, 136–141. [CrossRef] [PubMed]
12. Kim, J.; Park, H.D.; Chung, S. Microfluidic approaches to bacterial biofilm formation. *Molecules* **2012**, *17*, 9818–9834. [CrossRef] [PubMed]
13. Kumar, A.; Alam, A.; Rani, M.; Ehtesham, N.Z.; Hasnain, S.E. Biofilms: Survival and defense strategy for pathogens. *Int. J. Med. Microbiol.* **2017**, *307*, 481–489. [CrossRef] [PubMed]
14. Liu, J.; Ling, J.-Q.; Zhang, K.; Wu, C.D. Physiological properties of *Streptococcus mutans* UA159 biofilm-detached cells. *FEMS Microbiol. Lett.* **2013**, *340*, 11–18. [CrossRef] [PubMed]
15. Cardoso, J.G.; Iorio, N.L.; Rodrigues, L.F.; Couri, M.L.; Farah, A.; Maia, L.C.; Antonio, A.G. Influence of a Brazilian wild green propolis on the enamel mineral loss and *Streptococcus mutans'* count in dental biofilm. *Arch. Oral Biol.* **2016**, *65*, 77–81. [CrossRef] [PubMed]
16. Majeed, A.; Grobler, S.R.; Moola, M.H. The pH of various tooth-whitening products on the South African market. *SADJ* **2011**, *66*, 278–281. [PubMed]
17. Papaleo, E.; Tiberti, M.; Invernizzi, G.; Pasi, M.; Ranzani, V. Molecular determinants of enzyme cold adaptation: Comparative structural and computational studies of cold- and warm-adapted enzymes. *Curr. Protein Pept. Sci.* **2011**, *12*, 657–683. [CrossRef] [PubMed]
18. Russo, R.; Giordano, D.; Riccio, A.; Prisco, G.D.; Verde, C. Cold-adapted bacteria and the globin case study in the Antarctic bacterium *Pseudoalteromonas haloplanktis* TAC125. *Mar Genom.* **2010**, *3*, 125–131. [CrossRef] [PubMed]

19. Ma, C.; Ni, X.; Chi, Z.; Ma, L.; Gao, L. Purification and characterization of an alkaline protease from the marine yeast *Aureobasidium pullulans* for bioactive peptide production from different sources. *Mar. Biotechnol.* **2007**, *9*, 343–351. [CrossRef] [PubMed]
20. Cai, R.; Lu, M.; Fang, Y.; Jiao, Y.; Zhu, Q.; Liu, Z.; Wang, S. Screening, production, and characterization of dextranase from *Catenovulum* sp. *Ann. Microbiol.* **2014**, *64*, 147–155. [CrossRef]
21. Fukumoto, J.; Tsuji, H.; Tsuru, D. Studies on mold dextranases 1. *Penicillium luteum* dextranase: Its production and some enzymatic properties. *J. Biochem.* **1971**, *69*, 1113–1121. [CrossRef] [PubMed]
22. Wu, D.T.; Zhang, H.B.; Huang, L.J.; Hu, X.Q. Purification and characterization of extracellular dextranase from a novel producer, *Hypocrea lixii* F1002, and its use in oligodextran production. *Process Biochem.* **2011**, *46*, 1942–1950. [CrossRef]
23. Zhang, Y.Q.; Li, R.H.; Zhang, H.B.; Wu, M.; Hu, X.Q. Purification, characterization, and application of a thermostable dextranase from *Talaromyces pinophilus*. *J. Ind. Microbiol. Biotechnol.* **2017**, *44*, 1–11. [CrossRef] [PubMed]
24. Birol, H.; Damjanovic, D.; Setter, N. The purification and characterization of a dextranase from *Lipomyces starkeyi*. *Eur. J. Biochem.* **1989**, *183*, 161–167.
25. Wynter, C.V.A.; Chang, M.; Jersey, J.D.; Patel, B.; Inkerman, P.A.; Hamilton, S. Isolation and characterization of a thermostable dextranase. *Enzym. Microb. Technol.* **1997**, *20*, 242–247. [CrossRef]
26. Pleszczyńska, M.; Szczodrak, J.; Rogalski, J.; Fiedurek, J. Hydrolysis of dextran by *Penicillium notatum* dextranase and identification of final digestion products. *Mycol. Res.* **1997**, *101*, 69–72. [CrossRef]
27. Arnold, W.N.; Nguyen, T.B.P.; Mann, L.C. Purification and characterization of a dextranase from *Sporothrix schenckii*. *Arch. Microbiol.* **1998**, *170*, 91–98. [CrossRef] [PubMed]
28. Khalikova, E.F.; Usanov, N.G. An insoluble colored substrate for dextranase assay. *Appl. Biochem. Microbiol.* **2002**, *38*, 89–93. [CrossRef]
29. Khalikova, E.; Susi, P.; Usanov, N.; Korpela, T. Purification and properties of extracellular dextranase from a *Bacillus* sp. *J. Chromatogr. B-Anal. Technol. Biomed. Life Sci.* **2003**, *796*, 315–326. [CrossRef]
30. Shimizu, E.; Unno, T.; Ohba, M.; Okada, G. Purification and characterization of an isomaltotriose-producing endo-dextranase from a *Fusarium* sp. *Biosci. Biotechnol. Biochem.* **1998**, *62*, 117–122. [CrossRef] [PubMed]
31. Barrett, J.F.; Barrett, T.A.; Rd, C.R. Purification and partial characterization of the multicomponent dextranase complex of *Streptococcus sobrinus* and cloning of the dextranase gene. *Infect. Immun.* **1987**, *55*, 792–802. [PubMed]
32. Jin, H.L.; Nam, S.H.; Park, H.J.; Kim, Y.M.; Kim, N.; Kim, G.; Seo, E.S.; Kang, S.S.; Kim, D. Biochemical characterization of dextranase from *Arthrobacter oxydans* and its cloning and expression in *Escherichia coli*. *Food Sci. Biotechnol.* **2010**, *19*, 757–762.
33. Zohra, R.R.; Aman, A.; Ansari, A.; Haider, M.S.; Qader, S.A. Purification, characterization and end product analysis of dextran degrading endodextranase from *Bacillus licheniformis* KIBGE-IB25. *Int. J. Biol. Macromol.* **2015**, *78*, 243–248. [CrossRef] [PubMed]
34. Hattori, A.; Ishibashi, K.; Minato, S. The purification and characterization of the dextranase of *Chaetomium gracile*. *Agric. Biol. Chem.* **1981**, *45*, 2409–2416. [CrossRef]
35. Tsuru, D.; Hiraoka, N.; Fukumoto, J. Studies on mold dextranases IV. Substrate specificity of *Aspergillus carneus* dextranase. *J. Biochem.* **1972**, *71*, 653–660. [PubMed]
36. Sugiura, M.; Ito, A. Studies on dextranase. III. Action patterns of dextranase from *Penicillium funiculosum* on substrate and inhibition on hydrolysis reaction by substrate analogues. *Chem. Pharm. Bull.* **1974**, *22*, 1593–1599. [CrossRef] [PubMed]
37. Simonson, L.G.; Liberta, A.E.; Richardson, A. Characterization of an extracellular dextranase from *Fusarium moniliforme*. *Appl. Microbiol.* **1975**, *30*, 855–861. [PubMed]
38. Kim, Y.M.; Seo, M.Y.; Kang, H.K.; Atsuo, K.; Kim, D. Construction of a fusion enzyme of dextransucrase and dextranase: Application for one-step synthesis of isomalto-oligosaccharides. *Enzym. Microb. Technol.* **2009**, *44*, 159–164. [CrossRef]
39. Esawy, M.A.; Mansour, S.H.; Ahmed, E.F.; Hassanein, N.M.; El Enshasy, H.A. Characterization of Extracellular Dextranase from a Novel Halophilic *Bacillus subtilis* NRC-B233b a Mutagenic Honey Isolate under Solid State Fermentation. *E-J. Chem.* **2012**, *9*, 1494–1510. [CrossRef]
40. Huang, X.; Zhang, K.; Deng, M.; Ram, E.; Liu, C.; Zhou, X.; Cheng, L.; Ten Cate, J.M. Effect of arginine on the growth and biofilm formation of oral bacteria. *Arch. Oral Biol.* **2017**, *82*, 256. [CrossRef] [PubMed]

41. Li, S.; Hao, J.; Sun, M. Cloning and characterization of a new cold-adapted and thermo-tolerant iota-carrageenase from marine bacterium *Flavobacterium* sp. YS-80-122. *Int. J. Biol. Macromol.* **2017**, *102*, 1059–1065. [CrossRef] [PubMed]
42. Crespim, E.; Zanphorlin, L.M.; de Souza, F.H.M.; Diogo, J.A.; Gazolla, A.C.; Machado, C.B.; Figueiredo, F.; Sousa, A.S.; Nobrega, F.; Pellizari, V.H.; et al. A novel cold-adapted and glucose-tolerant GH1 beta-glucosidase from *Exiguobacterium antarcticum* B7. *Int. J. Biol. Macromol.* **2016**, *82*, 375–380. [CrossRef] [PubMed]
43. Margesin, R.; Schinner, F. Properties of cold-adapted microorganisms and their potential role in biotechnology. *J. Biotechnol.* **1994**, *33*, 1–14. [CrossRef]
44. Yano, A.; Kikuchi, S.; Yamashita, Y.; Sakamoto, Y.; Nakagawa, Y.; Yoshida, Y. The inhibitory effects of mushroom extracts on sucrose-dependent oral biofilm formation. *Appl. Microbiol. Biotechnol.* **2010**, *86*, 615–623. [CrossRef] [PubMed]
45. Zhang, Y.M.; Liu, B.L.; Zheng, X.H.; Huang, X.J.; Li, H.Y.; Zhang, Y.; Zhang, T.T.; Sun, D.Y.; Lin, B.R.; Zhou, G.X. Anandins A and B, two rare steroidal alkaloids from a marine *Streptomyces anandii* H41-59. *Mar. Drugs* **2017**, *15*, 355. [CrossRef] [PubMed]
46. Lenkkeri, A.-M.H.; Pienihakkinen, K.; Hurme, S.; Alanen, P. The caries-preventive effect of xylitol/maltitol and erythritol/maltitol lozenges: Results of a double-blinded, cluster-randomized clinical trial in an area of natural fluoridation. *Int. J. Paediatr. Dent.* **2012**, *22*, 180–190. [CrossRef] [PubMed]
47. Arruda, A.O.; Kannan, R.S.; Inglehart, M.R.; Rezende, C.T.; Sohn, W. Effect of 5% fluoride varnish application on caries among school children in rural Brazil: A randomized controlled trial. *Community Dent. Oral Epidemiol.* **2012**, *40*, 267–276. [CrossRef] [PubMed]
48. Laemmli, U.K. Cleavage of structural proteins during the assembly of the head of bacteriophage T4. *Nature* **1970**, *227*, 680–685. [CrossRef] [PubMed]
49. Hild, E.; Brumbley, S.M.; O'Shea, M.G.; Nevalainen, H.; Bergquist, P.L. A *Paenibacillus* sp. dextranase mutant pool with improved thermostability and activity. *Appl. Microbiol. Biotechnol.* **2007**, *75*, 1071–1078. [CrossRef] [PubMed]
50. Bradford, M.M. A rapid and sensitive method for the quantitation of microgram quantities of protein utilizing the principle of protein-dye binding. *Anal. Biochem.* **1976**, *72*, 248–254. [CrossRef]
51. Cardoso, S.N.; Cavalcante, T.T.; Araújo, A.X.; dos Santos, H.S.; Albuquerque, M.R.; Bandeira, P.N.; Da, C.R.; Cavada, B.S.; Teixeira, E.H. Antimicrobial and antibiofilm action of Casbane Diterpene from Croton nepetaefolius against oral bacteria. *Arch. Oral Biol.* **2012**, *57*, 550–555.
52. Tao, R.; Tong, Z.; Lin, Y.; Xue, Y.; Wang, W.; Kuang, R.; Wang, P.; Tian, Y.; Ni, L. Antimicrobial and antibiofilm activity of pleurocidin against cariogenic microorganisms. *Peptides* **2011**, *32*, 1748–1754. [CrossRef] [PubMed]

© 2018 by the authors. Licensee MDPI, Basel, Switzerland. This article is an open access article distributed under the terms and conditions of the Creative Commons Attribution (CC BY) license (http://creativecommons.org/licenses/by/4.0/).

Article

Identification of 2-keto-3-deoxy-D-Gluconate Kinase and 2-keto-3-deoxy-D-Phosphogluconate Aldolase in an Alginate-Assimilating Bacterium, *Flavobacterium* sp. Strain UMI-01

Ryuji Nishiyama, Akira Inoue and Takao Ojima *

Laboratory of Marine Biotechnology and Microbiology, Faculty of Fisheries Sciences, Hokkaido University, Hakodate, Hokkaido 041-8611, Japan; nsym2480rj@eis.hokudai.ac.jp (R.N.); inouea21@fish.hokudai.ac.jp (A.I.)
* Correspondence: ojima@fish.hokudai.ac.jp; Tel./Fax: +81-138-40-8800

Academic Editor: Antonio Trincone
Received: 28 October 2016; Accepted: 8 February 2017; Published: 14 February 2017

Abstract: Recently, we identified an alginate-assimilating gene cluster in the genome of *Flavobacterium* sp. strain UMI-01, a member of *Bacteroidetes*. Alginate lyase genes and a 4-deoxy-L-erythro-5-hexoseulose uronic acid (DEH) reductase gene in the cluster have already been characterized; however, 2-keto-3-deoxy-D-gluconate (KDG) kinase and 2-keto-3-deoxy-6-phosphogluconate (KDPG) aldolase genes, i.e., *flkin* and *flald*, still remained uncharacterized. The amino acid sequences deduced from *flkin* and *flald* showed low identities with those of corresponding enzymes of *Saccharophagus degradans* 2-40^T, a member of *Proteobacteria* (Kim et al., Process Biochem., 2016). This led us to consider that the DEH-assimilating enzymes of *Bacteroidetes* species are somewhat deviated from those of *Proteobacteria* species. Thus, in the present study, we first assessed the characteristics in the primary structures of KDG kinase and KDG aldolase of the strain UMI-01, and then investigated the enzymatic properties of recombinant enzymes, recFlKin and recFlAld, expressed by an *Escherichia coli* expression system. Multiple-sequence alignment among KDG kinases and KDG aldolases from several *Proteobacteria* and *Bacteroidetes* species indicated that the strain UMI-01 enzymes showed considerably low sequence identities (15%–25%) with the *Proteobacteria* enzymes, while they showed relatively high identities (47%–68%) with the *Bacteroidetes* enzymes. Phylogenetic analyses for these enzymes indicated the distant relationship between the *Proteobacteria* enzymes and the *Bacteroidetes* enzymes, i.e., they formed distinct clusters in the phylogenetic tree. recFlKin and recFlAld produced with the genes *flkin* and *flald*, respectively, were confirmed to show KDG kinase and KDPG aldolase activities. Namely, recFlKin produced 1.7 mM KDPG in a reaction mixture containing 2.5 mM KDG and 2.5 mM ATP in a 90-min reaction, while recFlAld produced 1.2 mM pyruvate in the reaction mixture containing 5 mM KDPG at the equilibrium state. An in vitro alginate-metabolizing system constructed from recFlKin, recFlAld, and previously reported alginate lyases and DEH reductase of the strain UMI-01 could convert alginate to pyruvate and glyceraldehyde-3-phosphate with an efficiency of 38%.

Keywords: alginate degradation; 4-deoxy-L-erythro-5-hexoseulose uronic acid (DEH) metabolism; *Bacteroidetes*; *Proteobacteria*; *Flavobacterium*; 2-keto-3-deoxy-D-gluconate (KDG) kinase; 2-keto-3-deoxy-6-phosphogluconate (KDPG) aldolase; alginate-derived products

1. Introduction

Alginate is an acidic heteropolysaccharide comprising two kinds of uronic acid, β-D-mannuronate and α-L-guluronate [1–3]. This polysaccharide exists as a structural material in cell-wall matrices of brown algae and biofilms of certain bacteria. Since alginate solution shows high viscosity and forms an elastic gel upon chelating Ca^{2+}, it has long been used as viscosifier and gelling agent in the fields of food and pharmaceutical industries. Alginate oligosaccharides produced by alginate lyases have also been recognized as functional materials since they exhibit various biological functions; e.g., promotion of root growth in higher plants [4,5], acceleration of growth rate of *Bifidobacterium* sp. [6], and promotion of penicillin production in *Penicillium chrysogenum* [7]. Anti-oxidant [8], anti-coagulant [9], anti-inflammation [10], and anti-infectious disease [11] are also bioactivities of alginate oligosaccharides. Recently, 4-deoxy-L-erythro-5-hexoseulose uronic acid (DEH), an end reaction product of alginate lyases, was proven to be available as a carbon source for ethanol fermentation by the genetically modified microbes [12–14]. Furthermore, 2-keto-3-deoxyaldonic acids like 2-keto-3-deoxy-D-gluconate (KDG) and 2-keto-3-deoxy-6-phosphogluconate (KDPG), which are intermediates in alginate metabolism, have been expected as leading compounds for antibiotics, antiviral agents, and other drugs and medicines [15]. Thus, such alginate-derived products are regarded as promising materials in various practical applications.

Alginate-degrading enzymes have been investigated in many organisms such as soil bacteria [16–21], marine bacteria [22–29], marine gastropods [30–33], and seaweeds [3,34]. Endolytic and exolytic alginate lyases split glycosyl linkages of alginate via β-elimination mechanism producing unsaturated oligosaccharides and monosaccharide, where a double bond is introduced between C4 and C5 of the newly formed non-reducing terminus [35]. Unsaturated monosaccharide, the end product of alginate lyases, is spontaneously [20] and/or enzymatically [36] converted to an open chain form, DEH, and further converted to KDG by the NAD(P)H-dependent DEH reductase. The KDG is phosphorylated to KDPG by KDG kinase and then split to pyruvate and glyceraldehyde-3-phosphate (GAP) by KDPG aldolase. The alginate-derived pyruvate and GAP are finally metabolized by Kreb's cycle. Bacterial alginate lyases have been identified in many species, e.g., *Sphingomonas* sp. [16,17], *Flavobacterium* sp. [26,27], *Saccharophagus* sp. [22,23], *Vibrio* sp. [29], and *Pseudomonas* sp. [20,21]. *Sphingomonas* sp. strain A1 possesses four kinds of alginate lyases, A1-I–IV, whose sequential action completely depolymerizes alginate to DEH [16,17]. *Flavobacterium* sp. strain UMI-01 also possesses four kinds of alginate lyases, FlAlyA, FlAlyB, FlAlyC and FlAlex, whose cooperative action efficiently degrades alginate to DEH [27]. Meanwhile, *Saccharophagus degradans* strain 2-40^T possesses two kinds of alginate lyases, Alg7D and Alg17C, which degrade alginate to unsaturated disaccharide and DEH [22,23]. The alginate-derived DEH is reduced to KDG by NAD(P)H-dependent DEH reductases as described above. Recently, this enzyme was identified in *Sphingomonas* sp. strain A1 [18,19], *Flavobacterium* sp. strain UMI-01 [28], *S. degradans* strain 2-40^T [24], *Vibrio splendidus* 12B01 [13], and marine gastropod *Haliotis discus hannai* [37]. The bacterial DEH reductases were classified under short-chain dehydrogenases/reductases (SDR) superfamily, while the gastropod enzyme was identified as a member of the aldo-keto reductase (AKR) superfamily. Information about alginate lyases and DEH reductases has been continuously accumulated; however, KDG kinase and KDPG aldolase have not been so well investigated.

Under these circumstances, DEH reductase, KDG kinase, and KDPG aldolase were recently characterized in *S. degradans* 2-40^T, a member of the phylum *Proteobacteria* [25]. The combined action of these enzymes could convert DEH to pyruvate and GAP in vitro. On the other hand, we also found the existence of alginate-assimilating gene cluster in the genome of *Flavobacterium* sp. strain UMI-01, a member of the phylum *Bacteroidetes* [27,28]. The endolytic and exolytic alginate lyase genes, *flalyA* and *flalyB*, and a DEH reductase gene, *flred*, are located in operon A, and KDG kinase-like gene *flkin* (GenBank accession number, BAQ25538) and KDPG aldolase-like gene *flald* (GenBank accession number, BAQ25539) are in operon B (Figure 1). The alginate lyases and DEH reductase of this bacterium have already been characterized [26–28]; however, KDG kinase and KDPG aldolase have not been identified yet. The amino acid sequences deduced from *flkin* and *flald* showed only 19% and 22% identities, respectively, with those of the corresponding enzymes from *S. degradans* 2-40^T [25]. These low sequence identities suggest that the properties of *Flavobacterium* (*Bacteroidetes*) enzymes may be somewhat different from those of *Saccharophagus* (*Proteobacteria*) enzymes. Therefore, in the present study, we first characterized the primary structures of KDG kinase and KDPG aldolase, FlKin and FlAld, of the strain UMI-01 compared with those of other bacterial enzymes. Then, we investigated enzymatic properties of proteins encoded by *flkin* and *flald* using recombinant enzymes, recFlKin and recFlAld. Furthermore, we constructed an in vitro alginate-metabolizing system using recFlKin and recFlAld, along with recombinant alginate lyases and DEH reductase of this bacterium to confirm that this enzyme system can produce pyruvate and GAP from alginate in vitro.

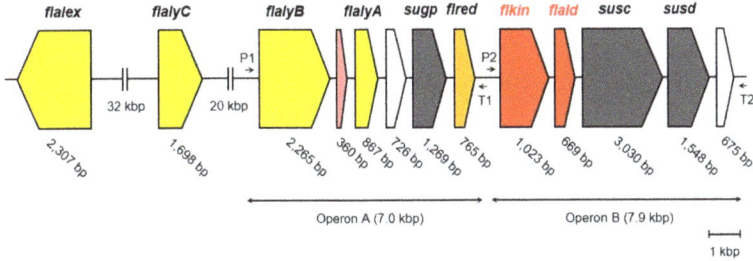

Figure 1. Alginate-assimilating enzyme genes in the genome of *Flavobacterium* sp. strain UMI-01. Yellow, alginate-lyase genes; pink, KdgF-like protein gene; white, transcriptional regulator-like protein genes; gray, membrane transporter-like genes; orange, 4-deoxy-L-erythro-5-hexoseulose uronic acid (DEH) reductase gene; red, 2-keto-3-deoxy-D-gluconate (KDG) kinase-like gene and 2-keto-3-deoxy-6-phosphogluconate (KDPG) aldolase-like gene. Arrows P1 and P2 and arrows T1 and T2 indicate predicted promoters and terminators, respectively.

2. Results

2.1. Characteristics in the Primary Structures of FlKin and FlAld

Deduced amino acid sequences of *flkin* and *flald* were compared with those of KDG kinases and KDPG aldolases from several *Proteobacteria* and *Bacteroidetes* species. Enzymes from two *Archaea* species are also included in the comparison of KDG kinases. FlKin showed considerably low amino acid identity (15%–26%) with KDG kinases from *Proteobacteria* species, i.e., *Escherichia coli* (GenBank accession number, WP_024175791) [38], *Serratia marcescens* (GenBank accession number, ABB04497) [39], and *S. degradans* 2-40^T (GenBank accession number, ABD82535) [25], and archaea, i.e., *Sulfolobus solfataricus* (GenBank accession number, WP_009991690) [40–42] and *Thermus thermophiles* (GenBank accession number, WP_011229211) [43] (Figure 2). Meanwhile, the sequence of FlKin showed relatively high identities (47%–68%) with the enzymes from *Bacteroidetes* species, i.e., *Gramella forsetii* KT0803 (GenBank accession number, CAL66135), *Dokdonia* sp. MED134 (GenBank accession number, WP_016501275), and *Lacinutrix* sp. 5H-3-7-4 (GenBank accession number, AEH01605). However, substrate-recognition residues of KDG kinase, which were identified in the *S. solfataricus* enzymes [42], i.e., Gly34, Tyr90, Tyr106, Arg108, Arg166, Asp258, and Asp294, were entirely conserved in FlKin as Gly34, Tyr89, Tyr104, Arg106, Arg169, Asp280, and Asp317, respectively. FlAld also showed low amino acid identity (22%–25%) with KDPG aldolases from *Proteobacteria* species such as *E. coli* (GenBank accession number, WP_000800517) [44,45], *Zymomonas mobilis* (GenBank accession number, S18559) [44], *Pseudomonas putida* (GenBank accession number, WP_016501275) [44,46], and *S. degradans* 2-40^T (GenBank accession number, ABD80644) [25] (Figure 3). Meanwhile, the sequence identities between FlAld and enzymes from other *Bacteroidetes* species such as *G. forsetii* KT0803 (GenBank accession number, KT0803), *Dokdonia* sp. MED134 (GenBank accession number, WP_013749799), and *Lacinutrix* sp. 5H-3-7-4 (GenBank accession number, AEH01606) were 61%–65%. Catalytic residue Lys133 and substrate-recognition residues, Glu45, Arg49, Thr73, Pro94 and Phe135 identified in the *E. coli* enzyme [45], were conserved in FlAld except for the substitution of Thr73 by Ser. Phylogenetic analyses for KDG kinases and KDPG aldolases (Figure 4A,B) suggested that the *Bacteroidetes* enzymes are somewhat phylogenetically deviated from the *Proteobacteria* (and *Archaea*) enzymes. Therefore, we decided to examine if FlKin and FlAld of the strain UMI-01 actually possess KDG kinase and KDPG aldolase activities.

Figure 2. Multiple alignment for amino acid sequences of FlKin and other KDG kinases. Closed triangles indicate substrate-recognition residues of KDG kinase from *Sulfobolus solfataricus* [42]. FlKin, KDG kinase from *Flavobacterium* sp. strain UMI-01 (GenBank accession number, BAQ25538); Lacin, KDG kinase-like protein from *Lacinutrix* sp. 5H-3-7-4 (GenBank accession number, AEH01605); Dokdo, KDG kinase-like protein from *Dokdonia* sp. MED134 (GenBank accession number, WP_013749800); Grame, KDG kinase-like protein from *Gramella forsetii* KT0803 (GenBank accession number, CAL66135); Sacch, KDG kinase from *Saccharophagus degradans* 2-40^T (GenBank accession number, ABD82535) [25]; Esche, KDG kinase from *Escherichia coli* (GenBank accession number, WP_024175791) [38]; Serra, KDG kinase from *Serratia marcescens* (GenBank accession number, ABB04497) [39]; Sulfo, KDG kinase from *Sulfolobus solfataricus* (GenBank accession number, WP_009991690) [40–42]; Therm, KDG kinase from *Thermus thermophiles* (GenBank accession number, WP_011229211) [43].

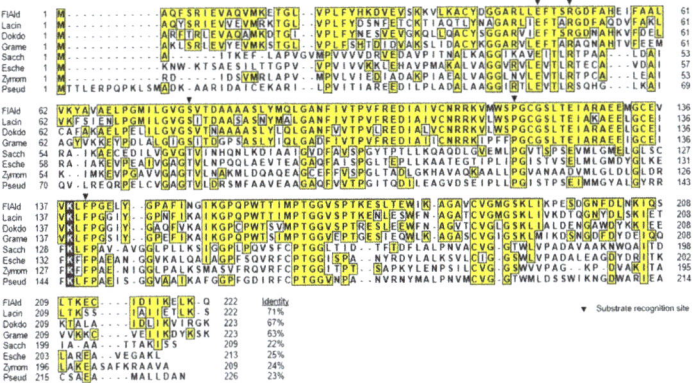

Figure 3. Multiple alignment for amino acid sequences of FlAld and other KDPG aldolases. Gray box and closed triangles indicate catalytic and substrate-recognition residues of KDPG aldolase from *E. coli* [44,45], respectively. FlAld, KDPG aldolase from *Flavobacterium* sp. strain UMI-01 (GenBank accession number, BAQ25539); Lacin, KDPG aldolase-like protein from *Lacinutrix* sp. 5H-3-7-4 (GenBank accession number, AEH01606); Dokdo, KDPG aldolase-like protein from *Dokdonia* sp. MED134 (GenBank accession number, WP_013749799); Grame, KDPG aldolase-like protein from *G. forsetii* KT0803 (GenBank accession number, CAL66136); Sacch, KDPG aldolase from *S. degradans* 2-40^T (GenBank accession number, ABD80644) [25]; Esche, KDPG aldolase from *E. coli* (GenBank accession number, WP_000800517) [44,45]; Zymom, KDPG aldolase from *Zymomonas mobilis* (GenBank accession number, S18559) [44]; Pseud, KDPG aldolase from *Pseudomonas putida* (GenBank accession number, WP_016501275) [44,46].

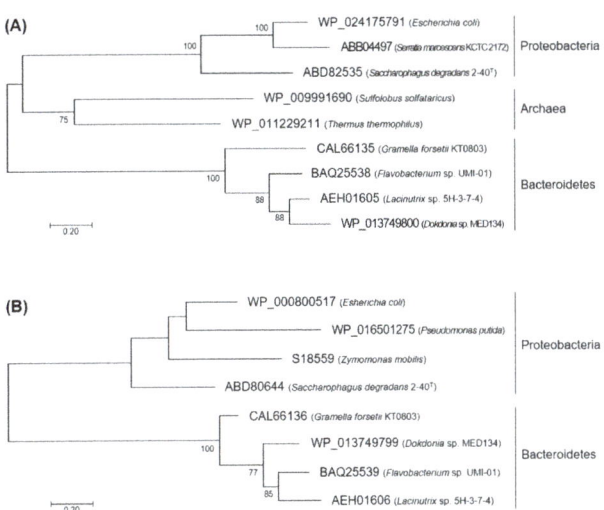

Figure 4. Phylogenetic trees for KDG kinases and KDPG aldolases. Phylogenetic analyses were carried out using amino acid sequences of KDG kinases from *Proteobacteria*, *Archaea* and *Bacteroidetes* species (**A**) and KDPG aldolases from *Proteobacteria* and *Bacteroidetes* species (**B**). Amino acid sequences of KDG kinases and KDPG aldolases were retrieved from the draft or complete genome data deposited in GenBank. Accession numbers for enzyme sequences along with the bacterial species are indicated in the right of each branch. Bootstrap values above 50% are indicated on the root of branches. Scale bar indicates 0.20 amino acid substitution.

2.2. Production of recFlKin and recFlAld, and Their Reaction Products

Coding regions of *flkin* and *flald* were amplified by PCR with specific primers listed in Table 1, cloned into pCold vector and expressed in *E. coli* BL21 (DE3). The recombinant enzymes were purified by Ni-NTA affinity chromatography. Molecular masses of recFlKin and recFlAld estimated by SDS-PAGE were 39 kDa and 26 kDa, respectively (Figure 5). These values were consistent with the calculated molecular masses of these enzymes, i.e., 39,391 Da and 25,808 Da, which include 8 × Gly + 8 × His-tag [26].

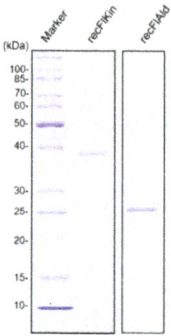

Figure 5. SDS-PAGE for recFlKin and recFlAld. Recombinant enzymes were purified Ni-NTA affinity chromatography and subjected to 0.1% SDS–10% polyacrylamide-gel electrophoresis. Proteins in the gel were stained by Coomassie Brilliant Blue R-250. Marker, molecular weight markers (Protein Ladder Broad Range, New England Biolabs, Ipswich, MA, USA).

Table 1. Primers used for amplification of *flkin* and *flald* genes.

Primer Name	Nucleotide Sequence
recFlKin-F	5′-AGGTAATACACCATGAAAAAAGTAGTCACTTTTGG-3′
recFlKin-R	5′-CACCTCCACCGGATCCTCTTGAAACTTTTCCTGAAA-3′
recFlAld-F	5′-ATGTAATACACCATGGCTCAATTTTCAAGAATAGA-3′
recFlAld-R	5′-CACCTCCACCGGATCCTTGTTTTAACTCTTTAATGA-3′

The recFlKin was allowed to react with KDG in the presence of ATP. TLC analysis suggested that the reaction product was KDPG (Figure 6A). Then, the molecular mass of the reaction product was determined by matrix-assisted laser desorption ionization-time of flight mass spectrometer (MALDI-TOF) mass spectrometry (Figure 7A,B). The 257 m/z peak was considered to be that of KDPG (MW = 258), and the 279 m/z peak was considered to be that of a sodium-salt form of KDPG. These results indicate that the reaction product of recFlKin is KDPG. Thus, we concluded that the protein encoded by *flkin* is KDG kinase. Here, it should be noted that the peak intensities of KDPG were considerably low. This was ascribable to the low ionization level of KDPG. Therefore, we attempted to improve the signal intensity of KDPG using other matrices, e.g., 2,5-dihydroxybenzoic acid and α-cyano-4-hydroxycinnamic acid. Unfortunately, signal intensity of KDPG was not improved much. We still need to investigate the suitable conditions for the detection of KDPG.

Reaction products of recFlAld were also analyzed by TLC (Figure 6B). recFlAld produced two kinds of reaction products with different mobility on TLC. According to their mobility, they were regarded as pyruvate and GAP. The staining intensity of pyruvate was significantly low compared with that of GAP. This difference was ascribable to the difference in the reactivity between pyruvate and GAP with 2,4-dinitrophenylhydrazine (DNP). Namely, GAP showed much higher reactivity with DNP than pyruvate. Then, the reaction products of recFlAld were subjected to MALDI-TOF mass spectrometry. The 87 m/z and 169 m/z peaks corresponding to pyruvate (MW = 88) and GAP

(MW = 170), respectively, were observed. The peak intensity of GAP was small (Figure 7C,D). This appeared to be due to the decomposition of GAP during the mass spectrometric analysis. Thus, we may conclude that recFlAld is the KDPG aldolase that splits KDPG to pyruvate and GAP.

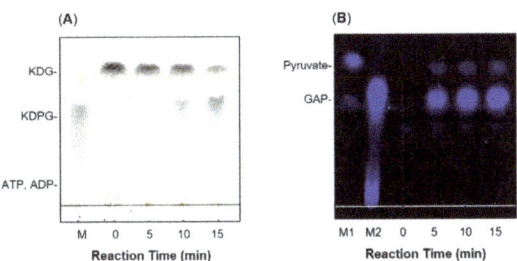

Figure 6. Thin-layer chromatography (TLC) analyses for reaction products of recFlKin and recFlAld. (**A**) Reaction products produced by recFlKin. The reaction products were visualized by spraying 10% (*v*/*v*) sulfuric acid in ethanol followed by heating at 130 °C for 10 min. M, standard KDPG; (**B**) Reaction products of recFlAld. The reaction products were visualized with 0.5% (*w*/*v*) 2,4-dinitrophenylhydrazine (DNP)–20% (*v*/*v*) sulfuric acid. The color was graphically inverted to ease the recognition of spots. M1, standard pyruvate; M2, standard glyceraldehyde-3-phosphate (GAP). Stained materials near the original position are GAP oligomers.

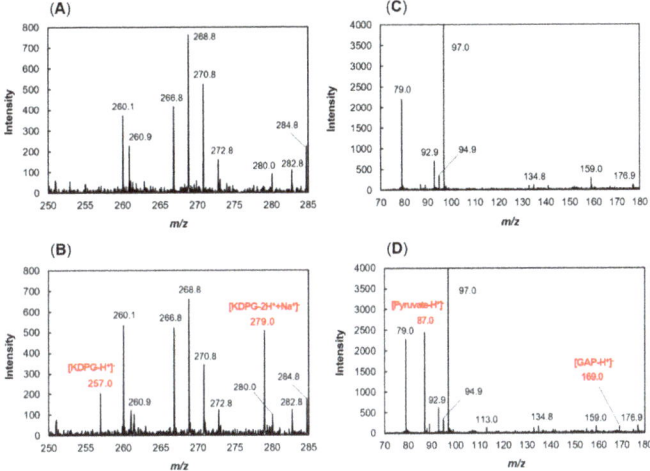

Figure 7. Mass spectrometry for reaction products of recFlKin and recFlAld. The reaction products prepared as in Section 4.10 were subjected to matrix-assisted laser desorption ionization-time of flight mass spectrometer (MALDI-TOF) mass spectrometry, and analyzed by negative-ion mode. (**A**,**B**) KDG before and after the recFlKin reaction, respectively; (**C**,**D**) KDPG before and after the recFlAld reaction, respectively. Reaction products are indicated with red letters along with molecular masses above the peaks.

2.3. Enzymatic Properties of recFlKin and recFlAld

We first investigated the kinetic parameter for recFlAld, since recFlAld was necessary for the KDG kinase assay. In the present study, the kinase activity was assayed by quantifying the pyruvate produced from KDPG by the action of recFlAld. KDPG-derived pyruvate was determined by the lactate dehydrogenase (LDH)–NADH system as described in Section 4.6. In the equilibrium state of recFlAld reaction, pyruvate concentration reached 1.2 mM. Since the KDPG concentration was originally 5 mM, that in the equilibrate state was regarded as 3.8 mM. From these values the equilibrium constant (K_{eq}) and $\Delta G°$ were calculated to be 3.8×10^{-1} M and +0.57 kcal/mol, respectively. This indicated that the equilibrium position of KDPG–aldolase reaction is slightly shifted toward the KDPG side. Next, we determined the reaction rate of recFlAld by the LDH–NADH method. By this method, the specific activity of recFlAld was estimated to be 57 U/mg at pH 7.4 and 30 °C. Coexistence of LDH–NADH in the reaction mixture could extend the aldolase reaction longer time by decreasing pyruvate concentration in the reaction equilibrium.

Next, KDG kinase activity of recFlKin was determined by using recFlAld and LDH–NADH. recFlKin was allowed to react with KDG in the presence of ATP at 30 °C and the reaction was terminated by heating at 100 °C for 3 min at the reaction times 1, 15, and 30 min. The KDPG produced in the reaction mixture was then split to pyruvate and GAP by recFlAld, and the pyruvate was quantified by the LDH–NADH system. At reaction time 90 min, recFlKin was found to produce 1.7 mM KDPG from 2.5 mM KDG at ~70% efficiency with the specific activity 0.72 U/mg. recFlKin showed an optimal temperature and pH at around 50 °C and 7.0, respectively, and was stable at 40 °C for 30 min.

2.4. Construction of In Vitro Alginate-Metabolizing System Using Recombinant Enzymes

In the present study, we identified *flkin* and *flald* in the genome of strain UMI-01 as KDG kinase and KDPG aldolase gene, respectively. Since alginate lyases and DEH reductase in this strain have already been characterized [26–28], here we examined if the sequential action of these alginate-degrading and -assimilating enzymes could convert alginate to pyruvate and GAP in vitro. Namely, recombinant alginate lyases (recFlAlyA, recFlAlyB, and recFlAlex) [26,27], DEH reductase (recFlRed) [28], KDG kinase (recFlKin), and KDPG aldolase (recFlAld) were allowed to react alginate in various combinations, and each reaction product was analyzed by TLC (Figure 8) and quantified by thiobarbituric acid (TBA) and LDH–NADH methods (Table 2). As shown in Figure 8, alginate was almost completely degraded to DEH by the simultaneous actions of recFlAlyA, recFlAlyB, and recFlAlex. The DEH was also almost completely reduced to KDG by recFlRed. Furthermore, a major part of the KDG was converted to KDPG by recFlKin, and the band of KDPG became faint by the reaction of recFlAld. This indicated the splitting of KDPG to pyruvate and GAP by the action of recFlAld. Accordingly, the sequential action of recombinant enzymes was considered to be capable of converting alginate to pyruvate and GAP in vitro. Then, the yields of intermediates in each reaction step were quantified by TBA and LDH–NADH methods (Table 2). Concentrations of the unsaturated oligo-alginates, DEH, KDG, KDPG, and pyruvate (and GAP), were determined to be 4.2 mM, 9.8 mM, 9.8 mM, 8.1 mM, and 3.8 mM, respectively. Since the initial concentration of alginate (0.2% (w/v)) corresponds to 10 mM monosaccharide, the yields of DEH and KDG were estimated to be ~100%, and the yields of KDPG and pyruvate were estimated to be ~80% and ~40%, respectively. These results indicated that high-value intermediates such as KDPG could be produced from alginate with fairly high efficiency by the recombinant enzymes of the strain UMI-01 in vitro.

Table 2. Quantification of reaction products produced by the recombinant enzymes.

Enzymes	Substrates/Products	Concentration (mM)	Yield (%)
None	Alginate [a]	10 [a]	-
recFlAlyA	Oligoalginates	4.2 ± 0.06	-
recFlAlyA + recFlAlyB + recFlAlex	DEH	9.8 ± 0.34	98
recFlAlyA + recFlAlyB + recFlAlex + recFlRed	KDG	9.8 ± 1.0	98
recFlAlyA + recFlAlyB + recFlAlex + recFlRed + recFlKin	KDPG	8.1 ± 0.54	81
recFlAlyA + recFlAlyB + recFlAlex + recFlRed + recFlKin + recFlAld	Pyruvate (and GAP)	3.8 ± 0.33	38

[a] 0.2% (w/v) sodium alginate theoretically corresponds to 10 mM monosaccharide.

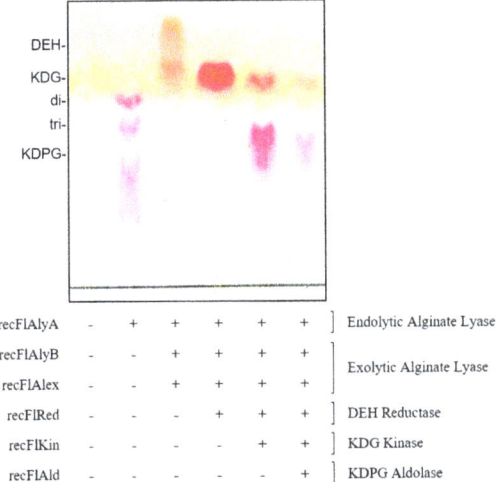

Figure 8. Construction of in vitro alginate-metabolizing system using recombinant enzymes. Alginate was allowed to react with recFlAlyA, recFlAlyB, recFlAlex, recFlRed, recFlKin, and recFlAld in various combinations at 25 °C for 12 h. The reaction products were subjected to TLC and detected by staining with 4.5% TBA. Presence and absence of each enzyme is indicated with '+' and '−', respectively. Detailed conditions are shown under Section 4.

3. Discussion

3.1. Alginate-Metabolizing Enzymes of Flavobacterium sp. Strain UMI-01

In the present study, *flkin* and *flald* in the genome of *Flavobacterium* sp. strain UMI-01 were confirmed to be the enzyme genes encoding KDG kinase and KDPG aldolase. The recombinant enzymes, recFlKin and recFlAld, showed KDG kinase and KDPG aldolase activity although low sequence identities were shown to the corresponding enzymes from other bacteria and archaea (Figures 2–4). Consequently, these genes, along with previously reported alginate lyase and DEH reductase genes were confirmed to be the genes responsible for alginate metabolism of this bacterium. The alginate-metabolizing pathway of this strain is summarized as in Figure 9. The alginate lyases degrade polymer alginate to unsaturated monomer (DEH) in the periplasmic space [24,25]. DEH reductase, KDG kinase and KDPG aldolase convert DEH to pyruvate and GAP in the cytosol. Therefore, DEH produced in the periplasmic space should be incorporated to the cytosol by certain transportation system(s). Such DEH transporters in this strain have not been identified yet; however, sugar permease-like gene *sugp* and membrane transporter-like genes *susc* and *susd* were found in the operons A and B, respectively (see Figure 1). Thus, the putative permease and transporters are also indicated in Figure 9. Another problem is how the expressions of alginate-metabolizing genes are regulated. We recently noticed that expression levels of alginate lyases were significantly low in the absence of alginate but strongly increased by the addition of alginate to the medium. This indicates that the expressions of alginate-metabolic enzymes are up-regulated by alginate. We are now searching regulatory genes for alginate-metabolizing enzyme genes in the UMI-01 strain genome.

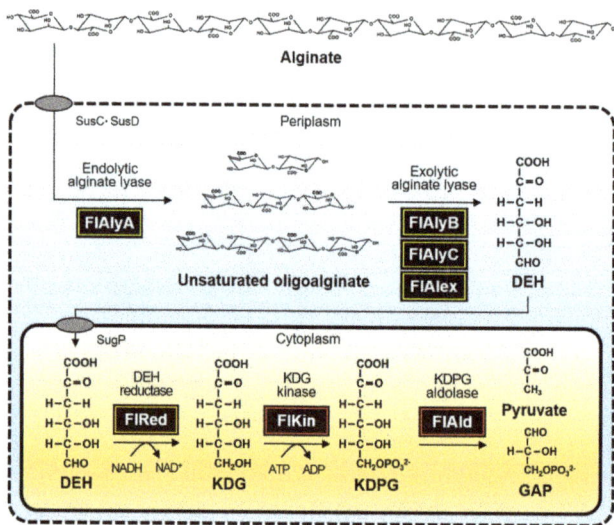

Figure 9. Alginate-metabolic system of *Flavobacterium* sp. strain UMI-01.

3.2. Properties of recFlKin and recFlAld

KDG kinase and KDPG aldolase are known to be the enzymes included in Entner–Doudoroff (ED) pathway. This pathway distributes over bacteria and archaea and play important roles in the metabolisms of glucuronate and glucose. In this pathway, KDG kinase phosphorylates KDG to KDPG, and KDPG aldolase split KDPG to pyruvate and GAP. Optimal temperature and pH of recFlKin were 50 °C and ~7.0, which were similar to those of KDG kinase from the bacteria *S. marcescens* [39]. While thermal stability of recFlKin was considerably low compared with the enzymes from archaea *S. tokodaii* [47] and *S. solfataricus* [40], e.g., these enzymes were stable up to 60–70 °C. recFlAld acts only on KDPG unlike archaea aldolases which split both KDG and KDPG [48,49]. Primary structures of bacterial aldolases showed low identity with those of archaea enzymes. The amino acid sequence of FlAld showed only 22%–25% identity with respect to *Proteobacteria* enzymes, while it showed 61%–65% identity with the *Bacteroidetes* enzymes. This suggests that somewhat deviated function between the *Proteobacteria* enzymes and *Bacteroidetes* enzymes. However, less different properties were found in recFlAld. Reverse reaction of bacterial aldolases was shown to be useful for the production of KDPG from pyruvate and GAP and also various compounds from pyruvate and aldehydes [44]. Our preliminary experiments also indicated that recFlAld could produce KDPG from pyruvate and GAP (data not shown, but see Section 2.3). Thus, recFlAld is also considered to be useful for producing novel compounds from pyruvate and various aldehydes.

3.3. Construction of In Vitro Alginate-Metabolizing System

An in vitro alginate-metabolizing system was successfully constructed from the recombinant enzymes, recFlAlyA, recFlAlyB, recFlAlex, recFlRed, recFlKin, and recFlAld. Accordingly, various kinds of intermediates could be produced by this system (Figure 8 and Table 2). Recently, alginate-assimilating enzymes of *S. degradans* 2-40^T were used for the production of KDG, KDPG, GAP and pyruvate [24,25]. However, the reaction efficiency of KDG kinase of *S. degradans* 2-40^T appeared to be lower than that of our system. Namely, the major part of KDG in the reaction mixture remained to be unphosphorylated in the *S. degradans* 2-40^T system. On the other hand, recFlKin in our system could convert KDG to KDPG with ~80% efficiency. This difference in the reaction efficiency between *S. degradans* enzyme and recFlKin may be derived from the origin of this enzyme, namely, from *Proteobacteria* species or *Bacteroidetes* species. To confirm this, we have to directly compare the KDG kinase properties between the enzymes from *Proteobacteria* and *Bacteroidetes* in future.

3.4. Production of a High-Value Product KDPG from Alginate

KDPG is a valuable leading compound for novel drugs and medicines. Synthesis of KDPG has been attempted by several methods [44,48,50]. For example, KDPG was first produced from gluconate with archaea enzymes [48]. However, this method required high-temperature reaction since the archaea enzymes are thermophilic. Reverse reaction of KDPG aldolase was also used for the production of KDPG from pyruvate and GAP [44,50]. However, this method required GAP, a significantly expensive raw material. On the other hand, we could produce KDPG from a much cheaper material, alginate, using the enzymes from the strain UMI-01. High recovery of KDPG from alginate (~80%) also indicated the practical potentiality of this enzyme. Thus, *Flavobacterium* sp. strain UMI-01 was considered to be a useful enzyme source for the production of value-added materials from alginate.

4. Experimental Section

4.1. Materials

Sodium alginate (*Macrocystis pyrifera* origin) was purchased from Sigma-Aldrich (St. Louis, MO, USA). Alginate-assimilating bacteria, *Flavobacterium* sp. strain UMI-01, was cultivated at 25 °C in a mineral salt (MS) medium including 1% (w/v) sodium alginate as described in our previous report [26]. Cell lysate (crude enzyme) of this strain was extracted from cell pellets by freeze and thaw followed by sonication as described previously [28]. DEH was prepared by the digestion of sodium alginate with the crude enzyme and purified by SuperQ-650S (Tosoh, Tokyo, Japan) anion-exchange chromatography [28]. Standard KDG, KDPG, pyruvate, and GAP were purchased from Sigma-Aldrich. pCold I expression vector was purchased from TaKaRa (Shiga, Japan) and modified to the form that can add $8 \times$ Gly + $8 \times$ His-tag to the C-terminus of the expressed proteins [26]. *E. coli* DH5α and BL21 (DE3) were purchased from TaKaRa. Ni-NTA resin was purchased from Qiagen (Hilden, Germany). A TLC silica gel 60 plate was purchased from Merk KGaA (Darmstadt, Germany). TSKgel DEAE-2SW (4.6×250 mm) and Superdex peptide 10/300 GL were purchased from Tosoh Bioscience LLC (King of Prussia, PA, USA) and GE Healthcare (Little Chalfont, Buckinghamshire, UK), respectively. Lactate dehydrogenase (LDH; porcine heart origin) and NADH were purchased from Oriental Yeast Co., LTD. (Tokyo, Japan). ATP and 9-aminoacridine were purchased from Sigma-Aldrich. Other chemicals were purchased from Wako Pure Chemical Industries Ltd. (Osaka, Japan).

4.2. Phylogenetic Analysis for KDG Kinases and KDPG Aldolases

Phylogenetic analysis was carried out using the amino acid sequences of KDG kinases or KDPG aldolases from *Proteobacteria*, *Bacteroidetes* and *Archaea* currently available. *Bacteroidetes* enzymes used are from *Gramella forsetii* KT0803, *Lacinutrix* sp. 5H-3-7-4, and *Dokdonia* sp. MED134, which were reported to be located in the alginolytic gene cluster of each species [51]. These amino acid sequences were first aligned with the sequences of FlKin or FlAld by the ClustalW program, then aligned sequences were trimmed with GBlocks. Phylogenetic trees were generated by the maximum likelihood algorithm on the basis of the LG model implemented in the Molecular Evolutionary Genetics Analysis version 6.0 (MEGA 6) software. The bootstrap values were calculated from 1000 replicates.

4.3. Cloning, Expression, and Purification of Recombinant FlKin and FlAld

Genomic DNA of strain UMI-01 was prepared with ISOHAIR DNA extraction kit (Nippon Gene, Tokyo, Japan). Coding regions of *flkin* and *flald*, 1023 bp and 669 bp, respectively, were amplified by PCR using specific primers including restriction sites, *Nco*I and *Bam*HI, in the 5′-terminal regions (Table 1). Genomic PCR was performed in a medium containing 10 ng of genomic DNA, 0.2 µM each primer, and Phusion DNA polymerase (New England Biolabs, Ipswich, MA, USA). The reaction medium was preincubated at 95 °C for 2 min, and a reaction cycle of 95 °C for 10 s, 55 °C for 20 s, and 72 °C for 60 s was repeated 30 times. The PCR product was ligated to pCold I vector pre-digested by *NcoI* and *Bam*HI using In-Fusion cloning system (Clontech Laboratories, Mountain View, CA, USA). Insertion of the genes in the vector was confirmed by nucleotide sequencing with DNA sequencer 3130xl (Applied Biosystems, Foster, CA, USA). Recombinant enzymes, recFlKin and recFlAld, were expressed with the pCold I–*E. coli* BL21 (DE3) system. The transformed BL21 (DE3) was inoculated to 500 mL of 2× YT medium and cultivated at 37 °C for 16 h. Then, the temperature was lowered to 15 °C and isopropyl β-D-1-thiogalactopyranoside was added to make the final concentration of 0.1 mM. After 24-h induction, bacterial cells were harvested by centrifugation at 5000× g for 5 min and suspended in a buffer containing 10 mM imidazole-HCl (pH 8.0), 0.5 M NaCl, 1% (v/v) TritonX-100, and 0.01 mg/mL lysozyme. The suspension was sonicated at 20 kHz (30W) for a total of 4 min (30 s × 8 times with each 1 min interval) and centrifuged at 10,000× g for 10 min. The supernatant containing recombinant proteins was mixed with 1 mL of Ni-NTA resin and incubated for 30 min on ice with occasional suspension. The resin was set on a disposal plastic column (1 × 5 cm) and washed three times with 20 mL of 30 mM imidazole-HCl (pH 8.0)–0.5 M NaCl. The recombinant proteins adsorbed to the resin were eluted with 250 mM imidazole-HCl (pH 8.0)–0.5 M NaCl and collected as 1 mL fractions. The fractions containing the recombinant proteins were pooled and dialyzed against 20 mM Tris-HCl (pH 7.4)–0.1 M NaCl.

4.4. Preparation of KDG

KDG was prepared from alginate using the crude enzyme of the strain UMI-01 as follows; 0.5% (w/v) sodium alginate (50 mL) was digested at 30 °C with 1 mg/mL of the crude enzyme, which contains alginate lyases and other metabolic enzymes. NADH was added to the mixture to make the final concentration 10 mM to reduce DEH with DEH reductase contained in the crude enzyme. After 12 h, four volumes of −20 °C 2-propanol were added to terminate the reaction and the proteins and NADH precipitated were removed by centrifugation at 10,000× g for 10 min. The supernatant containing KDG was dried up in a rotary evaporator at 35 °C. The dried powder was dissolved in 50 mL of distilled water and subjected to a TOYOPEARL SuperQ-650S column (2.4 × 22 cm) equilibrated with distilled water. The absorbed KDG and trace amount of unsaturated disaccharide were separately eluted by a linear gradient of 0–0.2 M NaCl in distilled water (total 400 mL). Elution of KDG and unsaturated disaccharide was detected by TBA reaction. In this chromatography, KDG was eluted at around 80 mM NaCl, while disaccharides were eluted at around 120 mM. Approximately 90 mg of KDG was obtained from 0.25 g of sodium alginate.

4.5. Preparation of KDPG

KDPG was prepared from the KDG by using recFlKin. Namely, recFlKin was (final concentration 10 µg/mL) was added to the reaction mixture (10 mL) containing 2.5 mM KDG, 2.5 mM ATP, 5 mM $MgCl_2$, 20 mM Tris-HCl (pH 7.4), 100 mM KCl, and 1 mM dithiothreitol, and incubated at 40 °C for 3 h. The mixture was lyophilized, dissolved in 500 µL of distilled water and the supernatant was subjected to a Superdex peptide 10/300 GL column equilibrated with 0.1 M CH_3COONH_4. KDPG and KDG, which eluted together in this chromatography, were lyophilized, dissolved in 1 mL of distilled water, and subjected to HPLC (Shimadzu Prominence LC-6AD, Tokyo, Japan) equipped by TSKgel DEAE-2SW (Tosoh). KDG and KDPG were separately eluted at around 150 mM and 320 mM CH_3COONH_4 by the linear gradient of 0–0.4 M CH_3COONH_4. The amount of KDPG was quantified by the system comprising recFlAld and LDH–NADH using authentic KDPG as a standard. By the above procedure, 1.2 mg of KDPG was obtained from 4.5 mg of KDG.

4.6. Assay for KDPG Aldolase Activity

KDPG aldolase activity of recFlAld was assayed by the determination of pyruvate using a lactate dehydrogenase (LDH)–NADH coupling system [50]. Namely, the aldolase reaction was conducted at

30 °C in a reaction mixture containing 5 mM KDPG, 20 mM Tris-HCl (pH 7.4), 100 mM KCl, 1 mM DTT, and 1 µg/mL recFlAld in the presence of 0.2 mM NADH and 1 unit/mL LDH. The reaction rate was estimated from the decrease in the Abs 340 nm due to the oxidation of NADH accompanied by the reduction of pyruvate. One unit (U) of KDPG aldolase activity was defined as the amount of enzyme that produced 1 µmol of pyruvate per min.

4.7. Assay for KDG Kinase Activity

KDG kinase activity was assayed as follows. The reaction mixture containing 2.5 mM KDG, 2.5 mM ATP, 5 mM $MgCl_2$, 20 mM Tris-HCl (pH 7.4), 100 mM KCl, 1 mM DTT, and 10 µg/mL recFlKin was incubated at 30 °C. At reaction times, 1, 15, and 30 min, an aliquot (160 µL) of the reaction mixture was taken out and heated at 100 °C for 3 min to terminate the reaction. To the mixture, 240 µL of a buffer containing 84 mM Tris-HCl (pH 7.4), 167 mM KCl, 0.67 mM NADH, 2.5 µg/mL recFlAld, and 1 unit of LDH was added and the pyruvate released was determined by the LDH–NADH system. One unit (U) of KDG kinase activity was defined as the amount of enzyme that produced 1 µmol of KDPG per min. Temperature dependence of recFlKin was determined at 10–60 °C. Thermal stability of recFlKin was assessed by measuring the activity remaining after the incubation at 10–50 °C for 30 min. pH dependence of recFlKin was determined with reaction mixtures adjusted to pH 4.5–5.3 with 20 mM CH_3COONa buffer, pH 5.6–7.3 with 20 mM PIPES-NaOH buffer, pH 7.1–8.8 with 20 mM Tris-HCl buffer, and pH 9.1–9.7 with 20 mM glycine–NaOH buffer. The activity assay was conducted three times and the mean value was shown with standard deviation in each figure.

4.8. Construction of In Vitro Alginate-Metabolizing System from Recombinant Enzymes

An in vitro alginate-metabolizing system was constructed using recombinant alginate lyases (recFlAlyA, recFlAlyB, and recFlAlex) [26,27], recombinant DEH reductase (recFlRed) [28], and recFlKin and recFlAld prepared in the present study. Alginate-metabolizing reaction was conducted at 25 °C in a mixture containing 0.2% (w/v) sodium alginate, 10 mM NADH, 10 mM ATP, 10 mM $MgCl_2$, 20 mM sodium phosphate (pH 7.4), 1 mM DTT, and various combinations of recFlAlyA, recFlAlyB, recFlAlex, recFlRed, recFlKin, and recFlAld with each final concentration at 10 µg/mL, 10 µg/mL, 10 µg/mL, 2.5 µg/mL, 10 µg/mL, and 1 µg/mL, respectively. After 12-h reaction, unsaturated oligo-alginates, DEH, and KDG, were analyzed by TLC and TBA reaction [52]. KDPG and pyruvate concentrations were determined by the LDH–NADH reaction.

4.9. Determination of Unsaturated Sugars

Unsaturated sugars were determined by the TBA method [52]. The sample containing unsaturated sugars (150 µL) was mixed with 150 µL of 20 mM $NaIO_4$–0.125 M H_2SO_4 and allowed to react for 1 h on ice. Then, 100 µL of $NaAsO_2$–0.5 N HCl was added to the mixture and incubated for 10 min at room temperature. To the mixture, 600 µL of 0.6% (w/v) TBA was added and heated for 10 min at 100 °C. The unsaturated sugars were determined by measuring Abs 548 nm, adopting the absorption coefficient for DEH and KDG, $\varepsilon = 41 \times 10^3$ $M^{-1} \cdot cm^{-1}$, which we determined in the present study using KDG and DEH standards.

4.10. Thin-Layer Chromatography

TLC silica gel 60 plate was used for the analysis of the reaction products produced by recFlKin and recFlAld. The reaction product of recFlKin was prepared with a reaction mixture containing 2.5 mM KDG and 2.5 mM ATP and 200 µg/mL recFlKin. The reaction was carried out at 30 °C for 0–15 min and terminated by heating at 100 °C for 2 min. Four microliters of each reaction mixture was applied to a TLC plate. The reaction product was developed with 1-butanol:acetic acid:water = 2:1:1 (v:v:v) and detected by heating at 130 °C for 10 min after spraying 10% (w/v) sulfuric acid–90% (w/v) ethanol. The reaction product of recFlAld was prepared with a reaction mixture containing 5 mM KDPG and 1 µg/mL recFlAld. After the reaction at 30 °C for 0–15 min, six microliters of the reaction mixture were applied to TLC plate and developed with the same solvent as described above. The reaction product on the plate was detected with 0.5% (w/v) 2,5-dinitrophenylhydrazine (DNP)–20% (v/v) sulfuric acid–60% (v/v) ethanol. In case of unsaturated sugars, they were visualized with 4.5% (w/v) TBA after the periodic acid treatment.

4.11. Mass Spectrometry

Phosphorylation of KDG by recFlKin was detected by mass spectrometry. The KDG phosphorylated by recFlKin in the conditions described in Section 4.10 was mixed with 6.7 mg/mL 9-aminoacridine–methanol at 1:3 ($v:v$). One microliter of the mixture was applied to a sample plate and air-dried at room temperature. The sample was subjected to a matrix-assisted laser desorption ionization-time of flight mass spectrometer (MALDI-TOF-MS) (Proteomics Analyzer 4700, Applied Biosystems, Foster City, CA, USA) and analyzed in a negative-ion mode.

4.12. SDS-PAGE

SDS-PAGE was performed by the method of Porzio and Pearson [53] using 10% polyacrylamide gel. Proteins in the gel were stained with 0.1% (w/v) Coomassie Brilliant Blue R-250–50% (v/v) methanol–10% (v/v) acetic acid and the background of the gel was destained with 5% (v/v) methanol–7% (v/v) acetic acid.

4.13. Determination of Protein Concentration

Protein concentration was determined by the method of Lowry [54] using bovine serum albumin fraction V as a standard.

5. Conclusions

Enzymes responsible for the metabolism of alginate-derived DEH had not been well characterized in alginolytic bacteria. In the present study, KDG kinase-like gene *flkin* and KDPG aldolase-like gene *flald* in the genome of *Flavobacterium* sp. strain UMI-01 were investigated and the activities of the proteins encoded by these genes were assessed by using recombinant enzymes recFlKin and recFlAld. Analyses for reaction product of recFlKin and recFlAld indicated that these enzymes were KDG kinase and KDPG aldolase, respectively. Thus, the alginate metabolism of *Flavobacterium* sp. strain UMI-01 was considered to be achieved by the actions of FlKin and FlAld along with alginate lyases FlAlyA, FlAlyB and FlAlex, and DEH reductase FlRed. An in vitro alginate-metabolizing system was successfully constructed from the above enzymes. This system could convert alginate to pyruvate and GAP with 38% efficiency. This result indicates that the UMI-01 enzymes are available for the production of high-value materials like KDPG from alginate.

Acknowledgments: This study was supported in part by the Program for Constructing "Tohoku Marine Science Bases" promoted by Ministry of Education, Culture, Sports, Science and Technology, Japan.

Author Contributions: Ryuji Nishiyama took charge of designing the research and performed biochemical analysis. Akira Inoue performed cloning of *flkin* and *flald* genes and expression of recombinant enzymes. Ryuji Nishiyama, Akira Inoue and Takao Ojima took charge of preparation of manuscript.

Conflicts of Interest: The authors declare no conflict of interest.

References

1. Haug, A.; Larsen, B.; Smidsrod, O. Studies on the sequence of uronic acid residues in alginic acid. *Acta Chem. Scand.* **1967**, *21*, 691–704. [CrossRef]
2. Gacesa, P. Alginates. *Carbohydr. Polym.* **1988**, *8*, 161–182. [CrossRef]
3. Wong, T.Y.; Preston, L.A.; Schiller, N.L. Alginate lyase: Review of major sources and enzyme characteristics, structure-function analysis, biological roles, and applications. *Annu. Rev. Microbiol.* **2000**, *54*, 289–340. [CrossRef] [PubMed]

4. Tomoda, Y.; Umemura, K.; Adachi, T. Promotion of barley root elongation under hypoxic conditions by alginate lyase-lysate (A.L.L.). *Biosci. Biotechnol. Biochem.* **1994**, *58*, 202–203. [CrossRef] [PubMed]
5. Xu, X.; Iwamoto, Y.; Kitamura, Y.; Oda, T.; Muramatsu, T. Root growth-promoting activity of unsaturated oligomeric uronates from alginate on carrot and rice plants. *Biosci. Biotechnol. Biochem.* **2003**, *67*, 2022–2025. [CrossRef] [PubMed]
6. Akiyama, H.; Endo, T.; Nakakita, R.; Murata, K.; Yonemoto, Y.; Okayama, K. Effect of depolymerized alginates on the growth of bifidobacteria. *Biosci. Biotechnol. Biochem.* **1992**, *56*, 355–356. [CrossRef] [PubMed]
7. Ariyo, B.; Tamerler, C.; Bucke, C.; Keshavarz, T. Enhanced penicillin production by oligosaccharides from batch cultures of *Penicillium chrysogenum* in stirred-tank reactors. *FEMS Microbiol. Lett.* **1998**, *166*, 165–170. [CrossRef] [PubMed]
8. Trommer, H.; Neubert, R.H.H. Screening for new antioxidative compounds for topical administration using skin lipid model systems. *J. Pharm. Pharm. Sci.* **2005**, *8*, 494–506. [PubMed]
9. Khodagholi, F.; Eftekharzadeh, B.; Yazdanparast, R. A new artificial chaperone for protein refolding: Sequential use of detergent and alginate. *Protein J.* **2008**, *27*, 123–129. [CrossRef] [PubMed]
10. Mo, S.-J.; Son, E.-W.; Rhee, D.-K.; Pyo, S. Modulation of tnf-α-induced icam-1 expression, no and h2o2 production by alginate, allicin and ascorbic acid in human endothelial cells. *Arch. Pharm. Res.* **2003**, *26*, 244–251. [CrossRef] [PubMed]
11. An, Q.D.; Zhang, G.L.; Wu, H.T.; Zhang, Z.C.; Zheng, G.S.; Luan, L.; Murata, Y.; Li, X. Alginate-deriving oligosaccharide production by alginase from newly isolated *Flavobacterium* sp. LXA and its potential application in protection against pathogens. *J. Appl. Microbiol.* **2009**, *106*, 161–170. [CrossRef] [PubMed]
12. Enquist-Newman, M.; Faust, A.M.E.; Bravo, D.D.; Santos, C.N.S.; Raisner, R.M.; Hanel, A.; Sarvahowman, P.; Le, C.; Regitsky, D.D.; Cooper, S.R.; et al. Efficient ethanol production from brown macroalgae sugars by a synthetic yeast platform. *Nature* **2014**, *505*, 239–243. [CrossRef] [PubMed]
13. Wargacki, A.J.; Leonard, E.; Win, M.N.; Regitsky, D.D.; Santos, C.N.S.; Kim, P.B.; Cooper, S.R.; Raisner, R.M.; Herman, A.; Sivitz, A.B.; et al. An engineered microbial platform for direct biofuel production from brown macroalgae. *Science* **2012**, *335*, 308–313. [CrossRef] [PubMed]
14. Takeda, H.; Yoneyama, F.; Kawai, S.; Hashimoto, W.; Murata, K. Bioethanol production from marine biomass alginate by metabolically engineered bacteria. *Energy Environ. Sci.* **2011**, *4*, 2575. [CrossRef]
15. Miyake, H.; Yamaki, T.; Nakamura, T.; Ishibashi, H.; Fukuiri, Y.; Sakuma, A.; Komatsu, H.; Anso, T.; Togashi, K.; Umetani, H. Gluconate dehydratase. U.S. Patent 7,125,704 B2, 24 October 2006.
16. Yoon, H.J.; Hashimoto, W.; Miyake, O.; Okamoto, M.; Mikami, B.; Murata, K. Overexpression in *Escherichia coli*, purification, and characterization of *Sphingomonas* sp. A1 alginate lyases. *Protein Expr. Purif.* **2000**, *19*, 84–90. [CrossRef] [PubMed]
17. Hashimoto, W.; Miyake, O.; Momma, K.; Kawai, S.; Murata, K. Molecular identification of oligoalginate lyase of *Sphingomonas* sp. strain A1 as one of the enzymes required for complete depolymerization of alginate. *J. Bacteriol.* **2000**, *182*, 4572–4577. [CrossRef] [PubMed]
18. Takase, R.; Ochiai, A.; Mikami, B.; Hashimoto, W.; Murata, K. Molecular identification of unsaturated uronate reductase prerequisite for alginate metabolism in *Sphingomonas* sp. A1. *Biochim. Biophys. Acta Proteins Proteom.* **2010**, *1804*, 1925–1936. [CrossRef] [PubMed]
19. Takase, R.; Mikami, B.; Kawai, S.; Murata, K.; Hashimoto, W. Structure-based conversion of the coenzyme requirement of a short-chain dehydrogenase/reductase involved in bacterial alginate metabolism. *J. Biol. Chem.* **2014**, *289*, 33198–33214. [CrossRef] [PubMed]
20. Preiss, J.; Ashwell, G. Alginic acid metabolism in bacteria I. *J. Biol. Chem.* **1962**, *237*, 309–316. [PubMed]
21. Preiss, J.; Ashwell, G. Alginic acid metabolism in bacteria II. *J. Biol. Chem.* **1962**, *237*, 317–321. [PubMed]
22. Kim, H.T.; Ko, H.J.; Kim, N.; Kim, D.; Lee, D.; Choi, I.G.; Woo, H.C.; Kim, M.D.; Kim, K.H. Characterization of a recombinant endo-type alginate lyase (Alg7D) from *Saccharophagus degradans*. *Biotechnol. Lett.* **2012**, *34*, 1087–1092. [CrossRef] [PubMed]
23. Kim, H.T.; Chung, J.H.; Wang, D.; Lee, J.; Woo, H.C.; Choi, I.G.; Kim, K.H. Depolymerization of alginate into a monomeric sugar acid using Alg17C, an exo-oligoalginate lyase cloned from *Saccharophagus degradans* 2-40. *Appl. Microbiol. Biotechnol.* **2012**, *93*, 2233–2239. [CrossRef] [PubMed]
24. Takagi, T.; Morisaka, H.; Aburaya, S.; Tatsukami, Y.; Kuroda, K.; Ueda, M. Putative alginate assimilation process of the marine bacterium *Saccharophagus degradans* 2-40 based on quantitative proteomic analysis. *Mar. Biotechnol.* **2016**, *18*, 15–23. [CrossRef] [PubMed]

25. Kim, D.H.; Wang, D.; Yun, E.J.; Kim, S.; Kim, S.R.; Kim, K.H. Validation of the metabolic pathway of the alginate-derived monomer in *Saccharophagus degradans* 2-40^T by gas chromatography–mass spectrometry. *Process Biochem.* **2016**, *51*, 1374–1379. [CrossRef]
26. Inoue, A.; Takadono, K.; Nishiyama, R.; Tajima, K.; Kobayashi, T.; Ojima, T. Characterization of an alginate lyase, FlAlyA, from *Flavobacterium* sp. strain UMI-01 and its expression in *Escherichia coli*. *Mar. Drugs* **2014**, *12*, 4693–4712. [CrossRef] [PubMed]
27. Inoue, A.; Nishiyama, R.; Ojima, T. The alginate lyases FlAlyA, FlAlyB, FlAlyC, and FlAlex from *Flavobacterium* sp. UMI-01 have distinct roles in the complete degradation of alginate. *Algal Res.* **2016**, *19*, 355–362. [CrossRef]
28. Inoue, A.; Nishiyama, R.; Mochizuki, S.; Ojima, T. Identification of a 4-deoxy-l-erythro-5-hexoseulose uronic acid reductase, FlRed, in an alginolytic bacterium *Flavobacterium* sp. strain UMI-01. *Mar. Drugs* **2015**, *13*, 493–508. [CrossRef] [PubMed]
29. Badur, A.H.; Jagtap, S.S.; Yalamanchili, G.; Lee, J.-K.; Zhao, H.; Rao, C.V. Alginate lyases from alginate-degrading *Vibrio splendidus* 12B01 are endolytic. *Appl. Environ. Microbiol.* **2015**, *81*, 1865–1873. [CrossRef] [PubMed]
30. Shimizu, E.; Ojima, T.; Nishita, K. cDNA cloning of an alginate lyase from abalone, *Haliotis discus hannai*. *Carbohydr. Res.* **2003**, *338*, 2841–2852. [CrossRef] [PubMed]
31. Suzuki, H.; Suzuki, K.I.; Inoue, A.; Ojima, T. A novel oligoalginate lyase from abalone, *Haliotis discus hannai*, that releases disaccharide from alginate polymer in an exolytic manner. *Carbohydr. Res.* **2006**, *341*, 1809–1819. [CrossRef] [PubMed]
32. Rahman, M.M.; Inoue, A.; Tanaka, H.; Ojima, T. Isolation and characterization of two alginate lyase isozymes, AkAly28 and AkAly33, from the common sea hare *Aplysia kurodai*. *Comp. Biochem. Physiol. B Biochem. Mol. Biol.* **2010**, *157*, 317–325. [CrossRef] [PubMed]
33. Rahman, M.M.; Inoue, A.; Tanaka, H.; Ojima, T. cDNA cloning of an alginate lyase from a marine gastropod *Aplysia kurodai* and assessment of catalutically important residues of this enzyme. *Biochimie* **2011**, *93*, 1720–1730. [CrossRef] [PubMed]
34. Inoue, A.; Mashino, C.; Uji, T.; Saga, N.; Mikami, K.; Olima, T. Characterization of an eukaryotic PL-7 alginate lyase in the marine red alga *Pyropia yezoensis*. *Curr. Biotechnol.* **2015**, *4*, 240–248. [CrossRef]
35. Gacesa, P. Enzymic degradation of alginates. *Int. J. Biochem.* **1992**, *24*, 545–552. [CrossRef]
36. Hobbs, J.K.; Lee, S.M.; Robb, M.; Hof, F.; Bar, C.; Abe, K.T.; Hehemann, J.H.; McLean, R.; Abbott, D.W.; Boraston, A.B. KdgF, the missing link in the microbial metabolism of uronate sugars from pectin and alginate. *Proc. Natl. Acad. Sci. USA* **2016**, *113*, 6188–6193. [CrossRef] [PubMed]
37. Mochizuki, S.; Nishiyama, R.; Inoue, A.; Ojima, T. A novel aldo-keto reductase HdRed from the pacific abalone *Haliotis discus hannai*, which reduces alginate-derived 4-deoxy-L-erythro-5-hexoseulose uronic acid to 2-keto-3-deoxy-D-gluconate. *J. Biol. Chem.* **2015**, *290*, 30962–30974. [CrossRef] [PubMed]
38. Cynkin, M.A.; Ashwell, G. Uronic acid metabolism in bacteria. *J. Biol. Chem.* **1960**, *235*, 1576–1579. [PubMed]
39. Lee, Y.S.; Park, I.H.; Yoo, J.S.; Kim, H.S.; Chung, S.Y.; Chandra, M.R.G.; Choi, Y.L. Gene expression and characterization of 2-keto-3-deoxy-gluconate kinase, a key enzyme in the modified Entner-Doudoroff pathway of *Serratia marcescens* KCTC 2172. *Electron. J. Biotechnol.* **2009**, *12*. [CrossRef]
40. Kim, S.; Lee, S.B. Characterization of *Sulfolobus solfataricus* 2-keto-3-deoxy-D-gluconate kinase in the modified Entner-Doudoroff pathway. *Biosci. Biotechnol. Biochem.* **2006**, *70*, 1308–1316. [CrossRef] [PubMed]
41. Lamble, H.J.; Theodossis, A.; Milburn, C.C.; Taylor, G.L.; Bull, S.D.; Hough, D.W.; Danson, M.J. Promiscuity in the part-phosphorylative Entner-Doudoroff pathway of the archaeon *Sulfolobus solfataricus*. *FEBS Lett.* **2005**, *579*, 6865–6869. [CrossRef] [PubMed]
42. Potter, J.A.; Kerou, M.; Lamble, H.J.; Bull, S.D.; Hough, D.W.; Danson, M.J.; Taylor, G.L. The structure of *Sulfolobus solfataricus* 2-keto-3-deoxygluconate kinase. *Acta Crystallogr. Sect. D Biol. Crystallogr.* **2008**, *64*, 1283–1287. [CrossRef] [PubMed]
43. Ohshima, N.; Inagaki, E.; Yasuike, K.; Takio, K.; Tahirov, T.H. Structure of *Thermus thermophilus* 2-keto-3-deoxygluconate kinase: Evidence for recognition of an open chain substrate. *J. Mol. Biol.* **2004**, *340*, 477–489. [CrossRef] [PubMed]
44. Shelton, M.C.; Cotterill, I.C.; Novak, S.T.A.; Poonawala, R.M.; Sudarshan, S.; Toone, E.J. 2-Keto-3-deoxy-6-phosphogluconate aldolases as catalysts for stereocontrolled carbon-carbon bond formation. *J. Am. Chem. Soc.* **1996**, *118*, 2117–2125. [CrossRef]

45. Allard, J.; Grochulski, P.; Sygusch, J. Covalent intermediate trapped in 2-keto-3-deoxy-6-phosphogluconate (KDPG) aldolase structure at 1.95-A resolution. *Proc. Natl. Acad. Sci. USA* **2001**, *98*, 3679–3684. [CrossRef] [PubMed]
46. Bell, B.J.; Watanabe, L.; Lebioda, L.; Arni, R.K. Structure of 2-keto-3-deoxy-6-phosphogluconate (KDPG) aldolase from *Pseudomonas putida*. *Acta Crystallogr. D Biol. Crystallogr.* **2003**, *59*, 1454–1458. [CrossRef] [PubMed]
47. Ohshima, T.; Kawakami, R.; Kanai, Y.; Goda, S.; Sakuraba, H. Gene expression and characterization of 2-keto-3-deoxygluconate kinase, a key enzyme in the modified Entner-Doudoroff pathway of the aerobic and acidophilic hyperthermophile *Sulfolobus tokodaii*. *Protein Expr. Purif.* **2007**, *54*, 73–78. [CrossRef] [PubMed]
48. Ahmed, H.; Ettema, T.J.G.; Tjaden, B.; Geerling, A.C.M.; Oost, J.V.D.; Siebers, B. The semi-phosphorylative Entner-Doudoroff pathway in hyperthermophilic archaea: A re-evaluation. *Biochem. J.* **2005**, *390*, 529–540. [CrossRef] [PubMed]
49. Reher, M.; Fuhrer, T.; Bott, M.; Schönheit, P. The nonphosphorylative entner-doudoroff pathway in the thermoacidophilic euryarchaeon *Picrophilus torridus* involves a novel 2-Keto-3-deoxygluconate-specific aldolase. *J. Bacteriol.* **2010**, *192*, 964–974. [CrossRef] [PubMed]
50. Cotterill, I.C.; Shelton, M.C.; Machemer, D.E.W.; Henderson, D.P.; Toone, E.J. Effect of phosphorylation on the reaction rate of unnatural electrophiles with 2-keto-3-deoxy-6-phosphogluconate aldolase. *J. Chem. Soc. Trans.* **1998**, *1*, 1335–1341. [CrossRef]
51. Kabisch, A.; Otto, A.; Konig, S.; Becher, D.; Albrecht, D.; Schuler, M.; Teeling, H.; Amann, R.I.; Scheweder, T. Functional characterization of polysaccharide utilization loci in the marine *Bacteroidetes 'Gramella forsetii'* KT0803. *ISME J.* **2014**, *8*, 1492–1502. [CrossRef] [PubMed]
52. Weissbach, A.; Hurwitz, J. The Formation of 2-Keto-3-deoxyheptonie Acid in Extracts of *Escherichia coli* B. I. Identification. *J. Biol. Chem.* **1959**, *234*, 705–710. [PubMed]
53. Porzio, M.A.; Pearson, A.M. Improved resolution of myofibrillar proteins with sodium dodecyl sulfate-polyacrylamide gel electrophoresis. *Biochim. Biophys. Acta (BBA) Protein Struct.* **1977**, *490*, 27–34. [CrossRef]
54. Lowry, O.H.; Rosebrough, N.J.; Farr, A.L.; Randall, R.J. Protein measurement with the dolin phenol reagent. *J. Biol. Chem.* **1951**, *193*, 265–275. [PubMed]

© 2017 by the authors. Licensee MDPI, Basel, Switzerland. This article is an open access article distributed under the terms and conditions of the Creative Commons Attribution (CC BY) license (http://creativecommons.org/licenses/by/4.0/).

Article

Microbial Population Changes in Decaying *Ascophyllum nodosum* Result in Macroalgal-Polysaccharide-Degrading Bacteria with Potential Applicability in Enzyme-Assisted Extraction Technologies

Maureen W. Ihua [1], Freddy Guihéneuf [2], Halimah Mohammed [1], Lekha M. Margassery [1], Stephen A. Jackson [1], Dagmar B. Stengel [3], David J. Clarke [1,4] and Alan D. W. Dobson [1,5,*]

[1] School of Microbiology, University College Cork, Cork, Ireland; w.ihua@umail.ucc.ie (M.W.I.); halimahmoh8@gmail.com (H.M.); lekha513@gmail.com (L.M.M.); sjackson@ucc.ie (S.A.J.); david.clarke@ucc.ie (D.J.C.)
[2] Sorbonne Université, CNRS-INSU, Laboratoire d'Océanographie de Villefranche-sur-Mer (LOV), 06230 Villefranche-sur-mer, France; freddy.guiheneuf@invale.com
[3] Botany and Plant Science, School of Natural Sciences, Ryan Institute for Environmental, Marine and Energy Research, National University of Ireland Galway, Galway H91 TK33, Ireland; dagmar.stengel@nuigalway.ie
[4] APC Microbiome Institute, University College Cork, Cork T12 TY20, Ireland
[5] School of Microbiology, Environmental Research Institute, University College Cork, Cork T23 XE10, Ireland
* Correspondence: a.dobson@ucc.ie; Tel.: +353-4902743

Received: 22 February 2019; Accepted: 26 March 2019; Published: 29 March 2019

Abstract: Seaweeds are of significant interest in the food, pharmaceutical, and agricultural industries as they contain several commercially relevant bioactive compounds. Current extraction methods for macroalgal-derived metabolites are, however, problematic due to the complexity of the algal cell wall which hinders extraction efficiencies. The use of advanced extraction methods, such as enzyme-assisted extraction (EAE), which involve the application of commercial algal cell wall degrading enzymes to hydrolyze the cell wall carbohydrate network, are becoming more popular. *Ascophyllum nodosum* samples were collected from the Irish coast and incubated in artificial seawater for six weeks at three different temperatures (18 °C, 25 °C, and 30 °C) to induce decay. Microbial communities associated with the intact and decaying macroalga were examined using Illumina sequencing and culture-dependent approaches, including the novel ichip device. The bacterial populations associated with the seaweed were observed to change markedly upon decay. Over 800 bacterial isolates cultured from the macroalga were screened for the production of algal cell wall polysaccharidases and a range of species which displayed multiple hydrolytic enzyme activities were identified. Extracts from these enzyme-active bacterial isolates were then used in EAE of phenolics from *Fucus vesiculosus* and were shown to be more efficient than commercial enzyme preparations in their extraction efficiencies.

Keywords: *Ascophyllum nodosum*; algal cell wall degrading enzymes; enzyme-assisted extraction; ichip device

1. Introduction

Ascophyllum nodosum (L.) Le Jolis is a brown fucoid which is dominant along the intertidal rocky shores of the North Atlantic [1]. This brown macroalga is of great economic value as an important source of diverse bioactive compounds, many with valuable pharmaceutical, biomedical, and biotechnological potential [2]; which include ascophyllan, laminarin, alginates, and polyphenols [3–5].

Extracts from *A. nodosum* have for example been reported to possess potent anticoagulant activity [6], antitumor activity [7], anti-inflammatory activity [8] together with antiviral activity [9], as well as possessing the ability to improve rumen function in animals [10]. Seaweed extracts are also known to help alleviate the consequences of abiotic stress in crops, with extracts from *A. nodosum* being reported to act at the transcriptional level in the model plant *Arabidopsis thaliana* [11] and to also alleviate drought stress in this plant species [12]. Thus, seaweeds and their useful derivatives have become subject to extensive research in recent times.

Macroalgal surfaces are well known to provide a suitable substratum for the attachment of microbial colonizers, including fungi and bacteria, with bacterial densities reaching levels ranging from 10^2 to 10^7 cells cm^{-2} [13]. Organic substances secreted by the macroalga act as an important nutritional source for these microorganisms [14–16]. Epiphytic bacterial communities are in fact believed to be essential for normal morphological development in the algal host, with these bacteria producing chemicals which help to protect the macroalga from potential harmful secondary colonization by pathogenic microorganisms [17,18]. It is believed that some algal species may contain distinct associated bacterial communities, related to the composition of the algal surfaces and its exudates [19]. Several environmental and non-environmental factors have been shown to influence the composition and abundance of such epibacterial communities associated with seaweeds. In addition to seasonal and temporal variations, the physiological state of the macroalga has also been found to play a significant role in the structure of algal associated microbial communities. Recent studies on *Cladophora*, a filamentous green alga, revealed changes in the diversity and composition of its associated epibacterial communities during decay, to include species involved in nutrient recycling [20].

Marine bacteria are likely to produce cell-wall degrading enzymes as a mechanism to mobilize polymers for nutritional purposes when growing in a nutrient limited state, such as growth on decaying algae. It has been proposed that macroalgal-polysaccharide-degrading (MAPD) bacteria will increase in numbers on weakened or dead macroalgae, thus contributing to recycling of the algal biomass [21]. Several algal polysaccharide-degrading bacteria, which were taxonomically assigned to the *Flavobacteria* and γ-Proteobacteria classes have recently been isolated from the microflora of *A. nodosum* [21]. These bacteria displayed diverse hydrolytic activities and the subsequent functional screening of plurigenomic libraries from these bacteria resulted in the discovery of a range of novel hydrolytic enzymes [22]. Thus, it is clear that the diverse and complex bacterial communities associated with *A. nodosum* represent a potential source of novel hydrolytic enzymes with biotechnological applications that could include enzyme-assisted extraction (EAE) strategies and improvements in the yields of algal components with cosmeceutical, functional food, nutraceutical, and biopharmaceutical applications [23,24].

While seaweeds are rich in useful polysaccharides and other metabolites [25], extraction of such algal components can be problematic due to the complexity and rigidity of the algal cell wall. Brown algal cell walls consist of complex sulfated and branched polysaccharide bound to proteins and ions that hinder extraction efficiencies of algal-derived metabolites [26]. Chemical and mechanical processes are currently used for the extraction of bioactive compounds or fractions from algae; however, problems exist if these compounds are sensitive to extraction techniques which involve heat or the use of solvents [27]. New improved extraction processes including microwave, ultrasound, supercritical fluid, pressurized liquid, and particularly the use of enzyme-assisted extraction (EAE) processes [27], offer ecofriendly, faster, and more efficient alternatives to traditional methods [23,28,29]. Commercial algal cell wall polysaccharidases, such as xylanase, alcalase, viscozyme, neutrase, and agarase, can be used to hydrolyze algal cell wall carbohydrates and eliminate solubility barriers for algal bioactive compounds [27,30]. Enzyme-assisted extraction (EAE) has been successfully employed to produce extracts from the green seaweed *Ulva rigida* (formerly *Ulva armoricana*), which displayed antiviral and antioxidant properties [23]. Carbohydrates and bioactive compounds have also been extracted from the brown algae *Ecklonia radiata* and *Sargassum muticum*, respectively, using carbohydrate hydrolases and proteases in EAE strategies [31].

In this study, our approach was to monitor the overall composition and population dynamics of the microbial communities associated with *A. nodosum* as it decayed over a six-week period by incubating the alga in artificial seawater, using culture independent approaches, targeting the 16S rRNA gene. We also employed culture dependent approaches, including the ichip method [32] to isolate bacteria from the seaweed during the decay process. While bacteria have previously been isolated from macroalgae using traditional cultivation methods [21,33], this study presents the first use of the ichip to cultivate bacteria from decaying seaweed. In this way we hoped to expand the range of bacterial isolates to include previously uncultured microorganisms. Following isolation, bacterial strains were screened for their ability to produce a range of algal cell wall degrading polysaccharidases and were characterized by 16S rRNA DNA sequencing. A range of species from the genus *Bacillus*, together with a number of *Vibrio* species, were isolated, these displayed multiple hydrolytic enzyme activities including hydroxyethyl cellulase, lichenase, and pectinase activities. Extracts from these bacteria were then successfully employed in the EAE of phenolics from the seaweed *Fucus vesiculosus*.

2. Results

2.1. MiSeq Sequencing and Data Processing

The microbial communities of *A. nodosum* samples were analyzed based on decay period (week 0, week 2, week 4, and week 6) and incubating temperature (18 °C, 25 °C, and 30 °C) by Illumina MiSeq sequencing targeting the V3-V4 16S rRNA gene region. A combined total of 8,872,164 raw reads were obtained which, when quality filtered produced, 1,655,910 reads with an average length of 301 bp, and were analyzed using the QIIME version 1.9.1 (http://qiime.org/tutorials/illumina_overview_tutorial.html) workflow [34]. The number of reads obtained after quality filtering and operational taxonomic units (OTUs) together with species richness and diversity indices of the microbial communities are shown in Table 1. With respect to the two diversity indices calculated; Shannon and Chao1, sample 4_30 (week 4; 30 °C) ranked the highest and sample 6_18 (week 6; 18 °C) ranked the lowest in diversity. In addition, it is noteworthy that up to 4% of the OTUs observed in the undecayed *Ascophyllum* sample (T_0; week 0) and 41% in the decaying samples were unclassified. These unassigned OTUs are likely to represent macroalgal associated bacterial populations that are as yet unknown or not present in the SILVA version 123 database [35] used for taxonomy assignment.

Table 1. Observed OTUs and species richness and diversity estimates of *Ascophyllum*-associated metagenomic communities obtained using MiSeq sequencing of the 16S rRNA gene from the intact seaweed (T_0) and each of decaying *Ascophyllum nodosum* samples collected at three phases of the decay period (week 2, week 4 and week 6) at 18 °C (2_18, 4_18, 6_18), 25 °C (2_25, 4_25, 6_25), and at 30 °C (2_30, 4_30, 6_30).

Decay Period	Sample	No. of Reads after Quality Filtering	No. of OTUs (at 97% Sequence Identity)	Chao1 Richness	Shannon Index
Week 0 intact seaweed	T_0	178,699	1467	2293.5	10.1
Week 2 early decay phase	2_18	123,551	854	2072.1	9.5
	2_25	350,135	1476	3327.1	10.2
	2_30	138,445	724	1822.7	9.2
Week 4 mid decay phase	4_18	148,904	749	1901.7	9.3
	4_25	151,285	1130	2061.6	9.8
	4_30	139,737	1633	3490.4	10.4
Week 6 late decay phase	6_18	120,679	443	1148	8.6
	6_25	138,659	1202	2884.5	10.0
	6_30	165,816	1250	3062.8	10.0

2.2. Metagenomic Communities Associated with Intact Ascophyllum nodosum

A total of 1467 OTUs were identified among sequence reads derived from the intact undegraded seaweed (T_0) with approximately 96% of the OTUs been classified into one of 19 different phyla, the major ones being Proteobacteria (24.2%), Planctomycetes (22.5%), Actinobacteria (15.2%), Verrucomicrobia (15.1%), Cyanobacteria (6.7%), Bacteroidetes (4.8%), and Firmicutes (3.8%) (Figure 1; T_0). The species richness and diversity of the microbial community associated with the macroalga in its intact state as represented by the number of OTUs and number of bacterial phyla identified is also reflected by the Chao1 richness and Shannon diversity index calculated (Table 1).

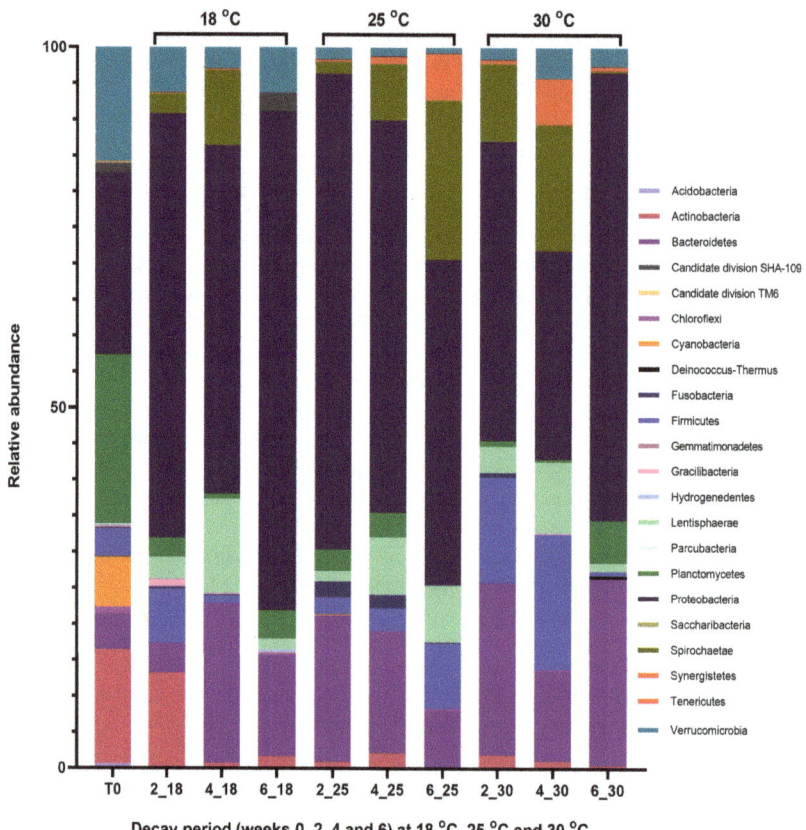

Figure 1. Relative abundances of bacterial phyla associated with intact (T_0) and decaying *Ascophyllum nodosum* at 2, 4, and 6 weeks of decay at 18 °C (2_18, 4_18, 6_18); 2, 4, and 6 weeks of decay at 25 °C (2_25, 4_25, 6_25), and 2, 4, and 6 weeks of decay at 30 °C (2_30, 4_30, 6_30) obtained from metagenomic 16S rRNA gene sequencing. The relative distribution of phyla in each group is represented as a percentage.

2.3. Metagenomic Communities Associated with Decaying Ascophyllum nodosum

2.3.1. Population Changes in the Seaweed Decaying at 18 °C

Within the first two weeks of the algal decay at 18 °C, the bacterial population associated with the intact seaweed, which was previously characterized by the prevalence of Proteobacteria (24.2%), Planctomycetes (22.5%) and Verrucomicrobia (15.1%), shifted towards a Proteobacteria-led population with a relative abundance of over 40% (Figure 1; 18 °C). This increase in the prevalence of Proteobacteria occurred with a corresponding decline in the relative abundances of some phyla including Planctomycetes (1.8%) and Verrucomicrobia (4.3%), amongst others. However, in the early decay period (at week 2), bacteria belonging to the phylum Spirochaetae which were not identified in the intact seaweed emerged, while other phyla such as Firmicutes, Fusobacteria, Lentisphaerae, and Gracilibacteria were found to be increasingly abundant, relative to their levels prior to the algal decay. The composition of the metagenomic communities associated with the decaying seaweed remained unchanged in the next phase of the algal decay process (at week 4), but changes in the relative abundances of some phyla occurred. At week 4, increase in the presence of bacterial groups classified as Bacteroidetes, Spirochaetae, and Lentisphaerae were more apparent while Proteobacteria levels declined. The bacterial population identified at the end of the decay process (week 6) was less diverse, with 443 OTUs identified and low diversity indices calculated (Table 1). Metagenomic results show that a number of bacterial phyla present in the preceding decay phases were not found at week 6. Some of these bacterial phyla which were not present include Firmicutes, Fusobacteria, Spirochaetae and Gracilibacteria.

2.3.2. Population Changes in the Seaweed Decaying at 25 °C

When compared to the bacterial population associated with fresh *Ascophyllum nodosum* samples, the macroalga allowed to decay at 25 °C experienced a sharp decrease in the relative abundances of bacteria belonging to the phyla Planctomycetes, Verrucomicrobia, Firmicutes and Actinobacteria amongst others in the early decay period. Metagenomic analysis of the 16S rRNA gene sequences revealed a steady shift in the composition and abundance of the microbial communities associated with the decaying seaweed. For example, while the phylum Spirochaetae was not identified prior to decay (T_0), bacteria classified as this phylum were identified at week 2 (0.9%) and increased in relative abundance throughout the decay period from 3.7% at week 4 to approximately 11% at week 6. Lentisphaerae, which was rarely identified in the undecayed macroalga, was also observed to follow a similar trend, by increasing in relative abundance as *A. nodosum* samples decayed at 25 °C. Levels of this phylum increased from 0.8% at week 2 to 3.8% and 4% at the mid and late phases of the decay period, respectively. Other phyla such as Fusobacteria and Bacteroidetes were observed to decrease consistently in their prevalence from week 2 to week 6 of the decay process (Figure 1; 25 °C). In general, *Ascophyllum nodosum* allowed to decay at 25 °C was dominated by bacteria recruiting to the phylum Proteobacteria.

2.3.3. Population Changes in the Seaweed Decaying at 30 °C

At 30 °C, the decaying macroalga was diversely comprised of bacteria belonging to different phyla (Figure 1; 30 °C). The seaweed-associated metagenomic population changed markedly upon decay, relative to the microbial communities found in the intact seaweed (described in Section 2.2) and some of the most notable changes observed occurred in the early decay phase. At week 2, the relative abundance of bacteria belonging to the phylum Actinobacteria was observed to have decreased from approximately 15% (found in week 0) to about 1%. This decrease in relative abundance of Actinobacteria was concurrent with a decline in the prevalence of bacteria belonging to the phyla Planctomycetes (0.4%) and Verrucomicrobia (0.9%) previously present at approximately 22% and 15%, respectively, in the intact seaweed. On the other hand, in the early decay phase, other phyla such as Bacteroidetes, Lentisphaerae, and Firmicutes were observed to increase to more than two-fold in their relative abundances in the decaying seaweed. The microbial population observed at the mid decay phase (week 4) did not differ greatly in its composition from the week 2 derived microbial population. However, the prevalence of Proteobacteria declined (22.4% early phase; 17.1% mid phase) as bacteria identified as Spirochaetae (5.8% early phase; 10.3% mid phase) and Synergistetes (0.3% early phase; 3.7% mid phase) increased in prevalence. Differences in the bacterial phyla present in the decaying seaweed became even more evident at the end of the decay period (week 6) when Proteobacteria regained dominance (43.1%) and the levels of Spirochaetae (0.2%) and Synergistetes (0.1%) declined.

2.4. Cultivable Surface Microbiota Associated with Intact A. nodosum

The cultivable epibacterial population of intact and decaying *A. nodosum* samples were assessed using the maceration cultivation method, which involves cutting the seaweed samples into smaller fine pieces. A total of 90 bacteria were isolated and taxonomically identified following 16S rDNA sequence analysis (Table 2) from the intact *Ascophyllum nodosum* sample (T_0) and were found to consist of bacteria belonging to the phyla Proteobacteria (46%), Bacteroidetes (43%) and Actinobacteria (11%) (Figure 2a). Members of the phylum Proteobacteria were largely dominated by the class γ–Proteobacteria (95%), with α-Proteobacteria and β-Proteobacteria being rarely isolated. In the total isolated bacterial population from the intact seaweed (T_0), eleven different genera were identified, the most abundantly represented being *Winogradskyella* (41%), *Marinobacter* (37%), *Microbacterium* (6%), and *Micrococcus* (6%) (Figure S1a; Supplementary information).

Table 2. Number of bacterial isolates cultured from intact (week 0) and decaying *Ascopyllum nodosum* samples (week 2, week 4, and week 6), incubated at different temperatures, using the maceration isolation method and the ichip device.

Week 0	Week 2	Week 4	Week 6	ichip Isolation	Incubating Temperature
	76	63	52	59	18 °C
90	35	47	70	76	25 °C
	63	67	53	89	30 °C

2.5. Cultivable Surface Microbiota Found on Decaying A. nodosum

The cultivable surface-attached microbiota of the intact seaweed differed greatly from the bacterial populations found on *Ascophyllum* samples allowed to decay at 18 °C, 25 °C, and 30 °C for six weeks (Figure 2). At the phylum level, while Proteobacteria maintained an overall dominance with over 70% relative abundance in the bacterial population associated with both the intact and decaying seaweed, members of the phylum Firmicutes which were not identified in the intact seaweed were present during the algal decay. Bacteria belonging to the phylum Bacteroidetes, which were prevalent in the intact macroalga (43%), represented only 3% of the total cultivable surface microbiota population in the decaying seaweed.

Similar bacterial phyla were present in the bacterial communities associated with the decaying seaweed in the three incubation flasks (18 °C, 25 °C, and 30 °C). However, distinct differences in the composition of the associated microbial populations found at the different decay periods and temperature are more evident at the genus level (Figure S1; Supplementary information). In the early decay phase, the bacterial population isolated from the macroalga decaying at 18 °C consisted of Proteobacteria (88%), Firmicutes (5%), Bacteroidetes (4%), and Actinobacteria (3%) (Figure 2b) and was diversely comprised of members of the genera; *Paracoccus* (71%), *Celeribacter* (7%), *Psychrobacter* (8%), *Bacillus* (5%), *Formosa* (4%), *Microbacterium* (3%), *Cobetia* (1%), and *Citricella* (1%) (Figure S1b; Supplementary information). Much lower diversity was observed in the bacterial communities isolated at week 2 from the 25 °C and 30 °C incubation flasks, with only members of the phyla Proteobacteria and Firmicutes being identified.

In the mid decay period (week 4), the genus *Celeribacter* dominated the bacterial population with a relative abundance of 95% and 100% in the 18 °C and 30 °C derived bacterial populations, respectively, while the 25 °C derived population at week 4 was comprised of more genera including *Paenisporosarcina* (51%), *Bacillus* (30%), *Celeribacter* (13%), *Paenibacillus* (2%) *Vibrio* (2%), and *Sporosarcina* (2%) (Figure S1c; Supplementary information).

Members of the bacterial phyla Proteobacteria (54%), Bacteroidetes (27%), and Firmicutes (19%) were identified in the bacterial population cultured at the end of the decay process (week 6) from the 18 °C incubation flask. These phyla were also present at 72%, 7%, and 21% relative abundances, respectively, in the 30 °C bacterial population (Figure 2d). In week 6, eleven distinct genera were identified at 18 °C, the most abundantly represented being *Paracoccus*, *Algoriphagus*, *Celeribacter*, *Bacillus*, and *Primorskyibacter* and four genera including *Celeribacter*, *Bacillus*, *Pseudozobellia*, and *Paracoccus* being observed in the 30 °C microbial community. The bacterial community associated with the seaweed decaying at 25 °C in the late decay phase differed from the bacterial communities isolated from the 18 °C and 30 °C microbial populations in this phase, with Proteobacteria (96%) dominating the dataset and three genera; *Celeribacter*, *Bacillus* and *Roseobacter* being isolated at 25 °C. Phylogenetic trees representing the bacteria cultured from both intact and decaying *Ascophyllum nodosum* samples using the maceration method are shown in Figures S2–S5 (Supplementary information).

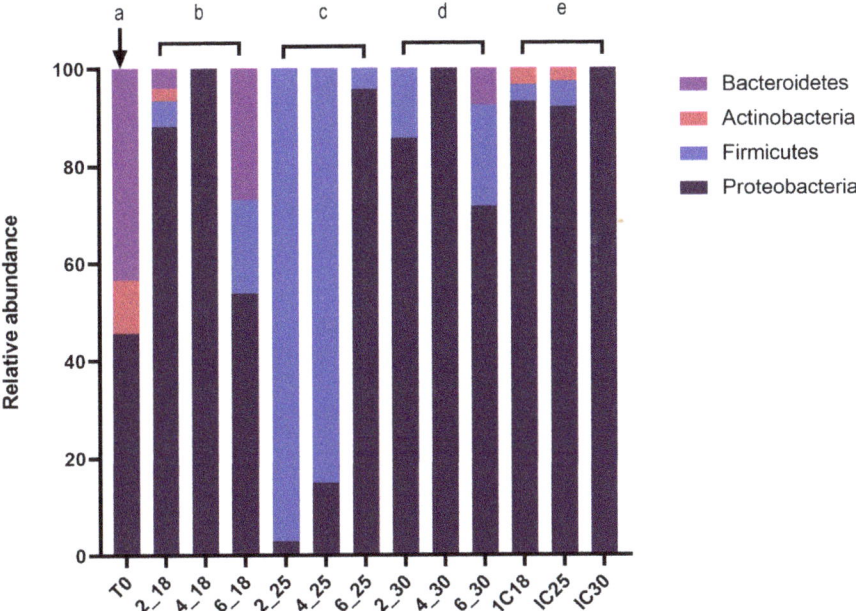

Figure 2. Relative abundances of bacterial phyla associated with the cultivable surface microbiota of (**a**) intact *Ascophyllum nodosum* and decaying *Ascophyllum nodosum* at 2, 4, and 6 weeks of decay at (**b**) 18 °C; 2_18, 4_18, 6_18, (**c**) 25 °C; 2_25, 4_25, 6_25, (**d**) 30 °C; 2_30, 4_30, 6_30 which were obtained by maceration culture isolation method and (**e**) obtained by ichip culture isolation method. 16S rRNA gene sequences were obtained from the bacterial isolates and taxonomic analyses were performed. The relative distribution of phyla in each group is represented as a percentage.

2.6. ichip Bacterial Isolation Method Applied to Decaying A. nodosum

In a bid to further analyze the cultivable bacteria present and potentially expand upon the range of bacterial isolates, the ichip device was also employed on the decaying seaweed samples. At week 4 of the decay period, the ichip device loaded with a cell–agar suspension prepared from the decaying seaweed was inoculated into each flask containing the alga which were decomposing at 18 °C, 25 °C, and 30 °C; with bacteria being recovered from the device following a further 2 weeks of incubation. A total of 224 bacteria (59 isolates from 18 °C, 76 and 89 isolates from 25 °C and 30 °C, respectively; Table 2) were isolated and taxonomically identified using 16S rRNA gene sequences. Taxonomic analysis of the cultivable microbial communities revealed the presence of three representative phyla—Proteobacteria, Actinobacteria, and Firmicutes—across the three different temperature groups. Proteobacteria dominated at the phylum level in the three datasets, comprising

93%, 92%, and 100% in the 18 °C, 25 °C, and 30 °C bacterial culture populations, respectively (Figure 2e). Low relative abundances of Actinobacteria and Firmicutes were observed in the 18 °C and 25 °C populations, with both phyla not being identified in the 30 °C bacterial population. The majority of bacterial isolates cultured from the 18 °C and 25 °C samples, which belong to the phylum Proteobacteria; were further classified as α–Proteobacteria, with only 9–11% recruiting to γ-Proteobacteria. In contrast, γ-Proteobacteria were found to dominate the microbial community isolated from the seaweed incubated at 30 °C, with a relative abundance of 98%. Thirteen distinct genera, including *Celeribacter*, *Paracoccus*, *Vibrio* and *Marinobacterium* were present in the total bacterial population isolated using the ichip device (Figure S1e; Supplementary information). The genus *Celeribacter* was present across all temperatures, at very high relative abundances in the 18 °C and 25 °C samples (64% and 83%, respectively) but at a much lower relative abundance at 30 °C (2%). In contrast, *Enterobacter*, which dominated at 30 °C, was not found to be present in either of the 18 °C or 25 °C derived microbial populations. The phylogenetic tree representing the bacteria cultured from decaying *Ascophyllum nodosum* samples using the ichip in situ cultivation method is shown in Figure S6 (Supplementary information).

2.7. Enzymatic Activities of A. nodosum Cultivable Surface Microbiota

2.7.1. Intact *Ascophyllum nodosum* Isolated Using the Maceration Method

Over 800 bacterial isolates cultured from intact (T_0) and decaying *Ascophyllum nodosum* samples using both the maceration and ichip isolation methods (Table 2) were screened for enzyme activity in plate assays containing hydroxyethyl cellulose, pectin, and lichenin as substrates. The cultivable surface microbiota community associated with the seaweed in its intact state was found not to produce any of the algal cell wall degrading enzymes examined under the conditions employed in this study, with none of the bacterial isolates testing positive on any of the plate assays used.

2.7.2. Decaying *Ascophyllum nodosum* Isolated Using the Maceration Method

The microbial population associated with the decaying seaweed isolated using the maceration method consisted of a total of 51 isolates (approximately 7%) with hydrolytic activity against at least one of the tested substrates (Table S1, Supplementary information). Of these enzyme active bacterial isolates, 65% belonged to the microbial community cultured from the decaying seaweed at week 2, another 10% belonged to the week 4 bacterial population, while 25% were cultured from week 6 and the majority of these MAPD bacteria were found to degrade lichenin (Table S1; Supplementary information). Bacteria belonging to the genus *Bacillus* (10%) represented one of the less abundant genera in the total microbial community associated with the decaying seaweed. However, among the bacteria cultured from the decaying seaweed using the maceration method, these *Bacillus* species were found to be the only producers of the algal cell wall polysaccharidases tested for in this study.

2.7.3. Decaying *Ascophyllum nodosum* Isolated Using the ichip Method

Approximately 5% of the ichip-derived microbial communities screened were identified as being positive for one or more of HE-cellulose, lichenin, and pectin degrading activities. None of the bacterial isolates from 30 °C displayed MAPD activity under the conditions tested in this study while less than 3% of the 25 °C derived population tested positive and 15% from the 18 °C bacterial population were enzyme active. All the enzyme active bacterial isolates cultured from 18 °C were identified as belonging to the *Vibrio* genus. These isolates were found to produce pectin degrading enzymes (Table S1; Supplementary information).

2.8. Enzyme-Assisted Extraction (EAE) of Total Phenolics from F. vesiculosus

We then compared the ability of an enzymatic bacterial supernatant (EBS) generated from the three isolates IC18_D7 (DSM 107285), IC18_D5 and ANT_0_A6 (DSM 107318) with ≥98% 16S rRNA gene sequence similarity to *Vibrio anguillarum* X0906, *Vibrio oceanisediminis* S37, and *Winogradskyella sp.* MGE_SAT_697, respectively, which we had selected as the best enzyme producers from our group of enzyme-active strains to perform enzyme-assisted extraction of phenolics from *Fucus vesiculosus*, and to compare their performance to commercially available enzyme preparations. Bacterial isolates IC18_D7 and IC18_D5 were shown to produce pectin degrading enzymes. ANT_0_A6 had previously been shown to produce good levels of amylase activity (data not shown). Results obtained from the EAE of the total phenolic compounds from *F. vesiculosus*, performed with or without commercial enzymes conducted at 50 °C, and with or without the enzymatic bacterial supernatants (EBS) conducted at 28 °C, are shown in Figure 3. The total phenolic content (TPC) of *F. vesiculosus* obtained by exhaustive solid-liquid extraction had previously been reported as 68.6 ± 8.3 mg $PE.g^{-1}$ DWB [36]. This content was thus considered as a reference value for TPC, corresponding to a yield of extraction of 100%. Although the highest TPC values were obtained using commercial enzymes, compared to the control (50 °C), the increase was only significant when xylanase was used on the larger biomass particles i.e., $0.5 < Ps < 2.5$ mm ($p = 0.021$). This TPC value of 35.6 ± 2.0 mg $PE.g^{-1}$ DWB, obtained using xylanase, was equivalent to an extraction yield of 52%. Using the enzymatic bacterial supernatants (EBS), the TPC values increased significantly for both particle sizes ($p < 0.01$), compared to the control (28 °C), reaching up to 44.8 ± 1.8 mg $PE.g^{-1}$ DWB (Ps < 0.5 mm) and 40.3 ± 1.7 mg $PE.g^{-1}$ DWB ($0.5 < Ps < 2.5$ mm), respectively. These TPC values correspond to extraction yields of 65% and 59%, respectively. The extraction yields were therefore increased by 10% using xylanase, while they increased by 11–13% using EBS, compared to their respective controls. Moreover, an increase in extraction temperature (control 28 °C vs control 50 °C) appeared to have an overall negative effect on the extraction yield for phenolics. These results indicate that cell-wall degrading enzyme preparations produced by the three bacterial isolates from *A. nodosum*, applied at 28 °C were more efficient than the commercial protease, cellulase and xylanase preparations in the extraction of total phenolics from *F. vesiculosus*.

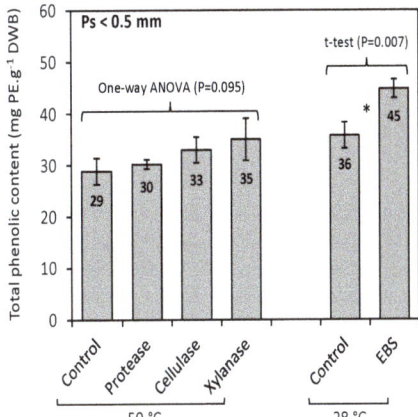
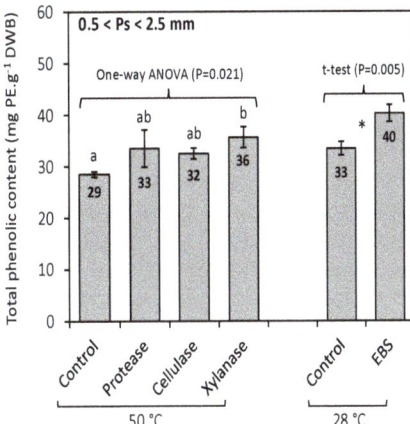

Figure 3. Enzymatic-assisted extraction of total phenolics from *Fucus vesiculosus* with commercial enzymes conducted at 50 °C, and with enzymatic bacterial supernatants (EBS) conducted at 28 °C. Control experiments without the addition of either commercial enzymes or EBS were conducted at 50 °C and 28 °C, respectively, under the same conditions. This experiment was undertaken using two different particle sizes (Ps) of ground biomass i.e., Ps < 0.5 mm, and $0.5 < Ps < 2.5$ mm. Total phenolic content (TPC) is expressed as milligram of phloroglucinol equivalents (PE) per gram of dry weight biomass (mg $PE.g^{-1}$ DWB). A one-way ANOVA was performed to assess significant differences ($p < 0.05$) between commercial enzymes and control (50 °C), results are arranged in increasing order: a < b; while a t-test was performed to determine significance differences ($p < 0.05$) between EBS and control (28 °C), asterisk (*) indicates a difference between both treatments.

3. Discussion

Macroalgal bioactive compounds are used in products to stimulate animal health or as functional food ingredients [28,37,38]. Phlorotannins exclusive to brown algae in high amounts (15% DW) have been shown to possess antidiabetic [39], antioxidant [40], and antiproliferative [41] effects. In addition, seaweed extracts are commonly used as biostimulants in agriculture [42] and have been proposed as a viable alternative protein crop for use in diets for monogastric livestock [43].

Seaweeds are well known to be associated with a diverse range of bacteria which colonize their nutrient-rich surfaces [13,21,22]. These bacteria are known to be a very good source of specific polysaccharidases, including pectinases, alginate lyases, carrageenanases, fucoidanases, and laminarinases [44] with several biotechnological applications. Some of these algal cell-wall degrading enzymes are produced to help mobilize polymers for nutritional purposes, for example, when growing in a nutrient limited state such as algal decay, and contribute to algal biomass recycling [20,21]. Thus, we reasoned that if *A. nodosum* was allowed to decay under controlled conditions at different temperatures, it should result not only in changes in the overall composition and dynamics of the bacterial communities present, but also in the isolation of bacteria that produce algal cell wall polysaccharidases, given the nutrient limited state to which they had been exposed, that might have potential application in EAE strategies.

The structure of the surface-attached bacterial population associated with intact and decaying *A. nodosum* incubated at 18 °C, 25 °C, and 30 °C was investigated in this study using both culture independent and culture dependent (traditional maceration and the in-situ cultivation based ichip device) approaches. The use of a next-generation sequencing approach (Illumina MiSeq) supplemented the 16rRNA gene-based approach employed on the cultured bacterial isolates. Given that the NGS approach circumvents the difficulties associated with the cultivation of bacteria from environmental samples and allows the identification of both cultivable and non-cultivable bacterial populations, it is not surprising that some phyla observed in the metagenomic communities of the macroalga were not identified in the total cultivable bacterial population. In particular, considering the isolation agar (SYP-SW) and the culture condition (72 h at 28 °C) employed in this study, it is highly unlikely that most phyla, including Planctomycetes, Spirochaetae, and Verrucomicrobia, which were found in the NGS dataset, would be recovered. These bacterial phyla would require a more targeted isolation strategy to be identified using various plate-based cultivation methods [45–48].

The ichip device, which has previously been reported to increase the microbial diversity of cultured bacterial isolates [49–51], was applied to potentially expand the range of bacterial isolates identified to include previously uncultured species. While the composition of the microbial communities derived from the ichip device did not differ greatly from the bacterial populations identified using the traditional approach (Figure 2), we recovered four potentially novel strains (IC25_B4, IC25_B12, IC25_C8, and IC25_G4) with 97% or less identity to their closest BLAST relative using the device. These bacterial isolates are currently being further characterized. The ichip device also resulted in the isolation of two strains (IC18_D5 and IC18_D7) identified as belonging to the *Vibrio* genus, extracts from which were subsequently utilized in the EAE of phenolics from *Fucus vesiculosus* and were found to be more efficient in the extraction process than current commercially available enzymes (Figure 3). This further demonstrates the utility of the ichip device as an important method to not only capture previously uncultivable bacteria, but also to recover bacteria with potential biotechnological applications [49,50].

Phylum-level analysis revealed that the structure of both the cultivable and metagenomic microbial communities found on the intact seaweed differed from that of the decaying macroalga, suggesting that the decay process plays a role in altering the algal associated microbial populations. However, these results should be interpreted with caution as our experiments were not conducted in replicates. Similar differences in the microbial community profiles associated with healthy and weakened bleached macroalgae have also been previously reported [52]. Although a causal link between such differences and the host condition has not been clearly established, it is known that host stress, such as bleaching and decay-related disruptions to the composition and abundance of its associated microbial consortium, can have detrimental effects on the host, causing diseases; for example, due to interferences with the seaweed–bacteria interactions that support algal development and host defense [18,53]. Chun et al. [20] suggest that microcosms which emerge as a result of the algal decay process may explain the differences in the bacterial populations associated with healthy and decaying algae. Decaying *Cladophora* samples have, for example, been shown to produce low oxygen and pH environments with increased ammonium-nitrogen levels. Subsequently, structural shifts in the microbial community towards bacterial groups better suited to thrive under such conditions were observed [20]. While the succession of oxygen concentration, pH and nutrient levels during the decay period was not monitored in this study, the structural shifts observed in the microbial communities with decay may be attributed to changes in the composition of the closed microcosm within the shake flasks.

Screening the cultivable surface microbiota of both the intact and decaying seaweed for the production of algal cell wall polysaccharide degrading enzymes revealed a number of MAPD bacteria. Bacteria belonging to the genus *Bacillus* which represented the major producers (>80%) of these hydrolytic enzymes were not identified in the bacterial population associated with *A. nodosum* in its healthy state but represented up to 10% of the surface microbiota communities isolated during the algal decay (Figure S1; Supplementary information). This marked difference in the composition and abundance of the microbial communities associated with the seaweed during its different physiological states (intact and decaying), mainly characterized by the emergence in the members of the enzymatically active *Bacillus* and *Vibrio* groups supports the hypothesis that nutrient limiting conditions such as algal decay is likely to promote the proliferation of MAPD producing bacteria [21]. However, while the number of MAPD isolates was not observed to steadily increase during decay as might be expected due to the weakened state of the seaweed, the few enzymatically active strains that we did identify during decay, such as the *Bacillus* and *Vibrio* species were efficient producers of the MAPD enzymes for which we tested (Table S1; Supplementary information).

Microorganisms are well-known to exhibit mutualism such that one or more individuals within a microbial population can gain from the collective characteristics expressed by its neighbors without expressing the trait itself [14,54,55]. Such phenotypically deficient bacteria may however possess the metabolic capability necessary to utilize nutrients provided by other members of the community [55]. A lack in the increase in the expected numbers of MAPD bacteria that we observed during the decay experiments may thus be explained by the efficiency of the less abundant enzymatically active strains who may be compensating for the inactivity of the dominant species by creating a pool of available nutrients thereby supporting the overall bacterial consortia present within the microcosm in the growth flasks, to which no nutrients had been added.

Finally, we assessed the ability of enzymatic bacterial supernatant (EBS) from a selected group of enzyme-active strains; IC18_D7, IC18_D5 and ANT$_0$_A6 with similarity to *Vibrio anguillarum* X0906, *Vibrio oceanisediminis* S37, and *Winogradskyella sp.* MGE_SAT_697, respectively, in the enzyme-assisted extraction of phenolics from *Fucus vesiculosus*. These enzyme preparations were shown to increase total phenolic content (TPC) extraction yields from *Fucus vesiculosus* by 11–13%, to levels which were greater than the extraction yields obtained using a commercially available xylanase (10%) (Figure 3). To our knowledge, this is the first study to report the application of macroalgal-derived bacterial culture extracellular supernatants in the enzyme-assisted extraction of phenolics from *Fucus vesiculosus*. Thus, it is clear that bacterial populations associated with *A. nodosum* are a good source of algal cell wall polysaccharide degrading enzymes with potential utility in EAE strategies. The isolation of macroalgal associated bacteria is frequently reported in the literature [20,21], with isolates being developed for use in various biotechnological applications, such as novel carrageenanases from *Flavobacteria* and γ-Proteobacteria isolated from *Ascophyllum nodosum* [21] and from *Pseudoalteromonas porphyrae* isolated from decayed seaweed [56] for potential biomedical and food applications, together with alginate lyase from *Zobellia galactanivorans* for biomass degradation [57]. Our study further demonstrates the potential utility of algal derived bacteria and their potential contribution to EAE based strategies aimed at the production of seaweed extracts for similar types of biotechnological applications.

4. Materials and Methods

4.1. Sampling

Ascophyllum nodosum samples were obtained in the intertidal zone at Rinville in Galway Bay, Ireland at 53°14′40″ North, 8°58′2″ West in late January, 2016. Approximately 2 kg of seaweed was sampled and packaged in sterile air-tight plastic bags and stored on dry ice at the sampling location. *Ascophyllum* samples were subsequently stored briefly at 4 °C in the laboratory before further analyses.

4.2. Experimental Design

Three sets of approximately 450 g of the seaweed were suspended in separate 950 mL sterile artificial seawater (3.33% w/v synthetic seawater salts Instant Ocean, Aquarium Systems, in distilled water), with each 2 L flask incubated at a different temperature (18 °C, 25 °C, and 30 °C) on a shaking platform at 125 rpm for a six-week period. The incubation flasks were single replicates ($n = 1$) per temperature treatment. Three separate ichip devices were subsequently inoculated four weeks after the initial incubation into each of the flasks under sterile conditions in a laminar flow hood (BioAir Safeflow 1.2—EuroClone, Pero, Italy), as previously described [32], and removed at the end of the decay period. To inoculate each ichip device, 1 mL suspension from the incubation flask containing the seaweed decaying at one of the three temperatures (18 °C, 25 °C, and 30 °C), with an estimated microbial density of 1.0×10^{12} cells mL^{-1} was diluted appropriately in sterile artificial seawater (3.33% w/v synthetic seawater salts Instant Ocean, Aquarium Systems, in distilled water) to attain an average of one cell per through-put hole in the iChip central plate (i.e., one cell per µL of inoculum) and suspended in molten 0.5% agar (Sigma Aldrich, Munich, Germany) solution. The cell-agar suspension was poured over the ichip central plate to allow cells that were immobilized within the suspension to be trapped in the small throughput holes on the device as the agar solidified. The device was then assembled and placed in the flask containing the decaying seaweed for another two weeks. Approximately 10 g of the fresh intact macroalga (T_0) was collected before incubation into the artificial seawater and 10 g of the decaying seaweed was collected from each incubating flask at two-week intervals (weeks 2, 4, and 6) during the decay period. All *Ascophyllum* samples were stored at -20 °C for further analyses.

4.3. 16S rRNA Gene Amplicon Library Preparation and MiSeq Sequencing

Metagenomic DNA was extracted from approximately 0.5 g of the intact seaweed (T_0) and 0.5 g each of decaying *Ascophyllum nodosum* samples collected at three phases of the decay period (week 2, week 4 and week 6) at 18 °C (2_18, 4_18, 6_18), 25 °C (2_25, 4_25, 6_25), and at 30 °C (2_30, 4_30, 6_30) as previously described [58]. PCR amplicon libraries were generated using forward (5′ *TCGTCGGCAGCGTCAGATGTGTATAAGAGACAG*CCTACGGGNGGCWGCAG-3′) and reverse (*GTCTCGTGGGCTCGGAGATGTGTATAAGAGACAG*GACTACHVGGGTATCTAATCC-3′) primers complementary to the V3-V4 16S rRNA gene region [59] with ligated Illumina adapter overhang sequences in italic text. This primer pair was identified as the most promising pair required for a good representation of bacterial diversity and has been successfully applied in a number of studies on a wide range of environments [59–62]. PCR amplification was performed under the following conditions: 98 °C for 30 s, followed by 30 cycles of denaturation (98 °C for 10 s), primer annealing (57 °C for 30 s), primer extension (72 °C for 30 s), and 72 °C for 5 min. PCR amplicons were purified using Agencourt AMPure XP beads (Beckman Coulter) according to the manufacturer's instructions and a subsequent reduced-cycle (8 cycles) reaction was performed to further attach unique dual eight-base Nextera XT multiplexing indices and sequencing adapters under similar cycling conditions. Index PCR products were purified using Agencourt AMPure XP beads (Beckman Coulter; Fisher Scientific, Dublin, Ireland) according to the manufacturer's instructions. All PCR reactions from each sample were performed in triplicates to minimize bias, replicate amplicons were pooled together and sequenced using the Illumina MiSeq platform by Macrogen (Seoul, Korea).

Scythe (v0.994 BETA) [63] and Sickle [64] programs were used to quality trim raw reads and remove adapter sequences. This service was provided by Macrogen Inc (Seoul, Korea) as part of a next generation sequencing package. Trimmed paired end reads were merged (using the join_paired_reads.py script with the fastq-join method [65]) in QIIME version 1.9.1 (QIIME.org) [34] and processed using standard QIIME version 1.9.1 protocols (http://qiime.org/tutorials/illumina_overview_tutorial.html). Briefly, a further quality step was applied by excluding reads with a Phred score less than 20 using the split_libraries_fastq.py QIIME script. The USEARCH algorithm [66] was used to remove chimeras and assign sequences to OTUs based on the SILVA database (version 123) (Max Plank Institute, Bremen, Germany) [35] at a threshold of 97% identity. Singletons were identified and filtered from the OTU table and the OTU table was CSS (cumulative sum scaling) normalized [67]. The taxonomy identified from the dataset was then represented through bar plots. Species diversity and richness within samples were calculated using alpha and beta diversity analyses (Chao1, Good's coverage, Shannon indices and principle coordinates analysis) using QIIME (version 1.9.1) (http://qiime.org/tutorials/illumina_overview_tutorial.html) scripts (alpha_diversity.py and beta_diversity_through_plots.py) [34].

4.4. Bacterial Isolation from Intact and Decaying A. nodosum Using Maceration Method

Surface-attached bacteria were isolated from the intact (T_0) and decaying seaweed samples collected at weeks 2, 4 and 6 of the decay period, each at three different temperatures (18 °C, 25 °C, and 30 °C) using the maceration method adapted from [68]. Briefly, approximately 0.5 g of the algal sample was cut into small pieces of about 1 cm^2 and suspended in 1 mL of sterile artificial seawater (3.33% w/v synthetic seawater salts Instant Ocean, Aquarium Systems, in distilled water) [69]. Serial dilutions of the suspension were plated on SYP-SW agar plates which consisted of soluble starch (Sigma Aldrich, Munich, Germany) 10 g L^{-1}; yeast extract (Sigma Aldrich, Germany) 4 g L^{-1}; peptone (Merck, Germany) 2 g L^{-1}; Instant Ocean (Aquarium Systems) 33.3 g L^{-1}; agar (Sigma Aldrich, Germany) 15 g L^{-1} and incubated at 28 °C for 72 h. The culture isolation procedure was conducted aseptically in a laminar flow hood (BioAir Safeflow 1.2—EuroClone, Pero, Italy). Individual colonies were selected and further streaked to isolate pure cultures which were grown at 28 °C overnight in SYP-SW medium and maintained in glycerol (20% w/v) stocks at −80 °C.

4.5. Bacterial Isolation from Decaying A. nodosum Using ichip Device

Three separate ichip devices were inoculated into each of the incubating flasks (18 °C, 25 °C, and 30 °C) at week 4 of the decay period and were removed at the end of the decay period (week 6). Macroalgal-associated bacteria from decaying *A. nodosum* were recovered from small throughput holes on the central plate of each ichip device and plated directly onto 96-well plates containing SYP-SW agar and incubated at 28 °C for 72 h. Individual colonies were selected and further streaked to isolate pure cultures which were grown at 28 °C overnight in SYP-SW medium and maintained in glycerol (20% w/v) stocks at −80 °C.

4.6. Taxonomic Identification of A. nodosum Cultivable Surface Microbiota Populations

Bacterial isolates recovered from both intact and decaying *A. nodosum* samples using both the traditional maceration method and the ichip method were taxonomically identified using 16S rRNA gene sequencing. Genomic DNA was extracted from the bacterial isolates grown overnight at 28 °C in SYP-SW medium using a modified Tris-EDTA boiling DNA extraction method [70]. Bacterial 16S rRNA PCR amplification was performed with the universal forward (8F; 5′-AGAGTTTGATCCTGGCTCAG-3′ or 27F; 5′-AGAGTTTGATCMTGGCTCAG-3′) and reverse (1492R; 5′-GGTTACCTTGTTACGACTT-3′) primers [71,72] under the following conditions: initial denaturation (95 °C for 30 s), followed by 35 cycles of denaturation (95 °C for 1 min), primer annealing (55 °C for 1 min), primer extension

(72 °C for 1 min) and a final primer extension step (72 °C for 5 min). PCR products were analyzed by gel electrophoresis on a 1% agarose gel and purified using a QIAquick PCR Purification Kit (Qiagen, Hilden, Germany) according to the manufacturer's instructions.

Sanger sequencing was performed on the amplified PCR products by GATC Biotech, (Konstanz, Germany) and Macrogen (Amsterdam, The Netherlands). Low quality 5' and 3' sequence ends were trimmed using FinchTV (http://www.geospiza.com/finchtv) depending on the data set. The BLAST program (NCBI) (https://blast.ncbi.nlm.nih.gov/Blast.cgi) was used to compare trimmed sequences against the GenBank database and closest relatives to the bacterial isolates were identified. 16S rRNA gene sequences were checked for chimeras using USEARCH algorithm [66] and the data sets were de-replicated using the Fastgroup database [73] and Avalanche NextGen Workbench version 2.30 (http://www.visualbioinformatics.com/html/)(bioinformatics.org) with a 99% cut-off value. Sequence alignment and phylogenetic tree construction were performed with MEGA (version 7) (Penn State University, PA, USA) [74]. The evolutionary history was inferred using the Neighbor-joining method [75].

4.7. Enzyme Screens

Bacterial isolates obtained from both intact *A. nodosum* and the decaying seaweed, using both the maceration method and the ichip device, were screened for the production of macroalgal cell wall degrading enzymes including pectinase, hydroxyethyl cellulase, and enzymes involved in lichenin degradation. Bacterial isolates were grown at 28 °C for 72 h on LB gellan gum (Sigma Aldrich, Munich, Germany) plates supplemented with the appropriate substrate, at a concentration of 0.2% (w/v) for pectin (Sigma Aldrich, Germany), 0.5% (w/v) for hydroxyethyl cellulose (Sigma Aldrich, Germany) and 0.05% (w/v) for lichenin (Megazyme). Enzymatic activities on lichenin and HE-cellulose were indicated by a surrounding zone of clearance upon flooding with Congo red solution (0.1% w/v Congo red in 20% v/v ethanol) for 30 min and wash with 1M NaCl for 5 min [76,77] while pectin supplemented plates were flooded with Lugol's iodine solution [78].

4.8. Enzyme-Assisted Extraction

Entire specimens of *Fucus vesiculosus* (2–3 kg fresh weight) were collected at low tide on 6 July 2016 at Finavarra, Co., Clare, Ireland (53°08′59″ North–9°08′09″ West). In the laboratory, on the day of collection, algal biomass was cleaned of epiphytes and rinsed in distilled water to remove excess salt. Samples were patted dry with tissue paper and stored at −20 °C. Then, the biomass was freeze-dried in a Labconco Freezone® freeze-dryer system (Labconco Corporation, Kansas City, MO 64132-2696, USA). Dried biomass was ground using a coffee-grinder and sieved to produce two types of powder, Ps < 0.5 mm and 0.5 < Ps < 2.5 mm), prior to subsequent enzymatic-assisted extraction (EAE) of total phenolic compounds. Three commercially available enzymes; cellulase (from *Aspergillus* sp., Sigma Aldrich, ≥1000 U/g), xylanase (from *Trichoderma* sp., Megazyme, 2.86 U/mL) and protease (from *Bacillus licheniniformis*, Sigma Aldrich, ≥2.4 U/g) were used for the hydrolysis of the seaweed. The potential of three bacterial strains (IC18_D7, IC18_D5 and ANT$_0$_A6 with ≥ 98% sequence similarity to *Vibrio anguillarum* X0906, *Vibrio oceanisediminis* S37 and *Winogradskyella* sp. MGE_SAT_697, respectively) isolated from *A. nodosum* and shown to be producers of algal cell wall degrading enzymes in this study was also tested. Bacterial isolates were grown overnight at 28 °C in SYP-SW medium and equal volumes of supernatants from the overnight cultures obtained by centrifugation at 4300× g for 10 min were used. Three sets of approximately 4 g dry weight of the crushed algae was incubated at 50 °C for 24 h on a shaking platform (185 rpm) with sodium acetate buffer (100 mL; 0.1 M; pH 5.2), each with 100 µL of one of the three commercial enzymes. Another set of the algal biomass was also incubated at 28 °C with a mixture of culture supernatants obtained from the bacterial isolates (enzymatic bacterial supernatants, EBS) to a final volume of 100 µL. All experiments were performed in triplicate and control experiments without the addition of either commercial enzymes or culture supernatants (EBS) were also conducted under the same conditions. The hydrolysate mixture from each experimental set was centrifuged at 4300× g for 10 min at 4 °C to eliminate the algal debris from the extract. The different extracts produced were freeze dried, weighed, and stored at -80 °C until further analysis for total phenolics.

4.9. Determination of Total Phenolic Content (TPC)

The total phenolic content (TPC) of the *F. vesiculosus* crude extracts was determined using a slightly modified version of the Folin–Ciocalteu assay [79] as described by [80]. A known amount of crude extract was re-suspended in methanol to a concentration of 1 mg.mL^{-1}. 100 µL of each crude extract was placed in a 1.5 mL Eppendorf tube along with 100 µL of methanol, 100 µL of Folin–Ciocalteu reagent (2N) and 700 µL of 20% sodium carbonate, to a final volume of 1 mL. Samples were vortexed and immediately afterwards placed in darkness to incubate for 20 min at room temperature. Samples were then centrifuged at 4300× g for 3 min before measuring the absorbance of the supernatant at 735 nm using a Cary UV50 Spectrophotometer and CaryWIN software (Varian Inc., Palo Alto, CA 94304, USA). A sample treated according to the same protocol, but where 100 µL methanol instead of 100 µL of crude extract (1 mg·mL^{-1}) was added, was used as a blank. Phloroglucinol was used as the external standard and a calibration curve was performed by serial dilution of a 2 mg mL^{-1} stock solution (10, 20, 50, 80, 120, 160 µg mL^{-1}). Total phenolic content (TPC) was expressed as milligram of phloroglucinol equivalents (PE) per gram of dry weight extract (mg PE g^{-1} DWE) or per gram of dry weight biomass (mg PE g^{-1} DWB) [36]. TPC quantification was performed in triplicate for each crude extract. The yield of extraction was calculated after exhaustive solid-liquid extraction (i.e., three successive extraction) of the total phenolics of 50 mg of *F. vesiculosus* freeze-dried ground biomass (Ps < 0.5 mm) using 80% methanol.

4.10. Accession Numbers

The metagenomic sequencing data (raw reads) was deposited in the European Nucleotide Archive (ENA) under the accession numbers ERR2608102 -ERR2608111. The 16S rRNA gene sequences for the bacterial isolates were deposited in GenBank under the accession numbers KY224981–KY225289, KY327837, KY327838, MG693225–MG693716, MG760723–MG760725, and MK480287–MK480325. Bacterial isolates IC18_D7 and ANT$_0$_A6 with enzymatic activities and applicability in EAE approach were deposited in DSMZ culture collection bank under the accession numbers DSM 107285 and DSM 107318, respectively.

4.11. Statistical Analysis

Prior to performing statistical analyses on data obtained by enzyme-assisted extraction (EAE) of total phenolics from *F. vesiculosus*, tests of normality were carried out with the Kolmogorov–Smirnov test for normal distributions and Levene's test for homogeneity of variance. A one-way ANOVA and post hoc Tukey's pairwise test was performed to assess significant differences ($p < 0.05$) between commercial enzymes and control (50 °C); and a *t*-test was applied to assess significance differences ($p < 0.05$) between EBS and control (28 °C). All data treatments and statistical analyses were performed using IBM SPSS Statistics V22.0 (IBM Corporation, Armonk, NY, USA).

5. Conclusions

In conclusion, we have demonstrated, using both metagenomic and culture based approaches, that changes occur in the composition and abundance of *A. nodosum*-associated epibiotic communities which are both time and temperature dependent and that the microflora of *A. nodosum* is composed of diverse and complex bacterial communities which produce a wide range of hydrolytic enzymes, some of which may be useful in future EAE based strategies in the agricultural, food, cosmeceutical, and pharmaceutical sectors.

Supplementary Materials: The following are available online at http://www.mdpi.com/1660-3397/17/4/200/s1; Table S1: Table showing enzymatic activities, Figure S1: *Ascophyllum* cultivable surface microbiota at genus classification level, Figures S2–S6; Phylogenetic trees of intact and decaying *A. nodosum* associated microbial communities.

Author Contributions: A.D.W.D., M.W.I. and D.S. conceived and designed the study; F.G., H.M. and M.W.I. performed the experiments and together with L.M.M. and S.A.J. analyzed the data. M.W.I., D.C., S.A.J., and A.D.W.D. wrote the manuscript.

Funding: This research was funded by the European Commission, PharmaSea project (www.pharma-sea.eu) (contract no. 312184), Marine Biotechnology ERA/NET, NEPTUNA project (contract no. PBA/MB/15/02), SMI-BIO project (15/F/698), Science Foundation Ireland (SSPC-2, 12/RC/2275), Department of Agriculture, Food and the Marine (DAFM) SMI-BIO project (15/F/698), DAFM, (FIRM 1/F009/MabS) and by the Beaufort Marine Research Award, part of the Sea Change Strategy and the Strategy for Science Technology and Innovation (2006–2012), with the support of The Marine Institute under the Marine Research Sub-Programme of the National Development Plan 2007–2013.

Conflicts of Interest: The authors declare that they have no conflict of interest. The funders had no role in the design of the study; in the collection, analyses, or interpretation of data; in the writing of the manuscript, or in the decision to publish the results.

References

1. Olsen, J.L.; Zechman, F.W.; Hoarau, G.; Coyer, J.A.; Stam, W.T.; Valero, M.; Åberg, P. The phylogeographic architecture of the fucoid seaweed *Ascophyllum nodosum*: An intertidal 'marine tree'and survivor of more than one glacial-interglacial cycle. *J. Biogeogr.* **2010**, *37*, 842–856. [CrossRef]
2. De Jesus Raposo, M.; de Morais, A.; de Morais, R. Emergent sources of prebiotics: Seaweeds and microalgae. *Mar. Drugs* **2016**, *14*, 27. [CrossRef] [PubMed]
3. Jiang, Z.; Okimura, T.; Yokose, T.; Yamasaki, Y.; Yamaguchi, K.; Oda, T. Effects of sulfated fucan, ascophyllan, from the brown Alga *Ascophyllum nodosum* on various cell lines: A comparative study on ascophyllan and fucoidan. *J. Biosci. Bioeng.* **2010**, *110*, 113–117. [CrossRef] [PubMed]
4. Nakayasu, S.; Soegima, R.; Yamaguchi, K.; Oda, T. Biological activities of fucose-containing polysaccharide ascophyllan isolated from the brown alga *Ascophyllum nodosum*. *Biosci. Biotechnol. Biochem.* **2009**, *73*, 961–964. [CrossRef] [PubMed]
5. Suleria, H.A.R.; Osborne, S.; Masci, P.; Gobe, G. Marine-based nutraceuticals: An innovative trend in the food and supplement industries. *Mar. Drugs* **2015**, *13*, 6336–6351. [CrossRef] [PubMed]
6. Adrien, A.; Dufour, D.; Baudouin, S.; Maugard, T.; Bridiau, N. Evaluation of the anticoagulant potential of polysaccharide-rich fractions extracted from macroalgae. *Nat. Prod. Res.* **2017**, *31*, 2126–2136. [CrossRef] [PubMed]
7. Abu, R.; Jiang, Z.; Ueno, M.; Isaka, S.; Nakazono, S.; Okimura, T.; Cho, K.; Yamaguchi, K.; Kim, D.; Oda, T. Anti-metastatic effects of the sulfated polysaccharide ascophyllan isolated from *Ascophyllum nodosum* on B16 melanoma. *Biochem. Biophys. Res. Commun.* **2015**, *458*, 727–732. [CrossRef] [PubMed]
8. Zhang, W.; Du, J.-Y.; Jiang, Z.; Okimura, T.; Oda, T.; Yu, Q.; Jin, J.-O. Ascophyllan purified from *Ascophyllum nodosum* induces Th1 and Tc1 immune responses by promoting dendritic cell maturation. *Mar. Drugs* **2014**, *12*, 4148–4164. [CrossRef]
9. Wang, W.; Wang, S.-X.; Guan, H.-S. The antiviral activities and mechanisms of marine polysaccharides: An overview. *Mar. Drugs* **2012**, *10*, 2795–2816. [CrossRef]
10. Belanche, A.; Ramos-Morales, E.; Newbold, C.J. In vitro screening of natural feed additives from crustaceans, diatoms, seaweeds and plant extracts to manipulate rumen fermentation. *J. Sci. Food Agric.* **2016**, *96*, 3069–3078. [CrossRef]
11. Goñi, O.; Fort, A.; Quille, P.; McKeown, P.C.; Spillane, C.; O'Connell, S. Comparative transcriptome analysis of two *Ascophyllum nodosum* extract biostimulants: Same seaweed but different. *J. Agric. Food Chem.* **2016**, *64*, 2980–2989. [CrossRef]
12. Santaniello, A.; Scartazza, A.; Gresta, F.; Loreti, E.; Biasone, A.; Di Tommaso, D.; Piaggesi, A.; Perata, P. *Ascophyllum nodosum* seaweed extract alleviates drought stress in *Arabidopsis* by affecting photosynthetic performance and related gene expression. *Front. Plant Sci.* **2017**, *8*, 1362. [CrossRef]
13. Bengtsson, M.M.; Sjøtun, K.; Øvreås, L. Seasonal dynamics of bacterial biofilms on the kelp Laminaria hyperborea. *Aquat. Microb. Ecol.* **2010**, *60*, 71–83. [CrossRef]

14. Egan, S.; Harder, T.; Burke, C.; Steinberg, P.; Kjelleberg, S.; Thomas, T. The seaweed holobiont: Understanding seaweed-bacteria interactions. *FEMS Microbiol. Rev.* **2013**, *37*, 462–476. [CrossRef]
15. Singh, R.P.; Shukla, M.K.; Mishra, A.; Reddy, C.; Jha, B. Bacterial extracellular polymeric substances and their effect on settlement of zoospore of *Ulva fasciata*. *Colloids Surf. B Biointerfaces* **2013**, *103*, 223–230. [CrossRef] [PubMed]
16. Steinberg, P.D.; De Nys, R.; Kjelleberg, S. Chemical cues for surface colonization. *J. Chem. Ecol.* **2002**, *28*, 1935–1951. [CrossRef] [PubMed]
17. Goecke, F.; Labes, A.; Wiese, J.; Imhoff, J.F. Chemical interactions between marine macroalgae and bacteria. *Mar. Ecol. Prog. Ser.* **2010**, *409*, 267–299. [CrossRef]
18. Singh, R.P.; Reddy, C. Seaweed-microbial interactions: Key functions of seaweed-associated bacteria. *FEMS Microbiol. Ecol.* **2014**, *88*, 213–230. [CrossRef] [PubMed]
19. Lachnit, T.; Fischer, M.; Künzel, S.; Baines, J.F.; Harder, T. Compounds associated with algal surfaces mediate epiphytic colonization of the marine macroalga *Fucus vesiculosus*. *FEMS Microbiol. Ecol.* **2013**, *84*, 411–420. [CrossRef] [PubMed]
20. Chun, C.L.; Peller, J.R.; Shively, D.; Byappanahalli, M.N.; Whitman, R.L.; Staley, C.; Zhang, Q.; Ishii, S.; Sadowsky, M.J. Virulence and biodegradation potential of dynamic microbial communities associated with decaying *Cladophora* in Great Lakes. *Sci. Total Environ.* **2017**, *574*, 872–880. [CrossRef] [PubMed]
21. Martin, M.; Barbeyron, T.; Martin, R.; Portetelle, D.; Michel, G.; Vandenbol, M. The cultivable surface microbiota of the brown alga *Ascophyllum nodosum* is enriched in macroalgal-polysaccharide-degrading bacteria. *Front. Microbiol.* **2015**, *6*, 1487. [CrossRef]
22. Martin, M.; Vandermies, M.; Joyeux, C.; Martin, R.; Barbeyron, T.; Michel, G.; Vandenbol, M. Discovering novel enzymes by functional screening of plurigenomic libraries from alga-associated *Flavobacteria* and *Gammaproteobacteria*. *Microbiol. Res.* **2016**, *186*, 52–61. [CrossRef]
23. Hardouin, K.; Bedoux, G.; Burlot, A.-S.; Donnay-Moreno, C.; Bergé, J.-P.; Nyvall-Collén, P.; Bourgougnon, N. Enzyme-assisted extraction (EAE) for the production of antiviral and antioxidant extracts from the green seaweed *Ulva armoricana* (Ulvales, Ulvophyceae). *Algal Res.* **2016**, *16*, 233–239. [CrossRef]
24. Kulshreshtha, G.; Burlot, A.-S.; Marty, C.; Critchley, A.; Hafting, J.; Bedoux, G.; Bourgougnon, N.; Prithiviraj, B. Enzyme-assisted extraction of bioactive material from *Chondrus crispus* and *Codium fragile* and its effect on herpes simplex virus (HSV-1). *Mar. Drugs* **2015**, *13*, 558–580. [CrossRef] [PubMed]
25. Leal, M.C.; Munro, M.H.; Blunt, J.W.; Puga, J.; Jesus, B.; Calado, R.; Rosa, R.; Madeira, C. Biogeography and biodiscovery hotspots of macroalgal marine natural products. *Nat. Prod. Rep.* **2013**, *30*, 1380–1390. [CrossRef] [PubMed]
26. Deniaud-Bouët, E.; Kervarec, N.; Michel, G.; Tonon, T.; Kloareg, B.; Hervé, C. Chemical and enzymatic fractionation of cell walls from Fucales: Insights into the structure of the extracellular matrix of brown algae. *Ann. Bot.* **2014**, *114*, 1203–1216. [CrossRef] [PubMed]
27. Kadam, S.U.; Tiwari, B.K.; O'Donnell, C.P. Application of novel extraction technologies for bioactives from marine algae. *J. Agric. Food Chem.* **2013**, *61*, 4667–4675. [CrossRef] [PubMed]
28. Joana Gil-Chávez, G.; Villa, J.A.; Fernando Ayala-Zavala, J.; Basilio Heredia, J.; Sepulveda, D.; Yahia, E.M.; González-Aguilar, G.A. Technologies for extraction and production of bioactive compounds to be used as nutraceuticals and food ingredients: An overview. *Compr. Rev. Food Sci. Food Saf.* **2013**, *12*, 5–23. [CrossRef]
29. Jeon, Y.J.; Wijesinghe, W.P.; Kim, S.K. Enzyme-assisted extraction and recovery of bioactive components from seaweeds. In *Handbook of Marine Macroalgae: Biotechnology and Applied Phycology*; John Wiley & Sons, Ltd.: Chichester, UK, 2011; pp. 221–228.
30. Nadar, S.S.; Rao, P.; Rathod, V.K. Enzyme assisted extraction of biomolecules as an approach to novel extraction technology: A review. *Food Res. Int.* **2018**, *108*, 309–330. [CrossRef] [PubMed]
31. Charoensiddhi, S.; Lorbeer, A.J.; Lahnstein, J.; Bulone, V.; Franco, C.M.; Zhang, W. Enzyme-assisted extraction of carbohydrates from the brown alga *Ecklonia radiata*: Effect of enzyme type, pH and buffer on sugar yield and molecular weight profiles. *Process Biochem.* **2016**, *51*, 1503–1510. [CrossRef]
32. Nichols, D.; Cahoon, N.; Trakhtenberg, E.; Pham, L.; Mehta, A.; Belanger, A.; Kanigan, T.; Lewis, K.; Epstein, S. Use of ichip for high-throughput in situ cultivation of "uncultivable" microbial species. *Appl. Environ. Microbiol.* **2010**, *76*, 2445–2450. [CrossRef] [PubMed]
33. Boyd, K.G.; Adams, D.R.; Burgess, J.G. Antibacterial and repellent activities of marine bacteria associated with algal surfaces. *Biofouling* **1999**, *14*, 227–236. [CrossRef]

34. Caporaso, J.G.; Kuczynski, J.; Stombaugh, J.; Bittinger, K.; Bushman, F.D.; Costello, E.K.; Fierer, N.; Pena, A.G.; Goodrich, J.K.; Gordon, J.I. QIIME allows analysis of high-throughput community sequencing data. *Nat. Methods* **2010**, *7*, 335. [CrossRef] [PubMed]
35. Quast, C.; Pruesse, E.; Yilmaz, P.; Gerken, J.; Schweer, T.; Yarza, P.; Peplies, J.; Glöckner, F.O. The SILVA ribosomal RNA gene database project: Improved data processing and web-based tools. *Nucleic Acids Res.* **2012**, *41*, D590–D596. [CrossRef]
36. Kenny, O.; Brunton, N.P.; Smyth, T.J. In vitro protocols for measuring the antioxidant capacity of algal extracts. In *Natural Products from Marine Algae*; Springer, Humana Press: New York, NY, USA, 2015; pp. 375–402.
37. Holdt, S.L.; Kraan, S. Bioactive compounds in seaweed: Functional food applications and legislation. *J. Appl. Phycol.* **2011**, *23*, 543–597. [CrossRef]
38. Pulz, O.; Gross, W. Valuable products from biotechnology of microalgae. *Appl. Microbiol. Biotechnol.* **2004**, *65*, 635–648. [CrossRef]
39. Lee, S.-H.; Jeon, Y.-J. Anti-diabetic effects of brown algae derived phlorotannins, marine polyphenols through diverse mechanisms. *Fitoterapia* **2013**, *86*, 129–136. [CrossRef]
40. Shibata, T.; Ishimaru, K.; Kawaguchi, S.; Yoshikawa, H.; Hama, Y. Antioxidant activities of phlorotannins isolated from Japanese Laminariaceae. In *Nineteenth International Seaweed Symposium*; Springer: New York, NY, USA, 2007; pp. 255–261.
41. Nwosu, F.; Morris, J.; Lund, V.A.; Stewart, D.; Ross, H.A.; McDougall, G.J. Anti-proliferative and potential anti-diabetic effects of phenolic-rich extracts from edible marine algae. *Food Chem.* **2011**, *126*, 1006–1012. [CrossRef]
42. Craigie, J.S. Seaweed extract stimuli in plant science and agriculture. *J. Appl. Phycol.* **2011**, *23*, 371–393. [CrossRef]
43. Angell, A.R.; Angell, S.F.; de Nys, R.; Paul, N.A. Seaweed as a protein source for mono-gastric livestock. *Trends Food Sci. Technol.* **2016**, *54*, 74–84. [CrossRef]
44. Martin, M.; Biver, S.; Steels, S.; Barbeyron, T.; Jam, M.; Portetelle, D.; Michel, G.; Vandenbol, M. Identification and characterization of a halotolerant, cold-active marine endo-β-1, 4-glucanase by using functional metagenomics of seaweed-associated microbiota. *Appl. Environ. Microbiol.* **2014**, *80*, 4958–4967. [CrossRef]
45. Dubinina, G.; Grabovich, M.; Leshcheva, N.; Gronow, S.; Gavrish, E.; Akimov, V. *Spirochaeta sinaica* sp. nov. a halophilic spirochaete isolated from a cyanobacterial mat. *Int. J. Syst. Evol. Microbiol.* **2015**, *65*, 3872–3877. [CrossRef]
46. Jeske, O.; Schüler, M.; Schumann, P.; Schneider, A.; Boedeker, C.; Jogler, M.; Bollschweiler, D.; Rohde, M.; Mayer, C.; Engelhardt, H. Planctomycetes do possess a peptidoglycan cell wall. *Nat. Commun.* **2015**, *6*, 7116. [CrossRef]
47. Lage, O.M.; Bondoso, J. Bringing Planctomycetes into pure culture. *Front. Microbiol.* **2012**, *3*, 405. [CrossRef]
48. Tanaka, Y.; Matsuzawa, H.; Tamaki, H.; Tagawa, M.; Toyama, T.; Kamagata, Y.; Mori, K. Isolation of novel bacteria including rarely cultivated Phyla, Acidobacteria and Verrucomicrobia, from the roots of emergent plants by simple culturing method. *Microbes Environ.* **2017**, *32*, 288–292. [CrossRef]
49. Berdy, B.; Spoering, A.L.; Ling, L.L.; Epstein, S.S. In situ cultivation of previously uncultivable microorganisms using the ichip. *Nat. Protoc.* **2017**, *12*, 2232. [CrossRef]
50. Kealey, C.; Creaven, C.; Murphy, C.; Brady, C. New approaches to antibiotic discovery. *Biotechnol. Lett.* **2017**, *39*, 805–817. [CrossRef]
51. Ling, L.L.; Schneider, T.; Peoples, A.J.; Spoering, A.L.; Engels, I.; Conlon, B.P.; Mueller, A.; Schäberle, T.F.; Hughes, D.E.; Epstein, S. A new antibiotic kills pathogens without detectable resistance. *Nature* **2015**, *517*, 455. [CrossRef] [PubMed]
52. Marzinelli, E.M.; Campbell, A.H.; Zozaya Valdes, E.; Vergés, A.; Nielsen, S.; Wernberg, T.; de Bettignies, T.; Bennett, S.; Caporaso, J.G.; Thomas, T. Continental-scale variation in seaweed host-associated bacterial communities is a function of host condition, not geography. *Environ. Microbiol.* **2015**, *17*, 4078–4088. [CrossRef] [PubMed]
53. Rosenberg, E.; Koren, O.; Reshef, L.; Efrony, R.; Zilber-Rosenberg, I. The role of microorganisms in coral health, disease and evolution. *Nat. Rev. Microbiol.* **2007**, *5*, 355. [CrossRef] [PubMed]
54. Hibbing, M.E.; Fuqua, C.; Parsek, M.R.; Peterson, S.B. Bacterial competition: Surviving and thriving in the microbial jungle. *Nat. Rev. Microbiol.* **2010**, *8*, 15. [CrossRef] [PubMed]

55. Thomas, T.; Evans, F.F.; Schleheck, D.; Mai-Prochnow, A.; Burke, C.; Penesyan, A.; Dalisay, D.S.; Stelzer-Braid, S.; Saunders, N.; Johnson, J. Analysis of the *Pseudoalteromonas tunicata* genome reveals properties of a surface-associated life style in the marine environment. *PLoS ONE* **2008**, *3*, e3252. [CrossRef] [PubMed]
56. Liu, G.-L.; Li, Y.; Chi, Z.; Chi, Z.-M. Purification and characterization of κ-carrageenase from the marine bacterium *Pseudoalteromonas porphyrae* for hydrolysis of κ-carrageenan. *Process Biochem.* **2011**, *46*, 265–271. [CrossRef]
57. Zhu, Y.; Thomas, F.; Larocque, R.; Li, N.; Duffieux, D.; Cladière, L.; Souchaud, F.; Michel, G.; Mcbride, M.J. Genetic analyses unravel the crucial role of a horizontally acquired alginate lyase for brown algal biomass degradation by *Zobellia galactanivorans*. *Environ. Microbiol.* **2017**, *19*, 2164–2181. [CrossRef]
58. Varela-Álvarez, E.; Andreakis, N.; Lago-Lestón, A.; Pearson, G.A.; Serrao, E.A.; Procaccini, G.; Duarte, C.M.; Marba, N. Genomic DNA isolation from green and brown algae (caulerpales and fucales) for microsatellite library construction 1. *J. Phycol.* **2006**, *42*, 741–745. [CrossRef]
59. Klindworth, A.; Pruesse, E.; Schweer, T.; Peplies, J.; Quast, C.; Horn, M.; Glöckner, F.O. Evaluation of general 16S ribosomal RNA gene PCR primers for classical and next-generation sequencing-based diversity studies. *Nucleic Acids Res.* **2013**, *41*, e1. [CrossRef] [PubMed]
60. Logue, J.B.; Langenheder, S.; Andersson, A.F.; Bertilsson, S.; Drakare, S.; Lanzén, A.; Lindström, E.S. Freshwater bacterioplankton richness in oligotrophic lakes depends on nutrient availability rather than on species–area relationships. *ISME J.* **2012**, *6*, 1127. [CrossRef] [PubMed]
61. Abrahamsson, T.R.; Jakobsson, H.E.; Andersson, A.F.; Björkstén, B.; Engstrand, L.; Jenmalm, M.C. Low diversity of the gut microbiota in infants with atopic eczema. *J. Allergy Clin. Immunol.* **2012**, *129*, 434–440. [CrossRef] [PubMed]
62. Herlemann, D.P.; Labrenz, M.; Jürgens, K.; Bertilsson, S.; Waniek, J.J.; Andersson, A.F. Transitions in bacterial communities along the 2000 km salinity gradient of the Baltic Sea. *ISME J.* **2011**, *5*, 1571. [CrossRef]
63. Buffalo, V. Scythe—A Bayesian Adapter Trimmer (version 0.994 BETA). 2014. Unpublished. Available online: https://github.com/vsbuffalo/scythe (accessed on 11 May 2017).
64. Joshi, N.; Fass, J. Sickle—A Windowed Adaptive Trimming Tool for FASTQ Files Using Quality. Available online: https://github.com/najoshi/sickle (accessed on 11 May 2017).
65. Aronesty, E. ea-utils: Command-Line Tools for Processing Biological Sequencing Data. 2011. Available online: https://expressionanalysis.github.io/ea-utils/ (accessed on 10 October 2017).
66. Edgar, R.C. Search and clustering orders of magnitude faster than BLAST. *Bioinformatics* **2010**, *26*, 2460–2461. [CrossRef]
67. Paulson, J.N.; Stine, O.C.; Bravo, H.C.; Pop, M. Differential abundance analysis for microbial marker-gene surveys. *Nat. Methods* **2013**, *10*, 1200. [CrossRef] [PubMed]
68. Santavy, D.; Willenz, P.; Colwell, R. Phenotypic study of bacteria associated with the caribbean sclerosponge, *Ceratoporella nicholsoni*. *Appl. Environ. Microbiol.* **1990**, *56*, 1750–1762.
69. Atkinson, M.; Bingman, C. Elemental composition of commercial seasalts. *J. Aquaricult. Aquat. Sci.* **1997**, *8*, 39–43.
70. Li, M.; Gong, J.; Cottrill, M.; Yu, H.; de Lange, C.; Burton, J.; Topp, E. Evaluation of QIAamp® DNA Stool Mini Kit for ecological studies of gut microbiota. *J. Microbiol. Methods* **2003**, *54*, 13–20. [CrossRef]
71. Turner, S.; Pryer, K.M.; Miao, V.P.; Palmer, J.D. Investigating deep phylogenetic relationships among cyanobacteria and plastids by small subunit rRNA sequence analysis 1. *J. Eukaryot. Microbiol.* **1999**, *46*, 327–338. [CrossRef]
72. Lane, D. 16S/23S rRNA sequencing. In *Nucleic Acid Techniques in Bacterial Systematic*; Stackebrandt, E., Goodfellow, M., Eds.; John Wiley and Sons: New York, NY, USA, 1991; pp. 115–175.
73. Seguritan, V.; Rohwer, F. FastGroup: A program to dereplicate libraries of 16S rDNA sequences. *BMC Bioinform.* **2001**, *2*, 9. [CrossRef]
74. Kumar, S.; Stecher, G.; Tamura, K. MEGA7: Molecular evolutionary genetics analysis version 7.0 for bigger datasets. *Mol. Biol. Evol.* **2016**, *33*, 1870–1874. [CrossRef] [PubMed]
75. Saitou, N.; Nei, M. The neighbor-joining method: A new method for reconstructing phylogenetic trees. *Mol. Biol. Evol.* **1987**, *4*, 406–425.

76. Walter, J.; Mangold, M.; Tannock, G.W. Construction, analysis, and β-glucanase screening of a bacterial artificial chromosome library from the large-bowel microbiota of mice. *Appl. Environ. Microbiol.* **2005**, *71*, 2347–2354. [CrossRef]
77. Wolf, M.; Geczi, A.; Simon, O.; Borriss, R. Genes encoding xylan and β-glucan hydrolysing enzymes in *Bacillus subtilis*: Characterization, mapping and construction of strains deficient in lichenase, cellulase and xylanase. *Microbiology* **1995**, *141*, 281–290. [CrossRef]
78. Soares, M.M.; Silva, R.D.; Gomes, E. Screening of bacterial strains for pectinolytic activity: Characterization of the polygalacturonase produced by *Bacillus* sp. *Rev. Microbiol.* **1999**, *30*, 299–303. [CrossRef]
79. Singleton, V.L.; Rossi, J.A. Colorimetry of total phenolics with phosphomolybdic-phosphotungstic acid reagents. *Am. J. Enol. Vitic.* **1965**, *16*, 144–158.
80. Kenny, O.; Smyth, T.; Hewage, C.; Brunton, N. Antioxidant properties and quantitative UPLC-MS analysis of phenolic compounds from extracts of fenugreek (*Trigonella foenum-graecum*) seeds and bitter melon (*Momordica charantia*) fruit. *Food Chem.* **2013**, *141*, 4295–4302. [CrossRef] [PubMed]

© 2019 by the authors. Licensee MDPI, Basel, Switzerland. This article is an open access article distributed under the terms and conditions of the Creative Commons Attribution (CC BY) license (http://creativecommons.org/licenses/by/4.0/).

Article

Characterization of the Specific Mode of Action of a Chitin Deacetylase and Separation of the Partially Acetylated Chitosan Oligosaccharides

Xian-Yu Zhu [1,2], Yong Zhao [1], Huai-Dong Zhang [1,3], Wen-Xia Wang [1], Hai-Hua Cong [2] and Heng Yin [1,*]

[1] Liaoning Provincial Key Laboratory of Carbohydrates, Dalian Institute of Chemical Physics, Chinese Academy of Sciences, Dalian 116023, China; zhuxy0721@126.com (X.-Y.Z.); zhaoyong_2019@163.com (Y.Z.); huaidongzhang@yahoo.com (H.-D.Z.); wangwx@dicp.ac.cn (W.-X.W.)
[2] College of Food Science and Engineering, Dalian Ocean University, Dalian 116023, China; haihuacong780@gmail.com
[3] Engineering Research Center of Industrial Microbiology, Ministry of Education; College of Life Sciences, Fujian Normal University, Fuzhou 350117, China
* Correspondence: yinheng@dicp.ac.cn; Tel./Fax: +86-0411-84379061

Received: 19 December 2018; Accepted: 16 January 2019; Published: 22 January 2019

Abstract: Partially acetylated chitosan oligosaccharides (COS), which consists of N-acetylglucosamine (GlcNAc) and glucosamine (GlcN) residues, is a structurally complex biopolymer with a variety of biological activities. Therefore, it is challenging to elucidate acetylation patterns and the molecular structure-function relationship of COS. Herein, the detailed deacetylation pattern of chitin deacetylase from *Saccharomyces cerevisiae*, *Sc*CDA$_2$, was studied. Which solves the randomization of acetylation patterns during COS produced by chemical. *Sc*CDA$_2$ also exhibits about 8% and 20% deacetylation activity on crystalline chitin and colloid chitin, respectively. Besides, a method for separating and detecting partially acetylated chitosan oligosaccharides by high performance liquid chromatography and electrospray ionization mass spectrometry (HPLC-ESI-MS) system has been developed, which is fast and convenient, and can be monitored online. Mass spectrometry sequencing revealed that *Sc*CDA$_2$ produced COS with specific acetylation patterns of DAAA, ADAA, AADA, DDAA, DADA, ADDA and DDDA, respectively. *Sc*CDA$_2$ does not deacetylate the GlcNAc unit that is closest to the reducing end of the oligomer furthermore *Sc*CDA$_2$ has a multiple-attack deacetylation mechanism on chitin oligosaccharides. This specific mode of action significantly enriches the existing limited library of chitin deacetylase deacetylation patterns. This fully defined COS may be used in the study of COS structure and function.

Keywords: chitin deacetylase; deacetylation patterns; chitooligosaccharides; separating; detecting

1. Introduction

Chitin, which consists of β-1,4-linked N-acetyl-D-glucosamine residues, is the main component of crustacean shells, such as shrimp, crab and shellfish [1,2]. Chitin, a renewable raw material whose annual production is about 10^{11} tons, is the second most abundant natural biopolymer after cellulose [3,4]. As a new type of functional material, chitin has attracted wide attention in various fields [5]. However, it is insoluble in water and most organic solvents, this property severely restricts its development and application [3]. On the other hand, chitosan, the deacetylation product of chitin, is soluble in dilute acid solution and has been widely used in agriculture, biomedicine, environmental science and other fields, as a plant inducer, biodegradable hydrogel and sewage treatment agent in antitumor drugs and in other green products [2,6–10]. Chitooligosaccharide (COS), the hydrolytic product of chitosan, has broader biological activities, such as immunological, antitumor, antioxidant and antibacterial activity [11–13]. Due to its water-soluble ability and broad biological activity, COS has attracted more attention than chitosan. The biological activity of COS are believed to be strongly dependent on the degree of polymerization (DP), the degree of acetylation (DA) and the pattern of acetylation (PA) [14]. Vander et al. reported that COS with different degrees of deacetylation is involved in the induction of phenylalanine ammonia lyase and peroxidase activities, both of which must be activated for lignin biosynthesis [15]. It has previously been observed that the specific recognition of the N-acetyl moiety allows AtCERK1 to distinguish chitin and chitosan, which then activate plant immune receptors and elicit a plant immune response [16].

In order to investigate the specific biological activity of COS in a particular acetylation pattern, COS with a completely known structure is required. However, chitosan and COS produced by chemical methods usually exhibit a randomized pattern of acetylation, making them difficult to control and predict their biological activity [17]. Moreover, chitosan produced by chemical methods requires high energy consumption and causes environmental pollution [1]. In contrast, chitin deacetylase (CDA, E.C. 3.5.1.41) is able to hydrolyse the N-acetamido groups of N-acetyl-D-glucosamine residues in chitin, chitin oligosaccharides, chitosan and chitosan oligosaccharides under mild conditions by a specific mode of action. Previous studies have identified CDAs from bacteria, fungi and insects, such as *Bacillus thuringiensis* [18], *Bacillus amyloliquefaciens* [19], *Colletotrichum gloeosporioides* [20] *Mucor rouxii* [21] *Aspergillus nidulans* [22] *Saccharomyces cerevisiae* [23] *Bombyx mori* [24], *Drosophila melanogaster* [25], *Encephalitozoon cuniculi* [26], *Mamestra configurata* [27]. Although some CDAs have been reported, the deacetylation patterns of deacetylases are poorly understood. CDA from different sources can modify their substrates in different ways: Some being specific for a single position [28], others show showing multiple-attack [29,30]. In addition, COS with specific deacetylation patterns can be produced by enzymatic deacetylation of chitin oligomers, but the diversity is limited by the available CDA.

Two genes encoding chitin deacetylases (CDA$_1$ and CDA$_2$) have been identified in *Saccharomyces cerevisiae* in previous reports. And these genes have been proved to be involved in the formation of the ascospores wall of *Saccharomyces cerevisiae* [31]. However, it is very interesting that the deletion of each gene will result in activity decrease of CDA, and the functions of the two genes cannot be replaced by each other [31]. Therefore, the deacetylation mechanisms of these two different chitin deacetylases may be different. However, detailed deacetylation mechanisms of chitin deacetylase from *Saccharomyces cerevisiae* have not been reported so far.

In this study, the chitin deacetylase (CDA$_2$) from *Saccharomyces cerevisiae* (ScCDA$_2$) with a specific mode of action has been characterized and a fast, convenient and online monitoring method has been developed that can be used to separate and detect partially acetylated chitosan oligosaccharides. Mass spectrometry sequencing showed that ScCDA$_2$ can hydrolyze N-acetamido groups rather than the reducing ends of chitin oligosaccharides, producing fully defined chitosan oligosaccharides by a multiple attack mode of action. Furthermore, ScCDA$_2$ is able to remove about 8% and 20% of the acetyl groups from crystalline chitin and colloidal chitin.

2. Results and Discussion

2.1. Bioinformatic Analysis and Expression of ScCDA$_2$

CDA belongs to the carbohydrate esterase family 4 (CE4) according to the classification of the CAZY database (www.cazy.org) [32]. The presence of divalent metal ions, such as Zn^{2+}, Ca^{2+} and Co^{2+}, have been proved to increase the catalytic activity and stability of the CDAs [30]. The *Colletotrichum lindemuthianum*'s CDA crystal structure indicates that there is a zinc-binding triad (His-His-Asp) around Zn^{2+} [33].

The sequence of ScCDA$_2$ aligned with deacetylase sequences from marine *Arthrobacter* (ArCE4A, 34%) [34], *Streptomyces lividans* (SlCE4, 33%) [35] and *Streptococcus pneumoniae* (SpPgdA, 29%) [36]

(Figure 1) [37]. The structure-based sequence alignments of *Ar*CE4A, *Sl*CE4 and *Sp*PgdA showed different levels of sequence identities according to their source from different genera and enabled identification the key residues that may contribute to catalysis function, including active site residues (Asp102, His250) and zinc-binding residues (Asp103, His149, His153) (Figure 1). Asp103, His149 and His153 form a zinc-binding triplet (His-His-Asp) around Zn^{2+}, which is similar to chitin deacetylase from *Colletotrichum lindemuthianum* [33], although the CDA sequence from *C. lindemuthianum* only has a 30% similarity to *Sc*CDA$_2$. The full-length open reading frame encoding the N-acetylglucosamine deacetylase sequence from *Saccharomyces cerevisiae* was successfully cloned and transformed into *Pichia pastoris* X-33 for highly efficient secretion expression (Figure 2). The molecular weight of *Sc*CDA$_2$, which was digested by N-glycosidase F (PNGase F), decreased by about 10 kDa. PNGase F is an amidase working by cleaving between the innermost GlcNAc and asparagine residues of high mannose, hybrid, and complex oligosaccharides from N-linked glycoproteins and glycopeptides. This results in a deaminated protein or peptide and a free glycan [38,39]. Therefore, there are N-glycosylation post-translational modifications in *Sc*CDA$_2$. Glycosylation is one of the most common post-translational modifications of proteins in fungi. It plays an important role in protein activity, thermal stability, proteolytic resistance, folding and secretion [40]. Mass spectrometry showed that *Sc*CDA$_2$ have N-glycosylation post-translational modification at positions Asn 181, Asn 199 and Asn 203 (Figure S5).

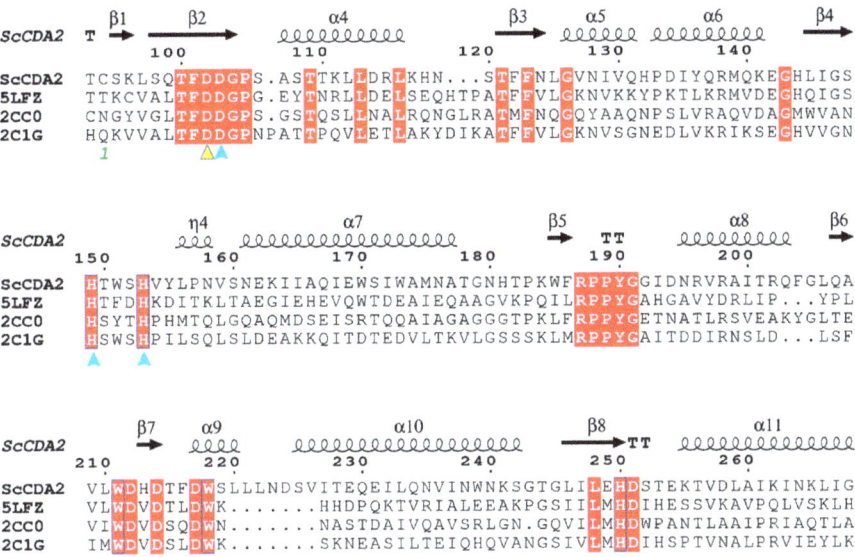

Figure 1. Structure-based on sequence alignments between four chitin deacetylases (CDAs). The sequence of chitin deacetylase from *Saccharomyces cerevisiae* (*Sc*CDA$_2$) was aligned with *Ar*CE4A sequences from a marine *Arthrobacter* species (PDB ID: 5LFZ), the *Sl*CE4 sequence from *Streptomyces lividans* (PDB ID: 2CC0) and the *Sp*PgdA sequence from *Streptococcus pneumoniae* (PDB ID: 2C1G). The conserved motifs are highlighted by a red background and the catalytic amino acids are marked with a yellow triangle. Amino acids capable of forming coordinate bonds with Zn^{2+} are marked with blue triangles. The symbol above the sequence represents the secondary structure, helices represent α-helices, and the arrow represents the beta fold.

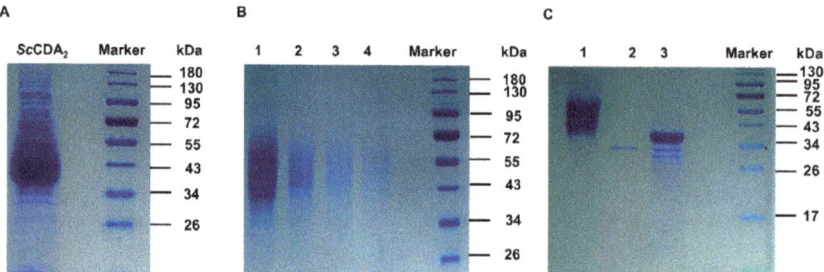

Figure 2. Analysis of molecular weight of ScCDA$_2$ protein by 12% SDS-PAGE. (**A**) ScCDA$_2$ crude enzyme; (**B**) purified ScCDA$_2$10 μM, 5 μM, 2.5 μM and 1.0 μM, marked as lanes 1, 2, 3 and 4, respectively; (**C**) PNGase F digestion confirmed that the enzyme is glycoprotein. Lane 1, ScCDA$_2$ before being digested by PNGase F; lane 2, PNGase F; lane 3, ScCDA$_2$ has been digested by PNGase F.

2.2. Homology Modeling and Substrate Binding Specificity of ScCDA$_2$

The crystal structures of several CDAs have already been determined, while CDA/substrate complex structure determination is less well defined and the interaction between the enzyme and the substrate is poorly understood. To study the characteristics of ScCDA$_2$ and chitin molecule interactions, we performed molecular docking to understand the binding mechanism of ScCDA$_2$. Homology modelling of ScCDA$_2$ (Figure 3A) revealed that the secondary structure consists of a conserved $(\alpha/\beta)_8$ folded barrel structure and six loops. The model was further evaluated for protein geometry by SAVES. Evaluation report shows that 97.3% residues in additional allowed regions and 85.57% of the residues have averaged 3D-1D score $\geq$0.2, and the quality factor is 91.2214, indicating that the structure quality was acceptable (Figure S1). The docking results (Figure 3B) show that chitin lies in the substrate-binding pocket which is surrounded by six loops, His250, Asp102, Asp103, His149 and His153. Asp103, His149 and His153 form a coordinate bond with Zn^{2+}, and the metal ion serves as a Lewis acid to assist the water affinity attack on the carbon atom on the amide bond. The adjacent His250 and Asp102 play a catalytic role through protonation, and the common action of these amino acids leads to the cleaving of the acetyl group. In addition, the structural superposition of ArCE4A (PDB ID: 5LFZ), SlCE4 (PDB ID: 2CC0), SpPgdA (PDB ID: 2C1G) and model of ScCDA$_2$ reveal that there are six conserved loop domains in ScCDA$_2$ (Figure 3C).

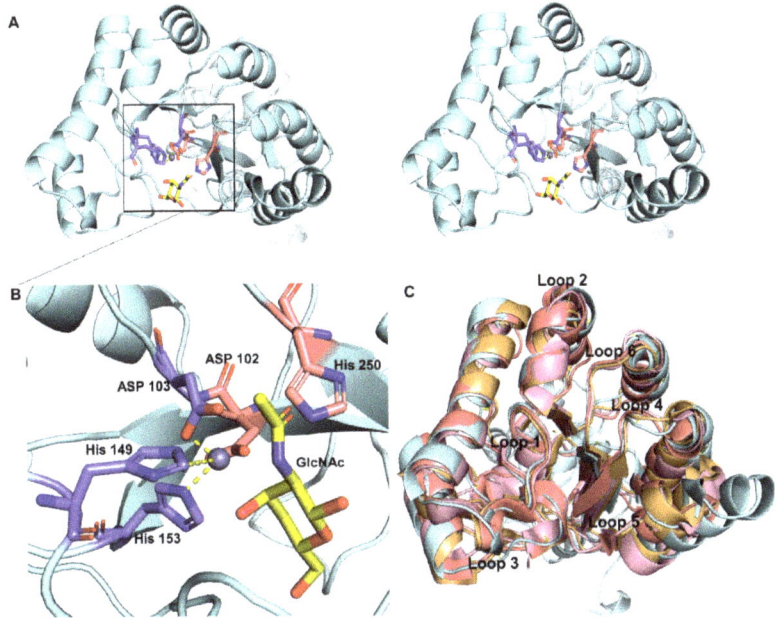

Figure 3. Catalytic binding mode resulting from homologous modelling and molecular docking. (**A**) The stereo view of the overall structure of ScCDA$_2$ (**B**) Highlights the binding pocket of ScCDA$_2$ docked with GlcNAc. The pink sticks represent a catalytic amino acid, and the blue sticks represent the amino acid that forms a coordinate bond with Zn^{2+}. The substrate GlcNAc is represented by a yellow stick. (**C**) Conservative loops were found through multiple structure superposition. The model of ScCDA2 was superposed with an ArCE4A structure from a marine *Arthrobacter* species (PDB ID: 5LFZ), a CE4 carbohydrate esterase structure from *Streptomyces lividans* (PDB ID: 2CC0) and an SpPgdA structure from *Streptococcus pneumoniae* (PDB ID: 2C1G). The conservative loops also have been marked.

2.3. Biochemical Characterization of ScCDA$_2$

The investigation of substrate specificity could provide important information for the potential applications of deacetylase. Using a coupled enzyme assay measure the amount of acetate released has been reported to be successfully applied to quantitatively determine the deacetylation activity of a recombinant chitin deacetylase [14]. When determining activity and substrate specific, interestedly, ScCDA$_2$ was observed that it is able to remove about 8% and 20% of the acetyl groups from crystalline chitin, alpha-chitin and beta-chitin (Figure 4). In addition, A$_n$ (A = GlcNAc; n = 1, 2, 3, 4, 5 or 6) as substrates also have been measured (Figure S3). To promote the application of ScCDA$_2$ in industry, more detailed physical and chemical properties characterization of CDA is essential. The optimal PH and metal ions of ScCDA$_2$ are pH = 8.0 and 50 °C when A4 was used as a substrate (Figure S4). When Co^{2+} is present, ScCDA$_2$ exhibits the maximum activity on A4. Despite the existence of a conserved zinc-binding triad in the ScCDA$_2$, biochemical data (Figure S4C) and structure-based on sequence alignments (Figure 1) indicate that ScCDA$_2$ as a metal-dependent metalloenzyme with a Co^{2+} dependence greater than Zn^{2+}. The peptidoglycan deacetylase from *Streptococcus pneumoniae* also shows that the peptidoglycan deacetylase is more metal-dependent on Co^{2+} than Zn^{2+}. Besides, the reported structures of two distinct acetylxylan esterases of CE4 from *Streptomyces lividans* and *Clostridium thermocellum*, in native and complex forms, show that the enzymes are sugar-specific and metal ion-dependent and possess a single metal (Zn^{2+}) center however with a chemical preference for Co^{2+} [35].

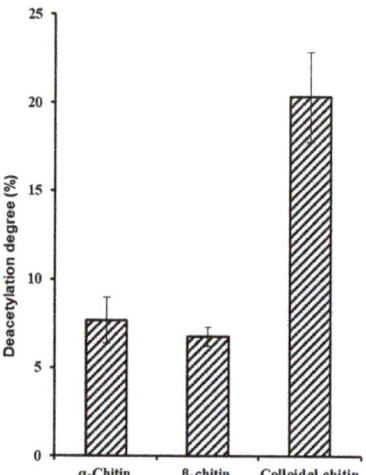

Figure 4. *Sc*CDA$_2$ substrate specificity on chitin. *Sc*CDA$_2$ activity on colloidal chitin, alpha-chitin and beta-chitin. 0.5 mg/mL substrates were incubated with 0. 75 μM *Sc*CDA$_2$ at 37 °C for 30 min. The data represent the mean SD values of the results from three independent experiments.

Most of the reported CDAs show only minimal activity or no activity on chitin *in vitro*. For example, CDA from *Cyclobacterium marinum* has been reported to be able to convert acetylglucosamine to glucosamine only with the cooperation of chitinase [17]. However, *Sc*CDA$_2$ can release up to 20.33% of acetyl groups from colloid chitin, as well as 9.16% and 7.29% of acetyl groups from insoluble alpha-chitin and Betabeta-chitin (Figure 4). Previous reported CDAs have no activity or low activity on insoluble chitin, which may be due to poor accessibility of chitin substrates [41]. However, the charge distribution on the surface of *Sc*CDA$_2$ indicates that *Sc*CDA$_2$ has an excessive negative charge in the region that interacts with the longer substrate, which may lead to enhanced substrate accessibility of *Sc*CDA$_2$ to chitin (Figure S2).

2.4. Isolation and Identification of Partially Acetylated Chitooligosaccharides

Due to its special biological activity, partially acetylated chitosan oligosaccharides have attracted wide interest, and these potential activities are significantly correlated with the degree of polymerization and degree of acetylation of chitooligosaccharides [14,42]. However, the method of preparing and isolating high-purity chitooligosaccharides is time consuming and labor intensive, which severely limits the large-scale production of partially acetylated chitooligosaccharides [43]. Much research into the separation of chitosan oligosaccharides has so far limited to the separation and identification of chitosan oligosaccharides of different degrees of polymerization [44–46]. As far as we know, the method for isolation and identification of partially acetylated chitosan oligosaccharides with a degree of polymerization of four has not been reported.

We have separated and identified the partially acetylated chitosan oligosaccharides with a degree of polymerization of 4. Chitin oligomers were deacetylated with recombinant *Sc*CDA$_2$ to form partially acetylated chitosan oligosaccharides. Three different partially acetylated chitosan oligosaccharides (A1D3, A2D2, A1D3) were obtained. These partially acetylated chitosan oligosaccharides were separated and detected by HPLC-ESI-MS (Figure 5).

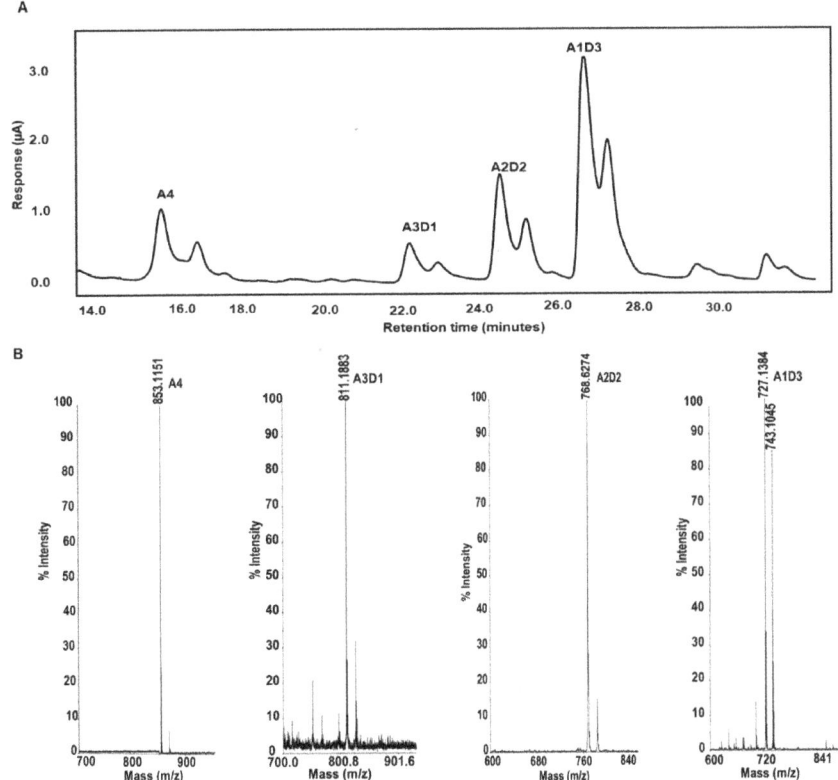

Figure 5. HPLC-ESI-MS analysis of chitin tetramer (A4) treated with ScCDA$_2$. (**A**) The target peak of the UHPLC-ESI chromatogram began to appear after 14 min, and the deacetylation peak was mainly concentrated between 20 and 26 min. (**B**) The m/z ratio in the MS spectrum corresponds to the mass of the substrate (A4; m/z 853.24), its mono-deacetylated products A3D1 (m/z 811.25), A2D2 (m/z 768.62) and A1D3 (m/z 727.13).

2.5. Partially Acetylated Chitooligosaccharides Production Processes

Exploring the partially acetylated chitooligosaccharides production process (simultaneously or in some order) is important to aid in understanding the action mode of CDA deacetylation. Therefore, the effects of enzyme concentration on the production process of partially acetylated COS have also been determined. As is shown in Figure 6, partially acetylated chitosan oligosaccharides (A1D3, A2D2, A1D3) are gradually produced according to the degree of deacetylation. With the amount of enzymes in the system increases, the types of enzyme reaction products gradually increase. From almost no product generation, to the production of the A3D1 and A2D2, the final substrate is completely consumed at the same time producing A1D3.

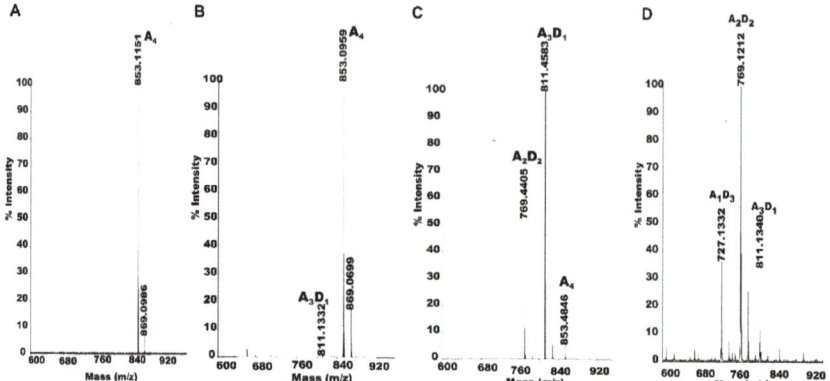

Figure 6. Partially acetylated chitooligosaccharides production processes. To explore the production processes of partially acetylated chitooligosaccharides 0.25 µM, 0.5 µM, 0.75 µM and 1 µm enzymes were incubated with A4 in 20 mM Tris-Cl buffer (pH 8.0) for 30 min. Then determined by MALDI-TOF-MS.

2.6. Specific Mode of Action of ScCDA$_2$ on A4

The partially acetylated chitooligosaccharide derivatized with a reducing amine showed molecular weights of 1005.96 Da (A3D1), 963.82 Da (A2D2) and 921.80 Da (A1D3) in MALDI-TOF mass spectrometry (Figure 7).

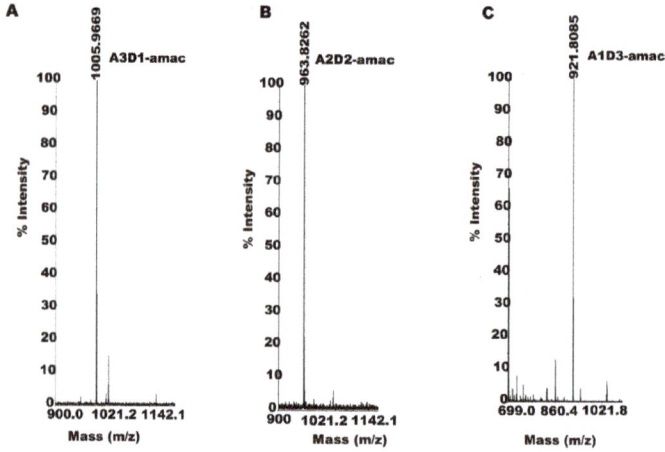

Figure 7. Analysis of partially acetylated COS by reductive amine derivatization with mass spectrometry. The reaction product generated after ScCDA$_2$ treatment (GlcNAc)$_4$ was labelled with AMAC and analyzed by MALDI-TOF-MS. (**A**). MS1 spectrum of A3D1 labelled with AMAC (m/z 1005.96). (**B**). MS1 spectrum of A2D2 labelled with AMAC m/z 963.82). (**C**). MS1 spectrum of A1D3 labelled with AMAC (m/z 921.80).

Then, by applying MALDI-TOF-MS analysis in MS2 mode, we were able to identify and analyze ScCDA$_2$'s partially acetylated products and determine specific acetylation pattern of partially acetylated chitooligosaccharides. As is shown in Figure 8, the A4 is first deacetylated to DAAA, ADAA and AADA (Figure 8A), followed by further deacetylation products to the intermediate product DDAA, DADA and ADDA (Figure 8B). Finally, the end product DDDA was obtained, due to the third deacetylation (Figure 8C). Therefore, deacetylation occurred mainly at the non-reducing end, and the acetyl at the reducing end was always present. No matter how we prolonged the reaction time or increased the concentration of the enzyme, the acetyl group at the reducing end could not be removed. After comparing the intermediate and end products generated by the deacetylation of the chitin tetramer, we found that the deacetylation occurred at any position except for the reducing end, indicating that *Sc*CDA$_2$ has a multiple attack mechanism like *Cl*CDA and *Sp*PgdA [33,36]. However, *Sc*CDA$_2$ cannot deacetylate at the reducing end to form completely deacetylated COS (DDDD). Therefore, the deacetylation pattern of *Sc*CDA$_2$ is significantly different from the reported CDA derived from *C. lindemuthianum, Mucor rouxii, Aspergillus nidulans, Vibrio cholera, Puccinia graminis*, etc [14,29,30,41,47]. A "subsite-capping model" has been proposed to explain the differentiation of the deacetylation process and product patterns of CDA [30]. This subsite-capping model states that the position and dynamics of loops play an important role in substrate preference and regioselectivity of deacetylation. So, the difference in the deacetylation mode of *Sc*CDA$_2$ may be due to the loop length, position and dynamic effects [47].

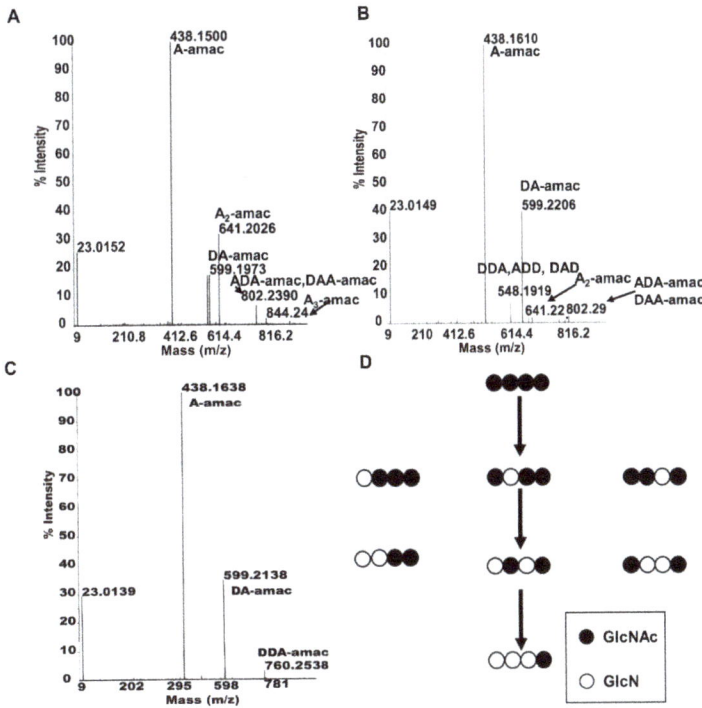

Figure 8. MALDI-TOF-MS2 determines the acetylation pattern of partially acetylated COS. (**A**) The MS$_2$ spectrum of A3D1 labelled with AMAC and the resulting ion fragments: A-amac, DA-amac, AA-amac, ADA-amac, DAA-amac, AAA-amac (*m/z* 438.15, 599.19, 641.20, 802.23, 844.24); so, the acetylation pattern of A3D1 is DAAA, ADAA and AADA. (**B**) MS$_2$ spectrum of A2D2 labelled with AMAC and the resulting ion fragments: A-amac, DDA, DAD, ADD, DA-amac, AA-amac and ADA-amac, (*m/z* 438.16, 548.19, 599.22, 641.22, 802.29); so, the acetylation pattern of A2D2 is DDAA, ADDA and DADA. (**C**) MS$_2$ spectrum of A1D3 labelled with AMAC, resulting in ion fragmentation of A-amac, DA-amac and DDA-amac (*m/z* 438.16, 599.21, 760.25); so, the acetylation pattern of A1D3 is DDDA. (**D**) The deacetylation process of *Sc*CDA$_2$ when A4 is used as a substrate.

3. Materials and Methods

3.1. Materials

Escherichia coli TOP10, plasmid pPICZαA, T4 DNA ligase and DNA polymerase were purchased from Takara Biotechnology (Dalian). *Pichia pastoris* X-33 was stored in our laboratory. Beta-chitin was purchased from Sigma (St. Louis, MO, USA). Alpha-chitin was purchased from Seikagaku (Tokyo, Japan) [33]. Colloidal chitin was prepared according to the previously reported method [48]. AMAC, chitin oligosaccharides (GlcNAc) n (n = 2–6, dimers to hexamers (A_2 to A_6; A, GlcNAc)) were purchased from Sigma-Aldrich (Munich, Germany). Unless otherwise noted, all reagents were analytical grade. Acetate release was measured using an acetate kit from R-Biopharm (Darmstadt, Germany) [14].

3.2. Cloning, Expression and Purification of ScCDA₂

The cda_2 gene from *Saccharomyces cerevisiae* S288c (GenBank: NP_013411.1) was amplified with upstream primer ScCDA₂-F:(5′-CATGCCATGGGAAGCTAATAGGGAAGATTTA-3′) and downstream primer: ScCDA₂-R (5′-CCGCTCGAGGGACAAGAATTCTTTTATGTAATC-3′). The target gene was digested and then ligated into a pPICZαA expression vector containing the N-terminal hexa histidine fusion tag coding region.

The cda_2 gene was recombined into a *Pichia pastoris* X-33 chromosome. Then the recombinant *Pichia pastoris* X-33 was induced by 0.5% methanol for 4 days, and methanol was added in batches every 24 h. The culture supernatant was collected by centrifugation at 8000× g, 4 °C for 20 min. The crude enzyme from the supernatant was concentrated using a 10 kDa ultrafiltration membrane. Then, the concentrated supernatant was purified by Ni-NTA Sepharose excel column (GE Healthcare). The pre-equilibrated buffer was subjected to Ni-NTA with buffer containing 20 mM PBS, pH 7.4, 300 mM NaCl and then washed with 50 mM PBS, pH 7.4, 300 mM NaCl, 20 mM imidazole. Finally, the target protein, eluted with 20 mM PBS, pH 7.4, 300 mM NaCl, 250 mM imidazole was obtained. Protein concentration was determined by using the Pierce™ BCA Protein Assay Kit (Thermo Fisher Scientific).

3.3. ScCDA₂ Activity Assay and Biochemistry Properties

The purified ScCDA₂ was studied to determine its enzymatic properties and deacetylation patterns. The colloidal chitin (water-soluble chitin), colloidal chitin and chitin oligomers dimers to hexamers (A_2 to A_6) were used as substrates [14]. The reaction mixture for ScCDA₂ enzyme activity assay containing 20 mM Tris-HCl buffer (pH8.0), including 1 mM $CoCl_2$, 0.5 mg/mL substrate and 0.75 µM purified soluble protein (ScCDA₂) or distilled water as a control was incubated at 37 °C for 30 min. The reaction was terminated by the addition of 10 µL 5% formic acid [14]. Determination of CDA activity by measuring the amount of acetate released by a coupled enzyme assay using an acetate assay kit [14]. The total reaction volume of the coupled enzyme reaction was 266 µL, which was measured spectrophotometrically at 340 nm [14].

In order to determine the optimum pH of ScCDA₂, protein in different buffers (final concentration 20 mM) was incubated at 37 °C for 30 min at the pH range of 4.0 to 10.0, in either sodium citrate disodium hydrogen phosphate buffer (pH 3.0–5.0), sodium phosphate dibasic sodium dihydrogen phosphate buffer (pH 6.0–7.0), Tris-HCl buffer (pH 8.0) or sodium carbonate sodium bicarbonate buffer (pH 9.0–10.0). The optimum temperature was determined in the 20 mM Tris-HCl buffer, at the optimum pH of 8.0, and each protein solution was incubated at 37 °C, 50 °C and 65 °C for 60 min. Subsequently, the remaining enzyme activity was measured using standard activity assays. The effects of different metal ions on enzyme activity were verified by adding 1 mM of different metal ion solutions (NaCl, NH_4Cl, $BaCl_2$, $CoCl_2$, $MnCl_2$, $ZnCl_2$, $CuCl_2$, $MgCl_2$ and $FeCl_3$) to the reaction mixtures [17].

3.4. Identification of ScCDA₂ Products by MALDI-TOF-MS

To determine the effect of different enzyme concentrations on the enzyme reaction products, four concentration gradients (0.25 µM, 0.5 µM, 0.75 µM, 1 µM) were prepared under 20 mM pH = 8.0 Tris-HCl. MS spectra were obtained using an Ultraflex™ ToF/ToF mass spectrometer (Bruker Daltonik GmbH, Bremen, Germany) to analyze the degree of acetylation, as previously described [49].

3.5. Preparation of Partially Acetylated COS

To analysis the deacetylation pattern of ScCDA$_2$, 20 mM Tris-HCl buffer (pH8.0), including 1 mM CoCl$_2$, 0.5 mg/mL substrate and 0.75 µM purified soluble protein (ScCDA$_2$) was incubated at 37 °C for 30 min. Then, 50 µL of the sample was injected into an X-Amide (4.6 mm × 250 mm) column for separation. The column was eluted with 0.3% formic acid and 50 mM ammonium formate buffer at a flow rate of 2 mL/min. The separated sample was analyzed by electrospray ionization mass spectrometry (ESI-MS).

3.6. Acetylation Pattern Analysis of COS

Reductive amine derivatisation of partially acetylated COS was performed as previously described [50]. In brief, 0.5 mg of the partially acetylated product was dissolved in 10 µL of 0.1 mol/L solution of 2-aminoacridone (AMAC) in acetic acid/DMSO (v/v, 3/17) and stirred manually for 30 s; then 10 µL of 1 M sodium cyanoborohydride solution was added and stirred for a further 30 s. The mixture was heated at 90 °C for 30 min, cooled to −20 °C and then completely freeze-dried. The dried sample was dissolved in 200 µL of methanol/water (v/v, 50/50) solution and sufficiently centrifuged at 12,000× g, 4 °C for 10 min. Then the supernatant was immediately analyzed by mass spectrometry or stored at −20 °C for one month. The method of mass spectrometry to detect the results of reductive amine derivatization was the same as the method of mass spectrometry detection of the enzyme reaction product mentioned previously [51]. MS^2 spectra were used to analyze the acetylation pattern of COS [52].

3.7. Homology Modelling and Molecular Docking

YASARA software (version 14.12.2) was used to build the homology model of ScCDA$_2$ with three crystal structures (PDB ID: 5LFZ, 2CC0 and 2C1G) as templates, the similarity between ScCDA$_2$ and templates is 34%, 33%, 29%, respectively [53], which are highly homologous to ScCDA$_2$, based on BLAST results using. The 3D structural model was visualized using VMD software (version 1.9.3, University of Illinois; Urbana–Champaign, IL, USA) [54]. The model was further evaluated for protein geometry by SAVES (http://services.mbi.ucla.edu/SAVES/), PROCHECK, ERRAT and VERIFY3D [55]. The chitin molecule structure was acquired from the zinc database (http://zinc.docking.org/). Molecular docking was performed using LeDock software (http://www.lephar.com/) with default parameters [56]. The dimensions of the binding box were set as 10 Å around the active site. The docking center was set at the Zn^{2+}. The number of binding poses of the ligand was 100. Finally, the docking pose that fulfilled the catalytic criteria was chosen as the initial conformation for analysis.

4. Conclusions

In this study, we firstly report the detailed deacetylation patterns of chitin deacetylase from *Saccharomyces cerevisiae* (ScCDA$_2$). Fully defined chitooligosaccharides (DAAA, ADAA, AADA, DDAA, DADA, ADDA and DDDA) have been produced by ScCDA$_2$ through multiple attack catalytic mechanisms. In addition, a fast, convenient and online monitoring method has been developed that can be used to separate and detect partially acetylated chitosan oligosaccharides. Enzymatic production of fully defined chitooligosaccharides and on-line monitoring and separation chitooligosaccharides, which solves the time-consuming and labor-intensive problem of isolating high purity chitooligosaccharides. This bio-enzymatic application could avoid the use of irritating chemicals and allows the production of functional chitosan and COS from crustacean waste chitin.

Supplementary Materials: The following are available online at http://www.mdpi.com/1660-3397/17/2/74/s1, **Figure S1:** The model was further evaluated for protein geometry by SAVES (A comprehensive measurement website for the quality of a protein structure). 97.3% Residues in additional allowed regions and 85.57% of the residues have averaged 3D-1D score $\geq$ 0.2, and the quality factor is 91.2214. **Figure S2:** Compare deacetylase charge distribution. These pictures show the surface charge distributions of chitin deacetylase from *Saccharomyces cerevisiae* (*Sc*CDA$_2$) and chitin deacetylase from *Aspergillus Nidulans* (*An*CDA, PDB ID: 2Y8U) calculated using ABPS (The Adaptive Poisson-Boltzmann Solver to generate electrostatic surface displayed) in VMD. Red represents a negative charge and blue represents a positive charge. **Figure S3:** *Sc*CDA$_2$ substrate specificity on chitin oligomers. 0.5 mg/mL chitin oligomers as substrates were incubated with 0. 75 μM *Sc*CDA$_2$ at 37 °C for 30 min. The data represents the mean SD values of the results from three independent experiments. **Figure S4:** Effects of pH, temperature and metal ion on enzyme activity. (**A**) The optimum pH was determined by incubating the 0.75 μM *Sc*CDA$_2$ with A4 chitin oligomer (0.5 mg/mL) for 60 min at pH 3–11 in universal buffer. (**B**) To obtain the optimal temperature, the enzyme (075 μmol) was incubated for 60 min at different temperatures in 50 mM Tris-HCl buffer (pH 8.0) containing chitin oligomer mixture (0.5 mg/mL) as a substrate. (**C**) Relative activity with different metal cations. Proteins were incubated with 1 mm metallized cations, and activity was determined in 50 mM Tris-HCl buffer (pH 8.0) using 0.5 mg/mL A4 as a substrate. **Figure S5:** Spectra of N-glycosylation of *Sc*CDA$_2$. Mass spectrometry showed that *Sc*CDA$_2$ have N-glycosylation post-translational modification at Asn 181, Asn 199 and Asn 203.

Author Contributions: General concept and design of studies: H.Y. Experimental concept and data analysis: X.-Y.Z., Y.Z., Experimental conduct and manuscript writing: X.-Y.Z. Manuscript review: H.Y., H.-D.Z., W.-X.W. and H.-H.C. Page: 12. Manuscript finalization: H.Y.

Funding: This research was supported by National Natural Science Foundation of China: (31670803, 31770847), National Key Research and Development Project of China: (2017YFD0200902), and H.Y was supported by Dalian city Innovative Support Program for High-Level Talents (2015R010), CAS Youth Innovation Promotion Association (2015144).

Acknowledgments: The authors would like to thank L.H.W. for the MALDI-TOF-MS analyses.

Conflicts of Interest: The authors declare no conflict of interest.

References

1. Kaur, S.; Dhillon, G.S. Recent trends in biological extraction of chitin from marine shell wastes: A review. *Crit. Rev. Biotechnol.* **2015**, *35*, 44–61. [CrossRef]
2. Yan, N.; Chen, X. Sustainability: Don't waste seafood waste. *Nature* **2015**, *524*, 155–157. [CrossRef] [PubMed]
3. Hamed, I.; Ozogul, F.; Regenstein, J.M. Industrial applications of crustacean by-products (chitin, chitosan, and chitooligosaccharides): A review. *Trends Food Sci. Technol.* **2016**, *48*, 40–50. [CrossRef]
4. Revathi, M.; Saravanan, R.; Shanmugam, A. Production and characterization of chitinase from Vibrio species, a head waste of shrimp Metapenaeus dobsonii (Miers, 1878) and chitin of Sepiella inermis Orbigny, 1848. *Adv. Biosci. Biotechnol.* **2012**, *3*, 392–397. [CrossRef]
5. Kumar, M.N.V.R. A review of chitin and chitosan applications. *React. Funct. Polym.* **2000**, *46*, 1–27. [CrossRef]
6. Chambon, R.; Despras, G.; Brossay, A.; Vauzeilles, B.; Urban, D.; Beau, J.M.; Armand, S.; Cottaz, S.; Fort, S. Efficient chemoenzymatic synthesis of lipo-chitin oligosaccharides as plant growth promoters. *Green Chem.* **2015**, *17*, 3923–3930. [CrossRef]
7. Zhang, X.D.; Xiao, G.; Wang, Y.Q.; Zhao, Y.; Su, H.J.; Tan, T.W. Preparation of chitosan-TiO2 composite film with efficient antimicrobial activities under visible light for food packaging applications. *Carbohydr. Polym.* **2017**, *169*, 101–107. [CrossRef] [PubMed]
8. Zhang, M.; Tan, T.W. Insecticidal and fungicidal activities of chitosan and oligo-chitosan. *J. Bioact. Compat. Polym.* **2003**, *18*, 391–400. [CrossRef]
9. Zhang, H.; Li, P.; Wang, Z.; Cui, W.W.; Zhang, Y.; Zhang, Y.; Zheng, S.; Zhang, Y. Sustainable Disposal of Cr(VI): Adsorption-Reduction Strategy for Treating Textile Wastewaters with Amino-Functionalized Boehmite Hazardous Solid Wastes. *ACS Sustain. Chem. Eng.* **2018**, *6*, 6811–6819. [CrossRef]
10. Yu, P.; Wang, H.-Q.; Bao, R.-Y.; Liu, Z.; Yang, W.; Xie, B.-H.; Yang, M.-B. Self-Assembled Sponge-like Chitosan/Reduced Graphene Oxide/Montmorillonite Composite Hydrogels without Cross-Linking of Chitosan for Effective Cr(VI) Sorption. *ACS Sustain. Chem. Eng.* **2017**, *5*, 1557–1566. [CrossRef]

11. Xu, G.; Liu, P.; Pranantyo, D.; Neoh, K.-G.; Kang, E.T. Dextran- and Chitosan-Based Antifouling, Antimicrobial Adhesion, and Self-Polishing Multilayer Coatings from pH-Responsive Linkages-Enabled Layer-by-Layer Assembly. *ACS Sustain. Chem. Eng.* **2018**, *6*, 3916–3926. [CrossRef]
12. Duri, S.; Harkins, A.L.; Frazier, A.J.; Tran, C.D. Composites Containing Fullerenes and Polysaccharides: Green and Facile Synthesis, Biocompatibility, and Antimicrobial Activity. *ACS Sustain. Chem. Eng.* **2017**, *5*, 5408–5417. [CrossRef]
13. Fang, Y.; Zhang, R.; Duan, B.; Liu, M.; Lu, A.; Zhang, L. Recyclable Universal Solvents for Chitin to Chitosan with Various Degrees of Acetylation and Construction of Robust Hydrogels. *ACS Sustain. Chem. Eng.* **2017**, *5*, 2725–2733. [CrossRef]
14. Naqvi, S.; Cord-Landwehr, S.; Singh, R.; Bernard, F.; Kolkenbrock, S.; El Gueddari, N.E.; Moerschbacher, B.M. A Recombinant Fungal Chitin Deacetylase Produces Fully Defined Chitosan Oligomers with Novel Patterns of Acetylation. *Appl. Environ. Microbiol.* **2016**, *82*, 6645–6655. [CrossRef]
15. Vander, P.; Km, V.R.; Domard, A.; Eddine, E.G.N.; Moerschbacher, B.M. Comparison of the ability of partially N-acetylated chitosans and chitooligosaccharides to elicit resistance reactions in wheat leaves. *Plant Physiol.* **1998**, *118*, 1353. [CrossRef]
16. Liu, T.; Liu, Z.; Song, C.; Hu, Y.; Han, Z.; She, J.; Fan, F.; Wang, J.; Jin, C.; Chang, J.; Zhou, J.-M.; Chai, J. Chitin-Induced Dimerization Activates a Plant Immune Receptor. *Science* **2012**, *336*, 1160–1164. [CrossRef]
17. Lv, Y.M.; Laborda, P.; Huang, K.; Cai, Z.P.; Wang, M.; Lu, A.M.; Doherty, C.; Liu, L.; Flitsch, S.L.; Voglmeir, J. Highly efficient and selective biocatalytic production of glucosamine from chitin. *Green Chem.* **2017**, *19*, 527–535. [CrossRef]
18. Hu, K.; Yang, H.; Liu, G.; Tan, H. Identification and characterization of a polysaccharide deacetylase gene from Bacillus thuringiensis. *Can. J. Microbiol.* **2006**, *52*, 935–941. [CrossRef]
19. Zhou, G.; Zhang, H.; He, Y.; He, L. Identification of a chitin deacetylase producing bacteria isolated from soil and its fermentation optimization. *Afr. J. Microbiol. Res.* **2010**, *4*, 2597–2603.
20. Pacheco, N.; Trombotto, S.; David, L.; Shirai, K. Activity of chitin deacetylase from Colletotrichum gloeosporioides on chitinous substrates. *Carbohydr. Polym.* **2013**, *96*, 227–232. [CrossRef]
21. Araki, Y.; Ito, E. Pathway of chitosan formation in mucor-rouxii—Enzymatic deacetylation of chitin. *Eur. J. Biochem.* **1975**, *55*, 71–78. [CrossRef] [PubMed]
22. Alfonso, C.; Nuero, O.M.; Santamaria, F.; Reyes, F. Purification of a heat-stable chitin deacetylase from aspergillus-nidulans and its role in cell-wall degradation. *Curr. Microbiol.* **1995**, *30*, 49–54. [CrossRef] [PubMed]
23. Martinou, A.; Koutsioulis, D.; Bouriotis, V. Expression, purification, and characterization of a cobalt-activated chitin deacetylase (Cda2p) from Saccharomyces cerevisiae. *Protein Expr. Purif.* **2002**, *24*, 111–116. [CrossRef] [PubMed]
24. Zhong, X.-W.; Wang, X.-H.; Tan, X.; Xia, Q.-Y.; Xiang, Z.-H.; Zhao, P. Identification and Molecular Characterization of a Chitin Deacetylase from Bombyx mori Peritrophic Membrane. *Int. J. Mol. Sci.* **2014**, *15*, 1946–1961. [CrossRef] [PubMed]
25. Wang, S.Q.; Jayaram, S.A.; Hemphala, J.; Senti, K.A.; Tsarouhas, V.; Jin, H.N.; Samakovlis, C. Septate-junction-dependent luminal deposition of chitin deacetylases restricts tube elongation in the Drosophila trachea. *Curr. Biol.* **2006**, *16*, 180–185. [CrossRef]
26. Brosson, D.; Kuhn, L.; Prensier, G.; Vivares, C.P.; Texier, C. The putative chitin deacetylase of Encephalitozoon cuniculi: A surface protein implicated in microsporidian spore-wall formation. *FEMS Microbiol. Lett.* **2005**, *247*, 81–90. [CrossRef]
27. Zhao, Y.; Park, R.-D.; Muzzarelli, R.A.A. Chitin Deacetylases: Properties and Applications. *Mar. Drugs* **2010**, *8*, 24–46. [CrossRef] [PubMed]
28. John, M.; Rohrig, H.; Schmidt, J.; Wieneke, U.; Schell, J. Rhizobium nodb protein involved in nodulation signal synthesis is a chitooligosaccharide deacetylase. *Proc. Natl. Acad. Sci. USA* **1993**, *90*, 625–629. [CrossRef] [PubMed]
29. Hekmat, O.; Tokuyasu, K.; Withers, S.G. Subsite structure of the endo-type chitin deacetylase from a Deuteromycete, Colletotrichum lindemuthianum: An investigation using steady-state kinetic analysis and MS. *Biochem. J.* **2003**, *374*, 369–380. [CrossRef]
30. Andres, E.; Albesa-Jove, D.; Biarnes, X.; Moerschbacher, B.M.; Guerin, M.E.; Planas, A. Structural basis of chitin oligosaccharide deacetylation. *Angew. Chem. Int. Ed. Engl.* **2014**, *53*, 6882–6887. [CrossRef] [PubMed]

31. Christodoulidou, A.; Bouriotis, V.; Thireos, G. Two sporulation-specific chitin deacetylase-encoding genes are required for the ascospore wall rigidity of Saccharomyces cerevisiae. *J. Biol. Chem.* **1996**, *271*, 31420–31425. [CrossRef] [PubMed]
32. Cantarel, B.L.; Coutinho, P.M.; Rancurel, C.; Bernard, T.; Lombard, V.; Henrissat, B. The Carbohydrate-Active EnZymes database (CAZy): An expert resource for Glycogenomics. *Nucleic Acids Res.* **2009**, *37*, D233–D238. [CrossRef] [PubMed]
33. Blair, D.E.; Hekmat, O.; Schuttelkopf, A.W.; Shrestha, B.; Tokuyasu, K.; Withers, S.G.; van Aalten, D.M.F. Structure and mechanism of chitin deacetylase from the fungal pathogen Colletotrichum lindemuthianum. *Biochemistry* **2006**, *45*, 9416–9426. [CrossRef]
34. Tuveng, T.R.; Rothweiler, U.; Udatha, G.; Vaaje-Kolstad, G.; Smalås, A.; Eijsink, V.G.H. Structure and function of a CE4 deacetylase isolated from a marine environment. *PLoS ONE* **2017**, *12*, e0187544. [CrossRef] [PubMed]
35. Taylor, E.J.; Gloster, T.M.; Turkenburg, J.P.; Vincent, F.; Brzozowski, A.M.; Dupont, C.; Shareck, F.; Centeno, M.S.J.; Prates, J.A.M.; Puchart, V.; et al. Structure and activity of two metal ion-dependent acetylxylan esterases involved in plant cell wall degradation reveals a close similarity to peptidoglycan deacetylases. *J. Biol. Chem.* **2006**, *281*, 10968–10975. [CrossRef] [PubMed]
36. Blair, D.E.; Schuttelkopf, A.W.; MacRae, J.I.; van Aalten, D.M.F. Structure and metal-dependent mechanism of peptidoglycan deacetylase, a streptococcal virulence factor. *Proc. Natl. Acad. Sci. USA* **2005**, *102*, 15429–15434. [CrossRef] [PubMed]
37. Schäffer, A.; Aravind, L.; Madden, T.; Shavirin, S.; Spouge, J.; Wolf, Y.; Koonin, E.; Altschul, S. Improving the accuracy of PSI-BLAST protein database searches with composition-based statistics and other refinements. *Nucleic Acids Res.* **2001**, *29*, 2994–3005. [CrossRef]
38. Tarentino, A.L.; Trimble, R.B.; Plummer, T.H., Jr. Enzymatic approaches for studying the structure, synthesis, and processing of glycoproteins. *Methods Cell Biol.* **1989**, *32*, 111–139. [PubMed]
39. Tarentino, A.L.; Plummer, T.H. Enzymatic deglycosylation of asparagine-linked glycans—Purification, properties, and specificity of oligosaccharide-cleaving enzymes from flavobacterium-meningosepticum. In *Guide to Techniques in Glycobiology*; Lennarz, W.J., Hart, G.W., Eds.; Elsevier Academic Press Inc.: San Diego, CA, USA, 1994; Volume 230, pp. 44–57.
40. Amore, A.; Knott, B.; Supekar, N.; Shajahan, A.; Azadi, P.; Zhao, P.; Wells, L.; Linger, J.; Hobdey, S.; Vander Wall, T.; et al. Distinct roles of N- and O-glycans in cellulase activity and stability. *Proc. Natl. Acad. Sci. USA* **2017**, *114*, 13667–13672. [CrossRef] [PubMed]
41. Liu, Z.; Gay, L.M.; Tuveng, T.R.; Agger, J.W.; Westereng, B.; Mathiesen, G.; Horn, S.J.; Vaaje-Kolstad, G.; van Aalten, D.M.F.; Eijsink, V.G.H. Structure and function of a broad-specificity chitin deacetylase from Aspergillus nidulans FGSC A4. *Sci. Rep.* **2017**, *7*, 1746. [CrossRef] [PubMed]
42. Li, K.C.; Xing, R.G.; Liu, S.; Qin, Y.K.; Li, P.C. Access to N-Acetylated Chitohexaose with Well-Defined Degrees of Acetylation. *Biomed. Res. Int.* **2017**, *2017*, 2486515. [CrossRef] [PubMed]
43. Wu, Y.X.; Lu, W.P.; Wang, J.N.; Gao, Y.H.; Guo, Y.C. Rapid and Convenient Separation of Chitooligosaccharides by Ion-Exchange Chromatography. In Proceedings of the 5th Annual International Conference on Material Science and Engineering, Xiamen, China, 20–22 October 2017; Aleksandrova, M., Szewczyk, R., Eds.; Iop Publishing Ltd.: Bristol, UK, 2018; Volume 275.
44. Cao, L.; Wu, J.; Li, X.; Zheng, L.; Wu, M.; Liu, P.; Huang, Q. Validated HPAEC-PAD Method for the Determination of Fully Deacetylated Chitooligosaccharides. *Int. J. Mol. Sci.* **2016**, *17*, 1699. [CrossRef] [PubMed]
45. Li, K.C.; Liu, S.; Xing, R.G.; Qin, Y.K.; Li, P.C. Preparation, characterization and antioxidant activity of two partially N-acetylated chitotrioses. *Carbohydr. Polym.* **2013**, *92*, 1730–1736. [CrossRef] [PubMed]
46. Li, K.C.; Xing, R.G.; Liu, S.; Li, R.F.; Qin, Y.K.; Meng, X.T.; Li, P.C. Separation of chito-oligomers with several degrees of polymerization and study of their antioxidant activity. *Carbohydr. Polym.* **2012**, *88*, 896–903. [CrossRef]
47. Aragunde, H.; Biarnes, X.; Planas, A. Substrate Recognition and Specificity of Chitin Deacetylases and Related Family 4 Carbohydrate Esterases. *Int. J. Mol. Sci.* **2018**, *19*, 30. [CrossRef] [PubMed]
48. Hirano, S.; Nagao, N. An Improved Method for the Preparation of Colloidal Chitin by Using Methanesulfonic-Acid. *Agric. Biol. Chem. Tokyo* **1988**, *52*, 2111–2112.

49. Wang, S.; Liu, C.; Liang, T. Fermented and enzymatic production of chitin/chitosan oligosaccharides by extracellular chitinases from Bacillus cereus TKU027. *Carbohydr. Polym.* **2012**, *90*, 1305–1313. [CrossRef]
50. Bahrke, S.; Einarsson, J.M.; Gislason, J.; Haebel, S.; Letzel, M.C.; Peterkatalinić, J.; Peter, M.G. Sequence analysis of chitooligosaccharides by matrix-assisted laser desorption ionization postsource decay mass spectrometry. *Biomacromolecules* **1900**, *3*, 696–704. [CrossRef]
51. Chen, M.; Zhu, X.; Li, Z.; Guo, X.; Ling, P. Application of matrix-assisted laser desorption/ionization time-of-flight mass spectrometry (MALDI-TOF-MS) in preparation of chitosan oligosaccharides (COS) with degree of polymerization (DP) 5–12 containing well-distributed acetyl groups. *Int. J. Mass Spectrom.* **2010**, *290*, 94–99. [CrossRef]
52. Lee, J.H.; Ha, Y.W.; Jeong, C.S.; Kim, Y.S.; Park, Y. Isolation and tandem mass fragmentations of an anti-inflammatory compound from Aralia elata. *Arch. Pharm. Res.* **2009**, *32*, 831–840. [CrossRef]
53. Krieger, E.; Koraimann, G.; Vriend, G. Increasing the precision of comparative models with YASARA NOVA—A self-parameterizing force field. *Proteins-Struct. Funct. Genet.* **2002**, *47*, 393–402. [CrossRef] [PubMed]
54. Humphrey, W.; Dalke, A.; Schulten, K. VMD: Visual molecular dynamics. *J. Mol. Graph. Model.* **1996**, *14*, 33–38. [CrossRef]
55. Eisenberg, D.; Luthy, R.; Bowie, J.U. VERIFY3D: Assessment of protein models with three-dimensional profiles. In *Macromolecular Crystallography, Pt B*; Carter, C.W., Sweet, R.M., Eds.; Academic Press: Cambridge, MA, USA, 1997; Volume 277, pp. 396–404.
56. Wang, Z.; Sun, H.; Yao, X.; Li, D.; Xu, L.; Li, Y.; Tian, S.; Hou, T. Comprehensive evaluation of ten docking programs on a diverse set of protein-ligand complexes: The prediction accuracy of sampling power and scoring power. *Phys. Chem. Chem. Phys.* **2016**, *18*, 12964–12975. [CrossRef] [PubMed]

© 2019 by the authors. Licensee MDPI, Basel, Switzerland. This article is an open access article distributed under the terms and conditions of the Creative Commons Attribution (CC BY) license (http://creativecommons.org/licenses/by/4.0/).

Article

Design and Synthesis of a Chitodisaccharide-Based Affinity Resin for Chitosanases Purification

Shangyong Li [1,*], Linna Wang [2], Xuehong Chen [1], Mi Sun [2] and Yantao Han [1,*]

[1] Department of Pharmacology, College of basic Medicine, Qingdao University, Qingdao 266071, China; chen-xuehong@163.com
[2] Key Laboratory of Sustainable Development of Marine Fisheries, Ministry of Agriculture, Yellow Sea Fisheries Research Institute, Chinese Academy of Fishery Sciences, Qingdao 266071, China; wlnwfllsy@163.com (L.W.); sunmi0532@yahoo.com (M.S.)
* Correspondence: lisy@qdu.edu.cn (S.L.); hanyt@qdu.edu.cn (Y.H.)

Received: 22 December 2018; Accepted: 15 January 2019; Published: 21 January 2019

Abstract: Chitooligosaccharides (CHOS) have gained increasing attention because of their important biological activities. Enhancing the efficiency of CHOS production essentially requires screening of novel chitosanase with unique characteristics. Therefore, a rapid and efficient one-step affinity purification procedure plays important roles in screening native chitosanases. In this study, we report the design and synthesis of affinity resin for efficient purification of native chitosanases without any tags, using chitodisaccharides (CHDS) as an affinity ligand, to couple with Sepharose 6B via a spacer, cyanuric chloride. Based on the CHDS-modified affinity resin, a one-step affinity purification method was developed and optimized, and then applied to purify three typical glycoside hydrolase (GH) families: 46, 75, and 80 chitosanase. The three purified chitosanases were homogeneous with purities of greater than 95% and bioactivity recovery of more than 40%. Moreover, we also developed a rapid and efficient affinity purification procedure, in which tag-free chitosanase could be directly purified from supernatant of bacterial culture. The purified chitosanases samples using such a procedure had apparent homogeneity, with more than 90% purity and 10–50% yield. The novel purification methods established in this work can be applied to purify native chitosanases in various scales, such as laboratory and industrial scales.

Keywords: chitosanases; adsorption analysis; affinity purification

1. Introduction

Chitosan, a natural cationic polysaccharide, has key roles in many biological processes, such as artificial skin, absorbable surgical sutures, and wound healing accelerators [1]. Chitosanase (EC 3.2.1.132) catalyzes the hydrolysis of β-1,4-linked glycosidic bond in the chitosan chain, releasing chitooligosaccharides (CHOS) as products [2]. In recent years, CHOS have increasingly gained more attention because of their important biological activities, such as anti-tumor [3,4], immuno-enhancing [5], anti-fungi [6], anti-bacterial [7], and anti-diabetic effect [8]. These activities are dependent on chemical structures and molecular sizes of the oligosaccharides [9].

It has been identified that chitosanases can be isolated from various organisms, including fungi, plants, and bacteria [10,11]. Based on the classification in Carbohydrate-Active enZYmes (CAZy) databases [12] (http://www.cazy.org), chitosanases are included into seven different glycoside hydrolase (GH) families: 3, 5, 7, 8, 46, 75, and 80. Currently, most known chitosanases belong to GH family 46, which comprise only chitosanases. Thus far, several chitosanases from different organisms have been identified [13–16]. According to the sequence alignments, chitosanases are classified into seven GH families: 3, 5, 7, 8, 46, 75, and 80 [11]. Among which, chitosanases, which belong to GH family 46, have been fully characterized, and several crystal structures of chitosanases have been determined (PDB codes: 1CHK, 1QGI, and 2D05). The catalytic mechanism of chitosanases from GH family 46 has been elucidated; it was identified to follow an "inverting" catalytic mechanism [12–14]. The 3-D crystal structure of chitosanase-substrate complex (CsnOU01) shows that the −2, −1, and +1 subsites of chitosanase from GH family 46 play a predominant role for the formation of hydrogen bond intermediate during substrate binding and catalysis [12,13]. Because chitosanases with special characteristics have potential applications in industry and biotechnology, a rapid and efficient purification method that can be used to purify chitosanases, from which their biochemical characteristics are determined, from different microorganisms is essential. According to the literature, the purification protocols of native chitosanases generally involve ultrafiltration, ammonium sulfate precipitation, salting out, hydrophobic interaction chromatography, ion-exchange chromatography, or gel filtration chromatography [13–16]. These traditional methods not only require a large number of steps, but are also time-consuming and difficult to scale up. Biomimetic affinity chromatography is potentially the most selective method in protein purification [17,18]. This technique requires lower number of steps, while resulting in higher yields, so can be considered economical. Therefore, it can be beneficial to develop a rapid and efficient one-step affinity purification protocol for chitosanase.

Affinity purification of enzymes often design affinity ligands that function as substrate analogues or specific inhibitors [16,18,19]. In this study, CHOS-based resin was synthesized by coupling chitosan-disaccharide (CHDS) to epoxy-activated Sepharose 6B using cyanuric chloride as a spacer. The resin was then used in the development of one-step affinity purification of chitosanases, in which three typical enzymes, GH families 46, 75, and 80, were purified. The developed purification protocol was highly efficient and resulted in high purity enzymes (more than 95% purity). The method was also applied to directly purify chitosanase from bacterial culture medium.

2. Results and Discussion

2.1. Design and Synthesis of CHOS-based Affinity Resin

Reversible inhibitor or substrate analogue are commonly used for the design of biomimetic affinity ligands of enzymes, due to the mild and efficient affinity values [17]. Chitooligosaccharides (CHOS) are natural cationic saccharides, while the catalytic domain of chitosanases is rich in acidic amino acids [12,13]; the acid-base interactions between the two molecules can provide affinity force during affinity purification. Immobilization of a ligand onto epoxy-activated resin can be achieved via nucleophilic groups (often is primary amine) presented in the ligand [14,19–22]. Because CHOS contains an amine group at the C-2 position of the sugar ring, we thus focused our efforts on the design of CHOS-based affinity resin for purification of chitosanase. The natural properties of affinity resins—in other words the selection of affinity ligands and spacers—have an important impact on the results of biomimetic affinity purification [19]. As reported previously, the final degradation products of the endo-type chitosanase are CHDS and chitotrisaccharide (CHTS), whereas that of the exo-type chitosanase is glucosamine [11]. To obtain affinity resin with optimal ligands, two types of CHOS-based ligands, CHDS-based and CHTS-based, were compared, and glucosamine was also chosen as a contrasted affinity ligand. The scheme for the synthesis protocol of CHDS-based Sepharose 6B is shown in Figure 1. The CHTS-based resin and glucosamine-based resin were synthesized from CHTS or glucosamine at the same concentration as that of CHDS (Figure 2A,B). The ninhydrin test was applied to examine the density of the free amino group (Table 1), and the linkage of cyanuric chloride to the amino groups. Purple color indicated the presence of free amino groups, and color disappearance indicated that cyanuric chloride had been linked to the amino groups. Through the change of purple color, almost all of the free amino groups linked to the cyanuric chloride. A ninhydrin test was also used to determine the coupling efficiencies and yields for CHDS, which showed an extra free amino group. About 16.8 μmol/ml free amino group was determined by ninhydrin test. The yields of the final affinity product were about 80.3%.

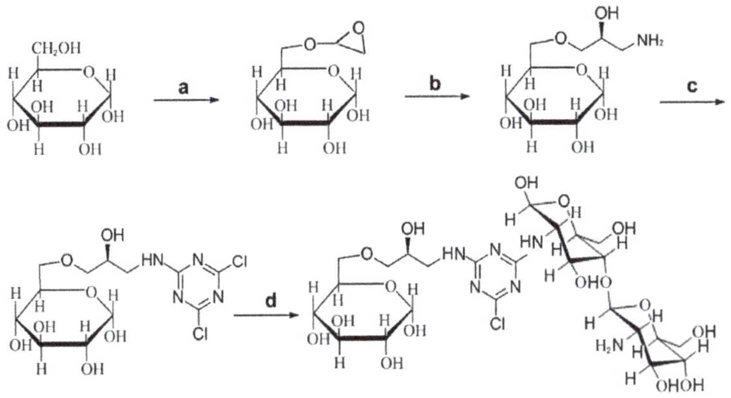

Figure 1. Synthesis protocol and scheme of the CHDS ligand coupled with active Sepharose 6B via cyanuric chloride spacer. Reagents and conditions: (**a**) epichlorohydrin, DMSO, NaOH aqueous solution, 2.5 h; (**b**) 35% saturated ammonia, overnight; (**c**) cyanuric chloride, 50% acetone, pH 7–8; (**d**), CHDS, sodium carbonate, 24 h.

Table 1. Ligand densities, desorption constant (K_d) and theoretical maximum absorption (Q_{max}) analysis of the affinity media.

Ligands	Spacer Arms	Ligand Density (μmol/mL)	K_d (μg/mL)	Q_{max} (mg/g)
Glucoamine	Cyanuric chloride	20.9	88.5	10.6
CHTS [a]	Cyanuric chloride	20.9	20.7	24.7
CHDS [b]	Cyanuric chloride	20.9	16.4	30.9
CHDS	5-atom spacer	41.8	24.2	24.4
CHDS	10-atom spacer	27.8	38.8	20.8

[a] CHTS represents chitosan trisaccharides; [b] CHDS represents chitosan disaccharides.

In this study, the typical GH family 46 chitosanase, CsnOU01, was chosen as the target protein in the determination of equilibrium adsorption of different affinity resins (Figure 3). The densities of free amino groups were determined by the ninhydrin test before the addition of the affinity ligand, giving equal ligand densities (Table 1). To find the optimal affinity ligand, control resins were synthesized from CHTS or glucosamine according to the method described above (Figure 2A,B). Equilibrium adsorption studies were performed to characterize the affinity value of CsnOU01 and these three affinity media (Figure 3A). The adsorption constant for CHDS-based resin was 16.4 μg/mL, which was notably lower than that for CHTS-based resin (20.7 μg/mL) and glucosamine-based resin (88.5 μg/mL). Additionally, the theoretical maximum absorption (Q_{max}) for the CHDS-based resin was significantly higher than that for other two types of resins (Table 1), indicating that the affinity of CHDS-based resin is high. Therefore, CHDS was chosen as the affinity ligand for further design and synthesis of affinity resins.

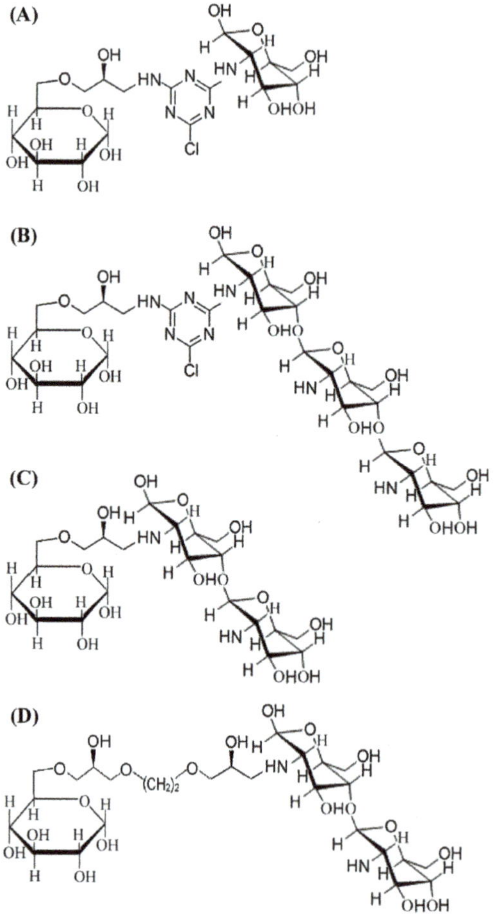

Figure 2. Schemes of four Sepharose 6B resins with different ligand and spacer. (**A**) Glucosamine ligand via cyanuric chloride spacer. (**B**) CHTS ligand via cyanuric chloride spacer. (**C**) CHDS ligand via 5-atom spacer arm. (**D**) CHDS ligand via 10-atom spacer arm.

To find the optimal spacer arm length, cyclic arm (cyanuric chloride) and linear arms (5-atom length and 10-atom length) were compared. Cyanuric chloride (2,4,6-trichloro-1,3,5-triazine) is a compound containing s-triazine (C_3N_3) ring, which can exert higher strength for ligand stabilization; it is widely used in the synthesis of affinity resin [23]. Figure 2C,D showed the corresponding scheme for the synthesis of resins, with spacers of 5-atom and 10-atom lengths, are shown in Figure 2C,D. According to the adsorption analysis (Figure 3B), CHDS ligand with cyclic spacer arm exhibited the highest desorption value (K_d, 16.4 µg/mL; Q_{max}, 30.9 mg/g), with an epoxy content (20.9 µmol/mL) lower than the content of 5-atom linear spacer and 10-atom linear spacer (Table 1). Therefore, cyanuric chloride was chosen as the optimal spacer arm. These observations indicate that in addition to CHDS ligand, the cyanuric chloride spacer arm is also important for the binding to chitosanase. Thus, in the larger scale of resin synthesis and chitosanases purification, CHDS was used as a ligand to couple with Sepharose 6B affinity resin through a spacer arm cyanuric chloride.

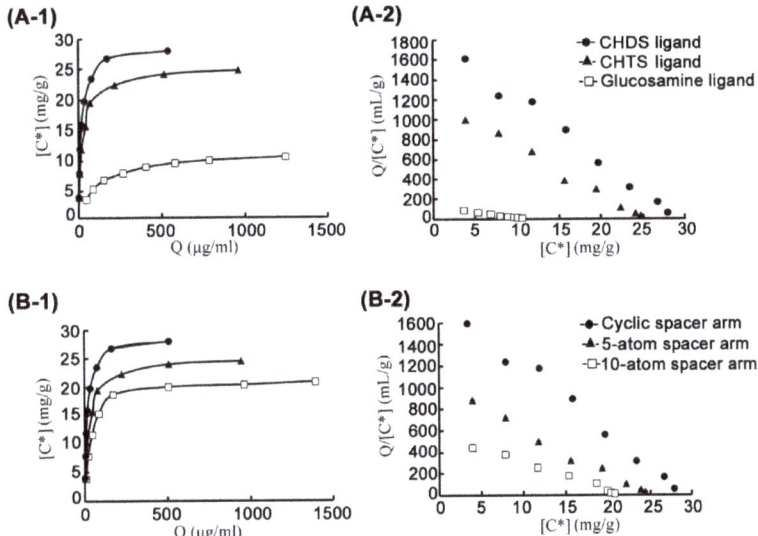

Figure 3. Adsorption analyses of GH family 46 enzyme CsnOU01. (**A**) Adsorption analysis of affinity resins with different ligands via cyanuric chloride as a spacer arm. (**B**) Adsorption analysis of affinity resins with CHDS as affinity ligand via different spacer arms. (**1**) Equilibrium adsorption of enzyme and affinity resin. (**2**) Plot describing the equilibrium of the absorption on the resin and the enzyme concentration in the liquid phase.

2.2. Affinity Purification of Chitosanases from Different GH Families

Three chitosanases (CsnOU01, Csn, and ChoA) from GH families 46, 75, and 80, respectively, were expressed in *E. coli* BL21(DE3) through pET-22b(+) system, with or without 6×His tag. After centrifugation at 12,000 rpm for 10 min, the supernatant containing strains without 6×His tag was loaded onto a 10 mL pre-equilibrated column and then washed with washing buffer (0.1 M Tris-HCl buffer, pH 8.0) until the eluate exhibited no detectable absorbance at 280 nm. Thereafter, the enzymes were purified by the established one-step purification using CHDS-based Sepharose 6B resin.

We tested different loading and elution conditions to optimize the yield of chitosanases (Supplementary Table S1). Chitosanases are stable at a pH range of 4.0–8.0, and Q_{max} values are usually determined at pH 8.0 with Tris-HCl buffer; therefore, 0.1 M Tris-HCl, pH 8.0 was used as the loading buffer. In previous elution process, non-target proteins were depleted by 0.1 M Tris-HCl, pH 8.0 containing 100 mM NaCl, in which the target protein was not eluted. In optimization of elution pH, elution buffers containing acetic acid buffers with different pH, ranging from 4.0–6.0, were used in the elution of chitosanases, and the results were compared (low pH buffers are known to favor the disruption of H-bond between chitosanases and affinity-based resin, especially for the substrate analogue-based resin). The highest protein yield was obtained at pH 5.4; 0.1 M acetic acid buffer, pH 5.4 was thus chosen as the elution buffer. In addition, the elution buffer was supplemented with 0.8 M NaCl to further deteriorate the affinity between the enzyme and the resin.

The established one-step purification method took a total time of as low as 10 min at the flow rate of 3 mL/min. With this simple and efficient affinity chromatography, CsnOU01 was purified with purities of ~98 and 98.7% according to SDS-PAGE (Figure 4(A-1)) and HPLC analysis (Figure 4(B-1)), respectively. The purification yielded CsnOU01 of 5.4 folds with the specific activity of 356.8 U/mg, and the molecular mass of the purified CsnOU01 was determined to be ~28 kDa, which was in good agreement with the theoretical molecular mass [24,25]. The bioactivity yield as a result of this affinity purification method was about 64.1% (Table 2). To determine whether or not the synthesized affinity resin has affinity for chitosanases from other GH families, Csn from GH family 75 [26] and ChoA from GH family 80 [13] were purified using the synthesized resin. As shown in Table 2, CHDS-based resin could efficiently purify the two chitosanases, with bioactivity recoveries of 45.2% and 40.8% for Csn and ChoA, respectively. The analysis by SDS-PAGE (Figure 4(A-2,A-3)) and HPLC using size-exclusion chromatography (Figure 4(B-2,B-3)) showed that both enzymes had purities of more than 95%.

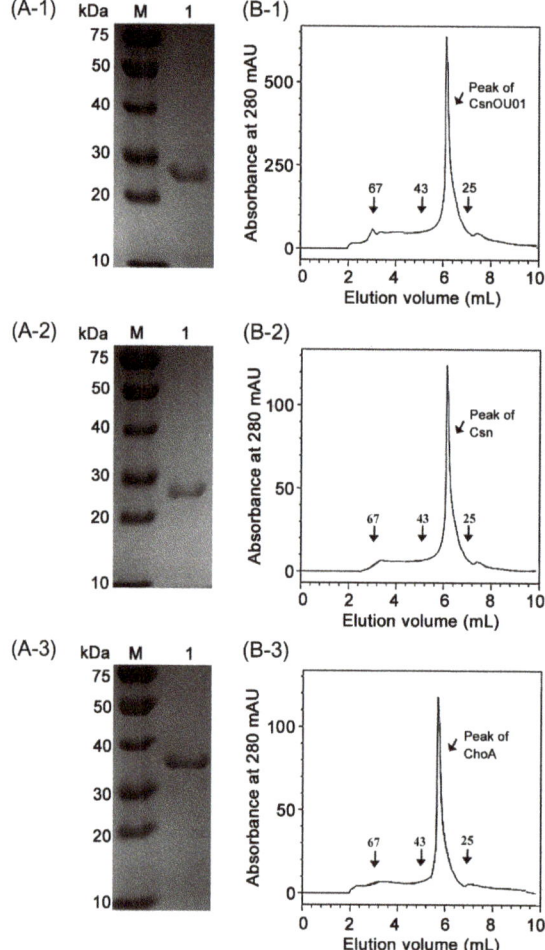

Figure 4. Purity analysis of purified chitosanases. (**A**) SDS-PAGE analysis of purified enzymes. *Lane M*, molecular mass standard protein marker. (**A-1**) the purified CsnOU01. (**A-2**) the purified Csn. (**A-3**) the purified ChoA. (**B**) HPLC analysis of the purified CsnOU01 (**1**), Csn (**2**), and ChoA (**3**) on a TSK 3000 SW column.

The traditional purification protocol towards CsnOU01, Csn, and ChoA was developed and shown in Supplementary Table S2. Here, we compared CHDS-based affinity protocol with the traditional methods, reporting in Table 2 all the different purification steps, activity yields, and specific activities of pure enzyme. The traditional protocol with multiple steps is expensive and leads to low recoveries. The specific activity of enzymes purified by the CHDS-based protocol and the traditional purification protocol is similar. However, the purity of CHDS-based affinity purification is higher than the traditional methods.

As a contrast, Ni-NTA Sepharose 6B resin was also used to purify the three recombinant chitosanases containing 6×His-tag, by immobilized metal affinity chromatography (IMAC). Even if the IMAC protocol led to an activity recovery higher than the CHDS-based affinity protocol, the specific activities are lower (Table 2). Because of all the obvious advantages of the CHDS-based affinity protocol, including one-step chromatography, no use of toxic imidazole, higher purity, and shorter times, this approach has the potential to be used for industrial applications of high purity chitosanase.

Table 2. Comparison of traditional, CHDS-based, and immobilized metal affinity chromatography (IMAC) affinity purification methods for three different chitosanases.

Enzymes	Purification Method	Activity Recovery (%)	Protein Purity (%)	Specific Activity (U/mg)
CsnOU01 (GH46)	CHDS-based protocol [a]	64.1	97.8	356.8
	Traditional protocol [c]	28.2	94.5	358.5
	IMAC protocol [b]	71.8	95.6	306.7
Csn (GH75)	CHDS-based protocol	45.2	96.3	664.6
	Traditional protocol [d]	10.5	93.2	682.7
	IMAC protocol	60.7	96.4	592.3
ChoA (GH80)	CHDS-based protocol	40.8	97.1	851.4
	Traditional protocol [e]	9.2	89.6	847.6
	IMAC protocol	63.4	90.3	727.8

[a] In the CHDS-based affinity purification protocol, enzymes without 6× his-tag were purified by CHDS-based medium; [b] in the Ni-NTA affinity purification protocol, enzymes with 6×his-tag were purified by Ni-NTA medium; [c] in the traditional protocol, CsnOU01 without 6×his-tag was purified by five steps, including ultrafiltration, ammonium sulfate precipitation, desalting, anion-exchange, and gel-filtration chromatography; [d] the traditional purification protocol of Csn was composed of three steps, including ammonium sulfate precipitation, desalting, and anion-exchange chromatography; [e] the traditional purification protocol of ChoA was composed of six steps, including ammonium sulfate precipitation, hydrophobic chromatography, desalting, anion-exchange chromatography, and two steps of gel-filtration chromatography.

2.3. Direct Purification of Chitosanases from Bacterial Culture Medium

In order to establish a rapid purification protocol for native chitosanase, nine bacterial strains with high chitosanase activity (more than 50 U/mL) were chosen as target strains in the affinity purification of chitosanase using the established CHDS-based materials. After bacterial culture was centrifuged at 10,000× g for 15 min, the supernatant was loaded onto 10 mL pre-equilibrated column and washed with washing buffer (0.1 M Tris-HCl buffer with 100 mM NaCl, pH 8.0) until the eluate showed no detectable absorbance at 280 nm. After that, the target protein was eluted with elution buffer (0.1 M acetic acid buffer, pH 5.6, 0.8 M NaCl). As shown in Figure 5, CHDS-based resin was able to purify chitosanases (with certain homogeneity) from culture medium.

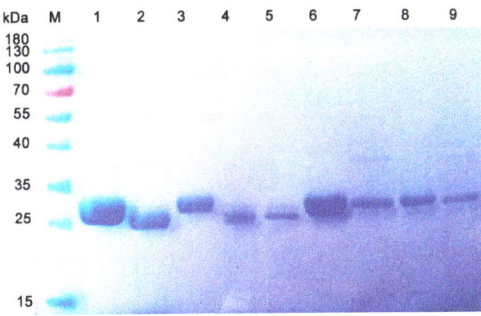

Figure 5. SDS-PAGE analysis of purified chitosanases direct from bacterial supernatant.

The activity recovery and purity of the purified enzymes were shown in Table 3. The purified chitosanase samples using such procedure had apparent homogeneity with more than 90% purity and 10–50% yield. Different strains showed different activity recoveries. This result may be caused by multi-factors. In this study, our affinity purification condition was only used for rapidly screening chitosanases. The optimal purification protocol toward the special enzyme needs further optimization. After characterizing these purified enzymes, the chitosanase from *Serratia* sp. QD07 showed high thermo-tolerant property and suitability for industrial usage (data not shown). The characterization data of this thermo-tolerant enzyme will be reported in the next paper.

Table 3. Affinity purification of chitosanases direct from bacterial supernatant.

Number	Bacterium	Activity Recovery (%)	Protein Purity (%)
1	*Bacillus* sp. QD08	49.2	95.4
2	*Bacillus* sp. QD102	40.6	96.2
3	*Bacillus* sp. QD72	41.1	98.1
4	*Paenibacillus* sp. QD03	10.5	92.2
5	*Mitsuaria* sp. QD129	10.7	90.5
6	*Mitsuaria* sp. QD130	39.5	96.7
7	*Renibacterium* sp. QD52	20.3	90.1
8	*Serratia* sp. QD07	12.7	97.8
9	*Flavobacterium* sp. QD28	11.6	91.9

Chitosanases that have special characteristics can potentially be applied in biotechnology industry and other fields. As has been described in the literature, the purification of native chitosanases (without His-tag) usually involves ultrafiltration, ammonium acetate precipitation, salting out, ion-exchange chromatography, and hydrophobic interaction chromatography [13–16]. These conventional methods usually involve a large number of steps, which are time-consuming and difficult to scale up. As far as we know, biomimetic affinity chromatography specially designed for native chitosanase has not been established. In this study, we developed the rapid and efficient affinity purification procedure, in which native chitosanase could be directly purified from supernatant of bacterial culture. The novel purification methods established in this work can be applied to screen and purify chitosanases, both in laboratory and industrial scales.

3. Materials and Methods

3.1. Materials

Chitosan, with degree of deacetylation (DDA) ≥ 95%, was purchased from Aladdin, China. To obtain chitodisaccharide (CHDS) and chitotrisaccharide (CHTS), chitosan (0.1%) was degraded by GH family 46 chitosanase CsnOU01 at a final enzyme concentration at 20 mg/mL for 6 h. CHDS and CHTS were purified from the degradation products of chitosan using a Biogel P-2 column (GE Healthcare, Madison, WI, USA). Briefly, 100 mg degradation product of CHDS and CHTS was loaded into the Biogel P-2 column (1.6 × 130 mm), using 0.2 M ammonium acid carbonate as mobile phase. The flow rate was 1 mL/min. Then, the effluent was collected every 1 min and the sugar content was determined by phenol sulfate method. Finally, oligosaccharides were collected and identified by TLC. Sigma-Aldrich (St. Louis, MO, USA) provided cyanuric chloride (2,4,6-trichloro-1,3,5-triazine) and glucosamine. Beijing Weishibohui Chromatography Technology Co., China, provided activated Sepharose 6B with two different spacer arm lengths (5-atom and 10-atom). Sinopharm Chemical Reagent (Shanghai, China) provided other analytical grade reagents.

3.2. Synthesis of Affinity Resins

CHDS-based affinity resins were synthesized according to our previous published method [20,21,27]. The synthesis scheme is shown in Figure 1. Originally, activated amino-sepharose resins were formed by modifying Sepharose 6B (100 g) using epichlorohydrin (Figure 1a). Briefly, Sepharose 6B (100 g) was first thoroughly washed with distilled water at a 10:1 ratio. After being drained and aired, Sepharose 6B was suspended in 50 mL activating solution (0.8 M NaOH aqueous solution, containing 25% DMSO and 10 mL epichlorohydrin) for 2 h at 40 °C. To form aminated Sepharose 6B, activated Sepharose 6B was suspended in 350 mL of distilled water and 35% saturated ammonia was added (150 mL) for mixing. The mixture was incubated for 6 h at 30 °C on a rotary shaker (Figure 1b). After that, cyanuric chloride (2,4,6-trichloro-1,3,5-triazine) was linked as a scaffold for the amino groups. The mixture was shaken in ice-salt bath, then 8 g of cyanuric chloride (44 mmol from Sigma-Aldrich; St. Louis, MO, USA), dissolved in 350 mL acetone, was slowly added (Figure 1c). About 100 mL NaOH aqueous solution (1 M) was slowly added to maintain the neutral pH. To clear away free cyanuric chloride, 50% (*v/v*) acetone was utilized to wash the resins. The ninhydrin test was applied to examine the density of free amino groups and the linkage of cyanuric chloride to the amino groups, according to the previously described procedure [18,20,21]. Briefly, a small aliquot of gel was smeared on filter paper, sprayed with ninhydrin solution (0.2%, *w/v*, in acetone), and heated briefly with a hair dryer. Purple color indicated the presence of free amino groups and color disappearance indicated that cyanuric chloride had been linked to the amino groups. Subsequently, dichlorotriazinylated Sepharose 6B resins were added with two-fold molar excess of CHDS dissolved in 2 M sodium carbonate and stirred for 24 h at room temperature (Figure 1d). The coupling efficiencies and yields for the CHDS ligands were also determined by ninhydrin test. Control resins were synthesized from chitotrisaccharide (CHTS) or glucosamine, according to the method described above (Figure 2A,B). Control resins with 5-atom or 10-atom spacer arms were also synthesized from CHDS-modified Sepharose 6B resins, according to the previously published method [21]. The schemes are shown in Figure 2C,D.

3.3. Expression and Purification of Three Typical Chitosanase

Currently, GH families 46, 75, and 80 comprise only chitosanases in the CAZy database [10,11,28]. Genes encoding three typical chitosanases, including CsnOU01 from GH family 46 (Genbank number ABM91442), Csn from GH family 75 (Genbank number AFG33049), and ChoA from GH family 80 (Genbank number BAA32084) were cloned into the pET22b (containing 6×His tag) vector and expressed in *Escherichia coli* BL21(DE3). The genes were optimized for *E. coli* and synthesized by BGI (Qingdao, China). The DNA fragment was digested to introduce *Nco* I and *Xho* I sites, then ligated into the *Nco* I and *Xho* I sites of plasmid pET22b. The recombinant plasmid was transferred into *E. coli* BL21 (DE3). Cells were cultured in LB medium containing 30 μg/mL ampicillin at 37 °C, until the OD600 reached 0.6. Afterwards, the expression of the target gene was induced by 0.1 mM isopropyl-β-thiogalactoside (IPTG) at 20 °C and 100 rpm for 18 h. These chitosanases with 6 his-tags were purified using a Ni-NTA Sepharose 6B column (GE Healthcare, Madison, WI, USA) at an AKTA avant 150 platform. After centrifugation for 10 min at 12,000 rpm, the supernatant was loaded into 10 mL equilibrated affinity column and washed with washing buffer (0.1 M Tris–HCl buffer, pH 7.6) until the elute exhibited no detectable absorbance at 280 nm. Then, the elution buffer 1 (0.02 M Tris-HCl, pH 7.6, with 10 mM imidazole) was used to deplete the impure protein. The target protein was eluted by elution buffer 2 (0.02 M Tris-HCl, pH 7.6, with 150 mM imidazole). The flow rate of the mobile phase was 3.0 mL/min. The concentrations of each elution peak were assayed by the Bradford method, using BSA as a standard. Chitosanses without 6 his-tag were purified by the traditional method and CHDS-based method, as shown in Section 3.5. Molecular weight and purity of the enzymes were confirmed by SDS-PAGE or HPLC with size-exclusion chromatography.

3.4. Calculation of Desorption Constant of Chitosanase

The characterization of the interactions between chitosanases and affinity resins was carried out using equilibrium adsorption study. Scatchard analysis model was used for analysis of the desorption constant (K_d) and the theoretical maximum adsorption capacity (Q_{max}) of different affinity resins [22,29]. Various concentrations of enzymes (10 mL, 0.1–0.9 mg/mL in 20 mM Tris-HCl buffer, pH 8.0) were combined with 5 g of each type of resin to reach the adsorption equilibrium in a shaken condition. Mixed culture was subsequently centrifuged at 1500× g for 5 min at 4 °C. Afterward, the residual activity of chitosanase and protein concentration in the supernatants was measured and analyzed according to the following Equation:

$$Q = \frac{Q_{max}[C*]}{K_d + [C*]} \quad (1)$$

where Q represents the amount of chitosanase adsorbed to the affinity resin (mg/g wet resin), Q_{max} represents the theoretical maximum absorption of chitosanase to the affinity resin (mg/g wet resin), [$C*$] represents the protein concentration of chitosanase in the mixed solution (mg/mL), and K_d represents the desorption constant. Scatchard plot represents one of the linearized forms of Equation (1). Equation (1) could be transformed into the following Equation (2):

$$\frac{Q}{[C*]} = \frac{Q}{K_d} + \frac{Q_{max}}{K_d} \quad (2)$$

According to the Scatchard model, a plot of $Q/[C*]$ against Q should yield a straight line. The batch adsorption of CsnOU01 towards the affinity medium showed that the respective correlation coefficient R^2 ranged from 0.921 to 0.991. These results indicate that the data fit well with the model.

3.5. Affinity Purification of Chitosanases Using CHDS-Based Resin

The synthesized CHDS-based affinity resins were pre-equilibrated with sample loading buffer (0.1 M Tris-HCl buffer, pH 8.0). Before sample was loaded, the supernatant of expression strains were centrifugated at 10,000× g for 10 min to remove the impurity. Approximately 100 mL samples were loaded onto 10 mL column with synthesized resins. Next, the column was washed with 30 mL washing buffer (0.1 M Tris-HCl buffer with 100 mM NaCl, pH 8.0). Elution buffers (0.1 M acetic acid buffer, pH 5.6, 0.8 M NaCl) were used for eluting the target proteins. The flow rate was 5.0 mL/min. As a control, the traditional purification protocols of three different chitosanses were also developed. The traditional purification protocol of CsnOU01 (GH46) was composed of five steps, including ultrafiltration with a Millipore Amicon® Ultra-10 kDa in 15 mL filter, 60% saturation ammonium sulfate precipitation in an ice-bath and still stirring for more than 2 h, desalting with a 5 mL desalting column (GE Healthcare, Madison, WI, USA), DEAE anion-exchange chromatography, and superdex 75 gel-filtration chromatography. The traditional purification protocol of Csn (GH75) was composed of three steps, including 60% saturation ammonium sulfate precipitation in an ice-bath and still stirring for more than 2 h, desalting with a 5 mL desalting column (GE Healthcare, Madison, WI, USA), and DEAE anion-exchange chromatography. The traditional purification protocol of ChoA (GH80) was composed of six steps, including 40% saturation ammonium sulfate precipitation in an ice-bath and still stirring for more than 2 h, phenyl hydrophobic chromatography (GE Healthcare, Madison, WI, USA), desalting with a 5 mL desalting column, DEAE anion-exchange chromatography, Superdex 75 gel-filtration chromatography, and Superdex 200 gel-filtration chromatography. Protein concentrations of elution peaks were determined by Bradford assay. The purified enzymes were analyzed by 10% SDS-PAGE and HPLC with size-exclusion chromatography.

3.6. Direct Affinity Purification of Chitosanases from Marine Bacteria

In our previous study, sixty-two strains of marine bacteria with chitosan degradation ability were isolated. Among these, twenty-three strains produced chitosanase, nine of which showed high chitosanase activity (>50 U/mL). In this study, the nine bacterial strains with high chitosanase activity were chosen as target strains in affinity purification of chitosanase using the CHDS-based resins. The bacteria were cultured in a medium (containing 30 g/L NaCl, 3 g/L $MgSO_4 \cdot 7H_2O$, 0.2 g/L $CaCl_2$, 0.1 g/L KCl, 0.02 g/L $FeSO_4$, 1.5 g/L Na_2HPO_4, 1 g/L NaH_2PO_4, and 2 g/L chitosan) at 25 °C for 48 h on a rotary shaker (speed, 150 rpm). After that, cultures were centrifuged at 10,000× g for 15 min to remove strains. Approximately 500 mL of supernatants was loaded onto a 10 mL pre-equilibrated column. Washing buffers (0.1 M Tris-HCl buffer with 100 mM NaCl, pH 8.0) were used to remove the uncombined proteins. After that, elution buffers (0.1 M acetic acid buffer, pH 5.6, 0.8 M NaCl) were used for eluting the target protein. The flow rate was 5.0 mL/min.

3.7. Assay of Enzyme Activity

The 3,5-dinitrosalicylic acid (DNS) method was used for assay of the enzyme activity of chitosanase. Briefly, 100 µL of enzyme was mixed with 900 µL of chitosan substrate (0.3% *w/v*, prepared in 50 mM sodium acetate buffer, pH 6.5). Reaction solution was incubated at 40 °C in a water bath for 10 min. Immediately, 750 µL of DNS solution was added into the reaction solution. After that, the mixtures were heating at 100 °C for 10 min. The reaction mixture was cooled down and then centrifuged at $10,000 \times g$ for 2 min to remove the remaining insoluble chitosan. The reducing sugars in the supernatant were analyzed at 520 nm. Each reaction was carried out in triplicate; standard deviation were calculated and used for analysis. D-glucosamine was used as standard. One unit of enzyme activity was defined as the amount of enzyme that releases 1 µmol D-glucosamine-equivalent reducing sugars per minute under the assay conditions.

3.8. Analysis of Protein Purity

Protein purity was determined by SDS-PAGE analysis. The purity of the purified chitosanase was estimated based on the intensity of the protein band using Gelpro Analyzer 3.2, a commonly used gel imaging analysis system. HPLC analysis was conducted using Agilent 1260, equipped with a TSK3000SW column (Tosoh Co., Tokyo, Japan), wherein protein was monitored by absorbance at 280 nm [27]. The mobile phase was 0.1 M PBS, pH 6.7, 0.1 M Na_2SO_4, 0.05% NaN_3. The flow rate was 0.6 mL/min.

4. Conclusions

In this study, a highly efficient affinity resin designed for chitosanase purification was synthesized and characterized. Among other purification protocols, the synthesized resins using CHDS to couple with Sepharose 6B resin via cyanuric chloride spacer were used in the direct purification of native chitosanase without any tags from bacterial culture. This protocol has several significant advantages, for instance, higher purity, fewer steps, and better activity recovery. Coupled with accessible reagents, efficacy, and time-saving procedures, this efficient affinity purification protocol can be a potentially important tool for screening native chitosanases that possible have unique characteristics.

Supplementary Materials: The following are available online at http://www.mdpi.com/1660-3397/17/1/68/s1, Table S1: Different loading and elution condition on the recovery yield and specific activity of CsnOU01. Table S2: The sum of traditional protocol for three different chitosanase.

Author Contributions: S.L. and Y.H. conceived the study. S.L. and L.W. performed the experiments. S.L., M.S., X.C., and Y.H. analyzed the data. S.L. wrote the paper. All the authors reviewed the manuscript.

Funding: This project was funded by Open Research Fund Program of Shandong Provincial Key Laboratory of Glycoscience and Glycotechnology (Ocean University of China) (KLGGOUC201703); National Natural Science Foundation of China (41376175).

Conflicts of Interest: The authors have declared that no competing financial interests exist.

References

1. Younes, I.; Rinaudo, M. Chitin and chitosan preparation from marine sources. Structure, properties and applications. *Mar. Drugs* **2015**, *13*, 1133–1174. [CrossRef] [PubMed]
2. Shinya, S.; Fukamizo, T. Interaction between chitosan and its related enzymes: A review. *Int. J. Biol. Macromol.* **2017**, *104*, 1422–1435. [CrossRef] [PubMed]

3. Liang, T.W.; Chen, Y.J.; Yen, Y.H.; Wang, S.L. The antitumor activity of the hydrolysates of chitinous materials hydrolyzed by crude enzyme from *Bacillus amyloliquefaciens* V656. *Process Biochem.* **2007**, *42*, 527–534. [CrossRef]
4. Shen, K.T.; Chen, M.H.; Chan, H.Y.; Jeng, J.H.; Wang, Y.J. Inhibitory effects of chitooligosaccharides on tumor growth and metastasis. *Food Chem. Toxicol.* **2009**, *47*, 1864–1871. [CrossRef] [PubMed]
5. Das, S.N.; Madhuprakash, J.; Sarma, P.V.; Purushotham, P.; Suma, K.; Manjeet, K.; Rambabu, S.; Gueddari, N.E.; Moerschbacher, B.M.; Podile, A.R. Biotechnological approaches for field applications of chitooligosaccharides (COS) to induce innate immunity in plants. *Crit. Rev. Biotechnol.* **2015**, *35*, 29–43. [CrossRef] [PubMed]
6. Oliveira, E.N., Jr.; El Gueddari, N.E.; Moerschbacher, B.M.; Peter, M.G.; Franco, T.T. Growth of phytopathogenic fungi in the presence of partially acetylated chitooligosaccharides. *Mycopathologia* **2008**, *166*, 163–174. [CrossRef] [PubMed]
7. Jeon, Y.J.; Park, P.J.; Kim, S.K. Antimicrobial effect of chitooligosacchardies produced by bioreactor. *Carbohydr. Polym.* **2001**, *44*, 71–76. [CrossRef]
8. Lee, H.W.; Park, Y.S.; Choi, J.W.; Yi, S.Y.; Shin, W.S. Antidiabetic effects of chitosan oligosaccharides in neonatal streptozotocin-induced noninsulin-dependent diabetes mellitus in rats. *Biol. Pharm. Bull.* **2003**, *26*, 1100–1103. [CrossRef]
9. Jung, W.J.; Park, R.D. Bioproduction of chitooligosaccharides: Present and perspectives. *Mar. Drugs.* **2014**, *12*, 5328–5356. [CrossRef]
10. Chavan, S.B.; Deshpande, M.V. Chitinolytic enzymes: An appraisal as a product of commercial potential. *Biotechnol. Prog.* **2013**, *29*, 833–846. [CrossRef]
11. Thadathil, N.; Velappan, S.P. Recent developments in chitosanase research and its biotechnological applications: A review. *Food Chem.* **2014**, *150*, 392–399. [CrossRef]
12. Lombard, V.; Golaconda Ramulu, H.; Drula, E.; Coutinho, P.M.; Henrissat, B. The carbohydrate-active enzymes database (CAZy) in 2013. *Nucl. Acids Res.* **2014**, *42*, 490–495. [CrossRef] [PubMed]
13. Park, J.K.; Shimono, K.; Ochiai, N.; Shigeru, K.; Kurita, M.; Ohta, Y.; Tanaka, K.; Matsuda, H.; Kawamukai, M. Purification, characterization, and gene analysis of a chitosanase (ChoA) from matsuebacter chitosanotabidus 3001. *J. Bacteriol.* **1999**, *181*, 6642–6649. [PubMed]
14. Shehata, A.N.; Abd El Aty, A.A.; Darwish, D.A.; Abdel Wahab, W.A.; Mostafa, F.A. Purification, physicochemical and thermodynamic studies of antifungal chitinase with production of bioactive chitosan-oligosaccharide from newly isolated *Aspergillus Griseoaurantiacus* KX010988. *Int. J. Biol. Macromol.* **2018**, *107*, 990–999. [CrossRef] [PubMed]
15. Yun, C.; Amakata, D.; Matsuo, Y.; Matsuda, H.; Kawamukai, M. New chitosan-degrading strains that produce chitosanases similar to ChoA of mitsuaria chitosanitabida. *Appl. Environ. Microbiol.* **2005**, *71*, 5138–5144. [CrossRef] [PubMed]
16. Wang, S.L.; Chen, S.J.; Wang, C.L. Purification and characterization of chitinases and chitosanases from a new species strain *Pseudomonas* sp. TKU015 using shrimp shells as a substrate. *Carbohydr. Res.* **2008**, *343*, 1171–1179. [CrossRef] [PubMed]
17. Labrou, N.E. Design and selection of ligands for affinity chromatography. *J. Chromatogr. B Anal. Technol. Biomed. Life Sci.* **2003**, *790*, 67–78. [CrossRef]
18. Mountford, S.J.; Daly, R.; Robinson, A.J.; Hearn, M.T. Design, synthesis and evaluation of pyridine-based chromatographic adsorbents for antibody purification. *J. Chromatogr. A* **2014**, *1355*, 15–25. [CrossRef]
19. Fasoli, E.; Reyes, Y.R.; Guzman, O.M.; Rosado, A.; Cruz, V.R.; Borges, A.; Martinez, E.; Bansal, V. Para-aminobenzamidine linked regenerated cellulose membranes for plasminogen activator purification: Effect of spacer arm length and ligand density. *J. Chromatogr. B Anal. Technol. Biomed. Life Sci.* **2013**, *930*, 13–21. [CrossRef] [PubMed]
20. Li, S.; Wang, L.; Xu, X.; Lin, S.; Wang, Y.; Hao, J.; Sun, M. Structure-based design and synthesis of a new phenylboronic-modified affinity medium for metalloprotease purification. *Mar. Drugs* **2017**, *15*. [CrossRef]
21. Li, S.; Wang, L.; Yang, J.; Bao, J.; Liu, J.; Lin, S.; Hao, J.; Sun, M. Affinity purification of metalloprotease from marine bacterium using immobilized metal affinity chromatography. *J. Sep. Sci.* **2016**, *39*, 2050–2056. [CrossRef] [PubMed]
22. Xin, Y.; Yang, H.; Xiao, X.; Zhang, L.; Zhang, Y.; Tong, Y.; Chen, Y.; Wang, W. Affinity purification of urinary trypsin inhibitor from human urine. *J. Sep. Sci.* **2012**, *35*, 1–6. [CrossRef] [PubMed]

23. Batra, S.; Bhushan, R. Amino acids as chiral auxiliaries in cyanuric chloride-based chiral derivatizing agents for enantioseparation by liquid chromatography. *Biomed. Chromatogr.* **2014**, *28*, 1532–1546. [CrossRef] [PubMed]
24. Lyu, Q.; Wang, S.; Xu, W.; Han, B.; Liu, W.; Jones, D.N.; Liu, W. Structural insights into the substrate-binding mechanism for a novel chitosanase. *Biochem. J.* **2014**, *461*, 335–345. [CrossRef] [PubMed]
25. Lyu, Q.; Shi, Y.; Wang, S.; Yang, Y.; Han, B.; Liu, W.; Jones, D.N.; Liu, W. Structural and biochemical insights into the degradation mechanism of chitosan by chitosanase OU01. *Biochim. Biophys. Acta* **2015**, *1850*, 1953–1961. [CrossRef] [PubMed]
26. Zhu, X.F.; Tan, H.Q.; Zhu, C.; Liao, L.; Zhang, X.Q.; Wu, M. Cloning and overexpression of a new chitosanase gene from *Penicillium* sp. D-1. *AMB Express* **2012**, *2*, 13. [CrossRef] [PubMed]
27. Li, S.; Wang, L.; Lin, S.; Yang, J.; Ma, Z.; Wang, Y.; Liu, J.; Hao, J.; Sun, M. Rapid and efficient one-step purification of a serralysin family protease by using a *p*-aminobenzamidine-modified affinity medium. *J. Sep. Sci.* **2017**, *40*, 1960–1965. [CrossRef] [PubMed]
28. Viens, P.; Lacombe-Harvey, M.E.; Brzezinski, R. Chitosanases from family 46 of glycoside hydrolases: From proteins to phenotypes. *Mar. Drugs.* **2015**, *13*, 6566–6587. [CrossRef]
29. Belenguer-Sapina, C.; Pellicer-Castell, E.; El Haskouri, J.; Guillem, C.; Simo-Alfonso, E.F.; Amoros, P.; Mauri-Aucejo, A. Design, characterization and comparison of materials based on beta and gamma cyclodextrin covalently connected to microporous silica for environmental analysis. *J. Chromatogr. A* **2018**, *1563*, 10–19. [CrossRef]

© 2019 by the authors. Licensee MDPI, Basel, Switzerland. This article is an open access article distributed under the terms and conditions of the Creative Commons Attribution (CC BY) license (http://creativecommons.org/licenses/by/4.0/).

Article

Novel Enzyme Actions for Sulphated Galactofucan Depolymerisation and a New Engineering Strategy for Molecular Stabilisation of Fucoidan Degrading Enzymes

Hang T. T. Cao [1,2], Maria D. Mikkelsen [1], Mateusz J. Lezyk [1], Ly M. Bui [2], Van T. T. Tran [2], Artem S. Silchenko [3], Mikhail I. Kusaykin [3], Thinh D. Pham [2], Bang H. Truong [2], Jesper Holck [1] and Anne S. Meyer [1,*]

[1] Protein Chemistry and Enzyme Technology, DTU Bioengineering, Department of Biotechnology and Biomedicine, Technical University of Denmark, Building 221, 2800 Kongens Lyngby, Denmark; caohang.nitra@gmail.com (H.T.T.C.); mdami@dtu.dk (M.D.M.); mateusz.lezyk@put.poznan.pl (M.J.L.); jesho@dtu.dk (J.H.)

[2] NhaTrang Institute of Technology Research and Application, Vietnam Academy of Science and Technology, 02 Hung Vuong Street, Nhatrang 650000, Vietnam; bminhly.nitra@gmail.com (L.M.B.); vanvvlnt@yahoo.com.vn (V.T.T.T.); ducthinh.nitra@gmail.com (T.D.P.); truonghaibangnt@gmail.com (B.H.T.)

[3] Laboratory of Enzyme Chemistry, G.B. Elyakov Pacific Institute of Bioorganic Chemistry, Far Eastern Branch, Russian Academy of Sciences, 159 100-Let Vladivostoku Ave., Vladivostok 690022, Russia; artem.silchencko@yandex.ru (A.S.S.); mik@piboc.dvo.ru (M.I.K.)

* Correspondence: asme@dtu.dk; Tel.: +45-45-252-600

Received: 30 September 2018; Accepted: 22 October 2018; Published: 1 November 2018

Abstract: Fucoidans from brown macroalgae have beneficial biomedical properties but their use as pharma products requires homogenous oligomeric products. In this study, the action of five recombinant microbial fucoidan degrading enzymes were evaluated on fucoidans from brown macroalgae: *Sargassum mcclurei*, *Fucus evanescens*, *Fucus vesiculosus*, *Turbinaria ornata*, *Saccharina cichorioides*, and *Undaria pinnatifida*. The enzymes included three endo-fucoidanases (EC 3.2.1.-GH 107), FcnA2, Fda1, and Fda2, and two unclassified endo-fucoglucuronomannan lyases, FdlA and FdlB. The oligosaccharide product profiles were assessed by carbohydrate-polyacrylamide gel electrophoresis and size exclusion chromatography. The recombinant enzymes FcnA2, Fda1, and Fda2 were unstable but were stabilised by truncation of the C-terminal end (removing up to 40% of the enzyme sequence). All five enzymes catalysed degradation of fucoidans containing α(1→4)-linked L-fucosyls. Fda2 also degraded *S. cichorioides* and *U. pinnatifida* fucoidans that have α(1→3)-linked L-fucosyls in their backbone. In the stabilised form, Fda1 also cleaved α(1→3) bonds. For the first time, we also show that several enzymes catalyse degradation of *S. mcclurei* galactofucan-fucoidan, known to contain α(1→4) and α(1→3) linked L-fucosyls and galactosyl-β(1→3) bonds in the backbone. These data enhance our understanding of fucoidan degrading enzymes and their substrate preferences and may assist development of enzyme-assisted production of defined fuco-oligosaccharides from fucoidan substrates.

Keywords: fucoidan; endo-fucoidanase; galactofucan; molecular stabilisation; *Sargassum mcclurei*; *Turbinaria ornata*

1. Introduction

Fucoidan polysaccharides are a family of sulphated, fucose-rich polysaccharides uniquely produced by brown marine macroalgae (seaweeds) and certain marine invertebrates, such as sea cucumbers [1,2]. In general, fucoidans, also known as fucose-containing sulphated polysaccharides (FCSPs), consist of a backbone of α-L fucosyl residues linked together by (1→3) and/or (1→4)-glycoside bonds. The bonds are organised in stretches of α(1→3) or of alternating α(1→3)- and α(1→4)-glycoside linkages, depending on the macroalgal origin of the fucoidan, i.e., the species, age, geographical origin, and collection time (season) [3]. The L-fucosyl residues may be sulphated ($-SO_3^-$) at position C2 and/or C4 (rarely at C3). Some fucoidans have fucose, galactose, glucuronic acid or other mono- and oligosaccharides as short branches [1,4,5]. Galactofucans are the most structurally diverse group of fucoidans that have been characterised from brown algae to date. The galactofucans have galactose residues in their backbone or in their branches; the position and number of these galactose residues depend on the type of algae [6,7].

The structural diversity of fucoidans or FCSPs is very high as both the sulphatation pattern and the backbone bond pattern of α(1→3) and α(1→4)-glycosidic bonds vary significantly depending on the fucoidan source. The fucoidan from *Fucus vesiculosus*, which is available commercially, is known to be made up of a backbone of repeating disaccharide units of α(1→3)- and α(1→4)-linked sulphated L-fucosyl residues (C2, C2/C3, C2/C4, C4 sulphatation) [8–10] (Figure 1). Fucoidan from *Fucus evanescens* has a similar L-fucosyl backbone of alternating α(1→4) and α(1→3) L-fucosyls with sulphate substitution at C2. An additional sulphate may occupy position 4 in some of the α(1→3)-linked fucosyls, and the remaining hydroxyl groups may be randomly acetylated [1] (Figure 1). In contrast, the bonds in the backbone of the fucoidan from *Undaria pinnatifida* and *Saccharina cichorioides* are exclusively α(1→3). The backbone U. pinnatifida fucoidan is moreover assumed to be rich in 2,4-disulphate substituted fucosyl residues and to contain some β(1→4)-linked galactosyl residues as branches [11] (Figure 1). Some fucoidans have even more complex backbone structures as is the case, e.g., for fucoidan from the brown macroalgae *Sargassum mcclurei* and *Turbinaria ornata* commonly found along the Pacific Ocean coastline of Vietnam. The S. mcclurei fucoidan is essentially a sulphated galactofucan polysaccharide having both α(1→3) and α(1→4) linked fucosyl residues, as well as galactosyl-β(1→3) links to fucosyl, and α(1→6) linkages from fucosyl to galactosyl in the reducing end of the backbone (Figure 1). The fucosyl residues in S. mcclurei fucoidan are moreover differentially sulphated at C2 and/or at C4 and some of the galactosyl moieties are sulphated at C6 [12] (Figure 1). Fucoidan extracted from T. ornata collected at Nha-Trang bay, Vietnam, also seems to be a galactofucan. The backbone of T. ornata fucoidan has thus been proposed to consist of α(1→3)-linked L-fucosyls with galactosyl branches (Fuc:Gal ≈ 3:1) and has been found to have a high sulphate content of about 25% with sulphate attached mostly at C2, and to a lesser extent at C4, of both the fucosyl and the galactosyl residues [13,14] (Figure 1). The biological function of fucoidans in brown macroalgae is uncertain, but fucoidans have long been known to exert beneficial biological activities including anti-tumorigenic, immune-modulatory, anti-inflammatory, anti-coagulant and anti-thrombotic effects, as demonstrated in vitro and in vivo [14–16]. Fucoidan from S. mcclurei, including the unique galactofucan structural moieties with sulphated α(1→3) L-fucosyl and α(1→4) linked galactosyl residues, have for example been shown to inhibit colony formation of DLD-1 human colon cancer cells in vitro [12], and crude, sulphated fucoidan products extracted from F. vesiculosus and Sargassum spp. are known to cause growth inhibition and apoptosis of melanoma B16 cells in vitro and to enhance the activity of natural killer cells in vivo in mice resulting in the specific lysis of YAC-1 cells (a murine T-lymphoma cell line sensitive to natural killer cells) [15]. However, the high molecular weight, irregular structure, and viscosity of fucoidans are an obstacle for providing homogeneous preparations for soluble and concentrate pharmaceutical use. One approach to solve this problem is to use enzymes that can depolymerise the fucoidans providing a preparation that is easier to handle and also with potentially bioactive properties.

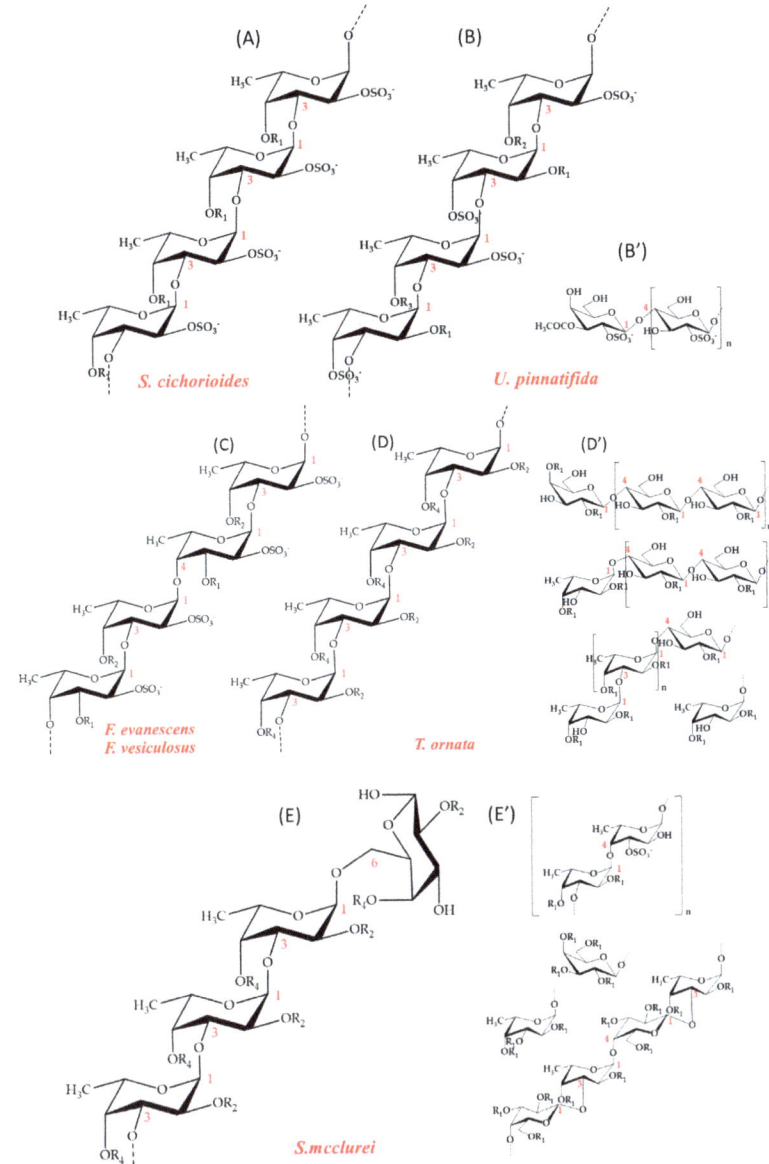

Figure 1. Representative fucoidan structures of brown macroalgae *Fucus evanescens*, *Fucus vesiculosus*, *Sargassum mcclurei*, *Turbinaria ornata*, *Saccharina cichorioides*, and *Undaria pinnatifida*: (**A**) main chain of *S. cichorioides* composed of α(1→3)-L-fucosyls; (**B**) main chain of *U. pinnatifida* fucoidan also composed of α(1→3)-L-fucosyls; (**B′**) branches of *U. pinnatifida* fucoidan [11]; (**C**) main chain of *F. evanescens* [1] and *F. vesiculosus* fucoidan [8–10], both composed of α(1→3)- and α(1→4)-linked L-fucosyls; (**D**) main chain of *T. ornata* fucoidan composed of α(1→3)-L-fucosyls [13,14]; (**D′**) branches of *T. ornata* of α(1→3)-L-fucosyls or of β(1→4)galactosyls and mixed fucosyl-galactosyls; (**E**) main chain of *S. mcclurei* fucoidan made up of mainly α(1→3)-L-fucosyls [12]; and (**E′**) branches or inserts in the main chain of *S. mcclurei* fucoidan. In all fucoidan structures: R_1: –H or –SO_3^-; R_2: –H, –SO_3^- or H_3COC–; R_3: SO_3^-, H_3COC– or branches; and R_4: SO_3^- or branches.

About 20 microorganisms, mainly marine bacteria, have been described that produce fucoidanases [17–21]. In addition, a few fucoidanases have been found in marine molluscs [22,23]. In 2006, the gene encoding a fucoidanase from the marine bacterium *Mariniflexile fucanivorans* SW5T was cloned and the recombinant enzyme named FcnA. A C-terminal truncated version of FcnA named FcnA2 was previously reported to exert endo α(1→4) action on fucoidan from *Pelvetia canaliculata* (a type of fucoidan encompassing both α(1→4) and α(1→3) fucosyl-linkages in the backbone) [24]. In 2002, the genes encoding for two endo-fucoidanases referred to as Fda1 and Fda2, from the marine bacterium *Alteromonas* sp. SN-1009 were sequenced and their use for degradation of sulphated fucoidan originating from the brown seaweed *Kjellmaniella crassifolia* (now called *Saccharina sculpera*) were patented [25]. In the patent, these enzymes were reported to catalyse cleavage of α(1→3)-glycosidic bonds in the *K. crassifolia* (*S. sculpera*) fucoidan [25]. FcnA, Fda1, and Fda2 all belong to the new glycoside hydrolase family GH107 in CAZy [26]. In 2017, two endo-fucoidanases, FFA1 and FFA2, from the marine bacterium *Formosa algae* (KMM 3553T) were characterised and also suggested to belong to GH family 107 [27,28]. The FFA2 enzyme was proposed to be a poly[(1→4)-α-L-fucoside-2-sulphate] glycano hydrolase [27]. Already in 2003 Sakai et al. reported the finding of a new type of extracellular endo-fucoidan-lyase activity from "*Fucobacter marina*" SA-0082, or more correctly *Flavobacterium* sp. SA-0082, which acted on sulphated fucoglucurono-mannan from *K. crassifolia* (*S. sculpera*) [29,30]. By sequence analyses, it was found that this lyase activity was apparently encoded by two separate coding regions. Recombinant expression of these two putative fucoidan degrading enzymes, referred to as FdlA and FdlB, respectively, showed that the two enzymes had about 56% amino acid sequence identity and both were claimed to act as (glucurono-) fucoidan lyases on *K. crassifolia* (*S. sculpera*) fucoidan [25].

The objective of this work was to compare the catalytic properties, notably the substrate degradation patterns, on different fucoidans of the three GH107 endo-fucoidanases (EC 3.2.1.-) referred to as FcnA2, Fda1, and Fda2, and the two enzymes previously reported to be endo-fucoglucuronomannan-lyases, referred to as FdlA and FdlB. The action of the enzymes on different fucoidan substrate structures was compared by assessing oligomer product profiles resulting after treatment with recombinantly produced enzymes on fucoidans originating from six different types of brown seaweeds: *Sargassum mcclurei*, *Turbinaria ornata*, *Fucus evanescens*, *Fucus vesiculosus*, *Saccharina cichorioides*, and *Undaria pinnatifida*. We also report stabilisation of the recombinantly produced enzymes by targeted gene truncation resulting in deletion of large parts of the C-terminal end of several of the enzymes.

2. Results

2.1. Recombinant Enzyme Expression

The enzymes FcnA2, FdlA and FdlB expressed well and the purified enzymes gave the expected band sizes as assessed by Sodium Dodecyl Sulfate Polyacrylamide Gel Electrophoresis (SDS-PAGE) (Figure 2A). The expression of recombinant Fda1 was high, but the protein remained in the cell debris after sonication (Figure S1). Several culture conditions for enzyme expression (temperature, medium, and isopropyl β-D-1-thiogalactopyranoside (IPTG) concentration) were tested to obtain a soluble enzyme, but without success. Fda2 expressed well, but migrated slower in the SDS-PAGE gel than expected (94 kDa). At present, the data do not allow any firm conclusions to be drawn regarding the cause of this retarded migration of the Fda2 protein, but high hydrophobicity and high levels of charged amino acids may cause anomalous SDS-PAGE migration as compared to the soluble, commercial protein standards [31]. For the enzymes FcnA2 and Fda2 more than one band was visible in both the SDS-PAGE gel and in the Western blot (Figure 2), suggesting spontaneous degradation rather than impurities from other proteins. This observation agrees with previously published data for recombinantly expressed FcnA2 reporting "co-elution" with other proteins, which could not be separated by anion exchange or SEC [24]. For the Fda2 enzyme, use of protease inhibitors such as

PMSF (36978) from Thermo Fisher Scientific (Waltham, MA, USA) and a protease inhibitor cocktail (P8849) from Sigma-Aldrich (Steinheim, Germany) during purification did not improve stability, corroborating that the degradation likely occurred during expression in *E. coli* cells or during the subsequent purification.

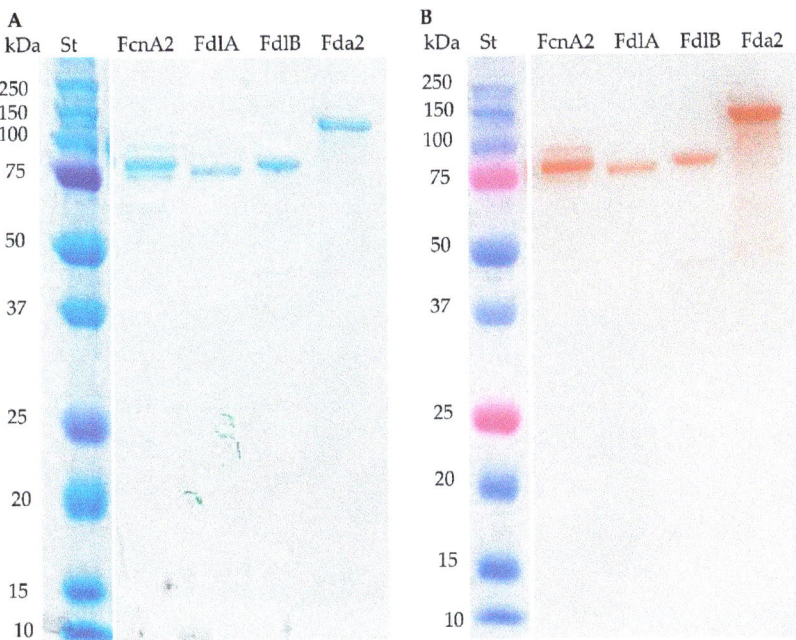

Figure 2. Purified recombinantly expressed fucoidan-modifying enzymes. (**A**) SDS-PAGE, and (**B**) Western blot of purified FcnA2, FdlA, FdlB, and Fda2. (St) is the protein plus molecular weight marker. The expected molecular weights of the recombinant enzymes FcnA2, FdlA, FdlB, and Fda2 were 87, 75, 76, and 94 kDa, respectively. The multiple bands seen for FcnA2 and Fda2, notably in the Western blot, indicate partial degradation of the proteins. Expression of recombinant Fda1 resulted in insoluble enzyme material which is not shown in this figure.

2.2. Substrate Specificity of the Recombinant Fucoidan-Degrading Enzymes

The six different fucoidan samples were treated with the purified enzymes FcnA2, Fda2, FdlA, and FdlB and the treatments produced different carbohydrate-polyacrylamide gel electrophoresis (C-PAGE) patterns with the six fucoidan samples (the expression of recombinant Fda1 resulted in insoluble enzymes, which is why there are no data for Fda1). The reactions were run for 24 h to ascertain maximal substrate degradation. Preliminary data using higher enzyme dosage or longer reaction time did not show higher extent of degradation except of the *S. mcclurei* fucoidan that gave more visible bands in the C-PAGE after a 48 h reaction (these data are discussed further below). Hence, the data obtained by C-PAGE showed both the selectivity and the maximal extent of fucoidan degradation obtainable for each set of enzyme and substrate. This means that it is presumed that the unreacted higher molecular weight polysaccharides do not contain structural units, i.e., backbone-stretches, linkages, substitutions or branches, attackable by the particular enzyme examined. The positive control standard (St) was the hydrolysate from the enzymatic reaction of the *Formosa algae* FFA2 on *F. evanescens* fucoidan, where the lowest band corresponds to a tetra-saccharide of (1→4)- and (1→3)-linked α-L-fucosyls with each fucosyl residue sulphated at C2 [27] (Figure 3). The data obtained by C-PAGE indicated more extensive degradation of the fucoidan substrates from *Sargassum* mcclurei (1), *Fucus vesiculosus* (2), and *Fucus evanescens* (3) than of substrates predominantly having α(1→3) glycoside bonds in their backbone structures, originating from *Turbinaria ornate* (4), *Saccharina cichorioides* (5), and *Undaria pinnatifida* (6), respectively (Figure 3). In general, the data obtained show that each enzyme produced differently sized sulphated oligomers in the C-PAGE chromatograms, suggesting that the different enzymes target different linkages and/or differently sulphated fucosyl residues. The results also suggest that all the enzymes were endo-acting as the enzymatic action left behind relatively high molecular weight fractions.

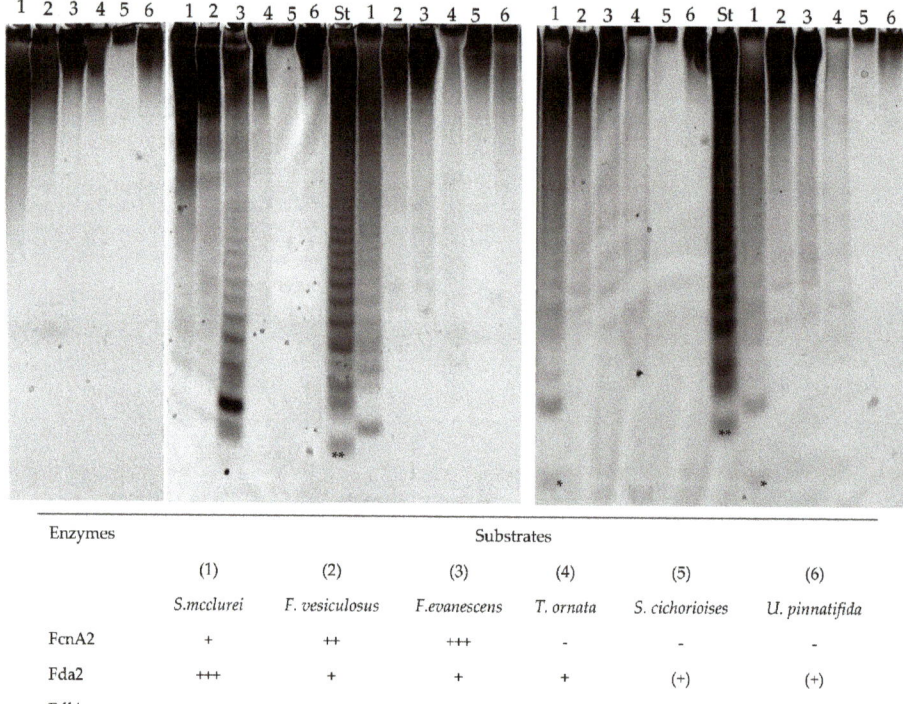

Enzymes	Substrates					
	(1)	(2)	(3)	(4)	(5)	(6)
	S.mcclurei	F. vesiculosus	F.evanescens	T. ornata	S. cichorioises	U. pinnatifida
FcnA2	+	++	+++	-	-	-
Fda2	+++	+	+	+	(+)	(+)
FdlA	+++	+	+	+	-	-
FdlB	+++	+	+	+	-	-

Figure 3. C-PAGE analysis of fucoidan degradation using purified enzymes: (**A**) substrate control with no enzyme; and (**B–E**) enzymatic products from reaction of FcnA2, Fda2, FdlA, and FdlB on different fucoidans, respectively (expression of recombinant Fda1 resulted in insoluble enzymes, which is why there are no data for Fda1): (1) *Sargassum mcclurei*; (2) *Fucus vesiculosus*; (3) *Fucus evanescens*; (4) *Turbinaria ornata*; (5) *Saccharina cichorioides*; and (6) *Undaria pinnatifida*. The extent of degradation is indicated with: (+++) highest, (++) medium, (+) lowest and (+) is positive activity resulting in a high molecular smear, while (−) is no activity. The standard (St) is the product profile of FFA2 treatment of fucoidan from *F. evanescens*. The lowest band (**) of the St is a tetra-saccharide of (1→4)- and (1→3)-linked α-L-fucosyls sulphated at every C2 with an approximate mass of 972 Da [27]. (*) indicates an enzymatic fucoidan degradation product of either lower mass or higher charge than the lowest St band (**) compound. The reaction time was 24 h.

2.3. FcnA2 Catalyses Cleavage of α(1→4) Fucosyl Bonds in Sulphated Fucoidan Backbones

The recombinantly expressed FcnA2 enzyme exerted highest activity on the fucoidan from *F. evanescens*, and the degradation of this substrate was much more profound than on *F. vesiculosus*, even though both substrates have similar alternating α(1→3) and α(1→4) glycoside bonds in the backbone. The degradation of fucoidan from *F. evanescens* was in agreement with previous data showing that FcnA2 is able to degrade fucoidan from *Pelvetia canaliculata* [24]. The fucoidans from *F. evanescens* and *P. canaliculata* presumably have less if any C2, C4 disulphates in the "−1" position of the α(1→4)-L-fucosyl linkage compared to the fucoidan substrate from *F. vesiculosus*, which likely contains more fucosyl residues with C2/C4 and even some with C2/C3 disulphatation than the *F. evanescens* fucoidan. The lesser degree of C2/C4 and C2/C3 disulphatation might be the reason for the *F. evanescens* fucoidan being more degraded than the *F. vesiculosus* fucoidan (Figure 3). Hence, FcnA2 most likely catalyses cleavage of (1→4)-α-glycosidic bonds between the −1 fucosyl residues having the sulphate group at C2, but not at both C2, C4. However, detailed structural elucidation of the fucoidan products and modelling of the substrate accommodation in the enzyme's active site are warranted to substantiate this hypothesis. The differences in the degradation of fucoidan from *F. evanescens* and *F. vesiculosus* thus indicate that differences in the sulphatation pattern or in other types of substitutions on the substrate backbones may influence the action of FcnA2 on these two *Fucus* sp. derived fucoidans. The data suggest that the presumed presence in *F. vesiculosus* of fucosyl residues with disulphate at C2, C4 (on either the −1 or +1 position of the α(1→4) glycoside bond) may retard the enzymatic action of FcnA2.

The smallest oligomers released from *F. evanescens* by FcnA2 also differed from those released by the FFA2 treatment of fucoidan from *F. evanescens* in the standard (st) (Figure 3). FFA2 catalyses the cleavage of (1→4)-α-glycosidic bonds in the *F. evanescens* fucoidan within the structural fragment [→3)-α-L-Fucp2S-(1→4)-α-L-Fucp2S-(1→]n but not in the fragment [→3)-α-L-Fucp2S,4S-(1→4)-α-L-Fucp2S-(1→]n. The difference in the oligomers released suggests that the sulphatation preferences of the FFA2 and FcnA2 may differ, which invites to further elucidation of the enzyme structures and detailed analyses and modelling of enzyme-substrate interactions. FcnA2 also catalysed degradation of the sulphated galacto-fucan fucoidan from *S. mcclurei* resulting in production of several low molecular weight bands in the C-PAGE (Figure 3B). The partial degradation is in agreement with the enzyme attacking α(1→4) linked (sulphated) L-fucosyl residues. Nevertheless, this enzymatic degradation of *S. mcclurei* fucoidan is a novel finding, as enzymatic modification of the *S. mcclurei* fucoidan has not previously been reported. The apparent lack of action of FcnA2 on the fucoidan from *T. ornata*, *S. cichorioides*, and *U. pinnatifida* suggests that FcnA2 does not catalyse cleavage of α(1→3) bonds between fucosyl residues, whereas the activity on the other three substrates supports the hypothesis that the enzyme attacks α(1→4) bonds between L-fucosyl residues as previously shown [24].

2.4. Fda2 Catalyses Cleavage of α(1→3) Fucosyl Bonds in Sulphated Fucoidan Backbones

Fda2 catalysed partial degradation of the galactofucan-rich fucoidan from *S. mcclurei* similar to the action of FcnA2 (Figure 3C). The C-PAGE results showed that this enzyme also exerted partial degradation of the fucoidans from *F. vesiculosus* and *F. evanescens* and had very low activity on the fucoidans rich in α(1→3) fucosyl linkages from *T. ornata*, *S. cichorioides*, and *U. pinnatifida*. The activity was very low, but still visible on the *S. cichorioides* fucoidan (with a smear at the top of the gel and weak bands in the lower part of the gel) and on the *U. pinnatifida* fucoidan (with a discernible smear at the top of the gel) (Figure 2C). The action of Fda2 on *S. mcclurei* fucoidan is a new finding which suggests that the Fda2 enzyme may be employed for controlled degradation of the complex galacto-fucan fucoidan from *S. mcclurei*. The activity of this enzyme on *S. mcclurei*, *F. evanescens* and *F. vesiculosus* together with the weak activity observed on substrates rich in α(1→3) fucosyl linkages corroborates previous claims of the action of Fda2 on α(1→3) bonded L-fucosyls in fucoidan [25]. Both Fda1 and Fda2 were previously shown to digest sulphated fucans from *K. crassifolia* (i.e., *S. sculpera*) with the backbone structure [3)-α-L-Fucp-(2OSO$_3$)-1→3-α-L-Fucp-(2,4OSO$_3$)-(1→] and to partially digest fucoidan from other brown algae of the order Laminariales, such as *Saccharina japonica*, *Lessonia nigrescens*, and *Ecklonia maxima* [32]. The data obtained further support the hypothesis that Fda2, despite its instability (Figure 2B), catalyses cleavage of α(1→3) fucosyl bonds in sulphated fucoidan backbones.

2.5. FdlA and FdlB Action

The FdlA and FdlB enzymes originating from *Flavobacterium* sp. SA-0081 (previously referred to as "*Fucobacter marina*") (Table 1) have been claimed to be specific for certain sulphated fuco-glucuronomannan (SFGM) structural fragments containing uronic acid and D-mannosyl α-linkages in fucoidan molecules [25]. The enzymes were purified from the SA-0082 strain and were shown to catalyse cleavage of SFGM fractions from the brown algae *Kjellmaniella crassifolia* (now *S. sculpera*) via a lyase mechanism cleaving the α-linkage between D-mannosyl and D-glucuronate in the SFGM fractions [33].

Table 1. Fucoidan-degrading enzymes, features, molecular weight and expression strains used.

Enzyme Name/GenBank No.	Organism	Features [a]	Length (aa) [b]	Expected MW (kDa)	*E. coli* Expression Strains
FcnA CAI47003.1	*Mariniflexile fucanivorans* SW5	nd	1007	nd	Nd
FcnA2	*Mariniflexile fucanivorans* SW5	His6 (N-term)	799	88	BL21 (DE3) pGro7 [c]
FcnAΔ229	*Mariniflexile fucanivorans* SW5	His10 (N-term)	720	80	BL21 (DE3) pGro7 [c]
Fda1 AAO00508.1	*Alteromonas* sp. SN-1009	His10 (N-term)	804	87	BL21 (DE3) pGro7 [c]
Fda1Δ145	*Alteromonas* sp. SN-1009	His10 (N-term) and His10 (C-term)	669	73	BL21 (DE3) pGro7 [c]
Fda1Δ395	*Alteromonas* sp. SN-1009	His10 (N-term) and His10 (C-term)	419	46	BL21 (DE3) pGro7 [c]
Fda2 AAO00509.1	*Alteromonas* sp. SN-1009	His10 (N-term)	868	94	BL21 (DE3) pGro7 [c]
Fda2-His	*Alteromonas* sp. SN-1009	His10 (N-term) and His10 (C-term)	878	95	BL21 (DE3) pGro7 [c]
Fda2Δ146	*Alteromonas* sp. SN-1009	His10 (N-term) and His10 (C-term)	732	80	BL21 (DE3) pGro7 [c]
Fda2Δ390	*Alteromonas* sp. SN-1009	His10 (N-term) and His10 (C-term)	488	53	BL21 (DE3) pGro7 [c]
FdlA AAO00510.1	*Flavobacterium* sp. SA-0082	His10 (N-term)	684	74	C41 (DE3)
FdlB AAO00511.1	*Flavobacterium* sp. SA-0082	His10 (N-term)	692	76	C41 (DE3)

nd, not determined in this study. [a] Wild type signal peptide had been removed for codon-optimised synthesised construct; [b] Includes his-tags; [c] groES-groEL chaperone expressed from the pGro7 plasmid.

In this study, FdlA and FdlB both exerted activity on the fucoidans from *S. mcclurei*, *F. vesiculosus*, and *F. evanescens*. Only weak action was observed on fucoidans from *T. ornata* and essentially no activity on *S. cichorioides* and *U. pinnatifida* was found (Figure 3D,E). Fucoidan preparations from *S. mcclurei*, *T. ornata*, *F. evanescens* and *F. vesiculosus* may contain low amounts of uronic acid and sometimes traces of mannose [12,14,33–35] but until now no data show that D-mannosyl and D-glucuronate are present in the backbone of these fucoidans. Moreover, no lyase activity was detected by monitoring absorbance at 232 nm, indicating that the degradation products did not include unsaturated uronic oligosaccharides. FdlA and FdlB most likely cleave α(1→4) fucosyl bonds in the backbone of these fucoidans, since lack of activity on fucoidan from *S. cichorioides* and from *U. pinnatifida* (and weak action on *T. ornata* fucoidan) indicate that FdlA and FdlB do not cleave the α(1→3) bonds in fucoidan.

The similar weak degradation of the fucoidan substrates from *F. vesiculosus* and *F. evanescens* by both enzymes, i.e., producing almost similar oligomer profiles in the C-PAGE, suggests a preference for rare or complex fucosyl-sulphatation (e.g., C2 and C4) in the *Fucus* fucoidan substrates. Such substrates may occur more abundantly in *S. mcclurei* fucoidan (Figure 2), and the enzyme most likely prefers to attack only α(1→4) fucosyl-bonds. Enzyme FdlB appeared to exert a more profound action on the *F. evanescens* substrate than FdlA. Interestingly, the action of the two enzymes on *S. mcclurei* galacto-fucan substrate produced a band that travelled further in the gel than the sulphated tetra-saccharide of the control, suggesting that both FdlA and FdlB are able to catalyse disintegration of sulphated fucoidan oligomers. Due to the high degree of depolymerisation, down to oligosaccharides of less than DP4 (Figure 3), and due to the high abundance of galactosyl residues in *S. mcclurei* fucoidan, we cannot rule out that FdlA and FdlB may cleave galactosyl-$\alpha(1\rightarrow 4)$ bonds (Figure 3D,E), and further analysis could confirm this conclusion.

2.6. Further Assessment of Sargassum mcclurei Fucoidan Degradation

C-PAGE and SEC of oligosaccharides released by the enzymes FcnA2, Fda2, FdlB and FdlA after extended reaction for 48 h, showed that each enzyme catalysed profound degradation of *S. mcclurei* fucoidan (Figure 4).

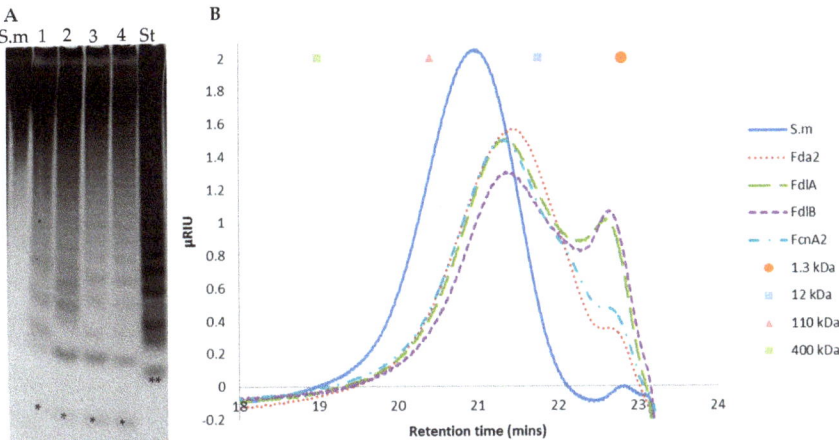

Figure 4. Degradation of *Sargassum mcclurei* fucoidan (S.m) by fucoidanase enzymes. (**A**) C-PAGE; and (**B**) size exclusion chromatography (SEC) of the products produced by: (1) FcnA2; (2) Fda2; (3) FdlA; and (4) FdlB on *S. mcclurei* fucoidan and molecular weight standards. The lowest band (**) of the standard (St), resulting from FFA2 treatment of fucoidan from *F. evanescens*, corresponds to a tetra-saccharide of $(1\rightarrow 4)$- and $(1\rightarrow 3)$-linked α-L-fucosyls with each fucosyl residue sulphated at C2; total mass has been calculated to be approximately 972 Da [27]. Reaction time was 48 h. (*) indicates an enzymatic fucoidan degradation product of either lower mass or higher charge than the lowest St band (**) compound.

In all cases, the smallest oligosaccharide ran further than the lowest of the standard, suggesting that the released oligosaccharides are either smaller or more charged, i.e., more sulphated, than the tetra-saccharide in the standard. The SEC profiles of the FcnA2 and Fda2 were similar, but the product profile differed from those of FdlA and FdlB which contained a smaller peak at around 22.5 min (corresponding to a molecular weight SEC standard of around 1.3 kDa), indicating that they acted slower if at all on certain fucoidan fragments <1.3 kD. Taken together with the C-PAGE results (Figure 4A), these data suggest that FdlA and FdlB exerted similar substrate attack preferences and left behind some oligomers around 1.3 kDa, whereas FcnA2 and Fda2 appeared to degrade the lower molar weight oligomers to a more significant extent.

2.7. New Construct of FcnA2

Western blot analysis of FcnA2 (Figure 2B) indicated that the spontaneous degradation of FcnA2 occurred from the C-terminal end, since the N-terminal His-tag was still present, making the protein visible in the Western blot. To avoid this degradation, a further truncation was made by removing an additional 80 amino acids from C-terminal end of the FcnA2 protein. This truncated enzyme was thus 229 amino acids shorter than the original FcnA enzyme and was called FcnAΔ229 (Table 1). FcnAΔ229 could be expressed very well and was purified with no apparent protein degradation, as illustrated by SDS-PAGE and Western blot analysis, giving the expected band size of 80 kDa (Figure 5A,B). FcnAΔ229 showed activity on the same substrates as FcnA2 (Figure 5C).

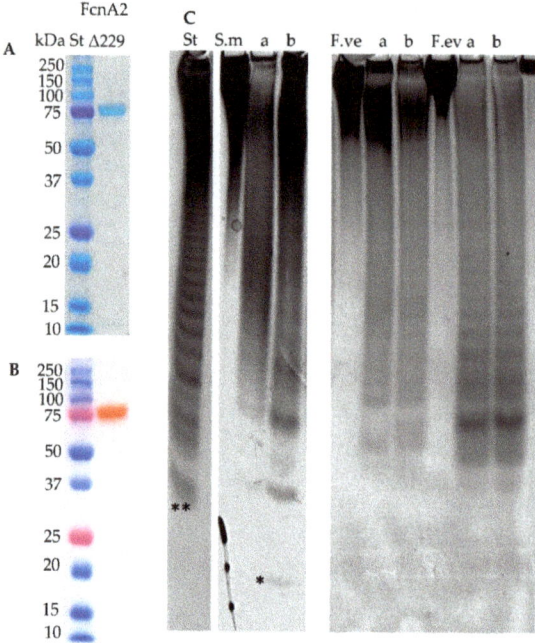

Figure 5. Purification and activity of enzyme FcnAΔ229. (**A**) SDS-PAGE indicating the expected molecular weight of 80 kDa and purity; (**B**) Western blot of purified FcnAΔ229. (St) is the protein plus molecular weight marker; and (**C**) enzyme activity by C-PAGE of (a) FcnA2 and (b) FcnAΔ229 on fucoidans from *S. mcclurei*, *F. vesiculosus* and *F. evanescens*. FcnA2 and FcnAΔ229 have similar profiles on *F. vesiculosus* and *F. evanescens* fucoidans. The reaction time was 24 h. The lowest band (**) of the standard (St), resulting from FFA2 treatment of fucoidan from *F. evanescens*, corresponds to a tetra-saccharide of (1→4)- and (1→3)-linked α-L-fucosyls with each fucosyl residue sulphated at C2; total mass has been calculated to be approximate 972 Da [27]. (*) An oligosaccharide of lower molecular weight or higher charge than the lowest band in the standard (**).

However, an oligosaccharide was released after 24 h that was running further than what was observed for FcnA2 (Figure 5C). This result indicated that the change in stability conferred by deletion of the 80 amino acids in FcnA2 in turn apparently enhanced substrate degradation, but the truncation did not confer any other apparent changes in the *S. mcclurei* degradation profile.

2.8. Stabilisation through C-Terminal Truncation of Fda1 and Fda2

The expression of Fda1 was high but the protein remained in the cell debris after sonication (Figure S1). By sequence analyses of Fda1 and Fda2, it was found that both enzymes contained two predicted Laminin G domains (IPR001791) (LamG domains) towards the C-terminal of each protein (Figure S2). Western blot analysis of Fda2 (Figure 2B) also indicated that enzyme destabilisation occurred via degradation from the C-terminal end as was observed for FcnA2. Hence, a strategy to stabilise the enzymes by deletion of the two predicted LamG domains in Fda1 and in Fda2 was developed and new constructs of Fda1, called Fda1Δ145 (one LamG domain deleted) and Fda1Δ395 (both LamG domains deleted), were prepared (Table 1). An additional his-tag was included with these new constructs to ensure better binding to the Ni^{2+} Sepharose column. Notably for Fda2, in addition to being highly unstable, substantial amounts of protein were lost during purification, presumably due to lack of binding to the column (data not shown). This new construct was called Fda2-His (Table 1). In addition, as for Fda1, new constructs devoid of either one or both of the two predicted LamG domains of Fda2 were also constructed. These Fda2 C-terminal deletion mutants were named Fda2Δ146 and Fda2Δ390 (Table S1 and Figure S2).

SDS-PAGE and Western blot analysis showed that all modified enzyme constructs expressed well. Some protein degradation was evident, but notably the double LamG deletion constructs, Fda1Δ395 and Fda2Δ390, appeared more stable than full length enzymes (Figure 6A,B). All the truncated enzymes exerted activity on *S. mcclurei* fucoidan, verifying the enzyme stabilisation strategy by LamG deletion (Figure 6C). Further study verified that Fda1Δ395 was stable but that degradation of the other truncated enzymes (Fda1Δ145, Fda2-C-His, Fda2Δ146, and Fda2Δ390) occurred already inside the *E. coli* cells, presumably via action of proteases in *E. coli*, recognising sites in the C-terminal of the enzymes, since degradation was evident in the *E. coli* cells before sonication (Figure S3).

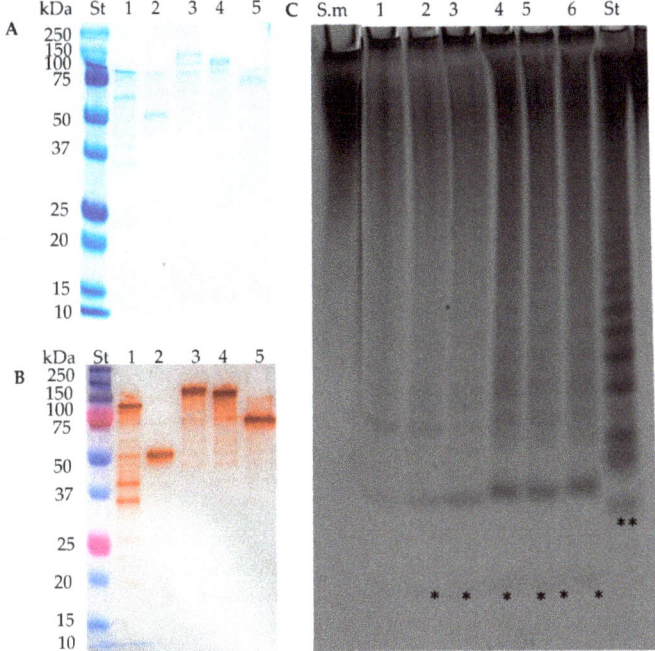

Figure 6. Purification and activity of Fda1 and Fda2 deletion mutants. (**A**) SDS-PAGE; and (**B**) Western blot of purified: (1) Fda1Δ145; (2) Fda1Δ395; (3) Fda2-C-His; (4) Fda2Δ146; and (5) Fda2Δ390. (St) is the protein plus molecular weight marker. The expected sizes of the proteins were 90, 50, 125, 110, 70 kDa respectively. (**C**) Enzymatic *S. mcclurei* fucoidan (S.m) degradation by C-PAGE: (1) Fda1Δ145; (2) Fda1Δ395; (3) Fda2-His; (4) Fda2Δ146; (5) Fda2Δ390; (6) Fda2; and the standard (St) resulting from FFA2 treatment of fucoidan from *F. evanescens*. The lowest band (**) of the standard (St), resulting from FFA2 treatment of fucoidan from *F. evanescens*, corresponds to a tetra-saccharide of (1→4)- and (1→3)-linked α-L-fucosyls with each fucosyl residue sulphated at C2; total mass has been calculated to be 972 Da [27]. (*) indicates an enzymatic fucoidan degradation product of either lower mass or higher charge than the lowest St band (**) compound. The reaction time was 48 h.

2.9. C-Terminally Truncated Fda1 Attacks α(1→3)-Linkages

The truncated Fda1 proteins Fda1Δ145 and Fda1Δ395 both catalysed degradation of most of the fucoidan substrates, although compared to the degradation achieved on the *S. mcclurei* fucoidan, the extent of degradation appeared to be lower (Figure 7). Both truncated enzymes produced comparable degradation patterns, releasing fucoidan oligo-saccharides that migrated to the same extent within the C-PAGE gels. Interestingly, Fda1 mutants were able to catalyse the degradation of fucoidans rich in α(1→3) fucosyl linkages from *T. ornata*, *S. cichorioides* and *U. pinnatifida* (Figure 7), indicating that the C-terminally truncated Fda1 enzymes attack α(1→3)-linkages as previously described [25]. Removal of up to 47% of the Fda1 sequence from the C-terminal thus resulted in a more stable enzyme that retain activity.

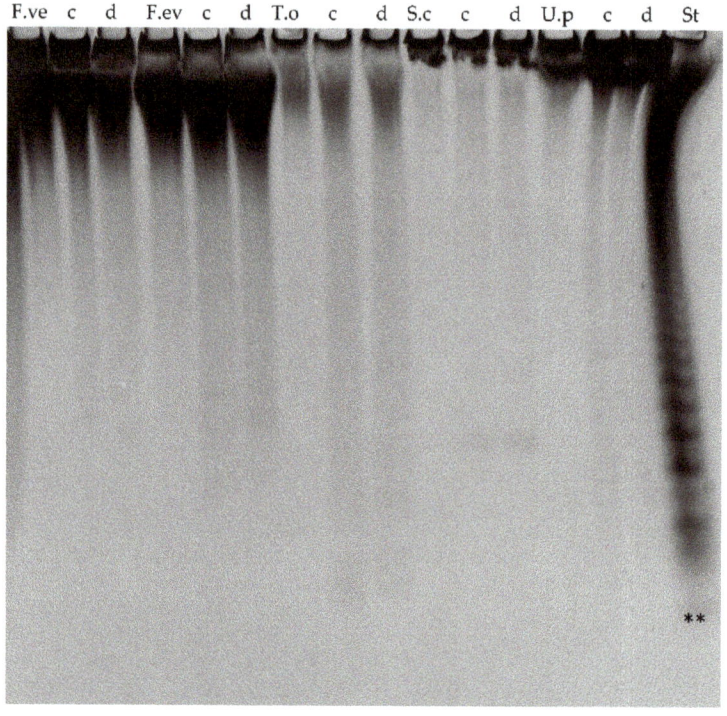

Figure 7. Enzyme activity of truncated Fda1 mutants by C-PAGE. Enzyme activity of (c) Fda1Δ145 and (d) Fda1Δ395 on fucoidans from *F. vesiculosus* (F.ve), *F. evanescens* (F.ev), *T. ornata* (T.o), *S. cichorioides* (S.c) and *U. pinnatifida* (U.p), and standard (st). Both enzymes show activity on all the tested substrates to a comparable degree. The lowest band (**) of the standard (St), resulting from FFA2 treatment of fucoidan from *F. evanescens*, corresponds to a tetra-saccharide of (1→4)- and (1→3)-linked α-L-fucosyls with each fucosyl residue sulphated at C2; total mass has been calculated to be 972 Da [27].

3. Discussion

This work showed that different microbially derived fucoidan-degrading enzymes exert activity on an array of different fucoidan substrates from brown macroalgae, even the very complex *S. mcclurei* fucoidan. FcnA2, Fda2, FdlA, and FdlB were found to degrade *S. mcclurei* fucoidan, with Fda2, FdlA and FdlB having particularly high activity on this fucoidan, which is known to contain sulphated galacto-fucan structural units and both α(1→4) and α(1→3) L-fucosyl linkages (Figure 1). FcnA2 and FcnA2Δ229 were more active than all the other enzymes on fucoidan from *F. evanescens* and they were also more active on fucoidan from *F. evanescens* than on fucoidan from *F. vesiculosus* suggesting an effect of the substrate sulphatation pattern or of other structural features of the substrate. Fda2 was the only enzyme that degraded fucoidans rich in α(1→3) L-fucosyl linkages, but FdlA and FdlB were also able to at least partially degrade the fucoidan from *T. ornata*. FdlA and FdlB were previously claimed to be lyases acting on manno-glucurono-linkages in fucoidan from *K. crassifolia* (i.e., *S. sculpera*). In the present work these enzymes were specifically found to act as endo-fucoidanases on fucoidans devoid of these types of bonds and did not produce any unsaturated 4–5 oligosaccharide uronides.

Enzyme stabilization was successfully achieved by targeted truncation of the C-terminal ends of FcnA2, Fda1 and Fda2. Interestingly, for FcnA2, the stabilisation by C-terminal truncation, to the enzyme variant FcnAΔ229, resulted in an enzyme which appeared able to foster more profound degradation of the *S. mcclurei* fucoidan than the parent enzyme. For Fda1 and Fda2, successful expression and stabilisation were attained by LamG domain deletion, in turn this stabilisation allowed us to show the ability of the otherwise unstable Fda1 to catalyse degradation of the *S. mcclurei* fucoidan. The data obtained have implications for use of these enzymes, including the stabilised versions, in fucoidan processing.

Enzymatically produced short sulphated fuco-oligosaccharides, with degree of polymerisation of 4–10, derived from *Sargassum horneri*, obtained via treatment with a recombinant GH family 107 endo-fucoidanase, FFA1 (originating from the marine bacterium *Formosa algae*), were recently reported unable to suppress growth of DLD-1 human colon cancer cells in vitro, whilst this ability, i.e., potential anti-cancerogenic activity, is significant for native fucoidan from *S. horneri* [28]. In contrast, enzymatically produced sulphated fucoidan products from *F. evanescens* have been reported to have a better effect than the corresponding native, higher molecular weight fucoidan, on the functional activity of innate immunity cells in vitro [36]. Partially depolymerised fucoidan fractions from *Saccharina cichorioides* exert strong inhibition of colony formation of colorectal carcinoma cells HT-29 in vitro [37]. It is not yet known whether specific structural units of fucoidan backbones or if particular sidechains or substitutions on fucoidans confer specific bioactivity functions. The results of the present work enable targeted production of defined fucoidan oligomer products. The availability of such homogenous fucoidan oligomers will permit rigorous research studies on the putative pharmaceutical functions of fucoidans of different structural configurations.

4. Materials and Methods

4.1. Fucoidan Substrates

Crude fucoidans from *Sargassum mcclurei*, *Fucus evanescens*, *Undaria pinnatifida*, and *Saccharina cichorioides* were extracted as described by Zvyagintseva et al. (1999) [38]. Fucoidan from *S. mcclurei* was purified further by ion-exchange chromatography [12]. *Turbinaria ornata* fucoidan was extracted as described by Thanh et al. (2013) [13]. Fucoidans from *F. evanescens*, *U. pinnatifida*, and *S. cichorioides* were purified as described by Kusaykin et al. (2006) [39]. Pure fucoidan from *Fucus vesiculosus* (F8190) was from Sigma-Aldrich (Steinheim, Germany).

4.2. Enzymes and Gene Constructs

Amino acid sequences for the five enzymes FcnA2, the C-terminal truncated version of FcnA (CAI47003.1) from *Mariniflexile fucanivorans* SW5 [24]; Fda1 (AAO00508.1) and Fda2 (AAO00509.1) from *Alteromonas* sp. SN-1009 (); FdlA (AAO00510.1) and FdlB (AAO00511.1) from *Flavobacterium* sp. SA-0082, were retrieved from GenBank (Table 1). The construct containing the gene encoding FcnA2 was designed to harbour an N-terminal 6xhistidine tag, while the gene constructs Fda1, Fda2, FdlA, and FdlB encoding the Fda1, Fda2, FdlA, FdlB proteins, respectively, were designed to harbour an N-terminal 10× histidine tag. The synthetic codon-optimised genes (optimised for *E. coli* expression), all devoid of their original signal peptide, were synthesised by GenScript (Piscataway, NJ, USA) and delivered as inserted either into the pET-45b(+) vector between the KpnI and PacI restriction sites (FcnA2) or into the pET-19b(+) plasmid vector between the NcoI and XhoI restriction sites (all other enzymes).

For FcnA2 C-terminal deletion of 80 amino acids of the enzyme equivalent to deletion of 229 amino acids of FcnA (GenBank No. CAI47003.1) was constructed, and the truncated protein was named FcnAΔ229 (Table 1).

Both Fda1 and Fda2 contain two predicted laminin G (LamG) domains in the sequence. Deletion mutants devoid of one or both predicted LamG domains were constructed by PCR amplification of the codon-optimised genes, each with an additional C-terminal 10× histidine tag, using CloneAmp HiFi polymerase premix (Takara Bio USA Inc., Mountain View, CA, USA) (primer sequences are listed in Table S1). For Fda1, the truncated proteins were named Fda1Δ145 and Fda1Δ395, as 145 and 395 amino acids had been removed from the C-terminal end, respectively. Analogously, for Fda2, the truncated versions were named Fda2Δ146 and Fda2Δ390, indicating that 146 and 390 amino acids, respectively, had been removed from the C-terminal. The construct of Fda2-His was done by adding 10× histidine tag at the C-terminal end. After PCR amplification, products were digested with BsaI and XhoI and ligated into the pET19b (+) vector between the NcoI and XhoI sites. Positive clones were confirmed by DNA sequencing.

The *Escherichia coli* strain DH5α (Invitrogen® Life Technologies, Thermo Fisher Scientific, Waltham, MA, USA), was used as plasmid propagation host. BL21 (DE3) and C41 (DE3) (also from Invitrogen® Life Technologies) were used as expression hosts for the fucoidan-degrading enzymes (Table 1). Protein expression was done as described below.

4.3. Production of Recombinant Enzymes

Expression of FcnA2 and FcnAΔ229 was performed in *E. coli* (BL21 (DE3) harbouring the Pch2 (pGro7) plasmid. Overnight cultures grown at 37 °C with agitation (180 rpm) in lysogeny broth (LB) medium containing 100 µg mL^{-1} ampicillin and 34 µg mL^{-1} chloramphenicol were used to inoculate 500 mL LB containing 100 µg mL^{-1} ampicillin, 34 µg mL^{-1} chloramphenicol and 0.05% arabinose. The inoculated LB was incubated at 37 °C with 180 rpm shaking until cultures reached 0.6-0.8 OD$_{600}$. Enzyme expression was induced with 1mM IPTG for 20 h at 20 °C and 180 rpm.

Expression of Fda1, Fda1Δ145, Fda1Δ395, Fda2, Fda2-His, Fda2Δ146, and Fda2Δ390 was also performed in *E. coli* (BL21 (DE3) with Pch2 (pGro7)). Overnight cultures grown at 37 °C and 180 rpm in LB medium containing 100 µg mL^{-1} ampicillin and 34 µg mL^{-1} chloramphenicol were used to inoculate 500 mL auto-induction media containing 0.6% Na$_2$HPO$_4$, 0.3% KH$_2$PO$_4$, 2%, tryptone, 5% yeast extract, 5% NaCl, 0.6% glycerol, 0.05% glucose, 0.2% lactose, 0.05% arabinose, 100 µg mL^{-1} ampicillin and 34 µg mL^{-1} chloramphenicol. Cells were grown at 20 °C, 180 rpm for 20 h. Expression of FdlA, FdlB was performed in *E. coli* (C41 (DE3)). Overnight cultures grown at 37 °C and 180 rpm in LB medium containing 100 µg mL^{-1} ampicillin were used to inoculate 500 mL LB containing 100 µg mL^{-1} ampicillin and were grown at 37 °C and 180 rpm until cultures reached 0.6-0.8 OD$_{600}$. The expression of the recombinant fucoidanases was induced with 1 mM IPTG during cell growth for 20 h at 20 °C and 180 rpm.

Cells were harvested by centrifugation at 5000× *g* for 20 min and 4 °C and the pellet was re-suspended in binding buffer (20 mM Tris-HCl buffer, 250 mM NaCl, 20 mM imidazole, pH 7.5) before being disrupted by UP400S Ultrasonic processor (Hielscher, Teltow, Germany)with 0.5 cycle and 100% amplitude. Cell debris was pelleted by centrifugation (20,000× *g*, 20 min at 4 °C). The supernatant obtained by centrifugation was then filtered through a 0.45 µm filter and applied to a 5 mL Ni^{2+} Sepharose HisTrap HP column (GE Healthcare, Uppsala, Sweden) which was equilibrated with binding buffer using an Äkta purifier (GE Healthcare, Uppsala, Sweden). The resin was washed 3 times with 20 mM Tris-HCl buffer, 250 mM NaCl, and 20 mM imidazole at pH 7.5 and proteins were eluted by a linear gradient of elution buffer (20 mM Tris-HCl buffer, 250 mM NaCl, and 20–500 mM imidazole, pH 7.5). The eluted fractions were analysed by sodium dodecyl sulphate–polyacrylamide gel electrophoresis (SDS-PAGE) and Western blotting as described below to assess the purity and homogenous fractions were pooled. Protein content was determined by the Bradford assay [40] with bovine serum albumin as standard.

4.4. SDS-PAGE

The homogeneity and molecular weight of the recombinantly expressed proteins were estimated by (SDS–PAGE) electrophoresis according to the Laemmli protocol [41]. Electrophoresis was performed in 12% acrylamide gels with the addition of 4× Laemmli loading-buffer, to 40 µg of crude protein and 5 µg purified protein and 5 mM DTT. The analysis of total intracellular proteins was conducted by using the biomass from 300 µL culture with 100 µL of 4× Laemmli loading-buffer, 10 µL of samples were loaded on the 12% acrylamide gels. The Protein Plus molecular weight marker (Bio-Rad Laboratories, Hercules, CA, USA) with molecular weights of 10–250 kDa was used as standard.

4.5. Western Blot Analysis of Proteins

Total intracellular protein, crude enzymes (40 µg) and pure enzymes (5 µg) were separated using 12% acrylamide gels with the addition of 4× Laemmli loading-buffer. Separated proteins were transferred onto a PVDF blotting membrane (GE Healthcare No. 1060022) and blotted in Tris-glycine pH 8.3 running buffer at 100 V for 45 min, after activation of the membrane in 96% ethanol for around 10 s. The membrane was blocked with 2% skim milk in 0.01 M TBS (Tris-based sodium chloride pH 7.6) buffer containing 0.1% Tween 20 (TBS_T buffer) for 60 min. The membrane was then incubated in TBS_T buffer with monoclonal anti-polyhistidine peroxidase conjugated antibody (Sigma-Aldrich, Steinheim, Germany) at 1:10.000 dilutions in a total volume of 30 mL for 1 h. The membrane was washed in TBS_T buffer 3 × 10 min and TBS with 0.1% Tween20 for 20 min. The bound antibodies were detected by horse radish peroxidase using the AEC Kit (Sigma-Aldrich, Steinheim, Germany) according to manufacturer's protocol.

4.6. Carbohydrate–Polyacrylamide Gel Electrophoresis (C-PAGE)

Reaction mixtures containing 0.5 µg/µL enzyme solution in 20 mM Tris–HCl buffer pH 7.4, 250 mM NaCl and 10 mM $CaCl_2$ (buffer A) and 1% weight/volume fucoidan in buffer A were incubated at 35 °C for 24–48 h. Each reaction mixture (10 µL) was mixed with 5 µL loading buffer (20% glycerol and 0.02% phenol red). Samples (5 µL) were electrophoresed at 400 V through a 20% (w/v) 1 mm thick resolving polyacrylamide gel with 100 mM Tris-borate buffer pH 8.3 for 1 h. Gel staining was performed in two steps, first with a solution containing 0.05% alcian blue 8GX (Panreac, Barcelona, Spain) in 2% acetic acid for 45 min and then with 0.01% O-toluidine blue (Sigma-Aldrich, Steinheim, Germany) in 50% aqueous ethanol and 1% acid acetic. The hydrolysate standard was obtained after enzymatic reaction of FFA2 on *Fucus evanescens* fucoidan [27].

4.7. SEC Analysis

High Performance Size Exclusion Chromatography was performed using an Ultimate iso-3100 SD pump with a WPS-3000 sampler (Dionex, Sunnyvale, CA, USA) connected to an RI-101 refractive index detector (Shodex, Showa Denko K.K., Tokyo, Japan). One hundred microliters of three times diluted reaction mixtures were loaded on a Shodex SB-806 HQ GPC column (300 × 8 mm) equipped with a Shodex SB-G guard column (50 mm × 6 mm) (Showa Denko K.K., Tokyo, Japan). Elution was performed with 100 mM sodium acetate pH 6 at a flow rate of 0.5 mL/min at room temperature. Pullunan standards were used as references.

Supplementary Materials: The following are available online at http://www.mdpi.com/1660-3397/16/11/422/s1, Figure S1: Recombinant expression of Fda1 in *E. coli*. (A) SDS-PAGE; and (B) Western blot of: (1) Autoinduced cells; (2) the cell debris (after sonication and protein extraction); and (3) crude extract after sonication and centrifugation. (St) is the protein plus molecular weight marker; Figure S2: Predicted protein domain structures of Fda1 and Fda2. Domains were predicted using NCBI conserved domain database (cdd) search tool and both proteins were found to contain two predicted LamG (Laminin G) superfamily domains. In Fda1, the domains span from position 429 to 574 aa and from 670 to 809 aa. For Fda2, the domains span from 496 to 641 aa and from 737 to 876 aa. Arrows indicate the points of truncation. Deletion mutants were named according to deletion from the C-terminal end, i.e., Fda1Δ145, Fda1Δ395, Fda2Δ146, and Fda2Δ390; Figure S3. (A) SDS-PAGE; and (B) Western blot of induced cells of: (1) Fda1; (2) Fda1Δ145; (3) Fda1Δ395; (4) Fda2-His; (5) Fda2; (6) Fda2Δ146; and (7) Fda2Δ390. St is the protein plus molecular weight marker (Bio-Rad Laboratories, Hercules, CA, USA). Table S1: Primers for constructing C-terminal deletion mutants.

Author Contributions: H.T.T.C., M.D.M., M.J.L., and A.S.M. designed the research, analysed and interpreted the data, and prepared the manuscript. A.S.S. and M.I.K. prepared the fucoidan from *F. evanescens*, *S. cichorioides* and *U. pinnatifida* and enzyme activity standard for C-PAGE. L.M.B., V.T.T.T., T.D.P. and B.H.T. prepared the *S. mcclurei* and *T. ornata* fucoidans, and J.H. contributed the SEC analyses. All authors have read and approved the final manuscript.

Acknowledgments: This work was supported by grants from Vietnam Academy of Science and Technology (VAST.ĐA47.12/16-19, VAST.HTQT.NGA.06/16-17), the Fucosan Interreg Germany–Denmark project, and the Seaweed Biorefinery Research Project in Ghana (SeaBioGha) supported by Denmark's development cooperation (Grant DANIDA-14-01DTU), The Ministry of Foreign Affairs of Denmark.

Conflicts of Interest: All authors declare no conflicts of financial or non-financial interest.

References

1. Bilan, M.I.; Grachev, A.A.; Ustuzhanina, N.E.; Shashkov, A.S.; Nifantiev, N.E.; Usov, A.I. Structure of a fucoidan from the brown seaweed *Fucus evanescens* C. Ag. *Carbohydr. Res.* **2002**, *337*, 719–730. [CrossRef]
2. Yu, L.; Ge, L.; Xue, C.; Chang, Y.; Zhang, C.; Xu, X.; Wang, Y. Structural study of fucoidan from sea cucumber acaudina molpadioides: A fucoidan containing novel tetrafucose repeating unit. *Food Chem.* **2014**, *142*, 197–200. [CrossRef] [PubMed]
3. Usov, A.I.; Bilan, M.I. Fucoidans—Sulfated polysaccharides of brown algae. *Russ. Chem. Rev.* **2009**, *78*, 785–799. [CrossRef]
4. Li, B.; Lu, F.; Wei, X.; Zhao, R. Fucoidan: Structure and bioactivity. *Molecules* **2008**, *13*, 1671–1695. [CrossRef] [PubMed]
5. Ale, M.T.; Mikkelsen, J.D.; Meyer, A.S. Important determinants for fucoidan bioactivity: A critical review of structure-function relations and extraction methods for fucose-containing sulfated polysaccharides from brown seaweeds. *Mar. Drugs* **2011**, *9*, 2106–2130. [CrossRef] [PubMed]
6. Pham, T.D.; Bui, L.M.; Tran, V.T.T.; Le, A.L.; Ermakova, S.P.; Zvyagintseva, T.N. Fucoidans from brown seaweeds collected from Nha Trang Bay: Isolation, structural characteristics, and anticancer activity. *VietNam J. Chem.* **2013**, *51*, 539–545.
7. Shevchenko, N.M.; Anastyuk, S.D.; Gerasimenko, N.I.; Dmitrenok, P.S.; Isakov, V.V.; Zvyagintseva, T.N. Polysaccharide and lipid composition of the brown seaweed *Laminaria gurjanovae*. *Russ. J. Bioorgan. Chem.* **2007**, *33*, 88–98. [CrossRef]
8. Holtkamp, A.D. *Isolation, Characterisation, Modification and Application of Fucoidan from Fucus vesiculosus Dissertation*; Südwestdeutscher Verlag für Hochschulschriften: Braunschweig, Germany, 2009.
9. Chizhov, A.O.; Dell, A.; Morris, H.R.; Haslam, S.M.; McDowell, R.A.; Shashkov, A.S.; Nifant'ev, N.E.; Khatuntseva, E.A.; Usov, A.I. A study of fucoidan from the brown seaweed *Chorda filum*. *Carbohydr. Res.* **1999**, *320*, 108–119. [CrossRef]
10. Holtkamp, A.D.; Kelly, S.; Ulber, R.; Lang, S. Fucoidans and fucoidanases-focus on techniques for molecular structure elucidation and modification of marine polysaccharides. *Appl. Microbiol. Biotechnol.* **2009**, *82*, 1–11. [CrossRef] [PubMed]
11. Anastyuk, S.D.; Shevchenko, N.M.; Nazarenko, E.L.; Dmitrenok, P.S.; Zvyagintseva, T.N. Structural analysis of a fucoidan from the brown alga *Fucus evanescens* by MALDI-TOF and tandem ESI mass spectrometry. *Carbohydr. Res.* **2009**, *344*, 779–787. [CrossRef] [PubMed]
12. Pham, T.D.; Menshova, R.V.; Ermakova, S.P.; Anastyuk, S.D.; Ly, B.M.; Zvyagintseva, T.N. Structural characteristics and anticancer activity of fucoidan from the brown alga *Sargassum mcclurei*. *Mar. Drugs* **2013**, *11*, 1456–1476. [CrossRef]
13. Thanh, T.T.T.; Tran, V.T.T.; Yuguchi, Y.; Bui, L.M.; Nguyen, T.T. Structure of fucoidan from brown seaweed *Turbinaria ornata* as studied by electrospray ionization mass spectrometry (ESIMS) and small angle X-ray scattering (SAXS) techniques. *Mar. Drugs* **2013**, *11*, 2431–2443. [CrossRef] [PubMed]

14. Ermakova, S.P.; Menshova, R.V.; Anastyuk, S.D.; Malyarenko, O.S.; Zakharenko, A.M.; Thinh, P.D.; Ly, B.M.; Zvyagintseva, T.N. Structure, chemical and enzymatic modification, and anticancer activity of polysaccharides from the brown alga *Turbinaria ornata*. *J. Appl. Phycol.* **2015**, *28*, 2495–2505. [CrossRef]
15. Ale, M.T.; Maruyama, H.; Tamauchi, H.; Mikkelsen, J.D.; Meyer, A.S. Fucoidan from *Sargassum* sp. and *Fucus vesiculosus* reduces cell viability of lung carcinoma and melanoma cells in vitro and activates natural killer cells in mice in vivo. *Int. J. Biol. Macromol.* **2011**, *49*, 331–336. [CrossRef] [PubMed]
16. Lapikova, E.S.; Drozd, N.N.; Tolstenkov, A.S.; Makarov, V.A.; Zvyagintseva, T.N.; Shevchenko, N.M.; Bakunina, I.U.; Besednova, N.N.; Kuznetsova, T.A. Inhibition of thrombin and factor Xa by *Fucus evanescens* fucoidan and its modified analogs. *Bull. Exp. Biol. Med.* **2008**, *146*, 328–333. [CrossRef] [PubMed]
17. Rodriguez-Jasso, R.M.; Mussatto, S.I.; Sepúlveda, L.; Agrasar, A.T.; Pastrana, L.; Aguilar, C.N.; Teixeira, J.A. Fungal fucoidanase production by solid-state fermentation in a rotating drum bioreactor using algal biomass as substrate. *Food Bioprod. Process.* **2013**, *91*, 587–594. [CrossRef]
18. Wu, Q.; Ma, S.; Xiao, H.; Zhang, M.; Cai, J. Purification and the secondary structure of fucoidanase from *Fusarium* sp. LD8. Evidence-based Complement. *Altern. Med.* **2011**. [CrossRef]
19. Sakai, T.; Ishizuka, K.; Shimanaka, K.; Ikai, K.; Kato, I. Structures of oligosaccharides derived from *Cladosiphon okamuranus* fucoidan by digestion with marine bacterial enzymes. *Mar. Biotechnol.* **2003**, *5*, 536–544. [CrossRef] [PubMed]
20. Furukawa, S.; Fujikawa, T.; Koga, D.; Ide, A. Purification and some properties of exo-type fucoidanases from *Vibrio* sp. N-5. *Biosci. Biotechnol. Biochem.* **1992**, *56*, 1829–1834. [CrossRef]
21. Kusaykin, M.I.; Silchenko, A.S.; Zakharenko, A.M.; Zvyagintseva, T.N. Fucoidanases. *Glycobiology* **2015**, *26*, 3–12. [CrossRef] [PubMed]
22. Silchenko, A.S.; Kusaykin, M.I.; Zakharenko, A.M.; Menshova, R.V.; Khanh, H.H.N.; Dmitrenok, P.S.; Isakov, V.V.; Zvyagintseva, T.N. Endo-1,4-fucoidanase from Vietnamese marine mollusk *Lambis* sp. which producing sulphated fucooligosaccharides. *J. Mol. Catal. B Enzym.* **2014**, *102*, 154–160. [CrossRef]
23. Kitamura, K.; Matsuo, M.; Yasuj, T. Enzymic degradation of Fucoidan by fucoidanase from the Hepatopancreas of *Patinopecten yessoensis*. *Biosci. Biotechnol. Biochem.* **1992**, *56*, 490–494. [CrossRef] [PubMed]
24. Colin, S.; Deniaud, E.; Jam, M.; Descamps, V.; Chevolot, Y.; Kervarec, N.; Yvin, J.C.; Barbeyron, T.; Michel, G.; Kloareg, B. Cloning and biochemical characterization of the fucanase FcnA: Definition of a novel glycoside hydrolase family specific for sulfated fucans. *Glycobiology* **2006**, *16*, 1021–1032. [CrossRef] [PubMed]
25. Takayama, M.; Koyama, N.; Sakai, T.; Kato, I. Enzymes Capable of Degrading a Sulfated-Fucose-Containing Polysaccharide and Their Encoding Genes. U.S. Patent No US 6,489,155 B1, 3 December 2002.
26. Lombard, V.; Golaconda Ramulu, H.; Drula, E.; Coutinho, P.M.; Henrissat, B. The carbohydrate-active enzymes database (CAZy) in 2013. *Nucleic Acids Res.* **2014**, *42*, 490–495. [CrossRef] [PubMed]
27. Silchenko, A.S.; Ustyuzhanina, N.E.; Kusaykin, M.I.; Krylov, V.B.; Shashkov, A.S.; Dmitrenok, A.S.; Usoltseva, R.V.; Zueva, A.O.; Nifantiev, N.E.; Zvyagintseva, T.N. Expression and biochemical characterization and substrate specificity of the fucoidanase from *Formosa algae*. *Glycobiol. Adv.* **2017**, 1–10. [CrossRef]
28. Silchenko, A.S.; Rasin, A.B.; Kusaykin, M.I.; Kalinovsky, A.I.; Miansong, Z.; Changheng, L.; Malyarenko, O.; Zueva, A.O.; Zvyagintseva, T.N.; Ermakova, S.P. Structure, enzymatic transformation, anticancer activity of fucoidan and sulphated fucooligosaccharides from *Sargassum horneri*. *Carbohydr. Polym.* **2017**, *175*, 654–660. [CrossRef] [PubMed]
29. Sakai, T.; Kimura, H.; Kojima, K.; Shimanaka, K.; Ikai, K.; Kato, I. Marine bacterial sulfated fucoglucuronomannan (SFGM) lyase digests brown algal SFGM into trisaccharides. *Mar. Biotechnol.* **2003**, *5*, 70–78. [CrossRef] [PubMed]
30. Sakai, T.; Kimura, H.; Kato, I. Purification of sulfated fucoglucuronomannan lyase from bacterial strain of Fucobacter marina and study of appropriate conditions for its enzyme digestion. *Mar. Biotechnol.* **2003**, *5*, 380–387. [CrossRef] [PubMed]
31. Rath, A.; Deber, C.M. Correction factors for membrane protein molecular weight readouts on sodium dodecyl sulfate-polyacrylamide gel electrophoresis. *Anal. Biochem.* 2013. [CrossRef] [PubMed]
32. Sakai, T.; Kawai, T.; Kato, I. Isolation and characterization of a fucoidan-degrading marine bacterial strain and its fucoidanase. *Mar. Biotechnol.* **2004**, *6*, 335–346. [CrossRef] [PubMed]
33. Sakai, T.; Kimura, H.; Kato, I. A marine strain of Flavobacteriaceae utilizes brown seaweed fucoidan. *Mar. Biotechnol.* **2002**, *4*, 399–405. [CrossRef] [PubMed]

34. Zvyagintseva, T.N.; Shevchenko, N.M.; Chizhov, A.O.; Krupnova, T.N.; Sundukova, E.V.; Isakov, V.V. Water-soluble polysaccharides of some far-eastern brown seaweeds. Distribution, structure, and their dependence on the developmental conditions. *J. Exp. Mar. Biol. Ecol.* **2003**, *294*, 1–13. [CrossRef]
35. Nishino, T.; Nishioka, C.; Ura, H.; Nagumo, T. Isolation and partial characterization of a novel amino sugar-containing fucan sulfate from commercial *Fucus vesiculosis* fucoidan. *Carbohydr. Res.* **1994**, *255*, 213–224. [CrossRef]
36. Kuznetsova, T.A.; Smolina, T.P.; Besednova, N.N.; Silchenko, A.S.; Imbs, T.I.; Ermakova, S.P. Effect of sulfated polysaccharides from brown alga *Fucus evanescens* and their enzymatic transformation product on functional activity of innate immunity cells. *Antibiot Khimioter.* **2016**, *61*, 10–14. (In Russian) [PubMed]
37. Anastyuk, S.D.; Shevchenko, N.M.; Usoltseva (Menshova), R.V.; Silchenko, A.S.; Zadorozhny, P.A.; Dmitrenok, P.S.; Ermakova, S.P. Structural features and anticancer activity in vitro of fucoidan derivatives from brown alga *Saccharina cichorioides*. *Carbohydr. Polym.* 2017. [CrossRef] [PubMed]
38. Zvyagintseva, T.N.; Shevchenko, N.M.; Popivnich, I.B.; Isakov, V.V.; Scobun, A.S.; Sundukova, E.V.; Elyakova, L.A. A new procedure for the separation of water-soluble polysaccharides from brown seaweeds. *Carbohydr. Res.* **1999**, *322*, 32–39. [CrossRef]
39. Kusaykin, M.I.; Chizhov, A.O.; Grachev, A.A.; Alekseeva, S.A.; Bakunina, I.Y.; Nedashkovskaya, O.I.; Sova, V.V.; Zvyagintseva, T.N. A comparative study of specificity of fucoidanases from marine microorganisms and invertebrates. *J. Appl. Phycol.* **2006**, *18*, 369–373. [CrossRef]
40. Bradford, M.M. A rapid and sensitive method for the quantitation of microgram quantities of protein utilizing the principle of protein-dye binding. *Anal. Biochem.* **1976**, *72*, 248–254. [CrossRef]
41. Laemmli, U.K. Cleavage of Structural Proteins during the Assembly of the Head of *Bacteriophage* T4. *Nature* **1970**, *227*, 680–685. [CrossRef] [PubMed]

© 2018 by the authors. Licensee MDPI, Basel, Switzerland. This article is an open access article distributed under the terms and conditions of the Creative Commons Attribution (CC BY) license (http://creativecommons.org/licenses/by/4.0/).

Article

A Comparative Study on Asymmetric Reduction of Ketones Using the Growing and Resting Cells of Marine-Derived Fungi

Hui Liu [1], Bi-Shuang Chen [1,2,3,*], Fayene Zeferino Ribeiro de Souza [4] and Lan Liu [1,2,3]

1. School of Marine Sciences, Sun Yat-Sen University, Guangzhou 510275, China; liuh229@mail.sysu.edu.cn (H.L.); cesllan@mail.sysu.edu.cn (L.L.)
2. Guangdong Provincial Key Laboratory of Marine Resources and Coastal Engineering, Guangzhou 510275, China
3. South China Sea Bio-Resource Exploitation and Utilization Collaborative Innovation Center, Sun Yat-Sen University, Guangzhou 510275, China
4. Departamento de Química, Faculdade de Ciências, UNESP, Bauru 17033-360, Brazil; faylittlefay@yahoo.com.br
* Correspondence: chenbsh23@mail.sysu.edu.cn; Tel.: +86-20-84725459

Received: 16 January 2018; Accepted: 3 February 2018; Published: 14 February 2018

Abstract: Whole-cell biocatalysts offer a highly enantioselective, minimally polluting route to optically active alcohols. Currently, most of the whole-cell catalytic performance involves resting cells rather than growing cell biotransformation, which is one-step process that benefits from the simultaneous growth and biotransformation, eliminating the need for catalysts preparation. In this paper, asymmetric reduction of 14 aromatic ketones to the corresponding enantiomerically pure alcohols was successfully conducted using the growing and resting cells of marine-derived fungi under optimized conditions. Good yields and excellent enantioselectivities were achieved with both methods. Although substrate inhibition might be a limiting factor for growing cell biotransformation, the selected strain can still completely convert 10-mM substrates into the desired products. The resting cell biotransformation showed a capacity to be recycled nine times without a significant decrease in the activity. This is the first study to perform asymmetric reduction of ketones by one-step growing cell biotransformation.

Keywords: growing cells; resting cells; asymmetric reduction; marine fungi; chiral alcohols

1. Introduction

The preparation of enantiomerically pure secondary alcohols is of ever-increasing significance because these intermediates are important building blocks for the production of chiral pharmaceuticals, flavours, agrochemicals and functional materials [1,2]. For example, (S)-1-(3,4-dichlorophenyl)ethan-1-ol (**2n**) is a versatile intermediate in the synthesis of sertraline [3,4], which is used to treat depression, obsessive-compulsive disorder, panic disorder, anxiety disorders, post-traumatic stress disorder, and premenstrual dysphoric disorder. Another optically active alcohol, ethyl (S)-4-chloro-3-hydroxybutanoate, is a key intermediate for the synthesis of 3-hydroxy-3-methylglutaryl coenzyme A (HMG-CoA) reductase inhibitors, which are the active ingredients of a cholesterol-lowering drug Lipitor [5], while 2-bromo-1-phenylethanol (**2a**) is the precursor for the synthesis of anti-depressants and α- or β-adrenergic drugs such as fluoxetine, tomoxetine, and nisoxetine [6,7].

While there are many synthetic approaches that furnish chiral alcohols in high enantiomeric excess, catalytic asymmetric reduction of prochiral carbonyl compounds offers several advantages because it can easily be applied to a wide range of substrates, the utilization of material is high, and waste and by-products can be drastically reduced [8,9]. Chemical methods using metal-complex catalysts have been employed [10–12]. Nevertheless, due to the presence of metals, most of the described methods require the use of either a cumbersome catalyst preparation or reductive/oxidative follow-up chemistry. Metal-free catalytic enantioselective reduction of prochiral ketones remains challenging.

Biocatalysis in asymmetric synthesis, as a result of their complex chiral constitution, are predominantly suited for the manufacture of optically pure stereoisomers [13–17]. Indeed, bioreduction of ketones using carbonyl reductases to chiral alcohols have been found and purified, such as S1 from *Candida magnoliae* [18], KaCR from *Kluyveromyces aestuarii* [19], and PsCR I from *Pichia stipitis* [20]. However, the high price of cofactors (approximately 1 g/485 euros), including Nicotinamide adenine dinucleotide (NADH) or Nicotinamide adenine dinucleotide phosphate (NADPH), is an impediment to the application of this approach. Therefore, efficient and cost-effective cofactor regeneration systems such as enzyme- and substrate-coupled systems must be developed [21]. Formate dehydrogenase (FDH) or glucose dehydrogenase (GDH) could be used as enzyme-coupled systems for the recycling of NAD^+ or $NADP^+$ [22], and 2-propanol can be used as a co-substrate because of its low cost and the feasibility of forcing the reaction towards completion by removing the acetone co-product under reduced pressure [23,24].

An easily operated and whole-cell system adaptable method is regarded as a viable "green" alternative synthetic approach [25,26] due to its unique advantages such as mild reaction conditions, environmental friendliness, regeneration of cofactors in situ, easy production and relatively low price; this method has therefore attracted great attention and been extensively investigated in recent years [27]. However, most whole-cell catalytic studies involve resting cells rather than growing cell biotransformation [28–30]. The resting cells are resuspended in buffer solution under non-growing conditions and are used as biocatalysts for the production of target compounds, which benefit from convenient downstream product separation. Growing cell biotransformation is the one-step process in which a certain amount of substrate was added to the medium and the target product was synthesized via one or several enzymatic reactions from the substrate during cell culture. Growing cell biotransformation is similar to microbial fermentation in its potential for industrial-scale production and shows significant advantages over resting cell biotransformation due to its ease of execution, which is a result of features such as simple operation steps, no need for cell preparation, and readiness for industrial scale production.

Therefore, in the present study, we report the results of a comparative study on the asymmetric reduction of a variety of aromatic ketones using growing and resting cells of marine-derived fungi that offers an alternative, highly enantioselective and minimally polluting route to important enantiomeric pure alcohols.

2. Results

To fully assess the potential of marine-derived fungi as biocatalysts for the enantioselective reduction of prochiral ketones, whole mycelia of 13 marine fungi (*Penicillium citrinum* GIM 3.458, *Penicillium citrinum* GIM 3.251, *Penicillium citrinum* GIM 3.100, *Aspergillus sclerotiorum* AS 3.2578, *Aspergillus sydowii* AS 3.7839, *Aspergillus sydowii* AS 3.6412, *Geotrichum candidum* GIM 2.361, *Geotrichum candidum* GIM 2.616, *Rhodotorula rubra* GIM 2.31, *Rhodotorula mucilaginosa* GIM 2.157, *Geotrichum candidum* AS 2.1183, *Geotrichum candidum* AS 2.498 and *Rhodotorula rubra* AS 2.2241) were screened for stereoselective reduction of 1-(3-bromophenyl)ethan-1-one **1b**. The screening reaction was initially performed with 10 mL of Na_2HPO_4-KH_2PO_4 buffer (100 mM, pH 6.0) containing glucose (0.5 g), 5 mM 1-(3-bromophenyl)ethan-1-one (**1b**), and 3 g of resting cells at 30 °C, due to its frequent use for described resting cell biotransformation [31–34]. The results are shown in Table 1. The absolute configuration of (S)-**2b** was determined by comparing the specific measured signs of rotation to those reported in the literature [35]. In a control reaction performed in parallel and involving only (**1b**), glucose and buffer (without resting cells), no yield of the desired product (S)-1-phenylpropan-1-ol [(S)-**2b**] was detected, indicating that no chemically catalysed reaction occurred; thus, the reaction was carried out by the active enzymes present in the marine-derived fungi. It must be emphasized that the complete genome of strain *Rhodotorula rubra* AS 2.2241 has been sequenced and annotated in our laboratory. Taking the conversion, enantioselectivity, and availability of the genome sequence into account, we decided to continue to use strain *R. rubra* AS 2.2241 for all further studies.

Table 1. Bioreduction of 1-(3-bromophenyl)ethan-1-one (**1b**) by marine-derived fungi.

Entry	Strains	Yield (%) [a]	Ee (%) [a]	Genome Sequences Available in Our Laboratory
1	*P. citrinum* GIM 3.458	25.3	99 (S)	No
2	*P. citrinum* GIM 3.251	5.5	53.7 (S)	No
3	*P. citrinum* GIM 3.100	39.2	99 (S)	No
4	*A. sclerotiorum* AS 3.2578	n.d.	n.d.	No
5	*A. sydowii* AS 3.7839	n.d.	n.d.	No
6	*A. sydowii* AS 3.6412	n.d.	n.d.	No
7	*G. candidum* GIM 2.361	75.9	99 (S)	No
8	*G. candidum* GIM 2.616	75.1	99 (S)	No
9	*R. rubra* GIM 2.31	99	99 (S)	No
10	*R. mucilaginosa* GIM 2.157	82	99 (S)	No
11	*G. candidum* AS 2.1183	44.3	99 (S)	No
12	*G. candidum* AS 2.498	77.7	99 (S)	No
13	*R. rubra* AS 2.2241	99	99 (S)	Yes [b]
control	no	n.d.	n.d.	-

[a] Yield and *ee* values were determined by HPLC equipped with a Chiracel OD-H chiral column; [b] Genome sequences were available in our laboratory, of which the genomic DNA libraries for the Illumina platform were generated and sequenced at BGI (Shenzhen, China). n.d. = not determined.

2.1. Reductions with Growing Cells

2.1.1. Optimization of Growing Cell Biotransformation

Substrate **1b** was added at the time of inoculation. From Table 2, it is evident that *R. rubra* AS 2.2241 could grow in the presence of substrate **1b** and could reduce substrate **1b** into the corresponding alcohol (S)-**2b**. The growth of *R. rubra* AS 2.2241 in the presence of substrate **1b** (entries 1–5) was less than in the control media without ketone (entry 2′). Thirty-five percent less growth was observed in the medium with 5 mM substrate **1b**. With 10 mM, 11.6 mM and 13.3 mM ketone in the cultivation medium, 60%, 84% and 93% less growth was observed, respectively. There was no growth of *R. rubra* AS 2.2241 with 15 mM ketone **1b** in the medium. It appears that ketone **1b** had an inhibitory effect on the growth of *R. rubra* AS 2.2241, and the inhibition increases with the increasing concentration of ketone **1b**. No transformation of ketone **1b** was observed in the flask that was not inoculated with *R. rubra* AS 2.2241 (control 1′). This indicated that ketone **1b** could not be reduced into alcohols without the organism. The second control contained only the RM1 medium without ketone **1b**, and the flask was inoculated (control 2′). No corresponding alcohol **2b** was observed, suggesting that alcohol **2b** was a metabolism-independent product, and this reaction was an enzyme-catalytic reduction. Heat-denatured cell preparations in control experiments (control 3′) clearly showed that no chemically catalysed reaction occurred, thus providing further support for the fact that the reaction is carried out by the active enzyme. Although the growth of *R. rubra* AS 2.2241 in the presence of ketone **1b** was lower, it could grow in the presence of ketone **1b** and reduce ketones **1b** into alcohols (S)-**2b**.

Table 2. Reduction of **1b** with growing cells of R. rubra AS 2.2241 in RM1 medium [a].

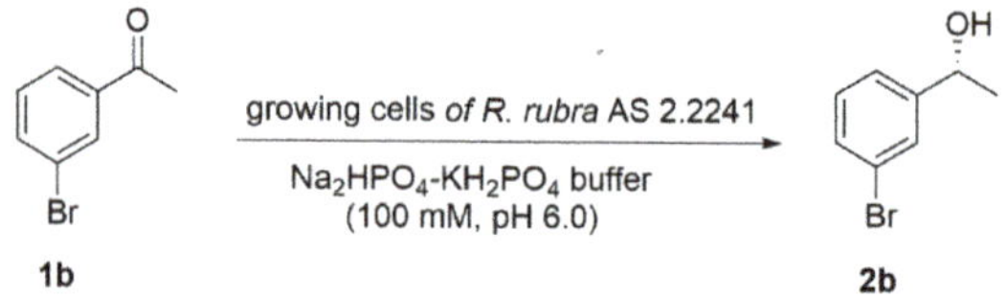

[Scheme: **1b** (3-bromophenyl methyl ketone, Br-C6H4-C(O)CH3) → **2b** (Br-C6H4-CH(OH)CH3) via growing cells of R. rubra AS 2.2241, Na$_2$HPO$_4$-KH$_2$PO$_4$ buffer (100 mM, pH 6.0)]

Entry	1b at 0 h (mM)	2b at 24 h (mM) [b]	2b at 48 h (mM) [b]	2b at 72 h (mM) [b]	2b at 96 h (mM) [b]	Cell Mass (Dry wt., g/L) [c]
1	5	2.36 (ee_S = 99)	3.06 (ee_S = 99)	4.95 (ee_S = 99)	4.07 (ee_S = 99)	4.61
2	10	0.42 (ee_S = 99)	8.33 (ee_S = 99)	9.9 (ee_S = 99)	6.21 (ee_S = 99)	2.81
3	11.6	n.c.	0.93 (ee_S = 99)	8.41 (ee_S = 99)	4.96 (ee_S = 99)	1.12
4	13.3	n.c.	n.c.	n.c.	n.c.	0.46
5	15	n.c.	n.c.	n.c.	n.c.	0
Control						
1'	10 (without inoculation) [d]	0	0	0	0	0
2'	0 (blank) [e]	0	0	0	0	7.16
3'	10 (with 5 g dry dead cells) [f]	0	0	0	0	n.d.

[a] 200 mL of RM1 medium, one loop of a single colony of R. rubra AS 2.2241, given amount of **1b**, cultured at 28 °C, 220 rpm for given times; [b] concentrations of **2b** were determined by HPLC analysis equipped with a Chiracel OD-H chiral column; [c] The cells were harvested by centrifugation at 4000 rpm and at 4 °C for 20 min; [d] This was used as a positive control for 1-(3-bromophenyl)ethan-1-one (**1b**); [e] This was done to check the growth of R. rubra AS 2.2241 in the absence of **1b**; [f] This was used to determine active enzymes by the cells; n.c. = no conversion; n.d. = not determined.

The next experiment was set up to determine the minimum growth of R. rubra AS 2.2241 required for the transformation of ketone **1b**. Here, different concentrations of ketones (5–15 mM) were added to the RM1 medium and inoculated with R. rubra AS 2.2241. The flasks were shaken (220 rpm) at 28 °C, and the samples were analysed at 24-h intervals. It was found that when the substrate concentration was smaller than 13.3 mM, the bioreduction reaction rate showed a clear trend towards an increase in product formation over time during the first 72 h of the reaction. The reaction velocity as well as product formation was low at the initial stage owing to low carbonyl reductase expression and then increased at the later stage, even though the desired product formation decreased over 72 h to 96 h. This decrease may be due to the alcohol dehydration, as was also found by Chandran and Das [36]. No significant changes in the product ee (almost above 99%) within the study were observed. When the substrate concentration was above 13.3 mM, cell growth and production of desired alcohols were not observed, suggesting the existence of substrate inhibition during the growing cell transformation. Therefore, the optimal substrate concentration in the medium was 10 mM. To date, yeasts, bacteria, fungi, and even plant tissues have been employed as biocatalysts for bio-reduction processes [28–30]. However, most of these biotransformations were performed with the resting cells of these microorganisms. Direct addition of substrates to the medium (one-pot) has significant potential for industrial scale production of chemicals because of the simplified cell preparation. Moreover, the total time of incubation was relatively short because growth and biotransformation occurred simultaneously. Although the substrates showed an inhibitory effect on the growth of R. rubra AS 2.2241, this growing cell biotransformation still showed good productivity in asymmetric reduction of tested ketones.

Substrate **1b** was added at different phases of growth. Addition of substrates at the beginning of the fermentation caused growth inhibition; therefore, substrate feeding (10 mM) was performed after induction. As illustrated in Figure 1, the 24-h-grown growing cells of R. rubra AS 2.2241 reduced ketone **1b** into alcohol (S)-**2b** within 24 h of incubation, while 48-h-grown growing cells also took 24 h to complete the transformation. In the case of 72-h-grown growing cells, the transformation was slightly smaller, and 84% of the added ketone **1b** was transformed after 24 h of incubation. Only 74% of ketone **1b** was transformed by 96-h-grown growing cells after 24 h of incubation. No further transformation was noticed with 96-h-grown cells with a prolonged period of incubation. In all cases, the initial ketones **1b** concentration was 10 mM. Thus, it is clear from these experiments that 24- and 48-h-grown growing cells were most active in transforming ketone **1b** to alcohol (*S*)-**2b**. This may be due to the presence of active enzymes after 24 and 48 h of growth. It has also been reported that maximum carbonyl reductase activity was obtained within 45–50 h of cultivation of *Acetobacter pasteurianus* GIM1.158 [37]. Poor enzyme activity was observed in 72-h-grown and 96-h-grown cultures, making the transformation of ketone **1b** inefficient. Poor enzyme activity may be due to the production of some metabolites that inactivated the enzyme. Transformation of ketone **1b** with growing cells used much more enzyme than that with the soluble enzyme system, but the time required for the transformation with the growing cells was not shorter. This is possible because the enzymatic system with the growing cells worked in an uncleaned environment (with cells, unused substrates and metabolites, etc.).

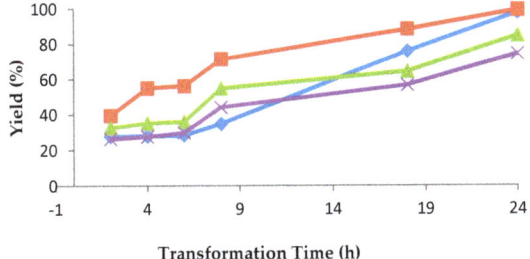

Figure 1. Effects of age of growing cells of *R. rubra* AS 2.2241 on the transformation of ketone **1b** (transformation was carried out with the growing cells of *R. rubra* AS 2.2241, and ketone **1b** was added at the different phases of growth: blue represents the data for **1b** added at 24 h of growth; red represents the data for **1b** added at 48 h of growth; green represents the data for **1b** added at 72 h of growth; purple represents the data for **1b** added at 96 h of growth).

2.1.2. Growing Cell Biotransformation of Ketones **1a–1n**

We next turned our attention to the substrates scope and the limitation of growing cells of strain *R. rubra* AS 2.2241 for the reduction of prochiral ketones **1a–1n**. Another 13 aromatic ketones that were structurally closely related to the main test substrate **1b** were tested (Figure 2) under the optimized conditions (addition of 10 mM substrates at the beginning of the fermentation and incubation for 72 h at 28 °C in RM1 medium buffer at pH 7.0), with the results presented in Table 3. Since the corresponding racemic reduction products (±)-β-phenylalcohols **2a–2m** (except **2b**) were not commercially available, they were obtained by the reduction of the aromatic ketones **1a**, **1c–1m** with sodium borohydride in methanol [38] and used as standard compounds for the analysis of the bioreduction products via chiral HPLC. The NMR spectra are in agreement with those reported in the literature [35,39–43].

For substrates (**1a–1i**) with an electron-withdrawing group such as –Br, –NO$_2$, –CF$_3$ in either *ortho*-, or *meta*-, or *para*-position (Table 3, entries 1–9), the reaction proceeded smoothly in all cases to yield the corresponding reduction products with more than 99% yield and 99% *ee*. By contrast, none of substrates (**1j**, **1k**, **1l**) with an electron-releasing group in the *para*-position were accepted by the enzyme (Table 3, entries 10–12), suggesting that an electron-withdrawing group might play an important role for the proper orientation of the ketones in the enzymes' active site. However, when substrates with electron-withdrawing groups in both ortho- and para-positions (**1m**) or in both meta- and para-positions (**1n**) were used, no desired reduction products were obtained, which is probably due to their more bulky structures. The absolute configurations of the reduction products (**2a–2i**) were determined by comparing the specific measured signs of rotation to those reported in the literature [35,39–43].

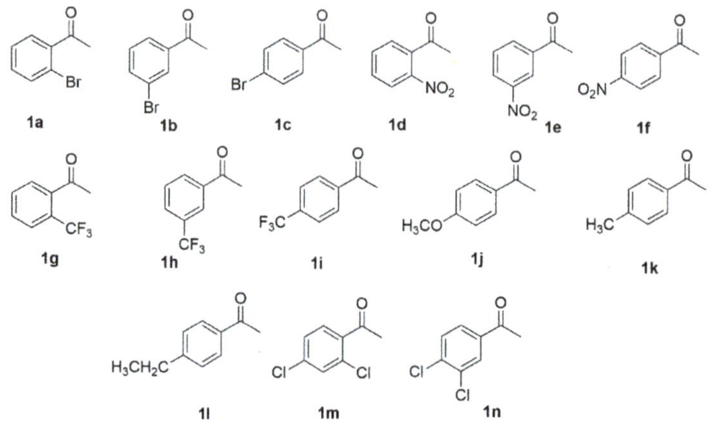

Figure 2. Structurally related aromatic ketones used for bioreduction by growing cells of R. rubra AS 2.2241.

Table 3. Stereoselective reduction of prochiral ketones with growing cells of R. rubra AS 2.2241 [a].

Entry	Substrates	Yield (%) [b]	Ee (%) [b]	Config. [c]
1	1a	99	99	S
2	1b	99	99	S
3	1c	99	99	S
4	1d	99	99	S
5	1e	99	99	S
6	1f	99	99	S
7	1g	99	99	S
8	1h	99	99	S
9	1i	99	99	S
10	1j	n.c.	n.c.	n.d.
11	1k	n.c.	n.c.	n.d.
12	1l	n.c.	n.c.	n.d.
13	1m	n.c.	n.c.	n.d.
14	1n	n.c.	n.c.	n.d.

[a] Reaction conditions: 200 mL of RM1 medium, one loop of a single colony of R. rubra AS 2.2241, 10 mM of substrates 1a–1n, inoculated at 28 °C, 220 rpm for 72 h; [b] Yield and ee were determined by chiral HPLC analysis equipped with a Chiracel OD-H chiral column; [c] The absolute configurations of the reduction products were determined by comparing the measured specific signs of rotation with those reported in the literature [35,39–43]; n.c. = no conversion; n.d. = not determined.

2.2. Reductions with Resting Cells

2.2.1. Optimization of Resting Cell Biotransformation

The transformation of ketone **1b** using the resting cells of different ages showed different types of transformation patterns. Here, the cells were grown under growth conditions in a nutrient medium for 24, 48, 72 and 96 h, and the cells were harvested from each phase and washed several times with a phosphate (0.1 M, pH 7.0) buffer. The strict separation of microbial growth and biotransformation offers many advantages: the washed cells were suspended to buffer (100 mM, pH 7.0) to transform ketone **1b** (10 mM) at 28 °C in an orbital shaker (220 rpm). The effects of the temperatures (in the range of 20–35 °C) and pH values (pH 6–8) on biotransformation were evaluated using the resting cells collected after 48 h of growth. As shown in Table 4, 24-h-grown resting cells gave the highest conversion (89% yield, 99% ee) at 25 °C with pH 7.0 buffered by Na_2HPO_4-KH_2PO_4 (100 mM) after 24 h. Therefore, these reaction conditions were used to further test the activity of the resting cells grown for 48 h, 72 h and 96 h. Gratifyingly, in each case, ketone **1b** was reduced into (*S*)-**2b** with good yields and excellent enantioselectivities. The 48-h-grown resting cells showed the slightly better activity than those obtained for the 72-h-grown and 96-h-grown resting cells, and the 48-h-grown resting cells were chosen for all further studies of the resting cell biotransformation.

Table 4. Reduction of **1b** with resting cells of *R. rubra* AS 2.2241 of different ages [a].

Entry	Types of Resting Cells	Reaction Conditions	Yield of 2b (%) [b]	Ee of 2b (%) [b]
1		25 °C, pH 6, 24 h	59	99 (S)
2		25 °C, pH 7, 24 h	89	99 (S)
3	24-h-grown	25 °C, pH 8, 24 h	52	99 (S)
4		20 °C, pH 7, 24 h	63	99 (S)
5		35 °C, pH 7, 24 h	41	99 (S)
6	48-h-grown	25 °C, pH 7, 24 h	99	99 (S)
7	72-h-grown	25 °C, pH 7, 24 h	81	99 (S)
8	96-h-grown	25 °C, pH 7, 24 h	77	99 (S)

[a] Reaction conditions: 3 g resting cells of *R. rubra* AS 2.224 of required ages, 10 mM of substrates **1b**, and 0.5 g glucose in 10 mL Na$_2$HPO$_4$-KH$_2$PO$_4$ buffer (100 mM, given pH-values), inoculated at given temperatures, 220 rpm for 24 h; [b] Yield and *ee* were determined using chiral HPLC instrument equipped with a Chiracel OD-H chiral column; [c] The absolute configurations of the reduction products were determined by comparing the specific signs of rotation measured with those reported in the literature [35,39–43].

One of the important considerations for industrial production is the capacity to recycle the catalyst. Therefore, experiments were performed to examine the recyclability of the resting cells of *R. rubra* AS 2.2241 for the reduction of **1b** as an example. For the results summarized in Figure 3, every reaction was performed in 10 mL of Na$_2$HPO$_4$-KH$_2$PO4 buffer (100 mM, pH 7.0) with 3 g of wet cells, 10 mM substrate and 0.5 g of glucose and shaken at 25 °C for 23 h. At the end of the reaction, the cells were centrifuged, washed twice with the same buffer [Na$_2$HPO$_4$-KH$_2$PO$_4$ buffer (100 mM, pH 7.0)] and reused for the next cycle under the same reaction conditions. Resting cells of strain *R. rubra* AS 2.2241 exhibited high activity and complete conversion for nine cycles (Figure 3). Only a slight decrease was observed in cycle 10, whereas almost no activity (2.6% yield of the desired product) was found in cycle 11. Remarkably, the reduction product *ee* showed no significant variation and remained above 99%. The high reusability of the whole cells of the strain *R. rubra* AS 2.2241 could be attributed to the expression of naturally immobilized ketone reductase in the cells, which is currently being implemented in our laboratory.

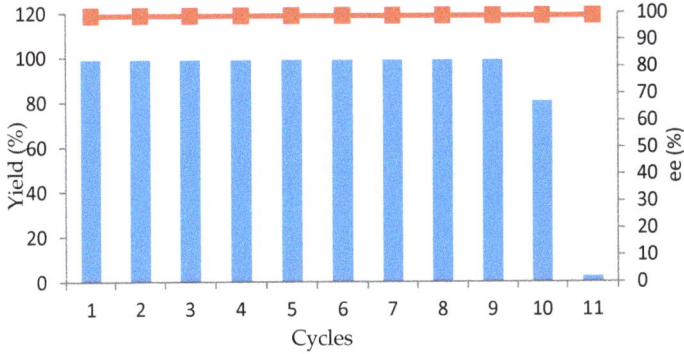

Figure 3. Reusability of resting cells of *R. rubra* AS 2.2241 of 48 h-grown age in the transformation of ketone **1b** (in each case the initial concentration of **1b** was 10 mM). Bar represents yields, square represents *ee* values. Yield and *ee* values were determined by chiral HPLC.

2.2.2. Resting Cell Biotransformation of Ketones 1a–1n

Under the condition optimized for the resting cell biotransformation, 14 substrates shown in Figure 2 were again used to test the substrate scope and limitations. As shown in Table 5, for most substrates (1a–1l, Table 5, entries, 1–12), the same results were obtained as those obtained with growing cell biotransformation. Surprisingly, substrates 1m and 1n were also converted by resting cells, and the corresponding products were obtained in 63% yield with 28% ee and 55% yield with 99% ee, suggesting that there might be different enzymes present in the resting cells which could be able to accept the substrates with bulky structure. The production of some metabolites during growing cell biotransformation inactivated the enzyme; therefore, there was no conversion for substrates 1m and 1n, as found by Banerjee [44] for the comparison of the activity of oxidases of *Curvularia lunata* in growing cell biotransformation and resting cell biotransformation.

Table 5. Stereoselective reduction of prochiral ketones with resting cells of R. rubra AS 2.2241 [a].

Entry	Substrates	Yield (%) [b]	Ee (%) [b]	Config. [c]
1	1a	99	99	S
2	1b	99	99	S
3	1c	99	99	S
4	1d	99	99	S
5	1e	99	99	S
6	1f	99	99	S
7	1g	99	99	S
8	1h	99	99	S
9	1i	99	99	S
10	1j	n.c.	n.c.	n.d.
11	1k	n.c.	n.c.	n.d.
12	1l	n.c.	n.c.	n.d.
13	1m	63	28	S
14	1n	55	99	S

[a] Reaction conditions: 10 mL Na_2HPO_4-KH_2PO_4 buffer (100 mM, pH 7.0), 3 g wet cells of 48-h-grown age, 10 mM various aromatic ketones, 0.5 g glucose, 25 °C, 24 h; [b] Yield and ee were determined by chiral HPLC instrument equipped with a Chiracel OD-H chiral column; [c] The absolute configurations of the reduction products were determined by comparing the specific signs of rotation measured with those reported in the literature [35,39–43]; n.c. = no conversion; n.d. = not determined.

3. Discussion

The development of biocatalysis requires novel biocatalysts in the form of isolated enzymes or whole cells [45], leading to a growing demand for robust and efficient biocatalysts. Fungi from marine environments are thoroughly adapted to surviving and growing under harsh conditions [46]. Such habitat-related characteristics are desirable features from a general biotechnological perspective and are of key importance to exploit a microorganism's enzymatic potential [47–50]. Indeed, fungi host novel enzymes showing optimal activities at extreme values of salt concentrations, pH and temperature, compared to enzymes isolated from terrestrial origins [51–53]. These advantages, in addition to their chemical and stereochemical properties and readily available sources (e.g., sea sources of enzymes represented by microorganisms or fungi, plants or animals; ease of growth), make marine enzymes ideal biocatalysts for fine chemistry and pharmaceutical sectors; these enzymes should be broadly explored [28–30,48,54–56].

Growing cells are under the growing conditions of proliferating and metabolically active cells, which are suspended in the medium of essential nutrients. Growing cell biotransformation is similar to the microbial fermentation, in which a certain amount of substrate was added at the time of inoculation or at regular phases of growth and was converted into the desired product via one or several enzymatic reactions during cell culture. The biotransformation is stable due to suitable growing environment for the cells. This one-step biotransformation showed substantial advantages such as the simultaneous growth and biotransformation, and simple operation steps, but usually substrate had an inhibition effect on the growing of cells which might be a limiting factor for growing cell biotransformation. Therefore, a straightforward approach for bioreduction of carbonyl compounds into the desired enantiomerically chiral alcohols using growing cells biotransformation still had not been described.

The resting cells refer to the group of active cells which are provided in buffer solution under non-growing conditions without any nutrients and are solely used as biocatalysts to yield many compounds [57]. Resting cell biotransformation show certain advantages over growing cell biotransformation, including no limit of substrate concentrations and convenient down downstream product separation. Besides, modelling and simulation of resting cell biotransformation processes had been successfully used to understand the investigated process, identify the limiting parameters, and optimize the reaction conditions. However, for industrial application, growing cells biotransformation showed more efficiency than that of resting cells biotransformation due to cumbersome catalyst preparation and storage.

Thus, we compared the productivity of marine fungi R. rubra AS 2.2241 from one-step growing cell biotransformation and two-step resting cell biotransformation in enantiomeric pure alcohols. Both of the reaction systems were carefully optimized for 10 mM-scale synthesis, resulting in good conversions and excellent enantioselectivities. Under the optimized conditions, R. rubra AS 2.2241 could grow in the presence of ketones **1a–1i** and could convert them into the corresponding alcohols (S)-**2a–2i**, offering an attractive alternative to the synthesis of enantiopure alcohols. Resting cells of different ages were also very effective in transforming ketones **1a–1i**, including **1m** and **1n**. The 48-h-grown resting cells could be reused for nine cycles without significant loss of activity while maintaining up to 99% *ee*. Further work will be focused on the elimination of substrate inhibition to improve the production and cell growth in growing cell biotransformation.

4. Materials and Methods

4.1. General

All chemicals were purchased from Sigma–Aldrich (Schnelldorf, Germany) and used without further purification unless specified otherwise. The culture media components were obtained from BD (Becton, Dickinson and Company, Bremen, Germany).

^{1}H and ^{13}C NMR spectra were recorded using a Bruker Advance 400 (Karlsruhe, Germany) (400 MHz and 100 MHz, respectively) instrument and internally referenced to residual solvent signals. Data for ^{1}H NMR are reported as chemical shift (d ppm), multiplicity (s = singlet, d = doublet, t = triplet, q = quartet, m = multiplet), integration, coupling constant (Hz) and assignment. Data for ^{13}C NMR are reported in terms of chemical shift. Optical rotations were obtained at 20 °C using a PerkinElmer 241 polarimeter (Shanghai, China) (sodium D line). Column chromatography was performed with a silica gel (0.060–0.200 mm, pore diameter ca. 6 nm) and mixtures of petroleum ether (PE) and ethyl acetate (EtOAc) as solvents. Thin-layer chromatography (TLC) was performed on 0.20 mm silica gel 60-F plates. Organic solutions were concentrated under reduced pressure with a rotary evaporator.

4.2. Chemical Synthesis of the Standard Racemic β-Phenylalcohols **2a–2f**, **2m** *and* **2n**

Ten mmol of NaBH$_4$ was added to a cooled (0 °C) solution of 2.5 mmol of each specific substrate (**1a–1f**, **1m** and **1n**) in 50 mL of methanol. After stirring for 10 min, the mixture was warmed to room temperature and stirred for another 3–4 h to complete the reduction. After quenching with 2 M HCl to pH 7.0, the mixture was extracted with EtOAc (50 mL × 3). The organic phases were washed with brine, dried over Na$_2$SO$_4$, filtered and concentrated in vacuum. The residue was purified by flash chromatography on silica gel (eluent: EtOAc/PE 1:20) to give the racemic alcohol **2a–2f**, **2m** and **2n** (see Supplementary Materials for NMR spectroscopic data).

4.3. Microorganisms

The marine fungi strains *Penicillium citrinum* GIM 3.458, *Penicillium citrinum* GIM 3.251, *Penicillium citrinum* GIM 3.100, *Aspergillus sclerotiorum* AS 3.2578, *Aspergillus sydowii* AS 3.7839, *Aspergillus sydowii* AS 3.6412, *Geotrichum candidum* GIM 2.361, *Geotrichum candidum* GIM 2.616, *Rhodotorula rubra* GIM 2.31, *Rhodotorula mucilaginosa* GIM 2.157, *Geotrichum candidum* AS 2.1183, *Geotrichum candidum* AS 2.498 and *Rhodotorula rubra* AS 2.2241 were isolated from a wide collection of isolates from marine sediments from Guangdong Province, China. All strains used in this study were deposited and commercially available at Guangdong Culture Collection Center or the China General Microbiological Culture Collection Center.

The marine fungi were maintained on agar plates at 4 °C and subcultured at regular intervals. The medium (RM1) used for cultivation contained glucose (15 g/L), peptone (5 g/L), yeast extract (grease, 5 g/L), disodium hydrogen phosphate (0.5 g/L), sodium dihydrogen phosphate (0.5 g/L), magnesium sulphate (0.5 g/L) and sodium chloride (10 g/L) and a final pH 7.0; this medium was sterilized at 115 °C in an autoclave for 25 min. A loop of a single colony was cut from the agar stock cultures and used to inoculate a given medium in an appropriate Erlenmeyer flask. This culture was shaken reciprocally (220 ppm) at 28 °C for given times.

4.4. Transformation with Growing Cells

Growing cells of *R. rubra* AS 2.2241 were used for the transformation of 1-(3-bromophenyl)ethanone (**1b**). 1-(3-bromophenyl)ethanone (**1b**) was added at the time of inoculation of the RM1 medium as described above by *R. rubra* AS 2.2241; in another set of experiments, 1-(3-bromophenyl)ethanone (**1b**) was added to the growing cells at different phases of growth, and the transformation continued.

4.4.1. **1b** Was Added at the Time of Inoculation

Different concentrations (5–15 mM) of 1-(3-bromophenyl)ethanone (**1b**) were added aseptically to 500-mL Erlenmeyer flasks containing 250 mL of sterile RM1 medium (pH 7.0). Three control flasks were prepared. The first control was prepared with 10 mM 1-(3-bromophenyl)ethanone in 250 mL of RM1 medium, and the flask was not inoculated with *R. rubra* AS 2.2241. This was used as a positive control for 1-(3-bromophenyl)ethanone (**1b**). The second control contained only the RM1 medium without 1-(3-bromophenyl)ethanone (**1b**), and the flask was inoculated with *R. rubra* AS 2.2241. This was done to check the growth of *R. rubra* AS 2.2241 in the absence of 1-(3-bromophenyl)ethanone (**1b**). The third control was prepared with 5 g of (dry weight) dead cells of *R. rubra* AS 2.2241 in RM1 medium with 10 mM 1-(3-bromophenyl)ethanone (**1b**). This was used to determine the active enzymes by the cells. This flask was not inoculated. All flasks were kept in a temperature-controlled (28 °C) orbital shaker at 220 rpm shaking speed. The samples were analysed after 24 h. The cells were removed by centrifugation at 4000 rpm and at 4 °C for 20 min. The collected cells were used for cell mass analysis. The supernatant (2 mL) was saturated with NaCl followed by extraction with 2×2 mL of HPLC eluent (*n*-hexane/*i*-PrOH = 95/5, *v*/*v*) by shaking for 5 min. The combined organic layer was dried over Na_2SO_4 and measured by HPLC for yield and *ee*.

4.4.2. **1b** Were Added at Different Phases of Growth

Several 500-mL Erlenmeyer flasks containing 250 mL of RM1 medium (pH 7.0) were inoculate with *R. rubra* AS 2.2241 and cultured at 28 °C on an orbital shaker (220 rpm). Next, 10 mM 1-(3-bromophenyl)ethanone (**1b**) was added to the growing cells after 24, 48, 72, 96 h of inoculation in separated flasks. For the transformation reaction, the flasks were again incubated under the same conditions. The samples were withdrawn at regular intervals and centrifuged, and the supernatant (2 mL) was saturated with NaCl followed by extraction with 2×2 mL of HPLC eluent

(n-hexane/i-PrOH = 95/5, v/v) by shaking for 5 min. The combined organic layer was dried over Na$_2$SO$_4$ and measured by HPLC for yield and *ee*.

4.5. General Methods for Growing Cells Biotransformation

All substrates **1a–1n** were added at the time of inoculation, and the reaction setup was the same as "**1b** was added at the time of inoculation". After a 72-h inoculation, the reaction was stopped by centrifugation at 4000 rpm and at 4 °C for 20 min to remove the cells. The supernatant (2 mL) was saturated with NaCl followed by extraction with 2 × 2 mL of HPLC eluent (n-hexane/i-PrOH = 95/5, v/v) by shaking for 5 min. The combined organic layer was dried over Na$_2$SO$_4$ and measured by HPLC for yield and *ee*.

4.6. Transformation with Resting Cells

R. rubra AS 2.2241 was cultivated in RM1 medium (pH 7.0) at 28 °C with 220 rpm speed. Cells in the culture age of 24 h, 48 h, 72 h and 96 h were harvested by centrifugation and washed twice with 100 mM Na$_2$HPO$_4$-KH$_2$PO$_4$ buffer (pH 7.0). Approximately 3 g of resting cells of 24-h-grown age was suspended in 10 mL of Na$_2$HPO$_4$-KH$_2$PO$_4$ buffer with the required pH (pH 6.0, 7.0 and 8.0) containing 0.5 g of glucose and 10 mM 1-(3-bromophenyl)ethanone (**1b**). The reaction mixtures were shaken (220 rpm) at the given temperatures (20 °C, 25 °C and 35 °C) for 24 h and stopped by centrifugation at 4000 rpm and at 4 °C for 20 min to remove the cells. The supernatant (2 mL) was saturated with NaCl followed by extraction with 2 × 2 mL of HPLC eluent (n-hexane/i-PrOH = 95/5, v/v) by shaking for 5 min. The combined organic layer was dried over Na$_2$SO$_4$ and measured by HPLC for yield and *ee*.

Resting cells of 48 h-, 72 h- and 96-h-grown ages were also suspended in 10 mL Na$_2$HPO$_4$-KH$_2$PO$_4$ buffer (pH 7.0, 100 mM) containing 0.5 g glucose and 10 mM of 1-(3-bromophenyl)ethanone (**1b**). The reaction mixtures were shaken (220 rpm) at 25 °C for 24 h and stopped by centrifugation at 4000 rpm and at 4 °C for 20 min to remove cells. The supernatant (2 mL) was saturated with NaCl followed by extraction with 2 × 2 mL of HPLC eluent (n-hexane/i-PrOH = 95/5, v/v) by shaking for 5 min. The combined organic layer was dried over Na$_2$SO$_4$ and measured by HPLC for yield and *ee*.

4.7. Recyclability

Reactions were carried out with 10 mM substrate **1b** in 10 mL of Na$_2$HPO$_4$-KH$_2$PO$_4$ buffer (100 mM, pH 7.0) containing 0.5 g of glucose and 3 g of resting cells of 48-h-grown age, shaken at 25 °C for 24 h. At the end of the reaction, the cells were centrifuged at 4000 rpm for 20 min to be separated from the reaction mixture, then washed by Na$_2$HPO$_4$-KH$_2$PO$_4$ buffer (100 mM, pH 7.0) and resuspended in 10 mL of the same buffer containing the same substrates and glucose. The reaction mixture (2 mL of supernatant separated from cells) was saturated with NaCl followed by extraction with 2 × 2 mL of HPLC eluent (n-hexane/i-PrOH = 95/5, v/v) by shaking for 5 min. The combined organic layer was dried over Na$_2$SO$_4$ and measured by HPLC for yield and *ee*.

4.8. General Methods for Resting Cell Biotransformation

Reactions were performed in 50-mL screw-capped glass vials to prevent evaporation of the substrate/product. Shaking was performed in a heated ground-top shaker at 25 °C with 220 rpm. Approximately 3 g of resting cells of 48-h-grown age (wet cells) were resuspended in 10 mL of Na$_2$HPO$_4$-KH$_2$PO$_4$ buffer (100 mM, pH 7.0) containing 0.5 g of glucose and 10 mM **1a–1n**. After 24 h, the reaction was stopped by centrifugation at 4000 rpm and at 4 °C for 20 min to remove cells. The supernatant (2 mL) was saturated with NaCl followed by extraction with 2 × 2 mL of HPLC eluent (n-hexane/i-PrOH = 95/5, v/v) by shaking for 5 min. The combined organic layer was dried over Na$_2$SO$_4$ and measured by HPLC for yield and *ee*.

4.9. Preparative-Scale Synthesis of Enantiomeric β-Phenylalcohols (S)-2a–2i, (S)-2m and (S)-2n by Resting Cells

For isolation and characterization of the bioreduction product, the reaction was performed on a preparative scale: 300 g resting cells of R. rubra AS 2.2241 were resuspended in 1000 mL of Na_2HPO_4-KH_2PO_4 buffer (100 mM, pH 7.0) with 50 g glucose and 10 mM of each substrate (1a–1i, 1m and 1n). The reaction mixture was incubated at 25 °C and shaken at 220 rpm for 24 h. The cells were removed by centrifugation and the supernatant was saturated with NaCl. The supernatant was extracted with EtOAc (1000 mL × 3). The organic phases were washed with brine, dried over Na_2SO_4, filtered and concentrated in vacuo. The residue was purified by flash chromatography on silica gel (eluent: EtOAc/PE 1:20) to give the enantiomerically pure alcohols 2a–2i, 2m, 2n. The isolated yield and ee of preparative-scale are comparable to those obtained from screening biotransformations. The spectroscopic data (^{1}H and ^{13}C NMR, and HPLC retention times) of enantiomerically alcohols 2a–2i, 2m and 2n are in agreement with those obtained for racemic forms, as described in the Supplementary Materials.

4.10. Assay Methods

Reaction products were analysed by chiral HPLC analysis using a Shimadzu LC-10AT VP series (Tokyo, Japan) and a Shimadzu SPD-M10Avp photo diode array detector (190–370 nm) with a Chiralcel AD-H column [eluent: n-hexane/i-PrOH (95:5, v/v), flow rate: 0.5 mL/min, column temperature 25 °C]. The yields (quantified using calibration curves) and product ee values of analytes were determined by chiral HPLC analyses according to the following retention time data: 1-(2-bromophenyl)ethanone (1a) [1a, 11.85 min; (R)-2a, 12.71 min (S)-2a, 13.23 min], 1-(3-bromophenyl)ethanone (1b) [1b, 10.66 min; (S)-2b, 16.39 min; (S)-2b, 17.26 min], 1-(3-bromophenyl)ethanone (1c) [1c, 11.21 min; (S)-2c, 16.83 min; (R)-2c, 17.99 min], 1-(2-nitrophenyl)ethanone (1d) [1d, 12.54; (R)-2d, 21.33 min; (S)-2d, 22.63 min], 1-(3-nitrophenyl)ethanone (1e) [1e, 21.14 min; (R)-2e, 26.25 min; (S)-2e, 27.06 min], 1-(3-nitrophenyl)ethanone (1f) [1f, 20.24 min; (S)-2f, 35.54 min; (R)-2f, 38.20 min], 1-(2-(trifluoromethyl)phenyl)ethanone (1g) [1g, 12.49 min; (R)-2g, 11.83 min; (S)-2g, 12.34 min], 1-(3-(trifluoromethyl)phenyl)ethanone (1h) [1h, 11.13 min; (S)-2h, 14.06 min; (R)-2h, 15.57 min], 1-(4-(trifluoromethyl)phenyl)ethanone (1i) [1i, 11.68 min; (S)-2i, 16.55 min; and (R)-2i, 17.79 min], 1-(2,4-dichlorophenyl)ethanone (1m) [1m, 15.09 min; (R)-2m, 19.09 min; (S)-2m, 21.26 min], 1-(3,4-dichlorophenyl)ethanone (1n) [1n, 12.39 min; (R)-2n, 12.79 min; and (S)-2n, 13.67 min].

The optical rotations of products isolated from the biocatalytic reactions were determined in a 1-dm cuvette using a Perkin-Elmer model 241 polarimeter and were referenced to the Na–D line. The experimental and reported data are listed below: (S)-2a, $[\alpha]_D^{20} = -62.4$ (c 1.00, CHCl$_3$) {(S)-1-(2-bromophenyl)ethanol $[\alpha]_D^{27} = -29.8$ (c 0.68, CHCl$_3$) [42]}; (S)-2b, $[\alpha]_D^{20} = -43.9$ (c 1.00, CHCl$_3$) {(S)-1-(3-bromophenyl)ethanol $[\alpha]_D^{20} = -27.6$ (c 1.00 in CHCl$_3$) [35]}; (S)-2c, $[\alpha]_D^{20} = -17.3$ (c 1.00, MeOH, {(S)-1-(4-bromophenyl)ethanol $[\alpha]_D^{21} = -20.6$ (c 1.07, MeOH) [39]}; (S)-2d, $[\alpha]_D^{20} = +64.9$ (c 0.1, MeOH) {(S)-1-(2-nitrophenyl)ethanone $[\alpha]_D^{25} = +18.5$ (c 0.23, MeOH) [40]}; (S)-2e, $[\alpha]_D^{20} = -88.5$ (c 0.1, CHCl$_3$) {(S)-1-(3-nitrophenyl)ethanone $[\alpha]_D^{25} = -20.5$ (c 1.0, CHCl$_3$) [40]}; (S)-2f, $[\alpha]_D^{20} = -67.2$ (c 0.1, CHCl$_3$) {(S)-1-(4-nitrophenyl)ethanone $[\alpha]_D^{25} = -20.5$ (c 1.2, CHCl$_3$) [40]}; (S)-2g, $[\alpha]_D^{20} = -43.1$ (c 0.1, MeOH) {(S)-1-(2-Trifluoromethylphenyl) ethanol $[\alpha]_D^{26} = -37.7$ (c 1.0, CH$_3$OH) [42]}; (S)-2h, $[\alpha]_D^{20} = -55.1$ (c 0.2, MeOH) {(S)-1-(3-(trifluoromethyl)phenyl)ethanol $[\alpha]_D^{25} = -21.9$ (c 1.40, MeOH) [35]}; (S)-2i, $[\alpha]_D^{20} = -62.2$ (c 0.2, CHCl$_3$) {(S)-1-(4-(trifluoromethyl)phenyl)ethanol $[\alpha]_D^{25} = -33.7$ (c 5.52, CHCl$_3$) [35]}; (S)-2m, $[\alpha]_D^{20} = -75.05$ (c 0.2, CHCl$_3$) {(S)-1-(2,4-dichlorophenyl)ethanol $[\alpha]_D^{25} = -52.4$ (c 0.55, CHCl$_3$) [43]}; (S)-2n, $[\alpha]_D^{20} = -30.0$ (c 0.3, CHCl$_3$) {(R)-1-(3,4-dichlorophenyl)ethanol $[\alpha]_D^{20} = +35.8$ (c 1.00, CHCl$_3$) [41]}.

Supplementary Materials: The following are available online at www.mdpi.com/1660-3397/16/2/62/s1. The NMR spectra and HPLC spectra of all the compounds are shown.

Acknowledgments: This study was funded by the Basic Research Program of Sun Yat-Sen University (Grant No. 17lgpy58), Natural Science Foundation of Guangdong Province (Grant No. 2017A030310232), the National Natural Science Foundation of China (Grant No. 41706148), China's Marine Commonweal Research Project (Grant No. 201305017), and the special financial fund of innovative development of marine economic demonstration project (Grant No. GD2012-D01-001).

Author Contributions: B.-S.C. and F.Z.R.d.S. designed the study. B.-S.C. wrote the paper. H.L. performed the enzymatic experiments. B.-S.C. and L.L. supervised the study. All authors made substantial contributions to the discussion of data and approved the final manuscript.

Conflicts of Interest: The authors declare no conflict of interest.

References

1. Lennon, L.C.; Ramsden, J.A. An efficient catalytic asymmetric route to 1-aryl-2-imidazol-1-yl-ethanols. *Org. Process Rev. Dev.* **2005**, *9*, 110–112. [CrossRef]
2. Shaikh, N.S.; Enthaler, S.; Junge, K.; Beller, M. Iron-catalyzed enantioselective hydrosilylation of ketones. *Angew. Chem. Int. Ed.* **2008**, *47*, 2497–2501. [CrossRef] [PubMed]
3. Barbieri, C.; Caruso, E.; D'Arrigo, P.; Fantoni, G.P.; Servi, S. Chemo-enzymatic synthesis of (R)- and (S)-3,4-dichlorophenylbutanolide intermediate in the synthesis of sertraline. *Tetrahedron Asymmetry* **1999**, *10*, 3931–3937. [CrossRef]
4. Quallich, G.J.; Woodall, T.M. Synthesis of 4-(S)-(3,4-dichlorophenyl)-3,4-dihydro-1(2H)-naphthalenone by SN2 cuprate displacement of an activated chiral benzylic alcohol. *Tetrahedron* **1992**, *48*, 10239–10248. [CrossRef]
5. Ma, S.K.; Gruber, J.; Davis, C.; Newman, L.; Gray, D.; Wang, A.; Grate, J.; Huisman, G.W.; Sheldon, R.A. A green-by-design biocatalytic process for atorvastatin intermediate. *Green Chem.* **2010**, *12*, 81–86. [CrossRef]
6. Xie, Y.; Xu, J.H.; Xu, Y. Isolation of a bacillus strain producing ketone reductase with high substrate tolerance. *Bioresour. Technol.* **2010**, *101*, 1054–1059. [CrossRef] [PubMed]
7. Zhu, D.; Chandrani, M.; Hua, L. 'Green' synthesis of important pharmaceutical building blocks: Enzymatic access to enantiomerically pure α-chloroalcohols. *Tetrahedron Asymmetry* **2005**, *16*, 3275–3278. [CrossRef]
8. Ma, Y.P.; Liu, H.; Chen, L.; Cui, X.; Zhu, J.; Deng, J.G. Asymmetric transfer hydrogenation of prochiral ketones in aqueous media with new water-soluble chiral vicinal diamine as ligand. *Org. Lett.* **2003**, *5*, 2103–2106. [CrossRef] [PubMed]
9. Sandoval, C.A.; Ohkuma, T.; Muñiz, K.; Noyori, R. Mechanism of asymmetric hydrogenation of ketones catalyzed by BINAP/1,2-diamine-ruthenium(II) complexes. *J. Am. Chem. Soc.* **2003**, *125*, 13490–13503. [CrossRef] [PubMed]
10. Noyori, R.; Ohkuma, T.; Kitamura, M.; Takaya, H.; Sayo, N.; Kumobayashi, H.; Akutagawa, S. Asymmetric hydrogenation of beta-keto carboxylic esters, a practical, purely chemical access to beta-hydroxy esters in high enantiomeric purity. *J. Am. Chem. Soc.* **1987**, *109*, 5856–5858. [CrossRef]
11. Noyori, R.; Ohkuma, T. Asymmetric catalysis by architectural and functional molecular engineering: Practical chemo- and stereoselective hydrogenation of ketones. *Angew. Chem. Int. Ed.* **2001**, *40*, 40–73. [CrossRef]
12. Noyori, R. Asymmetric catalysis: Science and opportunities (Nobel Lecture). *Angew. Chem. Int. Ed.* **2002**, *41*, 2008–2022. [CrossRef]
13. Brenna, E.; Fuganti, C.; Gatti, F.G.; Serra, S. Biocatalytic methods for the synthesis of enantioenriched odor active compounds. *Chem. Rev.* **2011**, *111*, 4036–4072. [CrossRef] [PubMed]
14. Choi, J.M.; Han, S.S.; Kim, H.S. Industrial applications of enzyme biocatalysis: Current status and future aspects. *Biotechnol. Adv.* **2015**, *33*, 1443–1454. [CrossRef] [PubMed]
15. Torrelo, G.; Hanefeld, U.; Hollmann, F. Biocatalysis. *Catal. Lett.* **2015**, *145*, 309–345. [CrossRef]
16. Albarránvelo, J.; Gonzálezmartínez, D.; Gotorfernández, V. Stereoselective biocatalysis: A mature technology for the asymmetric synthesis of pharmaceutical building blocks. *Biocatal. Biotransform.* **2018**, *36*, 102–130. [CrossRef]
17. Sheldon, R.A.; Woodley, J.M. Role of Biocatalysis in sustainable chemistry. *Chem. Rev.* **2018**, *118*, 801–838. [CrossRef] [PubMed]

18. Wada, M.; Kataoka, M.; Kawabata, H.; Yasohara, Y.; Kizaki, N.; Hasegawa, J.; Shimizu, S. Purification and characterization of NADPH-dependent carbonyl reductase, involved in stereoselective reduction of ethyl 4-chloro-3-oxobutanoate, from *Candida magnolia*. *Biotechnol. Biochem.* **1998**, *62*, 280–285. [CrossRef]
19. Yamamoto, H.; Mitsuhashi, K.; Kimoto, N.; Esaki, N.; Kobayshi, Y. Robust NADH-regenerator: Improved α-haloketone-resistant formate dehydrogenase. *Appl. Microbiol. Biotechnol.* **2005**, *67*, 33–39. [CrossRef] [PubMed]
20. Ye, Q.; Yan, M.; Xu, L.; Cao, H.; Li, Z.J.; Chen, Y.; Li, S.Y.; Ying, H.J. A novel carbonyl reductase from *Pichia stipitis* for the production of ethyl (S)-4-chloro-3-hydroxybutanoate. *Biotechnol. Lett.* **2009**, *31*, 537–542. [CrossRef] [PubMed]
21. Yasohara, Y.; Kizaki, N.; Hasegawa, J.; Wada, M.; Kataokac, M.; Shimizu, S. Stereoselective reduction of alkyl 3-oxobutanoate by carbonyl reductase from *Candida magnolia*. *Tetrahedron Asymmetry* **2001**, *12*, 1713–1718. [CrossRef]
22. Goldberg, K.; Schroer, K.; Lütz, S.; Liese, A. Biocatalytic ketone reduction—A powerful tool for the production of chiral alcohols—Part I: Processes with isolated enzymes. *Appl. Microbiol. Biotechnol.* **2007**, *76*, 237–248. [CrossRef] [PubMed]
23. Goldberg, K.; Edegger, K.; Kroutil, W.; Liese, A. Overcoming the thermodynamic limitation in asymmetric hydrogen transfer reactions catalyzed by whole cells. *Biotechnol. Bioeng.* **2006**, *95*, 192–198. [CrossRef] [PubMed]
24. Inoue, K.; Makino, Y.; Itoh, N. Production of (R)-chiral alcohols by a hydrogen-transfer bioreduction with NADH-dependent *Leifsonia* alcohol dehydrogenase (LSADH). *Tetrahedron Asymmetry* **2005**, *16*, 2539–2549. [CrossRef]
25. Milner, S.E. Recent trends in whole cell and isolated enzymes in enantioselective synthesis. *Arkivoc* **2012**, *2012*, 321–382.
26. Wachtmeister, J.; Rother, D. Recent advances in whole cell biocatalysis techniques bridging from investigative to industrial scale. *Curr. Opin. Biotechnol.* **2016**, *42*, 169–177. [CrossRef] [PubMed]
27. Wohlgemuth, R. Biocatalysis—Key to sustainable industrial chemistry. *Curr. Opin. Cell Biol.* **2010**, *21*, 713–724. [CrossRef] [PubMed]
28. Rocha, L.C.; Ferreira, H.V.; Pimenta, E.F.; Berlinck, R.G.S.; Rezende, M.O.O.; Landgraf, M.D.; Seleghim, M.H.R.; Sette, L.D.; Porto, A.L.M. Biotransformation of α-bromoacetophenones by the marine fungus *Aspergillus sydowii*. *Mar. Biotechnol.* **2010**, *12*, 552–557. [CrossRef] [PubMed]
29. Rocha, L.C.; Ferreira, H.V.; Luiz, R.F.; Sette, L.D.; Porto, A.L.M. Stereoselective bioreduction of 1-(4-methoxyphenyl)ethanone by whole cells of marine-derived fungi. *Mar. Biotechnol.* **2012**, *14*, 358–362. [CrossRef] [PubMed]
30. Rocha, L.C.; Seleghim, M.H.R.; Comasseto, J.V.; Sette, L.D.; Porto, A.L.M. Stereoselective bioreduction of α-azido ketones by whole cells of marine-derived fungi. *Mar. Biotechnol.* **2015**, *17*, 736–742. [CrossRef] [PubMed]
31. Chen, B.-S.; Yang, L.-H.; Ye, J.-L.; Huang, T.; Ruan, Y.-P.; Fu, J.; Huang, P.-Q. Diastereoselective synthesis and bioactivity of long-chain anti-2-amino-3-alkanols. *Eur. J. Med. Chem.* **2011**, *46*, 5480–5486. [CrossRef] [PubMed]
32. Chen, B.-S.; Hanefeld, U. Enantioselective preparation of (R) and (S)-3-hydroxycyclopentanone by kinetic resolution. *J. Mol. Catal. B Enzym.* **2013**, *85–86*, 239–242. [CrossRef]
33. Chen, B.-S.; Resch, V.; Otten, L.G.; Hanefeld, U. Enantioselective Michael addition of water. *Chem. Eur. J.* **2015**, *21*, 3020–3030. [CrossRef] [PubMed]
34. Chen, B.-S.; Liu, H.; de Souza, F.Z.R.; Liu, L. Organic solvent-tolerant marine microorganisms as catalysts for kinetic resolution of cyclic β-hydroxy ketones. *Mar. Biotechnol.* **2017**, *19*, 351–360. [CrossRef] [PubMed]
35. Kantam, M.L.; Laha, S.; Yadav, J.; Likhar, P.R.; Sreedhar, B.; Jha, S.; Bhargava, S.; Udayakiran, M.; Jagadeesh, B. An efficient copper-aluminum hydrotalcite catalyst for asymmetric hydrosilylation of ketones at room temperature. *Org. Lett.* **2008**, *10*, 2979–2982. [CrossRef]
36. Chandran, P.; Das, N. Role of plasmid in diesel oil degradation by yeast species isolated from petroleum hydrocarboncontaminated soil. *Environ. Technol.* **2012**, *33*, 645–652. [CrossRef] [PubMed]
37. Du, P.X.; Wei, P.; Lou, W.Y.; Zong, M.H. Biocatalytic anti-Prelog reduction of prochiral ketones with whole cells of *Acetobacter pasteurianus* GIM1.158. *Microb. Cell Fact.* **2014**, *13*, 84. [CrossRef] [PubMed]

38. De Oliveira, J.R.; Seleghim, M.H.R.; Porto, A.L.M. Biotransformation of methylphenylacetonitriles by Brazilian marine fungal strain *Aspergillus sydowii* CBMAI 934: Eco-friendly reactions. *Mar. Biotechnol.* **2014**, *16*, 156–160. [CrossRef] [PubMed]
39. Adachi, S.; Harada, T. Asymmetric mukaiyama aldol reaction of nonactivated ketones catalyzed by *allo*-threonine-derived oxazaborolidinone. *Org. Lett.* **2008**, *10*, 4999–5001. [CrossRef] [PubMed]
40. Barros-Filho, B.A.; de Oliveira, M.C.F.; Lemos, T.L.G.; de Mattos, M.C.; de Gonzalo, G.; Gotor-Fernandez, V.; Gotor, V. *Lentinus strigellus*: A new versatile stereoselective biocatalyst for the bioreduction of prochiral ketones. *Tetrahedron Asymmetry* **2009**, *20*, 1057–1061. [CrossRef]
41. Cheemala, M.N.; Gayral, M.; Brown, J.M.; Rossen, K.; Knochel, P. New paracyclophane phosphine for highly enantioselective ruthenium-catalyzed hydrogenation of prochiral ketones. *Synthesis* **2007**, *24*, 3877–3885. [CrossRef]
42. Martins, J.E.D.; Morris, D.J.; Wills, M. Asymmetric hydrogenation of ketones using Ir(III) complexes of N-alkyl-N′-tosyl-1,2-ethanediamine ligands. *Tetrahedron Lett.* **2009**, *50*, 688–692. [CrossRef]
43. Salvi, N.A.; Chattopadhyay, S. Asymmetric reduction of halo-substituted arylalkanones with *Rhizopus arrhizus*. *Tetrahedron Asymmetry* **2008**, *19*, 1992–1997. [CrossRef]
44. Banerjee, U.C. Transformation of rifamycin B with growing and resting cells of *Curvularia lunata*. *Enzym. Microb. Technol.* **1993**, *15*, 1037–1041. [CrossRef]
45. Liu, Z.; Weis, R.; Gliede, A. Enzymes from higher eukaryotes for industrial biocatalysis. *Food Technol. Biotechnol.* **2004**, *42*, 237–249.
46. Burton, S.G.; Cowan, D.A.; Woodley, J.M. The search for the ideal biocatalyst. *Nat. Biotechnol.* **2002**, *20*, 37–45. [CrossRef] [PubMed]
47. Trincone, A. Potential biocatalysts originating from sea environments. *J. Mol. Catal. B Enzym.* **2010**, *66*, 241–256. [CrossRef]
48. Trincone, A. Marine biocatalysts: Enzymatic features and applications. *Mar. Drugs* **2011**, *9*, 478–499. [CrossRef] [PubMed]
49. Trincone, A. Enzymatic processes in marine biotechnology. *Mar. Drugs* **2017**, *15*, 93. [CrossRef] [PubMed]
50. Nishiyama, R.; Inoue, A.; Ojima, T. Identification of 2-keto-3-deoxy-D-gluconate kinase and 2-keto-3-deoxy-D-phosphogluconate aldolase in an alginate-assimilating bacterium *Flavobacterium* sp. Strain UMI-01. *Mar. Drugs* **2017**, *15*, 37. [CrossRef] [PubMed]
51. Antranikian, G.; Vorgias, C.E.; Bertoldo, C. Extreme environments as a resource for fungi and novel biocatalysts. *Adv. Biochem. Eng. Biotechnol.* **2005**, *96*, 219–262. [PubMed]
52. Dionisi, H.M.; Lozada, M.; Olivera, N.L. Bioprospection of marine fungi: Biotechnological applications and methods. *Rev. Argent. Microbiol.* **2012**, *44*, 46–90.
53. Ferrer, M.; Golyshina, O.V.; Beloqui, A.; Golyshin, P.N. Mining enzymes from extreme environments. *Curr. Opin. Microbiol.* **2012**, *10*, 207–214. [CrossRef] [PubMed]
54. Liu, H.; de Souza, F.Z.R.; Liu, L.; Chen, B.-S. The use of marine-derived fungi for preparation of enantiomerically pure alcohols. *Appl. Microbiol. Biotechnol.* **2018**. [CrossRef] [PubMed]
55. De Vitis, V.; Guidi, B.; Contente, M.L.; Granato, T.; Conti, P.; Molinari, F.; Crotti, E.; Mapelli, F.; Borin, S.; Daffonchio, D.; et al. Marine fungi as source of stereoselective esterases and ketoreductases: Kinetic resolution of a prostaglandin intermediate. *Mar. Biotechnol.* **2015**, *17*, 144–152. [CrossRef] [PubMed]
56. Sarkar, S.; Pramanik, A.; Mitra, A.; Mukherjee, J. Bioprocessing data for the production of marine enzymes. *Mar. Drugs* **2010**, *8*, 1323–1372. [CrossRef] [PubMed]
57. De Carvalho, C.C. Enzymatic and whole cell catalysis: Finding new strategies for old processes. *Biotechnol. Adv.* **2011**, *29*, 75–83. [CrossRef] [PubMed]

© 2018 by the authors. Licensee MDPI, Basel, Switzerland. This article is an open access article distributed under the terms and conditions of the Creative Commons Attribution (CC BY) license (http://creativecommons.org/licenses/by/4.0/).

Article

Microbial Degradation of Amino Acid-Containing Compounds Using the Microcystin-Degrading Bacterial Strain B-9

Haiyan Jin [1,*], Yoshiko Hiraoka [2], Yurie Okuma [2], Elisabete Hiromi Hashimoto [2], Miki Kurita [2], Andrea Roxanne J. Anas [2], Hitoshi Uemura [3], Kiyomi Tsuji [3] and Ken-Ichi Harada [1,2,*]

1. Graduate School of Environmental and Human Science, Meijo University, Tempaku, Nagoya 468-8503, Japan
2. Faculty of Pharmacy, Meijo University, Tempaku, Nagoya 468-8503, Japan; y715o.c5@gmail.com (Y.H.); g0773315@ccalumni.meijo-u.ac.jp (Y.O.); elisabete.utfpr@gmail.com (E.H.H.); g0673319@ccalumni.meijo-u.ac.jp (M.K.); anasaroj@meijo-u.ac.jp (A.R.J.A.)
3. Kanagawa Prefectural Institute of Public Health, Shimomachiya, Chigasaki, Kanagawa 253-0087, Japan; uemura.aklt@pref.kanagawa.jp (H.U.); tsuji.df7@pref.kanagawa.jp (K.T.)
* Correspondence: 163891501@ccalumni.meijo-u.ac.jp (H.J.); kiharada@meijo-u.ac.jp (K.-I.H.); Tel./Fax: +81-52-839-2720 (H.J.); +81-52-834-8780 (K.-I.H.)

Received: 25 November 2017; Accepted: 30 January 2018; Published: 6 February 2018

Abstract: Strain B-9, which has a 99% similarity to *Sphingosinicella microcystinivorans* strain Y2, is a Gram-negative bacterium with potential for use in the degradation of microcystin-related compounds and nodularin. We attempted to extend the application area of strain B-9 and applied it to mycotoxins produced by fungi. Among the tested mycotoxins, only ochratoxin A was completely hydrolyzed to provide the constituents ochratoxin α and L-phenylalanine, and levels of fumonisin B1 gradually decreased after 96 h. However, although drugs including antibiotics released into the aquatic environment were applied for microbial degradation using strain B-9, no degradation occurred. These results suggest that strain B-9 can only degrade amino acid-containing compounds. As expected, the tested compounds with amide and ester bonds, such as 3,4-dimethyl hippuric acid and 4-benzyl aspartate, were readily hydrolyzed by strain B-9, although the sulfonamides remained unchanged. The ester compounds were characteristically and rapidly hydrolyzed as soon as they came into contact with strain B-9. Furthermore, the degradation of amide and ester compounds with amino acids was not inhibited by the addition of ethylenediaminetetraacetic acid (EDTA), indicating that the responsible enzyme was not MlrC. These results suggest that strain B-9 possesses an additional hydrolytic enzyme that should be designated as MlrE, as well as an esterase.

Keywords: microcystin-degrading bacteria; mycotoxin; protease; esterase; inhibitor

1. Introduction

Microcystins (MCs) are typical compounds produced by cyanobacteria, such as *Microcystis*, *Anabaena*, and *Planktothrix* [1]. They are cyclic heptapeptides showing potent hepatotoxicity and tumor-promoting activity [1]. In the environment, there are many bacteria which work to degrade such hazardous and harmful compounds. The first MC-degrading bacterium was isolated and identified as a *Sphingomonas* strain (ACM-3962) in 1994 [2]. Similar bacteria capable of degrading MC were reported by Dziga et al. [3]. As per their review [3], many MC-degrading microorganisms have been found and identified, and the corresponding genetic aspects with respect to MC degradation have been studied. However, in related published papers, no substrates other than MCs have been applied [4]. The purpose of the present study is to elucidate the inherent function and role of MC-degrading microorganisms in the aquatic environment.

Strain B-9, isolated from Lake Tsukui, Japan, exhibits degradation activity against MCs [1]. This strain belongs to the genus *Sphingosinecella* sp., and, based on the 16S rDNA sequence (GenBank accession no. AB084247), has 99% similarity to *Sphingosinecella microcystinivorans* strain Y2, a type of MC-degrading bacteria [5]. The *Sphingomonas* sp. strain ACM-3962 [2] was the first strain reported to degrade MCs. The cloning and molecular characterization of four genes from strain ACM-3962 revealed the presence of three hydrolytic enzymes (MlrA, MlrB, and MlrC), together with a putative oligopeptide transporter (MlrD) [6,7]. The three hydrolytic enzymes were putatively characterized as metalloproteases (MlrA and MlrC) or serine proteases (MlrB). The microcystinase MlrA catalyzes the initial ring opening of microcystin-LR (MC-LR) at the (2S,3S,8S,9S)-3-amino-9-methoxy-2,6,8-trimethyl-10-phenyldeca-4(E),6(E)-dienoic acid (Adda)-Arg peptide bond to give linearized MC-LR (Adda-Glu-Mdha-Ala-Leu-β-MeAsp-Arg). This further degrades to Adda-Glu-Mdha-Ala by MlrB, and the third enzyme, MlrC, hydrolyzes the tetrapeptide into smaller peptides and amino acids [6–8]. In terms of advancements, recombinant MlrA and MlrC have been prepared, and the degradation scheme has been almost completely verified [9,10]. The use of typical protease inhibitors, such as ethylenediaminetetraacetic acid (EDTA) and 1,10-phenanthroline, results in the accumulation of linear MC-LR and the tetrapeptide, which allows for the classification of the enzymes MlrA and MlrC as metalloproteases, [6,7]. Meanwhile, phenylmethylsulfonyl fluoride (PMSF) characterizes MlrB as a possible serine protease [6,7].

We extended the area of application of strain B-9 for bioremediation and applied it to the secondary fungal metabolites of mycotoxins that may have mutagenic, carcinogenic, cytotoxic, and endocrine-disrupting effects. These substances frequently contaminate agricultural commodities despite efforts for prevention, so successful detoxification methods are needed. The application of microorganisms to degrade mycotoxins is a possible strategy that shows potential for example in food and feed processing [11]; in antibiotics used in human and veterinary medicine (which can enter the environment via wastewater treatment plant effluents, hospitals, and processing plant effluents); in the application of agricultural waste and biosolids to fields; and in the case of leakage from waste-storage containers and landfills [12,13]. Such antibiotic pollution may facilitate the development and spread of antibiotic resistance [4].

Strain B-9 degrades MC-LR, the most toxic of the MCs, within 24 h [14,15]. After the discovery of strain B-9, we advanced our research on the degradation of the following compounds: non-toxic cyanobacterial cyclic peptides that are structurally different from MCs [16]; representative cyclic peptides (antibiotics) produced by bacteria [17]; physiologically-active cardiovascular and neuropeptides [18]; and the glucagon/vasoactive intestinal polypeptide (VIP) family of peptides [19]. The aforementioned experiments [17–19] confirmed that strain B-9 could degrade the tested peptides completely. During the application of strain B-9 to remove mycotoxins and drugs released in the aquatic environment, we obtained interesting results concerning the function of this strain. The purpose of this study was to demonstrate the additional hydrolytic enzymes (such as protease and/or esterase) of strain B-9.

2. Results

2.1. Mycotoxin and Drugs with Amide, Ester, and Sulfonamide Bonds

As already mentioned [14,15], strain B-9 can degrade the phycotoxins microcystin and nodularin, as well as non-toxic cyclic peptides and linear peptides. In this study, we applied strain B-9 to mycotoxins. We monitored the degradation behavior of the five mycotoxins using HPLC and LC/MS. No degradation was observed in zearalenone, deoxynivalenol, or patulin, while the peak of ochratoxin A completely disappeared (Figure 1). The peak of fumonisin B1 reduced to a certain extent after 96 h (Figure S1). Figure 1 shows (a) HPLC chromatograms; (b) total ion chromatograms, and selected ion monitoring (SIM) (c) at m/z 166.1 and (d) at m/z 257.0 of the reaction products of ochratoxin A by microbial degradation using strain B-9 at 0 h. Although ochratoxin A appeared at 18.7 min in the HPLC chromatogram, such a peak was not observed in the data at 0 h (Figure 1A). It was found that the peak at 20.28 min was derived from strain B-9 itself by comparison of the chromatograms of ochratoxin A and a mixture of the ochratoxin A and strain B-9 broth. Consequently, the peaks at 3.66 and 13.78 min were derived from ochratoxin A, which corresponded to phenylalanine and ochratoxin α, respectively (Figure 2). After 96 h, the latter was still observed, whereas the presence of the former was significantly reduced (Figure 1B).

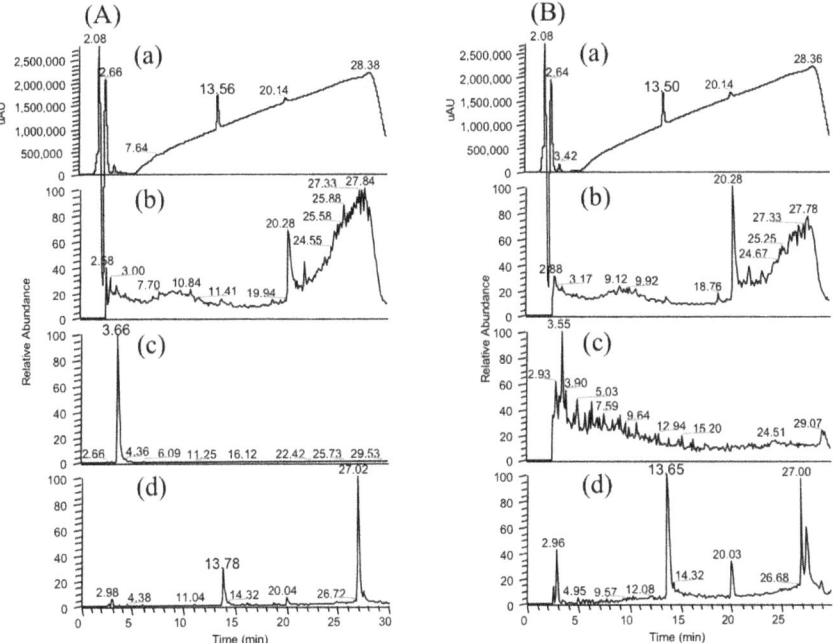

Figure 1. Degradation behavior of ochratoxin A at (**A**) 0 h and (**B**) 96 h after microbial degradation by strain B-9. (**a**) HPLC chromatograms; (**b**) total ion chromatograms, and selected ion monitoring (SIM) (**c**) at m/z 166.1 and (**d**) m/z 257.0 of the reaction products of ochratoxin A.

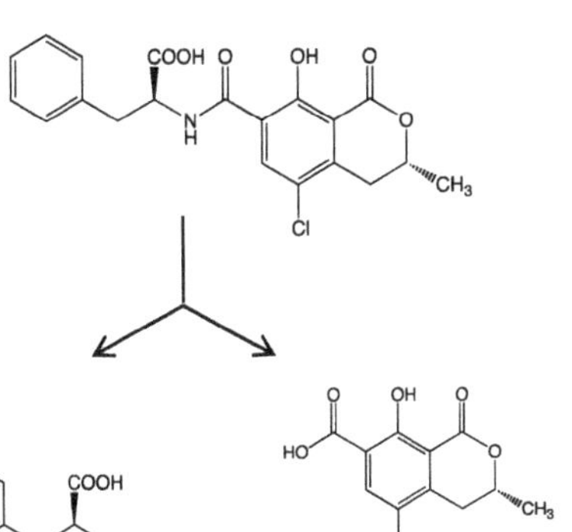

Figure 2. Microbial degradation of ochratoxin A (molecular weight (MW): 403.0) using strain B-9 to provide phenylalanine (MW: 165.1) and ochratoxin α (MW: 256.0).

We tried to degrade drugs including antibiotics released in the aquatic environment using strain B-9 and selected the following drugs with amide, ester, and sulfonamide bonds: oxytetracycline (OTC), sulphaminomethoxime, sulfadimethoxime, oseltamivir, crotamiton, N,N-diethyl-m-toluamide, and acetaminophen. Although these were treated in the same manner as the mycotoxins, no degradation was observed (data not shown). These results suggested that strain B-9 can degrade only amino acid-containing compounds.

2.2. Amino Acid-Containing Compounds (Amides, Esters, and Sulfonamides)

The following commercially available amino acid-containing compounds with different bonds were selected. Amides: 3,4-dimethylhippuric acid, D- and L-N-acetylphenylalanine, N-carbobenzoxy-L-phenylalanine-L-phenylalanine, and L-leucine-2-naphthylamide; esters: L-serine benzyl ester, and 4-benzyl L-aspartate; and sulfonamides: N-(p-toluenesulfonyl)-L-phenylalanine, and N-(1-naphthalenesulfonyl)-L-phenylalanine (Figure 3). L-Leucine-2-naphthylamide was treated in the same manner as the mycotoxins. The degradation proceeded smoothly and the starting material peak disappeared within 3 h (Figure 4a). A new peak was formed in the HPLC chromatogram. LC/MS showed that the starting material peak appeared after 18.13 min and the new peaks at 2.63 and 9.41 min were detected at 3 h (Figure 5). These peaks were determined to be 2-naphthylamine and leucine by selected ion chromatograms (SIM) at m/z 144.1 and m/z 132.03, respectively. The results indicated that L-leucine-2-naphthylamide was subjected to microbial degradation using strain B-9 to provide 2-naphthyl amine and leucine. As shown in Figure 4, the remaining amide compounds were also degraded and characteristic degradation behavior was observed. In addition, 3,4-dimethylhippuric acid was similarly degraded in the case of L-leucine-2-naphthylamide (Figure 4b). While N-acetyl-L-phenylalanine disappeared within 24 h, the D-amino acid derivative continued to appear at 96 h (Figure 4c,d). In the case of N-carbobenzoxy-L-phenylalanine-L-phenylalanine, the starting material peak disappeared as soon as it came into contact with strain B-9 (Figure 4e).

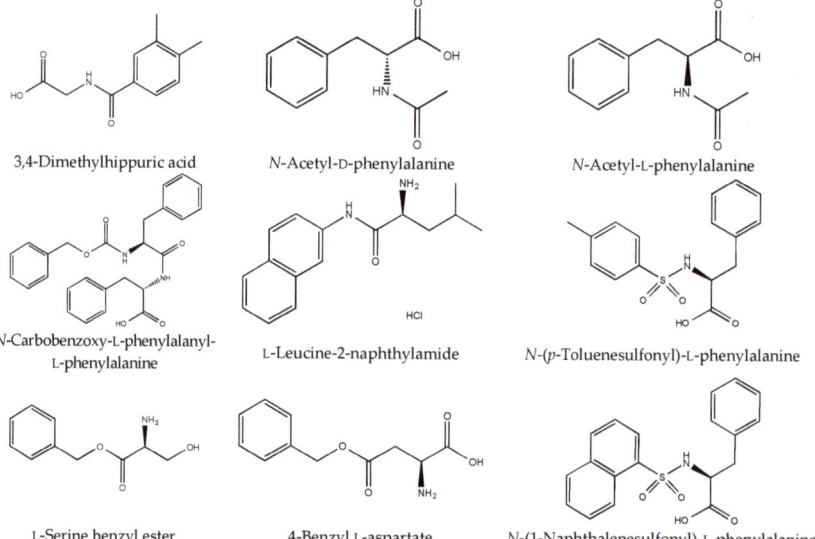

Figure 3. Structures of the tested amino acid-containing compounds. Amides: 3,4-dimethylhippuric acid, D- and L-N-acetylphenylalanines, N-carbobenzoxy-L-phenylalanine-L-phenylalanine, and L-leucine-2-naphthylamide; esters: serine benzyl ester, and 4-benzyl aspartate; and sulfonamides: N-(1-naphthalenesulfonyl)-phenylalanine, and N-(p-toluenesulfonyl)-L-phenylalanine.

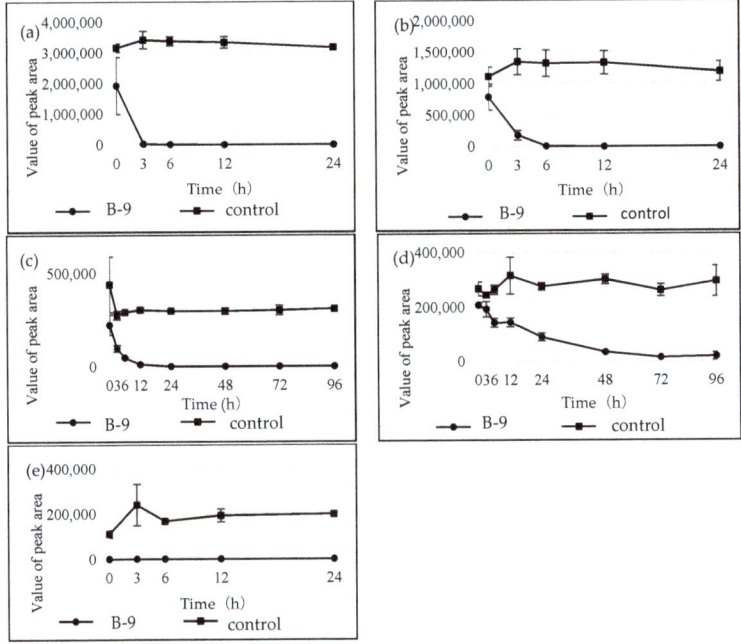

Figure 4. Time courses for the degradation of the tested compounds by strain B-9. (**a**) L-leucine-2-naphthylamide; (**b**) 3,4-dimethylhippuric acid; (**c**) N-acetyl-L-phenylalanine; (**d**) N-acetyl-D-phenyl alanine; and (**e**) N-carbobenzoxy-L-phenyl-L-phenylalanine.

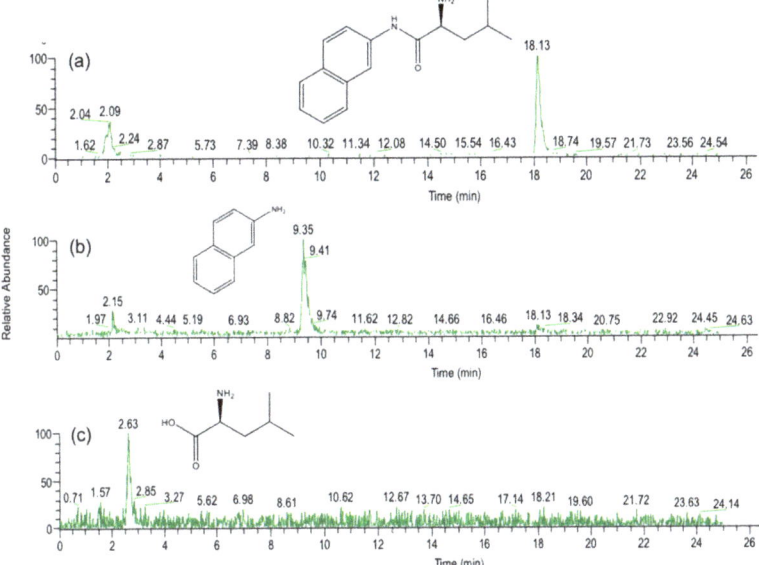

Figure 5. Selected ion chromatograms (SIMs) (**a**) at m/z 257.1 at 0 h; (**b**) at m/z 144.1 at 0 h; and (**c**) at m/z 132.03 at 3 h for reaction products of L-leucine-2-naphthylamide on microbial degradation using strain B-9.

There was a common degradation behavior of the tested ester compounds, in which the starting material peaks disappeared as soon as strain B-9 came into contact with the compounds, as shown in Figure 6a,b. The resulting benzyl alcohol continued to be detected by LC/MS during the experiment (data not shown). Figure 6c,d show the degradation behavior of the sulfonamide compounds. These compounds could not be degraded by strain B-9 during the experimental period.

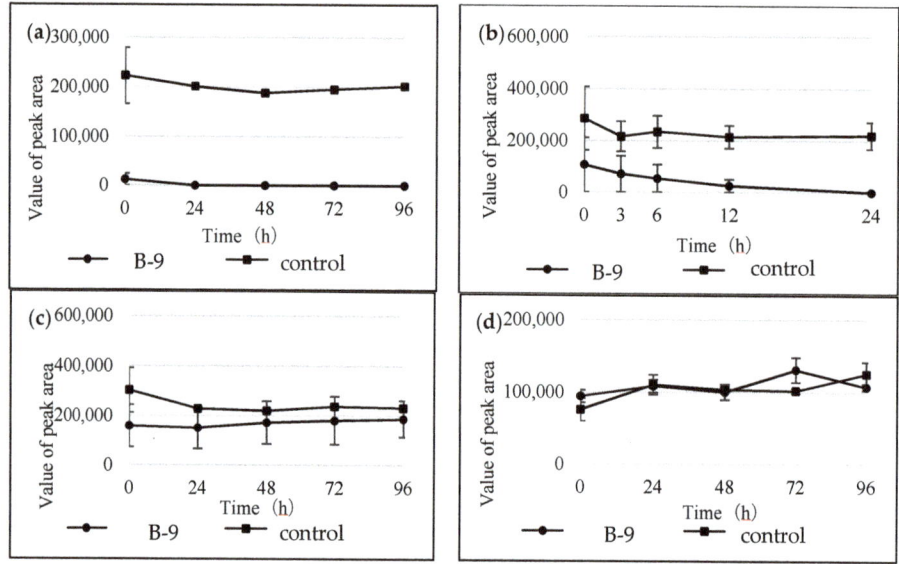

Figure 6. Time courses for the degradation of the tested compounds by strain B-9. (**a**) L-serine benzyl ester; (**b**) 4-benzyl L-aspartate; (**c**) N-(1-naphthalenesulfonyl)-L-phenylalanine; and (**d**) N-(p-toluenesulfonyl)-L-phenylalanine.

2.3. Inhibition of Hydrolysis of Amino Acid-Containing Compounds Using EDTA and PMSF

When inhibitors such as EDTA or PMSF were used at 10-mM concentrations to inhibit the degradation of MCs, the microbial degradation was effectively inhibited. Consequently, the concentration was set at 10 mM in this study. To check the inhibitory activity of the prepared solution, MC-LR was subjected to the microbial degradation in the presence or absence of the inhibitor. While MC-LR (22.7 min) and the resulting tetrapeptide (20.9 min) disappeared within 24 h in the HPLC chromatogram without the inhibitor (Figure S2A), the MC-LR disappeared and the tetrapeptide continued to appear even after 72 h in the HPLC chromatogram with the inhibitor (Figure S2B). These results were consistent with our previous findings that EDTA inhibits MlrC. Figure 7 shows the time course of the degradation using EDTA and the tested compounds. These were: (a) 3,4-dimethylhippuric acid; (b) N-carbobenzoxy-L-phenyl-L-phenylalanine; (c) L-leucine-2-naphthylamide; (d) 4-benzyl L-aspartate; and (e) L-serine benzyl ester. They showed a common degradation behavior in that the microbial degradation of the amide and ester compounds using strain B-9 was not inhibited by EDTA. These results indicated that the degradation mentioned above was not related to MlrC. In the case of PMSF, the following compounds tested positive: 3,4-dimethylhippuric acid; L-leucine-2-naphthylamide; and 4-benzyl-L-aspartate; while other compounds tested negative: N-carbobenzoxy-L-phenylalanine-L-phenylalanine and a-serine benzyl ester. No definitive conclusive information was obtained.

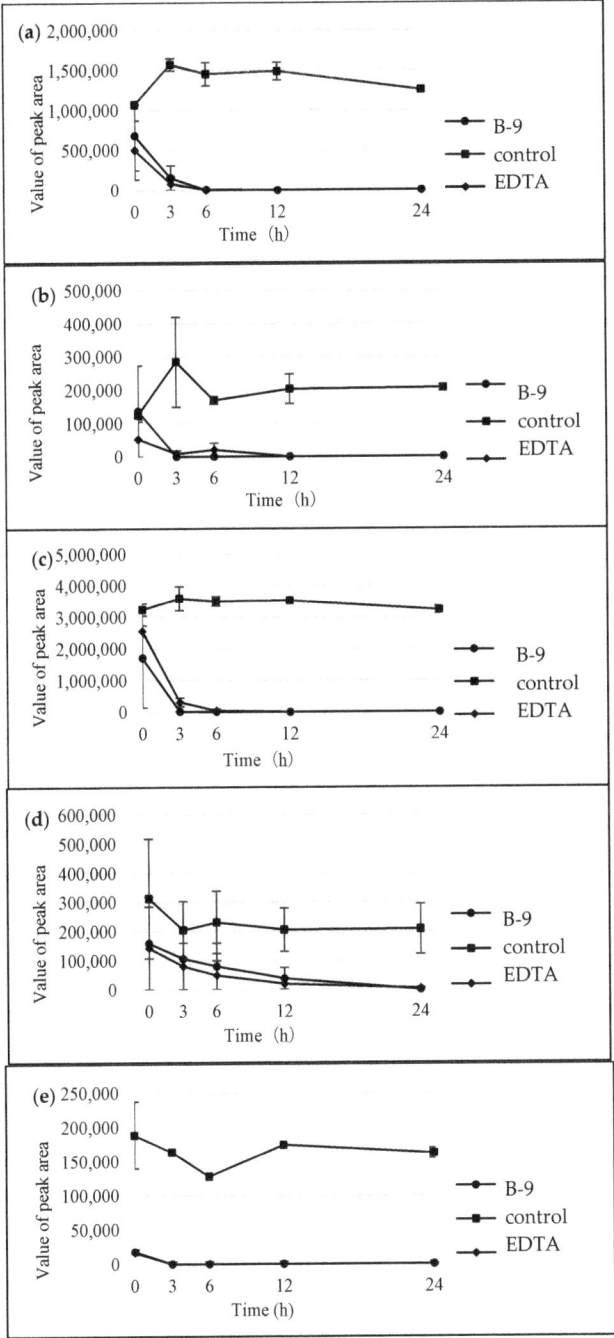

Figure 7. Time courses for the degradation of the tested compounds by B-9 in the presence of ethylenediaminetetraacetic acid (EDTA): (**a**) 3,4-dimethylhippuric acid; (**b**) N-carbobenzoxy-L-phenyl-L-phenylalanine; (**c**) L-leucine-2naphthyl amide; (**d**) 4-benzyl L-aspartate; and (**e**) L-serine benzyl ester.

3. Discussion

Strain B-9, which has 99% similarity to *Sphingosinicella microcystinivorans* strain Y2 [5], is Gram-negative and has, in several studies, shown promise for the degradation of MC-related compounds and nodularin [14,15]. Subsequently, we applied strain B-9 to other types of substrates, such as cyanobacterial peptides including depsipeptides [16], and bacterial cyclic peptides including depsipeptides [16], which are structurally different from the MCs and nodularin. Based on these results, the hydrolytic behavior using this strain is suggested as follows: (1) the reaction essentially occurs at a peptide bond in a cyclic peptide moiety to give a linearized peptide, which is more quickly hydrolyzed compared to their original ones; (2) strain B-9 primarily hydrolyzes an ester bond in a depsipeptide, in which the resulting peptides are further hydrolyzed; (3) a cyclic peptide is hydrolyzed at the acyclic part, and no further reaction occurs; and (4) the resulting linearized peptide is more quickly hydrolyzed compared to the cyclic one. In some cases, it is hard to detect the degraded peptides or amino acids due to rapid hydrolysis [16].

To confirm these observations and to further investigate the hydrolytic activities of the strain, we extended our study to physiologically active peptides such as neuropeptides and cardiovascular peptides [18]. The tested peptides were classified into two groups: (1) linear peptides, and (2) cyclic peptides with a loop formed by disulfide bond formation. The linearized peptides degraded faster than the loop-containing peptides because the loop formed by the disulfide bond was regarded as one of the degradation-resistant factors. Hydrolysis of the peptides occurred through the cleavage of various peptide bonds, and strain B-9 may bear similarities to the mammalian neutral endopeptidase (NEP) 24.11, a 94-kDa zinc metalloendoprotease widely distributed in humans and involved in the processing of peptide hormones due to its broad selectivity [20]. In a separate study, we observed the degradation behavior of the linear peptides—the glucagon/VIP family peptides (3200–5000 Da)—by strain B-9 in the absence or presence of two protease inhibitors, EDTA and PMSF. Consequently, we confirmed that one of the B-9 proteases (presumed to be MlrB), which is not inhibited by EDTA, cleaved bioactive peptides in the manner of an endopeptidase similar to NEP. Another protease, which is not inhibited by PMSF, corresponded to MlrC and cleaved the resulting middle-sized peptides to smaller peptides or amino acids [19].

In the present study, we attempted to extend the applications of strain B-9, applying it to mycotoxins produced by fungi. Among the tested mycotoxins, only ochratoxin A was completely hydrolyzed to provide the constituents ochratoxin α and L-phenylalanine (Figure 2), and fumonisin B1 levels gradually decreased to a certain extent after 96 h due to the formation of a new peak at 12.7 min (Figure S1). However, although drugs including antibiotics released into the aquatic environment were applied for microbial degradation using strain B-9, no degradation occurred. These results suggest that strain B-9 can only degrade amino acid-containing compounds. As expected, the tested compounds with amide and ester bonds were readily hydrolyzed by strain B-9, although the sulfonamides were not degraded (Figures 4 and 6). In particular, the ester compounds were characteristically and rapidly hydrolyzed as soon as they came into contact with strain B-9 (Figure 6). Furthermore, the degradations of the amide and ester compounds containing amino acids were not inhibited by the addition of EDTA, suggesting that the responsible enzyme is not MlrC.

It is understood that MlrC is found in the final stage in the microbial degradation of MC, catalyzing the degradation from linearized MC and tetrapeptide to smaller peptides and amino acids. Indeed, Dziga et al. reported the role of MlrC in MC degradation, in which linearized MC and tetrapeptides could be degraded to provide Adda by the cleavage of a peptide bond between Adda and Glu by a recombinant MlrC [14]. These results suggest that strain B-9 possesses an additional hydrolytic enzyme that should be designated as MlrE. Furthermore, the results of the present study suggest that strain B-9 possesses an esterase. As mentioned above, strain B-9 degraded depsipeptides such as aeruginosins and mikamycin A, in which the cleavage at the ester bond was predominant over that of other peptide bonds [16]. However, there may be a possibility that known proteases are responsible for the ester bond cleavage.

Since their discovery in 1994, it has been believed that MC-degrading microorganisms are only responsible for MC degradation [2]. Although many reports on MC-degrading microorganisms have appeared since then [3], few papers have described their substantial function and role in the aquatic environment. As reported by our group, one of the MC-degrading microorganisms, strain B-9, is applicable to structurally diverse peptide compounds, suggesting a different function. We should investigate the detailed function of each hydrolytic and transporter enzyme, as well as a system composed of these enzymes to understand their inherent roles under aquatic conditions.

4. Materials and Methods

4.1. Chemicals

As protease inhibitors, EDTA-2Na (purity: >99.5%) and PMSF (purity: >98.5%), were purchased from Dojindo Laboratories (Kumamoto, Japan) and Sigma-Aldrich Japan (Tokyo, Japan), respectively. Acetonitrile (ACN, LC/MS grade, purity: 99.8%), methanol (MeOH, LC/MS grade, purity: 99.7%), ethanol (EtOH, special grade, purity: 99.5%), formic acid (FA, LC/MS grade, purity: 99.5%), acetic acid (AcOH, LC/MS grade, purity: 99.5%), trifluoroacetic acid (TFA, special grade, purity: 98.0%), ammonium carbonate (special grade), and 28% ammonia solution (NH_4OH, special grade) were purchased from Wako Pure Chemical Industries, Ltd. (Osaka, Japan). Water used for the preparation of all the solutions was purified using a Milli-Q apparatus (Millipore, Billerica, MA, USA); LC/MS analysis used ultrapure water from Wako. The mycotoxins (ochratoxin A, fumonisin B1, zearalenone, deoxynivalenol, patulin) were purchased from Sigma (St. Louis, MO, USA). Drugs with amide, ester, and sulfonamide bonds and amino acid-containing compounds were obtained from the following companies: Aldrich and Sigma Japan (Tokyo, Japan), Nacalai Tesque (Kyoto, Japan), Tokyo Chemical Industry (Tokyo, Japan), and Wako Pure Chemical Industries, Ltd. (Osaka, Japan).

4.2. MC-Degrading Bacterium

Bacterial strain B-9, isolated from the surface water of Lake Tsukui, Kanagawa, Japan, was previously reported to degrade various MCs and nodularin [15]. This bacterium was inoculated into a flask containing 100 mL of Sakurai medium composed of peptone, yeast extract, and glucose, and incubated at 27 °C at 200 revolutions per minute (rpm) for 3 days.

4.3. Degradation of Tested Compounds

Two milligrams of the tested compounds was dissolved in 1 mL of EtOH and 50 µL of the solution was evaporated to dryness. One milliliter of the preincubated cell broth of strain B-9 (containing approximately 3×10^6 colony forming units (CFU) mL^{-1}) was added to the residue. The resulting solution was mixed, and then incubated at 27 °C for 5, 15, 30, 60, and 120 min. After incubation, 50 µL of each of these mixtures was added to 50 µL of MeOH containing 0.2% FA and filtered using an Ultrafree-MC membrane centrifuge-filtration unit (hydrophilic polytetrafluoroethylene (PTFA), 0.20 µm, Millipore, Bedford, MA, USA) to stop the degradation and to eliminate proteins. Each supernatant was then analyzed by HPLC and LC/MS.

4.4. Enzyme Inhibition

Enzyme inhibitors were prepared as follows: EDTA was prepared as a 200-mM stock solution in water and was used at an assay concentration of 10 mM. PMSF was prepared as a 200-mM stock solution in EtOH and was used at an assay concentration of 10 mM. The cell broth and required inhibitor were preincubated at 27 °C for 30 min.

4.5. High-Performance Liquid Chromatography

The degradation process was monitored by HPLC-photo diode array (PDA) at 220 or 254 nm. The system consisted of a pair of LC 10AD VP pumps, a DGU 12A degasser, a CTO 6A column oven, an SPD 10A VP photodiode array detector, and an SCL 10A VP system controller (Shimadzu, Kyoto, Japan). Five microliters of the filtered sample were loaded onto a TSK-gel Super ODS column (2.0 µm, 2.0 × 100 mm, TOSOH, Tokyo, Japan) at 40 °C. The mobile phase was 0.1% formic acid in water (A) and 0.1% formic acid in methanol (B). The gradient conditions were initially 40–90% B for 20 min, and the flow rate was 200 µL/min.

4.6. Liquid Chromatography/Ion Trap Mass Spectrometry

The sample, column, mobile phase, and gradient conditions were the same as those used for the HPLC analysis (12). The LC separation was performed using the Agilent 1100 HPLC system (Agilent Technologies, Palo Alto, CA, USA). Five microliters of the sample was filtrated using an Ultrafree-MC membrane centrifuge filtration unit (hydrophilic PTFE, 0.20 µm, Millipore, Bedford, MA, USA) and loaded onto a TSK-gel Super ODS column (2.0 µm, 2.0 × 100 mm, TOSOH, Tokyo, Japan) at 40 °C. The mobile phase was 0.1% formic acid in water (A) and 0.1% formic acid in acetonitrile (B). The flow rate was 200 µL/min with UV detection at 254 nm. The gradient conditions were initially 10–90% B for 40 min. The entire eluate was directed into the mass spectrometer, where it was diverted to waste 2.5 min after injection to avoid any introduction of salts into the ion source. The MS analysis was accomplished using a Finnigan LCQ Deca XP plus ITMS (Thermo Fischer Scientific, San Jose, CA, USA) equipped with an electrospray ionization (ESI) interface. The ESI conditions in the positive ion mode were as follows: capillary temperature 300 °C, sheath gas flow rate 35 (arbitrary unit), ESI source voltage 5000 V, capillary voltage 43 V, and tube lens offset 15 V. Various scan ranges were used according to the molecular weights of the tested compounds.

Supplementary Materials: The following are available online at http://www.mdpi.com/1660-3397/16/2/50/s1, Figure S1: Total ion chromatograms (a) and (b) and selected ion monitoring (SIM) at m/z 722.4 (c) and m/z 564.3 (d) of fumonisin B1 and a reaction product by microbial degradation using B-9 at 0 h (A) and 96 h (B), respectively, Figure S2: (A) HPLC chromatograms of MC-LR by B-9 without EDTA after 0, 6 and 24 h. (B) HPLC chromatograms of MC-LR by B-9 with EDTA after 0, 6 and 72 h.

Acknowledgments: K.-I.H. and H.J. gratefully acknowledge Tatsuko Sakai at the Analytical Services Center, Faculty of Pharmacy, Meijo University, for assistance and support.

Author Contributions: K.-I.H. and H.J. conceptualized the research. K.T., H.U. and Y.O. performed the degradation experiments of drugs including antibiotics with the supervision of K.-I.H.; E.H.H. and M.K. performed the experiments of mycotoxin degradation with the guidance of K.-I.H.; Y.H. and H.J. performed degradation of amide and esters with the guidance of K.-I.H.; M.K., Y.H. and H.J. ran the LC/MS and analyzed the LC/MS data under the supervision of A.R.J.A.; H.J. and K.-I.H. wrote the manuscript. All co-authors agreed to the contents of the paper.

Conflicts of Interest: The authors declare no conflict of interest.

References

1. Tsuji, K.; Asakawa, M.; Anzai, Y.; Sumino, T.; Harada, K.-I. Degradation of microcystins using immobilized microorganism isolated in an eutrophic lake. *Chemosphere* **2006**, *65*, 117–124. [CrossRef] [PubMed]
2. Jones, G.J.; Bourne, D.G.; Blakeley, R.L.; Doelle, H. Degradation of the cyanobacterial hepatotoxin microcystin by aquatic bacteria. *Nat. Toxins* **1994**, *2*, 228–235. [CrossRef] [PubMed]
3. Dziga, D.; Wasylewski, M.; Wladyka, B.; Nybom, S.; Meriluoto, J. Microbial degradation of microcystins. *Chem. Res. Toxicol.* **2013**, *26*, 841–852. [CrossRef] [PubMed]
4. Martinez, J.L. Antibiotics and antibiotic resistance genes in natural environments. *Science* **2008**, *321*, 365–367. [CrossRef] [PubMed]
5. Maruyama, T.; Park, H.D.; Ozawa, K.; Tanaka, Y.; Sumino, T.; Hamana, K.; Hiraishi, A.; Kato, K. *Sphingosinicella microcystinivorans* gen. nov., sp. nov., a microcystin-degrading bacterium. *Int. J. Syst. Evol. Microbiol.* **2006**, *56*, 85–89. [CrossRef] [PubMed]
6. Bourne, D.G.; Jones, G.J.; Blakeley, R.L.; Jones, A.; Negri, A.P.; Riddles, P. Enzymatic pathway for the bacterial degradation of the cyanobacterial cyclic peptide toxin microcystin, LR. *Appl. Environ. Microbiol.* **1996**, *62*, 4086–4094. [PubMed]

7. Bourne, D.G.; Riddles, P.; Jones, G.J.; Smith, W.; Blakeley, R.L. Characterisation of a gene cluster involved in bacterial degradation of the cyanobacterial toxin microcystin, LR. *Environ. Toxicol.* **2001**, *16*, 523–534. [CrossRef] [PubMed]
8. Hashimoto, E.H.; Kato, H.; Kawasaki, Y.; Nozawa, Y.; Tsuji, K.; Hirooka, E.Y.; Harada, K.-I. Further investigation of microbial degradation of microcystin using advanced Marfey's method. *Chem. Res. Toxicol.* **2009**, *22*, 391–398. [CrossRef] [PubMed]
9. Dziga, D.; Wasylewski, M.; Zielinska, G.; Meriluoto, J.; Wasylewski, M. Heterologous expression and characterization of microcystinase. *Toxicon* **2012**, *59*, 578–586. [CrossRef] [PubMed]
10. Dziga, D.; Wasylewski, M.; Szetela, A.; Bochenska, O.; Wladyka, B. Verification of the role of MlrC in microcystin biodegradation by studies using a heterologously expressed enzyme. *Chem. Res. Toxicol.* **2012**, *25*, 1192–1194. [CrossRef] [PubMed]
11. Cserháti, M.; Kriszt, B.; Krifaton, C.; Szoboszlay, S.; Háhn, J.; Tóth, S.; Nagy, I.; Kukolya, J. Mycotoxin-degradation profile of *Rhodococcus* strains. *Int. J. Food Microbiol.* **2013**, *166*, 176–185. [CrossRef] [PubMed]
12. Kümmerer, K. The presence of pharmaceuticals in the environment due to human use—Present knowledge and future challenges. *J. Environ. Manag.* **2009**, *90*, 2354–2366. [CrossRef] [PubMed]
13. Sarmah, A.K.; Meyer, M.T.; Boxall, A.B.A. A global perspective on the use, sales, exposure pathways occurrence, fate and effects of veterinary antibiotics (VAs) in the environment. *Chemosphere* **2006**, *65*, 725–759. [CrossRef] [PubMed]
14. Harada, K.-I.; Imanishi, S.; Kato, H.; Mizuno, M.; Ito, E.; Tsuji, K. Isolation of Adda from microcystin-LR by microbial degradation. *Toxicon* **2004**, *44*, 107–109. [CrossRef] [PubMed]
15. Imanishi, S.; Kato, H.; Mizuno, M.; Tsuji, K.; Harada, K.-I. Bacterial degradation of microcystins and nodularin. *Chem. Res. Toxicol.* **2005**, *18*, 591–598. [CrossRef] [PubMed]
16. Kato, H.; Imanishi, S.Y.; Tsuji, K.; Harada, K.-I. Microbial degradation of cyanobacterial cyclic peptides. *Water Res.* **2007**, *41*, 1754–1762. [CrossRef] [PubMed]
17. Kato, H.; Tsuji, K.; Harada, K.-I. Microbial degradation of cyclic peptides produced by bacteria. *J. Antibiot.* **2009**, *62*, 181–190. [CrossRef] [PubMed]
18. Kondo, F.; Okada, S.; Miyachi, A.; Kurita, M.; Tsuji, K.; Harada, K.-I. Microbial degradation of physiologically active peptides by strain B-9. *Anal. Bioanal. Chem.* **2012**, *403*, 1783–1791. [CrossRef] [PubMed]
19. Miyachi, A.; Kondo, F.; Kurita, M.; Tsuji, K.; Harada, K.-I. Microbial Degradation of linear peptides by strain B-9 of *Sphingosinicella* and its application in peptide quantification using liquid chromatography-mass spectrometry. *J. Biosci. Bioeng.* **2015**, *119*, 724–728. [CrossRef] [PubMed]
20. Stephenson, S.L.; Kenny, A.J. The hydrolysis of α-human atrial natriuretic peptide by pig kidney microvillar membranes is initiated by endopeptidase-24.11. *Biochem. J.* **1987**, *243*, 183–187. [CrossRef] [PubMed]

© 2018 by the authors. Licensee MDPI, Basel, Switzerland. This article is an open access article distributed under the terms and conditions of the Creative Commons Attribution (CC BY) license (http://creativecommons.org/licenses/by/4.0/).

Article

Optimization of Collagenase Production by *Pseudoalteromonas* sp. SJN2 and Application of Collagenases in the Preparation of Antioxidative Hydrolysates

Xinghao Yang [1,2,†], Xiao Xiao [1,†], Dan Liu [1], Ribang Wu [1], Cuiling Wu [1], Jiang Zhang [1], Jiafeng Huang [1], Binqiang Liao [1] and Hailun He [1,*]

1. School of Life Sciences, Central South University, Changsha 410013, China; 132511003@csu.edu.cn (X.Y.); 152501023@csu.edu.cn (X.X.); 162501022@csu.edu.cn (D.L.); 152511020@csu.edu.cn (R.W.); 152511020@csu.edu.cn (C.W.); jiangzhang915@csu.edu.cn (J.Z.); 1608110217@csu.edu.cn (J.H.); 162511005@csu.edu.cn (B.L.)
2. Hunan Bailin Biological Technology Incorporated Company, Changsha 410205, China
* Correspondence: helenhe@csu.edu.cn; Tel.: +86-0731-82650230
† These authors contributed equally to this work.

Received: 18 October 2017; Accepted: 29 November 2017; Published: 4 December 2017

Abstract: Collagenases are the most important group of commercially-produced enzymes. However, even though biological resources are abundant in the sea, very few of these commercially popular enzymes are from marine sources, especially from marine bacteria. We optimized the production of marine collagenases by *Pseudoalteromonas* sp. SJN2 and investigated the antioxidant activities of the hydrolysates. Media components and culture conditions associated with marine collagenase production by *Pseudoalteromonas* sp. SJN2 were optimized by statistical methods, namely Plackett–Burman design and response surface methodology (RSM). Furthermore, the marine collagenases produced by *Pseudoalteromonas* sp. SJN2 were seen to efficiently hydrolyze marine collagens extracted from fish by-products, and remarkable antioxidant capacities of the enzymatic hydrolysates were shown by DPPH radical scavenging and oxygen radical absorbance capacity (ORAC) tests. The final optimized fermentation conditions were as follows: soybean powder, 34.23 g·L^{-1}; culture time, 3.72 d; and temperature, 17.32 °C. Under the optimal fermentation conditions, the experimental collagenase yield obtained was 322.58 ± 9.61 U·mL^{-1}, which was in agreement with the predicted yield of 306.68 U·mL^{-1}. Collagen from Spanish mackerel bone, seabream scale and octopus flesh also showed higher DPPH radical scavenging rates and ORAC values after hydrolysis by the collagenase. This study may have implications for the development and use of marine collagenases. Moreover, seafood waste containing beneficial collagen could be used to produce antioxidant peptides by proteolysis.

Keywords: collagenase; fermentation optimization; collagen; *Pseudoalteromonas*; antioxidant peptides

1. Introduction

Collagen is the main component of the extracellular matrix. Collagen is the predominant constituent of skin, tendons and cartilage and is the main organic component of bones, teeth and corneas [1,2]. Collagen is not only a structural protein with high tensile strength, but also a protein that affects cell differentiation, migration and attachment. Collagen is an inexpensive and resourceful meat by-product that is used extensively as a food additive to increase the texture, water-holding capacity and stability of several food products. The main sources of collagen, such as bovine and porcine skin and bone, are derived from land-based animals. Recently, components of marine organisms, including fish skin, bone and scales, have received increasing attention as sources of collagen [3]. Additionally, with collagen being the most abundant protein in all higher organisms, collagenases have diverse biotechnological applications [4–6].

Collagenases, a class of proteases with high specificity for collagen, can cleave peptide bonds in collagen [7]. Collagenases are usually considered virulence factors and normally target the connective tissue in muscle cells and other organs. On the other hand, collagenases can be used to produce bioactive collagen peptides, which have been widely used in the pharmaceutical, food and cosmetic industries. To date, a number of collagenases of bacterial origin have been identified and characterized, such as MCP01 from *Pseudoalteromonas* sp. SM9913 [8], Col H from *Clostridium histolyticum* [9] and collagenases from *Vibrio alginolyticus* [10]. However, the productivity of most collagenases is not high enough for industrial application. Recently, marine bacteria have become known as important sources for the identification of novel enzymes. Compared with collagenases from land animals, marine collagenases usually have higher catalytic efficiency towards marine collagen from fish skin and bone [8]. Therefore, collagenases obtained from marine bacteria in particular have received attention owing to their diverse properties. Simultaneously, there have been attempts to increase the production of marine-derived collagenases and to optimize the fermentation conditions for the production of these enzymes [11].

Statistical methods such as factorial design, central composite design and response surface methodology (RSM) have been frequently used to optimize process parameters for the production of different kinds of enzymes [12]. Statistically-designed experiments are more effective than those designed using classical optimization strategies because they can be used to study many variables simultaneously with a low number of observations, saving time and expense. The Plackett–Burman design provides a way to rapidly screen for main variables with significant effects on a specified parameter from a large number of variables, and the information obtained can be retained for further optimization [13,14]. Response surface methodology (RSM) can identify interactions among various factors while requiring less experimental resources. Therefore, RSM, as a collection of statistical techniques that are useful for designing experiments, building models, evaluating the effects of different factors and searching for optimal conditions of studied factors for predictable responses, has been successfully applied in many areas of biotechnology, including the protein enzyme industry [15]. *Pseudoalteromonas* sp. SJN2, isolated from the inshore environment of the South China Sea, can produce extracellular collagenases that have high catalytic efficiency. The aim of this study was to use RSM to optimize the fermentation medium to increase collagenase production by *Pseudoalteromonas* sp. SJN2. In addition, marine collagen extracted from seafood by-products was digested by crude extracellular collagenases secreted by *Pseudoalteromonas* sp. SJN2, and the antioxidant activity of hydrolysates was measured.

2. Results

2.1. Purification and Enzymatic Properties of Ps sp. SJN2 Collagenases

Collagenases SJN2 was sequentially purified using ammonium sulfate precipitation, anion exchange and size exclusion chromatography; Col SJN2 purification is shown in Figure S1 and Table S1 (Supplementary Material). Currently, as a highly effective feed additive, enzymes used in the food industry are crude enzymes because the cost of pure enzymes is high and because the operation is tedious [16]. In addition, combinations of enzymes can result in a range of possible biological properties for the corresponding hydrolysates.

Compared with other strains isolated from the inshore environment of the South China Sea at the same time, *Pseudoalteromonas* sp. SJN2 collagenases, which appear as bright strips in Figure 1a Line 1, have some advantages in terms of the number and brightness of the strips, which suggest a strong ability to hydrolyze collagen.

The effects of ions on the enzymatic activities of *Ps* sp. SJN2 collagenases was measured, as shown in Figure 1b. Some ions (Zn^{2+}, Mg^{2+} and Ca^{2+}; at lower concentrations; red column) were seen to promote *Ps* sp. SJN2 collagenase activity or were less toxic to strain *Ps* sp. SJN2 than EDTA and Cd^{2+}.

Swelling of insoluble collagen after hydrolysis by *Ps* sp. SJN2 collagenases was observed (Figure 1c–g). With increasing enzyme hydrolysis time, the porosity of the insoluble collagen increased. As observed by SEM, the collagen structure in the control group remained compact and tough, while in the experimental group, the bulky collagen fiber bundles changed into small dispersed collagen fibers (Figure 1e), and the sub-fibers (Figure 1f) were exposed, which indicated the collagen-hydrolysis ability of *Ps* sp. SJN2 collagenases.

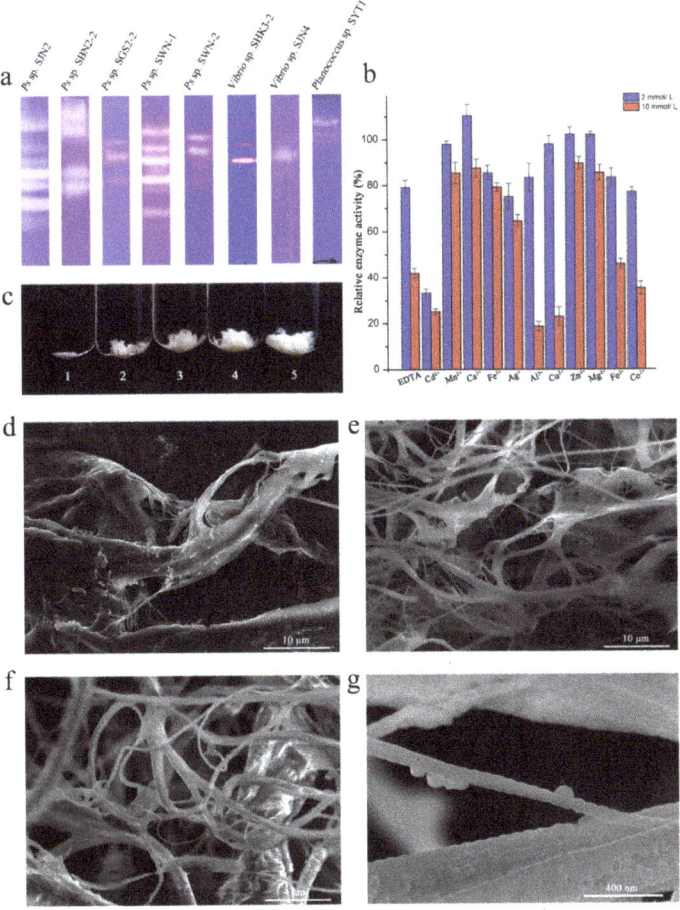

Figure 1. Enzymatic properties of *Pseudoalteromonas* sp. SJN2 collagenases. (**a**) Comparison of catalytic ability on gelatin of collagenases from *Ps* sp. SJN2, *Ps* sp. SBN2-2, *Ps* sp. SGS2-2, *Ps* sp. SWN-1, *Ps* sp. SWN-2, *Vibrio* sp. HK3-2, *Vibrio* sp. SJN4, and *Planococcus* sp. SYT1; (**b**) effect of ions against *Ps* sp. SJN2 collagenases' enzyme activity, red column for 2 mM, blue column for 10 mM; (**c**) swelling effect of *Ps* sp. SJN2 collagenase on bovine collagen type I: Tube 1: 200 µL 20 mM PBS treated for 24 h at 37 °C; Tubes 2–5: 200 µL *Ps* sp. SJN2 crude enzyme treated for 1 h, 5 h, 12 h and 24 h; (**d–g**) swelling of insoluble collagen was observed under SEM, the (**d**) control group with 8000× magnification and (**e–g**) the experimental group with 8000×, 20,000× and 250,000× magnification.

2.2. Catalytic Efficiency of Collagenases from Ps sp. SJN2

Compared with commercially available terrestrial Col H, collagenases from Ps sp. SJN2, as enzymes from marine sources, have a competitive edge in the degradation of collagen from marine-biological sources. Figure 2a shows the hydrolysis of fish skin collagen by collagenases from Ps sp. SJN2 into smaller molecular collagen-peptides than those produced using Col H, after a 1-min reaction at 45 °C with both collagenases having low concentrations (2 mg·mL^{-1} Col H and 0.2 mg·mL^{-1} crude enzymes from Ps sp. SJN2), which indicated that crude collagenases from Ps sp. SJN2 had higher catalytic efficiency than Col H. In the degradation of marine collagen, marine collagenases exhibit properties such as low enzyme concentrations and rapid catalysis, and the same results can be obtained for the degradation of collagen from fish scales (Figure 2b) and fish bone (Figure 2c). This result showed that crude collagenases from Ps sp. SJN2 could be more suitable for the digestion of marine collagen, which was seen to be spliced into small polypeptide fragments. Improving the production of collagenases from Ps sp. SJN2 would be useful for future application of this enzyme.

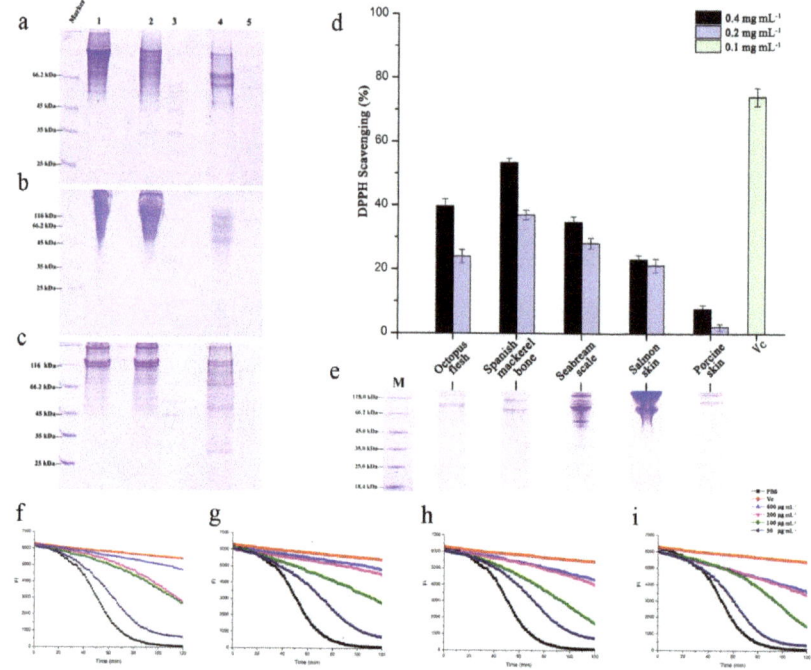

Figure 2. Differences between Col H and collagenases from Ps sp. SJN2 in the degradation of fish collagen and the antioxidative test. (a–c) Fish-collagen substrates from skin, scale and bone after hydrolysis are shown in Line 1: fish collagen; Line 2: fish collagen hydrolysed by Col H with a reaction time of 1 min; Line 3: Col H; Line 4: fish collagen hydrolysed by collagenases from Ps sp. SJN2 at a reaction time of 1 min; Line 5: collagenases from Ps sp. SJN2. (d,e) DPPH scavenging of five collagen peptides (d) and the extracted collagen (e). The reaction concentration of extracted collagen for the black column is 0.4 mg·mL^{-1}, and that for the blue column is 0.2 mg·mL^{-1}. The green is vitamin C at 0.1 mg·mL^{-1}, as a positive control. (f–i) ORAC assay for four kinds of marine collagen peptides: (f) octopus flesh, (g) seabream scale, (h) Spanish mackerel bone and (i) salmon skin. The red curve is vitamin C at 100 µg·mL^{-1}, as a positive control, while the black is PBS, as a negative control. For each substrate, the blue curve is 400 µg·mL^{-1} of hydrolysed collagen; the purple curve is 200 µg·mL^{-1}; the green curve is 100 µg·mL^{-1}; and the navy curve is 50 µg·mL^{-1}.

2.3. Initial Screening of Significant Fermentation Conditions

The effects of temperature, initial pH, seed inoculation, culture time, corn meal concentration, bran liquid concentration and soybean meal concentration, which are the seven variables associated with collagenase production in the Plackett–Burman design, are shown in Tables 1 and 2. A random experimental program was devised by using Design-Expert-8.0.6 software. Analysis of variance (ANOVA) results are listed in Table 3.

Table 1. Variables tested for collagenase production using Plackett–Burman designs and the levels for each variable.

Variables	Component	Unit	Lower Level (−1)	Higher Level (+1)
X_1	Temperature	°C	18.0	27.0
X_2	pH	-	7.5	8.5
X_3	Inoculum	μL	500.0	750.0
X_4	Culture time	d	4.0	6.0
X_5	Corn meal	g·L^{-1}	20.0	30.0
X_6	Bran liquid	mL·L^{-1}	10.0	15.0
X_7	Soybean powder	g·L^{-1}	20.0	30.0
X_8	Dummy variable	-	-	-
X_9	Dummy variable	-	-	-
X_{10}	Dummy variable	-	-	-
X_{11}	Dummy variable	-	-	-

Table 2. Plackett–Burman design matrix with corresponding results.

Run	X_1	X_2	X_3	X_4	X_5	X_6	X_7	X_8	X_9	X_{10}	X_{11}	Collagenase Activity U·mL^{-1}
1	−1	+1	−1	+1	+1	−1	+1	+1	+1	−1	−1	230.53
2	−1	−1	−1	+1	−1	+1	+1	−1	+1	+1	+1	235.50
3	−1	+1	+1	−1	+1	+1	+1	−1	−1	−1	+1	273.33
4	+1	−1	+1	+1	−1	+1	+1	+1	−1	−1	−1	162.23
5	+1	+1	+1	−1	−1	−1	+1	−1	+1	+1	−1	168.39
6	+1	−1	+1	+1	+1	−1	−1	−1	+1	−1	+1	127.82
7	+1	−1	−1	−1	+1	−1	+1	+1	−1	+1	+1	225.40
8	+1	+1	−1	−1	−1	+1	−1	+1	+1	−1	+1	202.29
9	−1	−1	−1	−1	−1	−1	−1	−1	−1	−1	−1	215.47
10	−1	−1	+1	−1	+1	+1	−1	+1	+1	+1	−1	218.72
11	+1	+1	−1	+1	+1	+1	−1	−1	−1	+1	−1	147.17
12	−1	+1	+1	+1	−1	−1	−1	+1	−1	+1	+1	173.70

Table 3. Variance analysis of the Plackett–Burman linear model.

Source	df	Collagenase Activity of Fermented Broth [a]				
		Sum of Squares	Coefficient Estimate	Mean Square	F-Value	p-value Probability > F
Model [a]	7	18,796.53	-	2685.22	9.770	0.0219 [b]
X_1	1	8213.77	−26.16	8213.77	29.880	0.0054 [b]
X_2	1	8.79	0.86	8.79	0.032	0.8668
X_3	1	1455.41	−11.01	1455.41	5.290	0.0829
X_4	1	4280.74	−18.89	4280.74	15.570	0.0169 [b]
X_5	1	356.35	5.45	356.35	1.300	0.3184
X_6	1	799.01	8.16	799.01	2.910	0.1634
X_7	1	3682.46	17.52	3682.46	13.400	0.0216 [b]
Residual	4	1099.59	-	274.90	-	-
Corrected Total	11	19,896.12	-	-	-	-

[a] $R^2 = 0.9447$. [b] Model terms are significant.

Mathematical analysis of this model shows that the model was significant (F-value = 9.77, p-value = 0.0219), with a complex correlation R^2 = 0.9447, indicating that the model can explain 94.47% of the experimental results. The multiple regression Equation (1) describes the mathematical relationship between collagenase production and fermentation conditions:

$$Y = 198.38 - 26.16X_1 + 0.86X_2 - 11.01X_3 - 18.89X_4 + 1.45X_5 + 8.16X_6 + 17.52X_7 \quad (1)$$

For every decrease in X_1, a decrease of 26.16 units in Y is predicted; a similar effect is predicted for the other variables. This relationship between Y and X was further used in the steepest ascent experiment. Table 3 indicates that the terms with p-values less than 0.05, i.e., X_1 (temperature, p = 0.0054), X_4 (culture time, p = 0.0169) and X_7 (soybean meal concentration, p = 0.0216), had relatively significant effects on collagenase yield. Therefore, the steepest ascent experiment was designed with the three significant variables mentioned above.

The path of steepest ascent was determined based on a decrease of 2.00 °C for X_1 (temperature), a decrease of 0.64 d for X_4 (culture time) and an increase of 3.00 g·L^{-1} for X_7 (soybean concentration). The movement was generated along the path until no improvement of Y (yield of collagenases from Ps sp. SJN2) occurred. After the first step of the coordinates, decreasing collagenase yield was observed, as shown in Table 4. Consequently, this combination (temperature, 16 °C; culture time, 3.36 d; and soybean concentration, 33 g·L^{-1}) was selected as the middle point (zero level) for a final optimal design by response surface methodology.

Table 4. Experimental details and results of the steepest ascent design with four steps approaching the response region.

Item	Temperature (°C)	Culture time (d)	Soybean Concentration (g·L^{-1})	Collagenase Activity (U·mL^{-1})
Slope [a]	−26.16	−18.89	17.52	-
Base [b]	18.00	4.00	30.00	-
Average [c]	22.50	5.00	25.00	-
Ratio [d]	0.017	0.034	0.034	-
Δ [e]	−2.00	−0.64	3.00	-
Base	18.00	4.00	30.00	224.66 ± 11.53
Base+Δ	16.00	3.36	33.00	279.18 ± 8.71
Base+2Δ	14.00	2.72	36.00	206.22 ± 26.47
Base+3Δ	12.00	2.08	39.00	195.09 ± 9.65

[a] Coefficient estimate in Table 3. [b] −1/+1 level in the Plackett–Burman design in Table 1. [c] Average value of the −1 level and +1 level in Table 1. [d] An appropriate ratio determined by the experimenter, based on experiential knowledge and laboratory conditions, to reasonably adjust the step coefficient in this model. [e] Step length, calculated as Equation (4).

2.4. Further Optimization by Response Surface Methodology

The fermentation conditions were further explored by response surface methodology based on central composite design. The design matrix and the corresponding results of RSM are shown in Tables 5 and 6, and the ANOVA results of the RSM are displayed in Table 7.

Table 5. Levels of significant variables for the response surface design.

Variables	Component	Unit	+1.68 Level	+1 Level	0 Level	−1 Level	−1.68 Level
A	Temperature	°C	19.36	18.00	16.00	14.00	12.64
B	Culture time	d	4.44	4.00	3.36	2.72	2.28
C	Soybean powder	g·L^{-1}	38.04	36.00	33.00	30.00	27.96

Table 6. The matrix of the response surface experiment and the corresponding results.

Run	Variables			Collagenase Activity (U·mL^{-1})
	A	B	C	
1	+1.68	0	0	251.13
2	+1	−1	+1	221.71
3	0	−1.68	0	187.44
4	+1	−1	−1	160.45
5	−1	+1	−1	123.73
6	0	0	0	278.35
7	0	0	0	282.97
8	+1	+1	−1	197.21
9	+1	+1	+1	302.56
10	0	+1.68	0	243.74
11	−1.68	0	0	182.53
12	0	0	0	278.07
13	0	0	0	287.85
14	0	0	−1.68	107.18
15	−1	+1	+1	214.38
16	0	0	+1.68	192.30
17	−1	−1	+1	182.51
18	−1	−1	−1	119.22
19	0	0	0	280.56
20	0	0	0	319.77

Table 7. Variance analysis of the response surface methodology.

Source	df	Collagenase Activity of Fermented Brotha [a]			
		Sum of Squares	Mean Square	F-Value	p-value Probability > F
Model	9	75,803.27	8422.59	35.19	0.0001 [b]
A	1	9356.35	9356.35	39.09	0.0001 [b]
B	1	4528.07	4528.07	18.92	0.0014 [b]
C	1	15,744.59	15,744.59	65.78	0.0001 [b]
AB	1	824.79	824.79	3.45	0.0931
AC	1	20.07	20.07	0.084	0.7781
BC	1	638.14	638.14	2.67	0.1335
A^2	1	8823.57	8823.57	36.87	0.0001 [b]
B^2	1	9139.01	9139.01	38.18	0.0001 [b]
C^2	1	33,848.58	33,848.58	141.43	0.0001 [b]
Residual	10	2393.38	239.34	-	-
Lack of Fit	5	1111.68	222.34	0.87	0.5601
Pure Error	5	1281.71	256.34	-	-
Corrected Total	19	78,196.66	-	-	-

[a] R^2 = 0.9694, C.V. = 7.01%. [b] Model terms are significant.

The adequacy of the model was assessed using ANOVA (Table 7). The coefficient of determination (R^2) was 0.9694 for collagenase production, indicating good agreement between the experimental and predicted values [17]. The results demonstrated that the 96.94% variability in the response experiment could be explained by this model. The R^2 is always between zero and 1.0, and a value closer to 1.0 indicates a stronger model and better response. A very low value of the coefficient of variation (C.V., 7.01%) indicated high reliability and precision of the experimental simulation. The F-value of the model was 35.19, which implied that the model was statistically significant. The smaller the p-value, the more significant the corresponding coefficient could be considered. The p-value of 0.0001, which is less than 0.05, suggested that the model terms were significant. These results indicated that temperature, culture time and soybean concentration have a direct relationship with collagenase production.

A multivariate regression model with interaction terms was characterized by the following regression Equation (2):

$$Y = 287.87 + 26.22A + 18.21B + 33.95C + 10.15AB + 1.58AC + 8.93BC - 24.74A^2 - 25.18B^2 - 48.46C^2 \quad (2)$$

where Y is the predicted collagenase production (U·mL^{-1}) and A, B and C are the values of temperature (°C), culture time (d) and soybean powder concentration (g·L^{-1}), respectively.

The three-dimensional response surface plots and two-dimensional contour plots were used to elucidate the relationship and interaction effects of the chosen fermentation conditions for maximal production of collagenases from Ps sp. SJN2, as shown in Figure 3. In each sketch, two variables linearly changed within the experimental range, while the other variable remained constant at the central point. For the two-dimensional contour plots, the shape of the contour described the interaction significance of the paired variables, and the elliptical appearance suggested an extremely significant interaction.

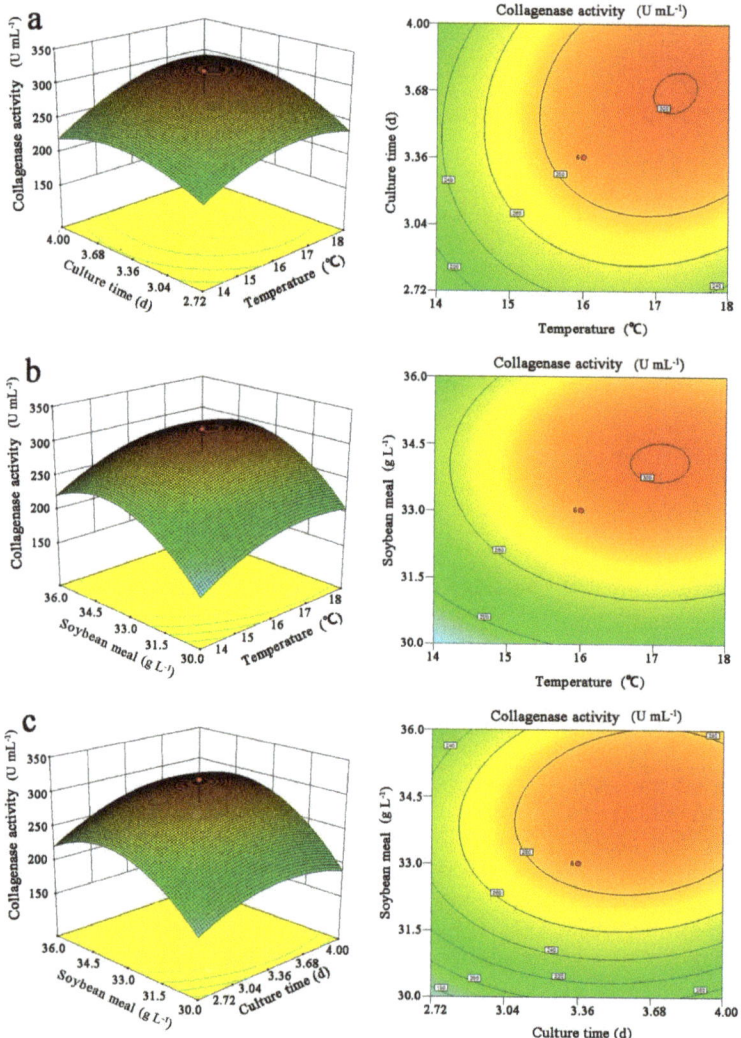

Figure 3. Response surface plots (left) and two-dimensional contour plots (right) of the effects of (**a**) culture time vs. temperature, (**b**) soybean concentration vs. temperature and (**c**) soybean concentration vs. culture time on the yield of collagenases from Ps sp. SJN2.

Figure 3a shows the response surface and the corresponding contour plot of culture time (A) vs. temperature (B), keeping soybean concentration (C) at the zero level. It can be noticed from the surface that the optimal yield of collagenases from Ps sp. SJN2 was observed when the culture time was approximately at the -1 level, while the temperature was nearly at the $+1$ level. The two-dimensional contour plot of culture time (A) vs. temperature (B) showed an elongated pattern, suggesting that the interaction between culture time and temperature has a significant effect on the yield of collagenases from Ps sp. SJN2. Similar profiles were also observed in Figure 3b,c.

The experimental data were fitted to Equation (2), and the optimal levels of the significant variables were determined to be as follows: culture time, 3.72 d; temperature, 17.32 °C; and soybean concentration, 34.23 g·L^{-1}; with a predicted maximum production of collagenases from Ps sp. SJN2 at 306.683 U·mL^{-1}.

2.5. Experimental Validation of the Model

The validation of the statistical model and the regression equation was conducted using the optimized conditions. The predicted response for collagenase production was 306.68 U·mL^{-1}, and the observed experimental value was 322.58 U·mL^{-1}. The experimental production was close to the predicted response, and the yield of collagenases from Ps sp. SJN2 increased 2.2-fold compared to the yield from the original fermentation scheme. The optimal fermentation scheme was further examined by the continuous determination of collagenase activity by bacterial biomass evaluation and gelatine-immersing zymography, as shown in Figure 4.

As shown in Figure 4a, the activity of the collagenases from Ps sp. SJN2 increased steadily as Ps sp. SJN2 grew in the logarithmic-growth phase, and the activity was highest when the biomass entered the stationary phase at approximately 72 h of cultivation time. Then, within approximately 24 h, the activity of the collagenases from Ps sp. SJN2 started to decrease, which may be explained by the decrease in biomass over the same period or by transformation of part of collagenases into other non-catalytic proteins. This finding affirmed that collagenase activity is related to biomass, which is characterized by the phenomenon that collagenase activity and biomass both exhibit maximum values at approximately 3.5 d, which is in agreement with the optimal model prediction.

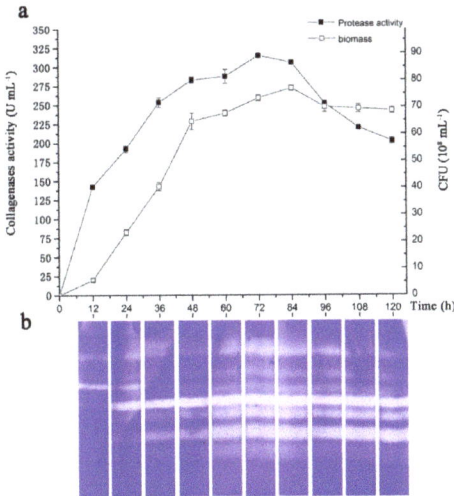

Figure 4. Time course of collagenase yield vs. biomass of Ps sp. SJN2 (a) and gelatine-immersing zymography (b).

In Figure 4b, the gelatine-substrate immersing zymography validation showed that when the proteases reach a maximum yield at 3.5 d, which was also proven by the time course experiment, there were eight bright strips, which represented activated zymogen and proteases (primarily metalloproteases and serine proteases; Figure S2, Supplementary Material) that might possess the ability to hydrolyze collagen, which proved that the increased activity of the crude enzyme under optimized conditions was caused by the increased yield. Furthermore, the visible strips of proteases with high molecular weight grew progressively darker over culture time, which might partly be due to enzyme autolysis into other mature forms.

2.6. Antioxidant Activity of the Hydrolysates

Five kinds of collagen were successfully extracted from seafood waste (salmon skin, seabream scale and Spanish mackerel fish bone), octopus flesh and porcine skin, as shown in Figure 2e. After hydrolysis, the collagen peptides were tested for potential antioxidant activity by DPPH scavenging (Figure 2d) and ORAC assays (Figure 2f–i).

The scavenging of nitrogen radicals released by DPPH was affected by the concentration of collagen peptides; the group of 0.4 mg·mL^{-1} reaction concentrations of collagen (black) has a higher rate than the group of 0.2 mg·mL^{-1} (blue). Furthermore, the collagen peptides from Spanish mackerel bone showed a strong scavenging rate, more than 50%, compared with the positive control, 0.1 mg·mL^{-1} vitamin C, at 73%, while the porcine collagen peptides are lower than 10%.

The ORAC assay was performed using four kinds of marine collagen peptides hydrolysed by collagenases from Ps sp. SJN2. The results are shown in Figure 2f–i: the red curve is the positive control, 100 µg·mL^{-1} vitamin C, and the black curve is the negative control, PBS (0.01 M). It was noticed that all the marine peptides have the capacity to absorb the oxygen radical released by AAPH (2,2′-Azobis(2-methylpropionamidine) dihydrochloride). When the collagen peptides were both at low concentrations (navy curve), the absorption capacities were also weak. The absorption capacity increased as the concentration of the collagen peptides increased, and a strong oxygen radical absorption capacity appeared at 400 µg·mL^{-1} (blue curve) for collagen from seabream scales (g) and octopus flesh (h). In addition, antioxidant activity of marine collagen-derived peptides have been widely reported in recent years, such as those from bluefin leatherjacket, tuna, yellowfin sole, Alaska pollock, halibut, round scad and Pacific hake [18], marine by-products of which have been named as sources of antioxidant peptides. Although the bioactive peptides are encrypted within the protein structure, enzymatic hydrolysis could be a useful method to obtain the peptides naturally in an environmentally-friendly, safe and efficient manner.

3. Materials and Methods

3.1. Microorganism

The strain *Pseudoalteromonas* sp. SJN2 used in this study was originally isolated from the inshore environment of the South China Sea (18°29′198″ N, 109°34′761″ E). *Ps* sp. SJN2 can produce certain extracellular collagenases. SJN2 was maintained on 2216E agar slants and stored at 4 °C. See the Supplementary Material for descriptions of the raw materials and inoculum preparation.

3.2. Fermentation

Bench-scale fermentation was performed in 250mL Erlenmeyer flasks containing different concentrations of raw solution (Supplementary Material) and media components, which were tested according to the statistical experimental design. Flasks were inoculated with seed culture and incubated at 16 °C for 96 h on a rotary shaker at 180 rpm. After fermentation, the crude enzyme was purified by centrifugation at 10,000× g for 10 min. Fermentation was carried out in triplicate, and the results represent the average of three trials.

3.3. Assay of Collagenase Activity and Substrate Immersing Zymography

The collagenolytic activities against bovine collagen were determined using the method provided by Worthington Biochemical Co. (Lakewood, NJ, USA). The reaction time was 5 h for bovine insoluble type I collagen fiber. For insoluble collagenases, one unit of enzyme releases 1 µmol of Leucine equivalents from collagen in 1 h. Substrate immersing zymography was performed as described in our previous studies. The bands from the SDS-PAGE gel were cut separately from each lane, immersed in each of the pre-warmed substrate solutions and incubated for reaction at 37 °C for 1 h. After washing, the gels were stained with 0.1% (w/v) Coomassie Brilliant Blue R-250 (Sangon, Shanghai, China) for 3 h and then destained with a solution containing 30% ethanol and 70% acetic acid, and destaining was complete when bands indicating proteolytic activity were clearly visible [19].

3.4. Screening of Significant Variables by Using Plackett–Burman Designs

The Plackett–Burman experimental design was used to screen the variables that significantly influenced collagenase production; the detailed scheme was designed by using Design-Expert Version 8.0.6 software (Stat-Ease Inc., Minneapolis, MN, USA). The Plackett–Burman design allows the evaluation of many variables, with 1–4 dummy variables reducing the error. Each variable was tested for two contrary levels: +1 as the high level and −1 as the low level; the level parameters should be appropriate selected such that the strain can survive under the conditions prescribed by each level. The Plackett–Burman experimental design is based on the first-order model as Equation (3) [13]:

$$Y = \beta_0 + \sum \beta_i x_i \quad (3)$$

where Y is the predicted response (collagenase activity), β_0 is the model intercept, β_i is the linear coefficient and x_i is the level of the independent variable. This model identifies the main parameters required for maximal collagenase production. A total of seven variables, namely temperature, initial pH of the medium, seed inoculation, incubation time, corn meal concentration, bran liquid concentration and soybean meal concentration, were chosen for the present study. The factors under investigation and the levels for each factor selected in the Plackett–Burman design are illustrated in Table 1, and the experimental matrix and results are presented in Table 2. The collagenase activity assay was carried out in duplicate, and the average value was calculated as the response Y.

3.5. Investigation of the Shifts in the Trends of the Significant Variables Using the Path of Steepest Ascent

Frequently, the initial estimate of the optimal conditions for the system is far from the actual optimal conditions. Before the final response surface analysis, the path of steepest ascent usually appears as a line through the center of the region of interest and is normal to the fitted surface contours [20]. Three significant independent variables were selected based on the results from the Plackett–Burman designs. The direction of the shift in the trend of each variable was determined by the positive or negative of the coefficient estimate listed in Table 3. The step length (Δ) of the path of steepest ascent was calculated by following Equation (4):

$$\Delta = (\text{Average} - \text{Base}) \times \text{Slope} \times \text{Ratio} \quad (4)$$

In this formula, Average is the average value of the −1 or +1 level of each variable and Slope is equal to the coefficient estimate ratio. The Ratio is an empirical value derived from pre-experimental data in order to obtain a rational numerical variable for operation.

Based on the physical and chemical characteristics of the bacteria, the steepest ascent model contained four steps. Specific experimental designs for the path of steepest ascent are shown in Table 4.

3.6. Optimization by Response Surface Methodology

Based on the significant variables chosen by the Plackett–Burman design experiment and the operating conditions selected by the steepest ascent experiment, response surface methodology was applied for the augmentation of collagenase production using a central composite design (CCD) [21]. There are five levels (−1.68, −1, 0, +1 and +1.68) estimated for each variable in the CCD design based on the coded values and actual values, as listed in Table 5. The significant variables and their 0-level values were as follows: temperature (16 °C), culture time (3.36 d) and soybean concentration (33.0 g·L^{-1}). The +1.68-level value was calculated as the value of the 0-level plus the product of the step unit multiplied by +1.68. A total of 20 combined experiments was performed in triplicate and repeated twice. The complete experimental plan, including the RSM design and collagenase production as the corresponding response value (Y), are shown in Table 6. The statistical software package Design-Expert Version 8.0.6 was used to analyze the design.

3.7. Statistical Analysis

The data obtained from the central composite designs with three factors (temperature, culture time and soybean powder concentration) were used to perform analysis of variance (ANOVA) with Design-Expert Version 8.0.6 software, as presented in Table 7. After completing the response surface methodology experiments and measuring the collagenase yield, the response surface regression procedure was used to fit the experimental results from the RSM to the following second-order polynomial regression Equation (5) [21,22]:

$$Y = \beta_0 + \sum_i \beta_i X_i + \sum_{ii} \beta_{ii} X_i^2 + \sum_{ij} \beta_{ij} X_i X_j \tag{5}$$

where Y is the predicted response value (collagenase production), β_0 is the center point of the system, β_i is the linear coefficient, β_{ii} is the quadratic coefficient and β_{ij} is the interaction coefficient, while X_i, X_i^2 and X_j are the linear, quadratic and interaction terms of the independent variables, respectively. The fitted model was then expressed as three-dimensional surface and contour plots to describe the relationship between the responses and the experimental levels of each of the variables studied.

3.8. Validation of the Optimization Model

The optimization model was validated by adjusting three parameters for each of the variables. A triplicate culture set was grown under experimental conditions derived from the optimization scheme, and collagenase production was compared to the predicted response value.

For further validation, a time-course experiment for collagenase production and biomass fluctuation was conducted, and gelatine-immersing zymography was also performed. Continuous fermentation was performed under the optimal conditions, and collagenase production and biomass were recorded [23,24]. Changes in extracellular collagenase secretion were detected by non-denaturing gel electrophoresis with gelatine as the immersing substrate of soluble collagen analogues [19]. The collagenase activity was measured, and the biomass was quantified using the spread plate method.

3.9. Purification of Ps sp. SJN2 Collagenases

The total extracellular collagenases from the fermentation broth were extracted by ammonium sulfate precipitation. The fermentation liquid was centrifuged at 10,000 rpm for 30 min at 4 °C. Ammonium sulfate was added slowly up to 40% (w/w), and the solution was allowed to stand overnight at 4 °C. The precipitate was collected and redissolved in water. The supernatant, containing active collagenases, was dialyzed in Tris-HCl buffer (pH 8.8, 20 mM) at 4 °C overnight.

The fraction containing active collagenases (Col SJN2) in the dialysis fluid was further purified on a HiTrap Capto DEAE (GE Healthcare, Boston, MA, USA) column with an NGC chromatography system (Bio-Rad, Hercules, CA, USA) after filtration through a 0.45-µm filter membrane. The column was equilibrated with Tris-HCl buffer (pH 8.8, 20 mM). Then, 5 mL of Col SJN2 were loaded onto the pre-equilibrated column at a flow rate of 1.0 mL·min^{-1} and washed with buffer Tris-HCl for 10 min. Then, elution was conducted using a linear gradient of 1 M NaCl (0–100%) at a flow rate of 2.0 mL·min^{-1}. Fractions were monitored at 280 nm, and the Col SJN2 fraction containing active collagenases was centrifuged in an ultrafiltration tube with a 3-kDa molecular weight cut-off (Millipore, Temecula, CA, USA) at 5000× g for 30 min at 4 °C. The portion of the Col SJN2 fraction with molecular weight higher than 3 kDa was further purified by size exclusion chromatography.

Col SJN2 was then loaded onto a Superdex-200 (GE Healthcare, Boston, MA, USA) column pre-equilibrated with distilled water. Elution was performed using distilled water at a flow rate of 0.5 mL·min^{-1}. Fractions were monitored at 280 nm, and the fractions with collagenase activity were stored.

3.10. Preparation of Collagen from Fishery By-Products

To study the enzymatic properties of collagenases [25] from *Ps* sp. SJN2, five kinds of native collagen were extracted from porcine skin, octopus flesh and fishery by-products (salmon fish skins, seabream fish scales and Spanish mackerel fish bones). Extraction methods are described in the Supplementary Material.

3.11. Enzymatic Hydrolysis and Properties of Ps sp. SJN2 Collagenase

Extracted collagen was digested with a crude protease of *Ps* sp. SJN2 at an enzyme-to-substrate ratio of 1:10 (v (mL)/w (mg)). The reaction was carried out at 45 °C in PBS (pH 7.0, 0.1 M) for 1 hour and stopped by incubation at 95 °C for 10 min. The resultant slurry was collected and centrifuged at 10,000× g at 4 °C for 10 min. The supernatant was collected and stored at 4 °C until further analysis.

Substrate-immersing zymography was conducted to compare the properties of the crude enzymes from *Ps* sp. SJN2, *Ps* sp. SBN2-2, *Ps* sp. SGS2-2, *Ps* sp. SWN-1, *Ps* sp. SWN-2, *Vibrio* sp. HK3-2, *Vibrio* sp. SJN4 and *Planococcus* sp. SYT1.

Additionally, the effects of different concentrations of ions on the enzymatic activity of *Ps* sp. SJN2 collagenase were tested. Different ions (Cd^{2+}, Mn^{2+}, Ca^{2+}, Fe^{3+}, Ag^+, Al^{3+}, Ba^{2+}, Cu^{2+}, Zn^{2+}, Mg^{2+}, Fe^{2+} and Co^{2+}) and the metal-chelating agent EDTA were added to the reaction buffer containing crude enzyme at concentrations of 2 mM and 10 mM, and the specific activity of the collagenase was measured.

The collagen-swelling effect was also observed under SEM. A total of 5 mg of type I insoluble collagen was mixed with 2 mL of 20 mM PBS (pH 8.5) containing 200 µL of *Ps* sp. SJN2 collagenase. The samples were incubated at 37 °C for 12 h with continuous stirring. The samples were observed using scanning electron microscopy (FEI, Hillsboro, OR, USA).

3.12. DPPH Radical Scavenging and ORAC Assay

A modified DPPH radical scavenging assay was performed [26,27]. Ethanolic solutions of DPPH (10^{-4}) and collagen hydrolysates were mixed in disposable Eppendorf tubes so that the final mass ratios of hydrolysates to DPPH were 1:5. The samples were sealed and incubated for 60 min in the dark at 37 °C, and the decrease in absorbance at 517 nm was measured against ethanol in 96-well plates using an Enspire spectrophotometer (Perkin Elmer, Waltham, MA, USA). Vitamin C was used as a positive control. All determinations were performed in triplicate.

The ORAC assay was performed based on the method described by Zulueta [28], with some modifications. In this system, AAPH is the source of free radicals that can attack the fluorescein and lead to fluorescence decay. The reaction was carried out in 75 mM phosphate buffer (pH 7.4). Sample solution (20 µL) and fluorescein (150 µL, 96 nM) were added into a 96-well plate and pre-incubated at 37 °C in the Enspire spectrophotometer. The reaction was initiated by adding 30 µL of AAPH (320 mM). The reaction was performed at 37 °C. The fluorescence intensity [29] was measured every 60 s for 120 cycles with excitation and emission wavelengths of 485 nm and 538 nm, respectively. The positive control was vitamin C, which was used at the same concentration as in the DPPH radical scavenging assay.

4. Conclusions

Compared with other native bacteria, the extracellular collagenases of *Ps* sp. SJN2, composed of proteases (e.g., metalloproteases and serine proteases) with different molecular weights, showed higher enzyme activity towards porcine, bovine and marine collagen. Given the increasing economic relevance of marine collagenases, this study was conducted to optimize a variety of fermentation parameters, including medium composition and culture conditions, for maximal collagenase production. The results indicated that the Plackett–Burman design and response surface methodology are effective and efficient approaches to improve collagenase production via optimization of *Ps* sp. SJN2 fermentation conditions. The maximum collagenase yield of 322.58 ± 9.61 U·mL^{-1} was achieved under the optimized conditions, which was in agreement with the production value of 306.68 U·mL^{-1} predicted by the model. Finally, for optimal fermentation, the cells were cultured at 17.32 °C for 3.72 d in a medium containing soybean 34.23 g·L^{-1}, corn meal 30 g·L^{-1} and bran liquid 15 mL·L^{-1} at an initial pH value of 8.5 and an inoculum volume of 1% (v/v). Significant improvement (2.2-fold) in the production of collagenases by *Ps* sp. SJN2 was accomplished. In addition, the marine collagen hydrolysed by collagenases from *Ps* sp. SJN2, especially those from Spanish mackerel fish bone and seabream fish scale, showed better prospects for application as antioxidant peptides. Above all, the optimized medium established in this study could provide a basis for further study of large-scale fermentation of *Ps* sp. SJN2 for marine collagenase yield improvement.

Supplementary Materials: The following are available online at www.mdpi.com/1660-3397/15/12/377/s1: Figure S1: Gelatin immersing zymography of Col SJN2, Figure S2: Crude enzyme zymography of *Ps* sp. SJN2, Table S1: Collagenases activity in purification process. Supporting information about the methods of inoculum preparation for fermentation and preparation of collagen from fishery by-products is provided in the Supplementary Material.

Acknowledgments: This work was supported by the National Natural Science Foundation of China (3140002, 31070061, 31370104, 21205142), the National Sparking Plan Project (2013GA770009), the Opening Foundation of the Chinese National Engineering Research Center for Control and Treatment of Heavy Metal Pollution (2015CNERC-CTHMP-07), the Open-End Fund for the Valuable and Precision Instruments of Central South University (CSUZC201640) and the Fundamental Research Funds for the Central Universities of Central South University (2015zzts273).

Conflicts of Interest: The authors declare no conflicts of interest.

References

1. Zhao, G.Y.; Chen, X.L.; Zhao, H.L.; Xie, B.B.; Zhou, B.C.; Zhang, Y.Z. Hydrolysis of insoluble collagen by deseasin mcp-01 from deep-sea *Pseudoalteromonas* sp. Sm9913: Collagenolytic characters, collagen-binding ability of c-terminal polycystic kidney disease domain, and implication for its novel role in deep-sea sedimentary particulate organic nitrogen degradation. *J. Biol. Chem.* **2008**, *283*, 36100–36107. [PubMed]
2. Silva, T.H.; Moreira-Silva, J.; Marques, A.L.; Domingues, A.; Bayon, Y.; Reis, R.L. Marine origin collagens and its potential applications. *Mar. Drugs* **2014**, *12*, 5881–5901. [CrossRef] [PubMed]
3. Blanco, M.; Vazquez, J.A.; Perez-Martin, R.I.; Sotelo, C.G. Hydrolysates of fish skin collagen: An opportunity for valorizing fish industry byproducts. *Mar. Drugs* **2017**, *15*, 131. [CrossRef] [PubMed]
4. Watanabe, K. Collagenolytic proteases from bacteria. *Appl. Microbiol. Biotechnol.* **2004**, *63*, 520–526. [CrossRef] [PubMed]
5. Duarte, A.S.; Correia, A.; Esteves, A.C. Bacterial collagenases—A review. *Crit. Rev. Microbiol.* **2016**, *42*, 106–126. [CrossRef] [PubMed]
6. Shekhter, A.B.; Balakireva, A.V.; Kuznetsova, N.V.; Vukolova, M.N.; Litvitsky, P.F.; Zamyatnin, A.A. Collagenolytic enzymes and their applications in biomedicine. *Curr. Med. Chem.* **2017**, *24*, 1–19. [CrossRef] [PubMed]

7. Ran, L.Y.; Su, H.N.; Zhao, G.Y.; Gao, X.; Zhou, M.Y.; Wang, P.; Zhao, H.L.; Xie, B.B.; Zhang, X.Y.; Chen, X.L.; et al. Structural and mechanistic insights into collagen degradation by a bacterial collagenolytic serine protease in the subtilisin family. *Mol. Microbiol.* **2013**, *90*, 997–1010. [CrossRef] [PubMed]
8. Zhao, G.Y.; Zhou, M.Y.; Zhao, H.L.; Chen, X.L.; Xie, B.B.; Zhang, X.Y.; He, H.L.; Zhou, B.C.; Zhang, Y.Z. Tenderization effect of cold-adapted collagenolytic protease mcp-01 on beef meat at low temperature and its mechanism. *Food Chem.* **2012**, *134*, 1738–1744. [CrossRef] [PubMed]
9. Matsushita, O.; Koide, T.; Kobayashi, R.; Nagata, K.; Okabe, A. Substrate recognition by the collagen-binding domain of clostridium histolyticum class I collagenase. *J. Biol. Chem.* **2001**, *276*, 8761–8770. [CrossRef] [PubMed]
10. Bassetto, F.; Maschio, N.; Abatangelo, G.; Zavan, B.; Scarpa, C.; Vindigni, V. Collagenase from *Vibrio alginolyticus* cultures: Experimental study and clinical perspectives. *Surg. Innov.* **2016**, *23*, 557–562. [CrossRef] [PubMed]
11. Vijayaraghavan, P.; Vincent, S.G. Statistical optimization of fibrinolytic enzyme production by *Pseudoalteromonas* sp. Ind11 using cow dung substrate by response surface methodology. *SpringerPlus* **2014**, *3*, 60. [CrossRef] [PubMed]
12. Haddar, A.; Fakhfakh-Zouari, N.; Hmidet, N.; Frikha, F.; Nasri, M.; Kamoun, A.S. Low-cost fermentation medium for alkaline protease production by *Bacillus Mojavensis* a21 using hulled grain of wheat and sardinella peptone. *J. Biosci. Bioeng.* **2010**, *110*, 288–294. [CrossRef] [PubMed]
13. Reddy, L.V.; Wee, Y.J.; Yun, J.S.; Ryu, H.W. Optimization of alkaline protease production by batch culture of *Bacillus* sp. Rky3 through plackett-burman and response surface methodological approaches. *Bioresour. Technol.* **2008**, *99*, 2242–2249. [CrossRef] [PubMed]
14. Yang, F.; Long, L.; Sun, X.; Wu, H.; Li, T.; Xiang, W. Optimization of medium using response surface methodology for lipid production by *Scenedesmus* sp. *Mar. Drugs* **2014**, *12*, 1245–1257. [CrossRef] [PubMed]
15. Gupta, R.; Beg, Q.K.; Lorenz, P. Bacterial alkaline proteases: Molecular approaches and industrial applications. *Appl. Microbiol. Biotechnol.* **2002**, *59*, 15–32. [PubMed]
16. He, H.L.; Liu, D.; Ma, C.B. Review on the angiotensin-i-converting enzyme (ace) inhibitor peptides from marine proteins. *Appl. Biochem. Biotechnol.* **2013**, *169*, 738–749. [CrossRef] [PubMed]
17. Yu, M.L.; Kim, J.S.; Wang, J.K. Optimization for the maximum bacteriocin production of *Lactobacillus brevis* df01 using response surface methodology. *Food Sci. Biotechnol.* **2012**, *21*, 653–659.
18. Aleman, A.; Gimenez, B.; Montero, P.; Gomez-Guillen, M.C. Antioxidant activity of several marine skin gelatins. *Lwt-Food Sci. Technol.* **2011**, *44*, 407–413. [CrossRef]
19. Liu, D.; Yang, X.; Huang, J.; Wu, R.; Wu, C.; He, H.; Li, H. In situ demonstration and characteristic analysis of the protease components from marine bacteria using substrate immersing zymography. *Appl. Biochem. Biotechnol.* **2015**, *175*, 489–501. [CrossRef] [PubMed]
20. Kleijnen, J.P.C.; Hertog, D.D.; Angün, E. Response surface methodology's steepest ascent and step size revisited: Correction. *Eur. J. Oper. Res.* **2002**, *170*, 664–666. [CrossRef]
21. Zhao, H.L.; Yang, J.; Chen, X.L.; Su, H.N.; Zhang, X.Y.; Huang, F.; Zhou, B.C.; Xie, B.B. Optimization of fermentation conditions for the production of the M23 protease pseudoalterin by deep-sea *Pseudoalteromonas* sp. CF6-2 with artery powder as an inducer. *Molecules* **2014**, *19*, 4779–4790. [CrossRef] [PubMed]
22. Liu, S.; Fang, Y.; Lv, M.; Wang, S.; Chen, L. Optimization of the production of organic solvent-stable protease by bacillus sphaericus ds11 with response surface methodology. *Bioresour. Technol.* **2010**, *101*, 7924–7929. [CrossRef] [PubMed]
23. Sahan, T.; Ceylan, H.; Sahiner, N.; Aktas, N. Optimization of removal conditions of copper ions from aqueous solutions by *Trametes versicolor*. *Bioresour. Technol.* **2010**, *101*, 4520–4526. [CrossRef] [PubMed]
24. Oskouiea, S.F.G.; Tabandehb, F.; Yakhchalib, B.; Eftekhara, F. Response surface optimization of medium composition for alkaline protease production by *Bacillus clausii*. *Biochem. Eng. J.* **2008**, *39*, 37–42. [CrossRef]
25. Ogawa, M. Biochemical properties of bone and scale collagens isolated from the subtropical fish black drum (pogonia cromis) and sheepshead seabream (*Archosargus probatocephalus*). *Food Chem.* **2004**, *88*, 495–501. [CrossRef]

26. Vazquez, J.A.; Blanco, M.; Massa, A.E.; Amado, I.R.; Perez-Martin, R.I. Production of fish protein hydrolysates from *Scyliorhinus canicula* discards with antihypertensive and antioxidant activities by enzymatic hydrolysis and mathematical optimization using response surface methodology. *Mar. Drugs* **2017**, *15*, 306. [CrossRef] [PubMed]
27. Wu, R.; Chen, L.; Liu, D.; Huang, J.; Zhang, J.; Xiao, X.; Lei, M.; Chen, Y.; He, H. Preparation of antioxidant peptides from salmon byproducts with bacterial extracellular proteases. *Mar. Drugs* **2017**, *15*, 4. [CrossRef] [PubMed]
28. Zulueta, A.; Esteve, M.J.; Frígola, A. ORAC and TEAC assays comparison to measure the antioxidant capacity of food products. *Food Chem.* **2009**, *114*, 310–316. [CrossRef]
29. Dávalos, A.; Carmen Gómezcordovés, A.; Bartolomé, B. Extending applicability of the oxygen radical absorbance capacity (orac−fluorescein) assay. *J. Agric. Food Chem.* **2004**, *52*, 48–54. [CrossRef] [PubMed]

© 2017 by the authors. Licensee MDPI, Basel, Switzerland. This article is an open access article distributed under the terms and conditions of the Creative Commons Attribution (CC BY) license (http://creativecommons.org/licenses/by/4.0/).

Article

Structure-Based Design and Synthesis of a New Phenylboronic-Modified Affinity Medium for Metalloprotease Purification

Shangyong Li [1,†], Linna Wang [1,†], Ximing Xu [2,†], Shengxiang Lin [3,*], Yuejun Wang [1], Jianhua Hao [1,4] and Mi Sun [1,4,*]

1. Key Laboratory of Sustainable Development of Marine Fisheries, Ministry of Agriculture, Yellow Sea Fisheries Research Institute, Chinese Academy of Fishery Sciences, 106 Nanjing Road, Qingdao 266071, China; lshywln@163.com (S.L.); wlnwfllsy@163.com (L.W.); wangyj@ysfri.ac.cn (Y.W.); haojh@ysfri.ac.cn (J.H.)
2. Institute of Bioinformatics and Medical Engineering, School of Electrical and Information Engineering, Jiangsu University of Technology, Changzhou 213000, China; ximing.xu@jsut.edu.cn
3. Laboratory of Oncology and Molecular Endocrinology, CHUL Research Center (CHUQ) and Laval University, 2705 Boulevard Laurier, Ste-Foy, Ville de Québec, QC G1V 4G2, Canada
4. Laboratory for Marine Drugs and Bioproducts, Qingdao National Laboratory for Marine Science and Technology, Qingdao 266237, China
* Correspondence: sheng-xiang.lin@crchul.ulaval.ca (S.L.); sunmi0532@yahoo.com (M.S.); Tel.: +1-418-654-2296 (S.L.); +86-532-8581-9525 (M.S.); Fax: +1-418-654-2761 (S.L.); +86-532-8581-9525 (M.S.)
† These authors contributed equally to this paper.

Academic Editors: Vassilios Roussis and Antonio Trincone
Received: 17 November 2016; Accepted: 21 December 2016; Published: 27 December 2016

Abstract: Metalloproteases are emerging as useful agents in the treatment of many diseases including arthritis, cancer, cardiovascular diseases, and fibrosis. Studies that could shed light on the metalloprotease pharmaceutical applications require the pure enzyme. Here, we reported the structure-based design and synthesis of the affinity medium for the efficient purification of metalloprotease using the 4-aminophenylboronic acid (4-APBA) as affinity ligand, which was coupled with Sepharose 6B via cyanuric chloride as spacer. The molecular docking analysis showed that the boron atom was interacting with the hydroxyl group of Ser176 residue, whereas the hydroxyl group of the boronic moiety is oriented toward Leu175 and His177 residues. In addition to the covalent bond between the boron atom and hydroxyl group of Ser176, the spacer between boronic acid derivatives and medium beads contributes to the formation of an enzyme-medium complex. With this synthesized medium, we developed and optimized a one-step purification procedure and applied it for the affinity purification of metalloproteases from three commercial enzyme products. The native metalloproteases were purified to high homogeneity with more than 95% purity. The novel purification method developed in this work provides new opportunities for scientific, industrial and pharmaceutical projects.

Keywords: metalloprotease; adsorption analysis; molecular docking; affinity purification; aminophenylboronic acid

1. Introduction

Proteases are enzymes that catalyze the hydrolysis of peptide bonds. Based on the mechanism of catalysis, proteases can be classified into six classes, including metallo, serine, aspartic, cysteine, glutamic, and threonine proteases [1]. Proteases are the most important industrial enzymes, accounting for more than 60% of the total enzyme market [2,3]. They have broad applications in the pharmaceutical, leather, food, and detergent industries [1–4]. Proteases play critical roles in normal biological processes; their unusual activities have been implicated in the development and progression of many diseases, e.g., fibrosis, arthritis, cancer, cardiovascular diseases, nephritis, and central nervous system disorders [5–7]. Among all of the six classes of proteases, only untagged serine proteases can be purified in one step using p-aminobenzamidine-modified affinity medium [8,9]. This simple procedure of affinity purification significantly accelerated the pharmaceutical application of many serine proteases [10–15].

Currently, there is no straightforward and efficient protocol for the purification of metalloproteases [14–16]. The traditional protocol that has multiple steps is expensive and results in low recovery [17–20]. Although some reports refer to the high-yield purification of metalloproteases (more than 90% purity) in one-step procedure, these protocols were based on immobilized metal affinity chromatography (IMAC) that has its disadvantages [21,22]. The first one is the use of high concentrations of imidazole and salt in the elution buffer of the IMAC procedure, which necessitates additional dialysis or a desalting step [23,24]. Also, it is well known that purification of a metalloprotein via a metal ion chelated by the resin in a similar manner results in the exchange of metal ion from resin with a metal ion from metalloprotein. This metal transfer causes a decrease in the stability of purified metalloprotein [21]. In addition, the use of chelating agents during purification has to be avoided as these compounds can remove the metal ion from the enzyme active site [23–25]. Therefore, design, synthesis and application of a new specific and efficient medium for the purification of metalloproteases are important tasks.

The structure-based design of the affinity ligand that serves as a specific inhibitor or substrate analogue is an efficient and commonly used approach in the affinity purification of enzymes [26–28]. An alkaline metalloprotease, MP (accession no. ACY25898) from marine bacterium *Flavobacterium* sp. YS-80-122, has been previously isolated in our laboratory [3]. This enzyme is a typical Zn-containing metalloprotease with antioxidant activity, and it has been commercially used as a detergent additive. The analysis of its crystal structure (PDB: 3U1R) [29] allowed suggestion of a novel affinity ligand that could reversibly bind to the active site and could be used for the affinity purification of the enzyme. Our preliminary virtual screening and experimental verification indicated that boronic acid derivatives (BADs) could reversibly inhibit the activity of MP [30]. Phenylboronate group, which can form a temporary covalent bond with any molecule that contains a 1,2-*cis*-diol group, is widely used in the affinity purification of 1,2-*cis*-diol-containing biomolecules such as glycoproteins, glycopeptides, nucleosides, and nucleic acids [31–35]. However, application of the resins modified by phenylboronate in the purification of metalloproteases has never been reported.

Here, the phenylboronate-modified resin was synthesized through the coupling of 4-aminophenylboronic acid (4-APBA) with epoxy-activated Sepharose 6B via cyanuric chloride spacer. The binding site and structure-activity relationship between 4-APBA-modified medium and MP were analyzed using molecular docking and adsorption determination, correspondingly. The synthesized medium was used for development of one-step affinity purification of metalloproteases. Three commercially available metalloproteases were efficiently purified with a high purity (more than 95%) using the protocol developed. Our research provides new opportunities for the development of industrial methods of metalloprotease purification.

2. Results and Discussion

2.1. Design and Synthesis of Affinity Medium for Metalloprotease Purification

Our initial virtual screening showed that some BADs could inhibit MP catalytic activity [30]. To confirm that BADs could inhibit MP, ten BADs were purchased or synthesized, and then their inhibitory effect was tested on MP. Surprisingly, three compounds were strong MP inhibitors with apparent K_i value of 0.8–1.2 μM. Thus, we focused our efforts on the design of BADs-based affinity medium for metalloprotease purification. Immobilisation of a ligand onto the epoxy-activated resin should be achieved via a nucleophilic group present in the ligand, often a primary amine [27].

Aminephenylboronic acid was chosen as an affinity ligand for our study because it was commercially available and had a favorable configuration for synthesis affinity medium.

The nature of the immobilized complex or, in other words, the choice of affinity ligand and spacer arm, has a major influence on the outcome of a biomimetic affinity of purification procedure [27–29,36]. To obtain an optimal affinity medium, two types of APBA-based ligands, 4-APBA, and 3-APBA, were tested. The affinity ligands were coupled with activated Sepharose 6B via cyanuric chloride spacer. To estimate the effect of the presence of a boron atom in the affinity ligand, another type of the affinity ligand lacking of boron atom (aniline ligand) was synthesized. The scheme for the synthesis of 4-APBA-modified Sepharose 6B is shown in Figure 1. To confirm the ligand structure, the medium was hydrolyzed with 6 M HCl, and then the resultant with molecular formula of C12H15BClN5O4 and molecular mass of 339.5. Because the chlorine on the triazine ring was unstable in acidic condition, the hydrolysis with 6 M HCl would replace the chlorine on the ligand with a hydroxyl group [37], thus the theoretical structure of the purified ligand should be with a molecular formula of C12H16BN5O5 and molecular mass of 321.1. The ligand may be broken into fragments as C9H10BN5O3 at cone voltage of 170 V and molecular mass of 247.01. As shown in Figure S1, the main peak, 247.02, showed good agreement with [M-C3H6O2-H]$^+$. The possible structures of chemicals in principal peaks are also shown in Figure S1. These results showed that the synthesized ligands had a good reliability.

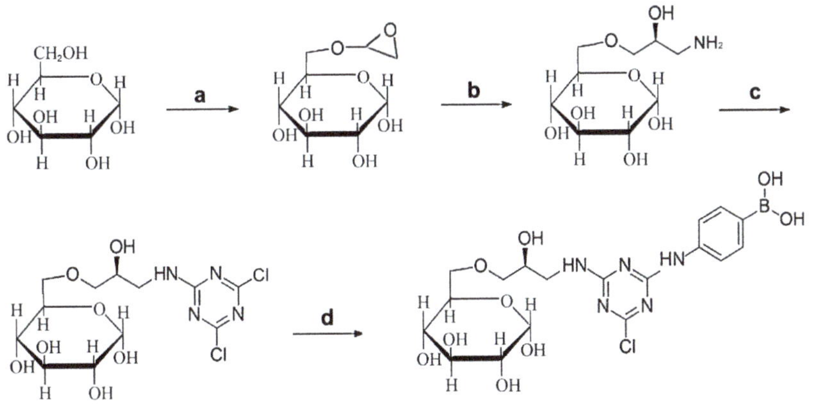

Figure 1. Synthesis protocol and scheme of the 4-APBA ligand coupled with actived Sepharose 6B via cyanuric chloride spacer. Reagents and conditions: (**a**) epichlorohydrin, DMSO, NaOH aqueous solution, 2.5 h; (**b**) 35% saturated ammonia, overnight; (**c**) cyanuric chloride, 50% acetone, pH 7–8; (**d**) 4-APBA, sodium carbonate, 24 h.

The 3-APBA-modified medium and aniline-modified medium were synthesized using the same concentration of 3-APBA or aniline as for 4-APBA (Figure 2A,B). The density of the free amino groups was determined by the ninhydrin test before the adding of the APBA ligands, giving equal ligand densities (tab:marinedrugs-15-00005-t001). Equilibrium adsorption studies were performed to characterize the affinity value of MP and these three affinity media (Figure 3A). Desorption constant for the 4-APBA-modified medium was 14.9 µg/mL which was significantly lower than that for the 3-APBA medium (21.5 µg/mL) and aniline medium (67.2 µg/mL). Meanwhile, the theoretical maximum absorption (Q_{max}) for the 4-APBA medium (29.6 mg/g) was significantly higher than it was for the other two media (24.9 mg/g and 10.6 mg/g, respectively) (tab:marinedrugs-15-00005-t001), indicating the high affinity of 4-APBA-modified Sepharose 6B towards MP. Therefore, 4-APBA was chosen as the affinity ligand for the further design and synthesis of affinity medium.

Table 1. Ligand densities, desorption constant (K_d) and theoretical maximum absorption (Q_{max}) analysis of the affinity media.

Ligands	Spacer Arms	Ligand Density (μmol/mL)	K_d (μg/mL)	Q_{max} (mg/g)
Aniline	Cyanuric chloride	20.9	67.2	10.6
3-APBA [a]	Cyanuric chloride	20.9	21.5	24.9
4-APBA	Cyanuric chloride	20.9	14.9	29.6
4-APBA	10-atom spacer	41.8	24.4	24.6
4-APBA	5-atom spacer	27.8	46.3	22.3

[a] APBA represents aminophenylboronic acid.

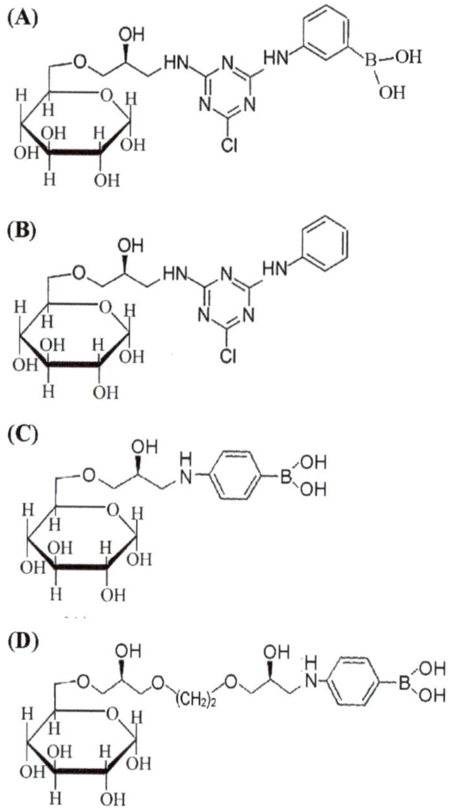

Figure 2. The scheme of four different affinity media. (**A**) 3-APBA ligand coupled with activated Sepharose 6B via cyanuric chloride spacer; (**B**) Aniline ligand coupled with activated Sepharose 6B via cyanuric chloride spacer; (**C**) 4-APBA ligand coupled with activated Sepharose 6B via 5-atom spacer arm; (**D**) 4-APBA ligand coupled with activated Sepharose 6B via 10-atom spacer arm.

To find the optimal spacer arm, two different lengths of linear arms (5-atom spacer and 10-atom spacer) and a cyclic arm (cyanuric chloride) were tested. Cyanuric chloride is a typical cyclic compound containing the s-triazine (C_3N_3) ring that could supply a higher mechanical strength for the ligand stabilization and was widely used in the affinity medium synthesis [38–41]. The scheme for the synthesis of media with the 5-atom spacer and the 10-atom spacer are shown in Figure 2C,D, correspondingly. In the adsorption analysis (Figure 3B), 4-APBA ligand with cyclic spacer arm exhibited the highest adsorption value, even though its epoxy content (20.9 µmol/mL) was lower than the content of 5-atom linear spacer (41.8 µmol/mL) and the 10-atom linear spacer (27.8 µmol/mL) (tab:marinedrugs-15-00005-t001). Thus, cyanuric chloride was chosen as the compound for generation of optimal spacer arm.

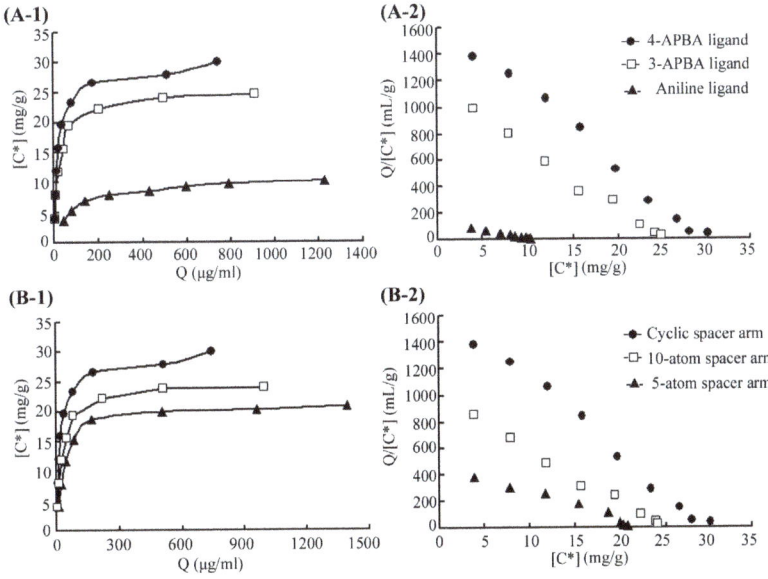

Figure 3. Adsorption analyses of different affinity media. (**A**) Adsorption analysis of affinity media with three different ligands via the same spacer arm (cyanuric chloride); (**B**) Adsorption analysis of affinity media with the same ligand (4-APBA) via three different spacer arms. (**1**) Equilibrium adsorption of metalloprotease (MP) on the affinity medium in a batch system (50 mM Gly-NaOH buffer, pH 8.6, 25 °C), (**2**) Plot describing the equilibrium of the absorption on the medium and the enzyme concentration in the liquid phase.

2.2. Binding Analysis for 4-APBA-Modified Medium and MP

Quite a few of studies show that boron-containing small molecules interacted with proteins through a covalent bond between the boron atom and the oxygen atom in the hydroxyl group of a serine [42]. In this study, the molecular docking analysis also indicated that the boron atom interacted with the hydroxyl group of Ser176 residue through covalent bonding, whereas the hydroxyl group of the boronic moiety is oriented toward Leu175 and His177 residues (Figure 4). We found that several secondary interactions could contribute to the stabilization of MP interaction with 4-APBA-modified medium. For example, the benzene ring of the 4-APBA ligand formed a π-π interaction with His171 residue of MP. In addition, the hydrogen bond between the s-triazine ring of the spacer and the molecule of water was observed, as well as the hydrogen bond between the hydroxyl group of an atom of the Ala128 residue. The aniline ligand bound with Sepharose 6B via cyanuric chloride also exhibited a low affinity (K_d, 67.2 µg/mL; Q_{max}, 10.6 mg/g) toward MP, implying that several secondary interactions can occur in addition to the interaction with the boronate ion.

Boronate affinity materials have gained increasing attention in recent years [31–33]. The mechanism involved is similar to other conventional boronate affinity chromatography. Moreover, other possible binding mechanisms were also exhibited in the molecular docking performance. One performance showed that it could be possible for Ser176 and His177 to interact with the hydroxyl groups of the boronic acid (not the boron atom) through hydrogen binding [31]. This binding mechanism relied on the hydrogen binding, which exhibited much lower affinity than the conventional binding. In the adsorption analysis, the aniline ligand exhibited a much lower affinity than APBA ligand with boronic acid, implying that the boronic acid was very important in the binding mechanism. The other possible performance is for the boron atom to coordinate with the water molecule through intermolecular B-N coordination [34]. The Ser176 residue was located in the bottom of the active-site pocket that had enough space for binding with a molecule larger than the molecule of water. Also, the 4-APBA ligand with 10-atom linear spacer showed a similar adsorption value with that for the 5-atom spacer, even though its epoxy content (27.8 µmol/mL) was smaller than the 5-atom spacer (41.8 µmol/mL) (tab:marinedrugs-15-00005-t001). This probably occurred because the longer spacer arm provided the larger spatial distance and thus provided a better accessibility of the Ser176 residue in the cavity of active site. Summarizing, the boron atom bound to MP by trapping the Ser176 hydroxyl group in the active site pocket.

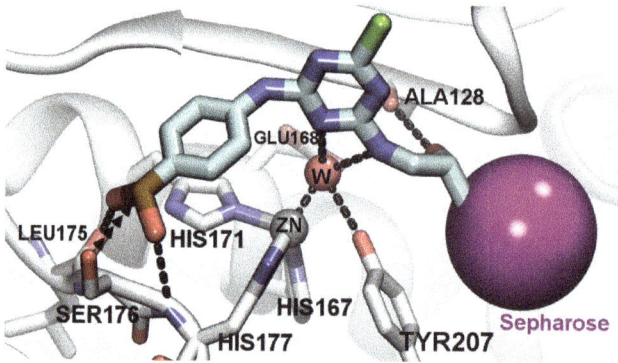

Figure 4. The binding mode of MP and the 4-APBA-modified medium. The atom force field maps were generated using Autogrid4 software for AutoDock4 (Zn); binding conformation was analyzed by Lamarckian Genetic Algorithm-Local Search combined algorithm with default searching parameter.

2.3. One-Step Affinity Purification of Commercial Metalloprotease Products

Three commercially available products (MP, DENIE-B LPS-P and ViscozymeL) containing metalloproteases were dissolved in the loading buffer to a final concentration of 10 mg/mL each. Then, the enzymes were purified by a one-step purification protocol using the 4-APBA-modified Sepharose 6B medium. We tested different loading and elution conditions to optimize the yield of metalloproteases. Almost all of the metalloproteases contained seven or eight calcium ions stabilizing their three-dimensional structure [29]. Thus, to obtain properly folded enzymes, both of the loading and elution buffers contained 1 mM $CaCl_2$. The 0.1 M Gly-NaOH buffer, pH 8.6, was chosen as the loading buffer because of the highest affinity of MP to the beads and stability of all three enzymes at this pH. Different acetic acid buffers (pH ranging from 4.0 to 6.0) were tested to select an optimal pH for MP elution, as low acidity favored the disruption of the H-bond interactions between MP and the medium. The highest protein yield was obtained at pH 5.4. Thus, 0.1 M acetic acid (pH 5.4) was chosen as the elution buffer. The SDS-PAGE analysis of the crude and purified metalloproteases is shown in Figure S2. The activity and purity of purified enzymes are shown in tab:marinedrugs-15-00005-t002.

Table 2. Comparison of affinity and traditional purification methods for three available metalloprotease products.

Enzymes	Purification Method	Activity Recovery (%)	Protein Purity (%)	Specific Activity (U/mg)	Time Requirement
MP	Affinity protocol [a]	64.1	98.8	95.6	~1 h
	Traditional protocol [b]	8.9	97.6	96.2	>48 h
DENIE-B LPS-P	Affinity protocol [a]	45.2	95.9	64.6	~1 h
	Traditional protocol [c]	10.7	91.4	62.3	>48 h
ViscozymeL	Affinity protocol [a]	37.8	97.1	51.4	~1 h
	Traditional protocol [d]	3.4	58.3	27.8	>96 h

[a] In the affinity protocol, enzymes were purified by 4-APBA-modified medium; [b] The traditional purification protocol of MP was composed of five steps, including ultrafiltration, ammonium sulfate precipitation, desalting, anion-exchange and gel-filtration chromatography; [c] The traditional purification protocol of DENIE-B LPS-P was composed of three steps, including ammonium sulfate precipitation, desalting and anion-exchange chromatography; [d] The traditional purification protocol of ViscozymeL was composed of six steps, including ammonium sulfate precipitation, hydrophobic chromatography, desalting, anion-exchange chromatography, and two steps of gel-filtration chromatography.

In our previous work, the five-step purification protocol for the purification of MP was developed. It included ammonium sulfate precipitation, desalting, anion-exchange and gel-filtration chromatography and took more than 48 h of work [3,21]. Here, we report a simple and efficient one-step MP purification procedure that takes less than one hour. Our protocol is based on the 4-APBA-modified Sepharose 6B medium that efficiently bound native MP from natural sources. Here we compared this one-step protocol with the previous reported methods, along with all the different purification steps, activity yields, specific activities and time requirement. According to the measurements of MP activity in the initial sample and purified protein, almost 64.1% of initial MP was purified, whereas only 8.9% of initial MP was recovered using the traditional protocol. The specific activity of MP purified by the APBA-modified protocol (95.6 U/mg) is similar with the value obtained through the traditional purification protocol (96.2 U/mg) and the IMAC protocol (94.8 U/mg). However, the purity of MP is different, being higher with the APBA-modified affinity purification (98.8%) with respect to the traditional and IMAC methods (92.5% and 94.7%, respectively). Even if the IMAC protocol results an activity recovery higher than the 4-APBA protocol, it is longer. Moreover, the APBA-modified affinity protocol avoids the use of toxic imidazole and the loss of metallic ions in the MP active pocket that reduce the enzyme stability. Based on all the positive features of this affinity protocol, such as one-step of chromatography, shorter times, and higher purity, it is clear there is potential in this approach for the industrial production of high-purity MP.

To determine whether our medium has an affinity value to metalloproteases from other sources, two other commercial metalloprotease products, DENIE-B LPS-P and ViscozymeL, were used for enzyme purification. DENIE-B LPS-P was an enzyme concentrate produced from *Bacillus subtilis* that was widely used in leather softening [43]. Based on the activity measurement in our study, only 10.7% of metalloprotease was purified from this commercial product using the traditional three-step purification protocol, including ammonium sulfate precipitation, desalting and anion-exchange chromatography on a Q Sepharose column. ViscozymeL was a cell wall degrading enzyme complex from *Aspergillus* sp., containing a wide range of carbohydrases and metalloprotease [44]. Traditional purification of the metalloprotease from ViscozymeL resulted in only less than 60% pure enzyme, which required six steps, including ammonium sulfate precipitation, hydrophobic chromatography, desalting, anion-exchange chromatography, and two steps of gel-filtration chromatography. Meanwhile, IMAC (Cu-IDA ligand) purification of these two protein products resulted in less than 60% purity of metalloproteases (data not shown). However, the 4-APBA-modified medium could efficiently purify metalloproteases from those two products (tab:marinedrugs-15-00005-t002). The activity recoveries of DENIE-B LPS-P and ViscozymeL were 45.2% and 37.8%, respectively. Meanwhile, when the purified enzymes were analyzed by HPLC with a TSK3000SW gel filtration column, both of them were more than 95% pure (Figure 5). To sum, our novel methodology had multiple advantages in comparison with all known techniques of metalloprotease purification.

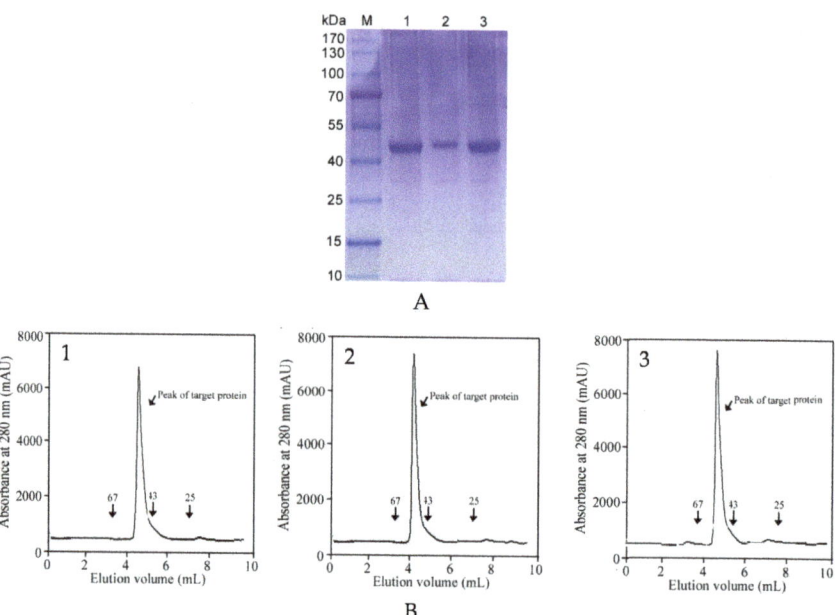

Figure 5. Purity analysis of three purified enzyme products. (**A**) SDS-PAGE (10.0%) analysis showed that the enzymes were purified to an apparent homogeneous population with a molecular mass of 48 kDa and the purity was more than 95%. *Lane M*, molecular mass standard protein marker; *Lane 1*, the purified MP; *Lane 2*, the purified DENIE-B LPS-P; *Lane 3*, the purified ViscozymeL; (**B**) HPLC analysis using the size exclusion by gel filtration of the purified MP (**1**), DENIE-B LPS-P (**2**) and ViscozymeL (**3**) on a TSK 3000SW column.

3. Materials and Methods

3.1. Materials

The dried powders of crude metalloprotease, MP, were yielded from marine bacterium *Flavobacterium* sp. YS-80-122. A commercial metalloprotease concentrate produced from *Bacillus subtilis*, DENIE-B LPS-P, was purchased from Denykem Ltd. (Shanghai, China). Cell wall degrading enzyme complex from *Aspergillus* sp., ViscozymeL, containing a wide range of carbohydrases and metalloprotease was obtained from Novozymes, Denmark. The 4-aminophenylboronic acid, 3-aminophenylboronic acid (3-APBA), aniline and cyanuric chloride (2,4,6-trichloro-1,3,5-triazine) were purchased from Sigma-Aldrich, St. Louis, MO, USA. Activated Sepharose 6B with two different spacer arm lengths (5-atoms, 10-atoms) were from Beijing Weishibohui Chromatography Technology Co., Beijing, China. All remaining reagents were of analytical grade (Sinopharm Chemical Reagent, Shanghai, China).

3.2. Synthesis of Affinity Medium

The affinity media were prepared according to the methods developed previously [45,46]. The scheme of the synthesis procedure is shown in Figure 1. Initially, Sepharose 6B was modified by epichlorohydrin to form activated amino-sepharose. Briefly, Sepharose 6B (100 g) was thoroughly washed with deionized water at a 1:10 ratio until the pH value of the eluate reached 7.0 and the beads were dried. To activate Sepharose 6B, the beads were resuspended in 50 mL of activating solution (1 M NaOH, 2.5 g DMSO, and 10 mL epichlorohydrin) followed by incubation at 40 °C for 2.5 h with shaking (Figure 1a). Then, 35% saturated ammonia (150 mL) was added to the activated Sepharose 6B resuspended in 350 mL distilled water. The beads were incubated overnight at 30 °C on a rotary 39 shaker to form aminated Sepharose 6B (Figure 1b). To attach cyanuric chloride to the amino groups of aminated Sepharose 6B, the beads were resuspended in 350 mL 50% (v/v) acetone in an ice-salt bath, and then 8 g of cyanuric chloride dissolved in 70 mL acetone was added with a flow rate of 0.5 mL/min in the shaking station. The neutral pH was maintained by simultaneous addition of 1 M NaOH. The beads were washed with 50% (v/v) acetone to remove the free cyanuric chloride (Figure 1c). The density of the free amino group was determined by the ninhydrin test in the following procedure: a small aliquot of beads was smeared on filter paper, sprayed with ninhydrin solution (0.2% (w/v) in acetone), and heated briefly with a hair dryer. The appearance of purple color indicated the presence of free amino groups, whereas the color disappearance indicated that cyanuric chloride had been linked to the amino groups [27]. Then, a twofold excess of 4-APBA dissolved in 2 M sodium carbonate was added to the dichlorotriazinylated Sepharose 6B beads. After 24 h of stirring at room temperature, the beads were filtered, washed well with water and stored in 0.02% (w/v) sodium azide (Figure 1d). To confirm the conformation of the 4-APBA ligand on the medium, 100 mg dried medium was incubated with 6 M HCl at boiling condition for 24 h, and HCl was removed by vacuum evaporation. The hydrolyzed chemical was purified and analyzed with ESI-MS (HP1100LC MSD, Agilent, San Francisco, CA, USA) according to the methods reported [37].

To generate control beads with 3-APBA and/or aniline, the affinity medium with 3-APBA or aniline instead of 4-APBA was synthesized according to the described method above (Figure 2A,B). To generate control beads with two different spacer arms, the 4-APBA-modified Sepharose 6B beads with 5-atom or 10-atom spacer arms were synthesized according to the published method [21,28,36]. The schemes for the generation of these beads are shown in Figure 2C,D. Briefly, 5 g of 4-APBA dissolved in 80 mL of 2 M sodium carbonate was added to the previously activated Sepharose 6B. After 24 h of incubation at room temperature with stirring, the beads were filtered, washed well with water and stored in 0.02% (w/v) sodium azide [13,14].

3.3. Adsorption Value Analysis

To characterize the interaction of MP with five different types of affinity media, an equilibrium adsorption study was performed. The constant of desorption (K_d) and the theoretical maximum adsorption capacity (Q_{max}) of these affinity media were analyzed according to the Scatchard analysis model [24,47]. Briefly, one milliliter of increasing concentrations of purified metalloprotease (0.1–0.9 mg/mL in 20 mM Gly-NaOH buffer, pH 8.6) was mixed with 0.5 g of each affinity medium and shaken for 2 h at 4 °C until the solution reached adsorption equilibrium. Then, the mixtures were centrifuged at 1500 g for 5 min. The protease activity and protein concentration were measured in the supernatants.

The analysis of equilibrium adsorption provided a relationship between the concentration of metalloprotease in the solution and the amount of enzyme absorbed on the affinity medium. The data obtained were analyzed using the Scatchard plot according to the following equation:

$$Q = \frac{Q_{max}[C^*]}{K_d + [C^*]}$$

Therein, Q is the adsorption amount of enzyme to the medium (mg/g), Q_{max} is the theoretical maximum of metalloprotease absorption to the affinity medium (mg/g), $[C^*]$ is the concentration of metalloprotease in solution (mg/mL), and K_d is the desorption constant.

3.4. Molecular Docking Analysis

The MP protein structure (PDB ID 3U1R) [26] was prepared by AutoDockTools (The Scripps Research Institute, San Diego, CA, USA). Briefly, hydrogens and gasteiger charge were added and waters were removed, except the water molecules bound to the zinc ion, which was treated as hydrogen acceptor. Ca^{2+} and Zn^{2+} were kept in the structure. The protein structure was then prepared according to the reference [48], using an improved zinc force field for AutoDock4 (Zn) (The Scripps Research Institute, San Diego, CA, USA). For the ligand, the sepharose part was not considered in molecular docking, as it was an inert polymeric support, and frequently used for coupling the "active" affinity ligands to the matrix. The ligand structure was built and minimized with Maestro (Schrödinger LLC., Cambridge, MA, USA). The type of boron atom was set to be sp^3 hybridization to mimic its binding with the hydroxyl group in Ser/Thr amino acids. Finally, the ligand was converted to the pdbqt format by AutoDockTools. The atom force field maps were generated using Autogrid4 software for AutoDock4 (Zn); binding conformation was searched by Lamarckian Genetic Algorithm-Local Search combined algorithm with default searching parameter. Fifty conformations were generated for further analysis. The representation was visualized with VMD 1.9.2 software (The Scripps Research Institute, San Diego, CA, USA) [49].

3.5. Traditional and Affinity Purification of Three Commercial Metalloproteases

Two grams of dried powder of three commercially available products containing three different metalloproteases, MP, DENIE-B LPS-P, and ViscozymeL, were dissolved in 50 mL sample loading buffer (0.1 M Gly-NaOH buffer, pH 8.6) each. The traditional purification protocol of MP was composed of five steps, including ultrafiltration, ammonium sulfate precipitation on 60% saturation, desalting, anion-exchange on a Q-sepharose column and gel-filtration chromatography on Sephacryl S-200 HR [3]. Meanwhile, the other two commercial metalloproteases were purified used traditional column purification protocol in this study. The traditional purification protocol of DENIE-B LPS-P was composed of three steps, including ammonium sulfate precipitation on 60% saturation, desalting and anion-exchange chromatography on a Q-sepharose column. The traditional purification protocol of ViscozymeL was composed of six steps, including ammonium sulfate precipitation on 40% saturation, hydrophobic chromatography on a phenyl column, desalting, anion-exchange chromatography on a diethylaminoethanol(DEAE)-sepharose column, and two step of gel-filtration chromatography on Sephacryl S-200 HR (GE Healthcare, Madison, WI, USA).

In the affinity purification protocol, the supernatant was loaded onto 10 mL pre-equilibrated column and washed with washing buffer (0.1 M Gly-NaOH buffer, pH 8.6) until the eluate exhibited no detectable absorbance at 280 nm. The target protein was eluted with elution buffer (0.1 M acetic acid buffer, pH 5.4). The flow rate of the mobile phase was 3.0 mL/min. The concentrations of each elution peak were measured by the Bradford method, using bovine serum albumin (BSA) as a standard. The purified enzyme was further characterized by 10% SDS-PAGE and high performance liquid chromatography (HPLC) analysis. The purification process was repeated more than five times.

3.6. Enzymatic Activity Assay

One hundred microliters of enzyme solution were mixed with 4.9 mL of casein solution (0.6% (w/v)) in 25 mM borate buffer, pH 10.0) and incubated at 25 °C for 10 min. The relative enzyme activity was measured using Folin-Ciocalteu's method [3,4]. One unit was defined as the amount of enzyme causing the release of 1 µg tyrosine per minute under the above conditions.

3.7. Protein Purity Analysis

SDS-PAGE analysis was carried out on a Mini-protean II system from Bio-Rad (Hercules, CA, USA). The purity of the purified proteases was calculated by a gel imaging analysis system (Gelpro Analyzer 3.2 (Thermo Fisher Scientific, Waltham, MA, USA) according to the integration of the lane darkness. HPLC (Agilent 1260, San Francisco, CA, USA) analysis was performed with a TSK3000SW gel filtration column (Tosoh Co., Tokyo, Japan) monitored at 280 nm [27]. The solvent phase was 0.1 M PBS, 0.1 M Na_2SO_4, 0.05% NaN_3, pH 6.7. The flow rate was 0.6 mL/min.

4. Conclusions

In this study, an affinity medium more efficient for metalloprotease purification than other currently available techniques was designed, synthesized and experimentally characterized. Testing the adsorption properties of five designed resins, the Sepharose 6B media coupled with 4-APBA via cyanuric chloride spacer was selected for the purification of native metalloproteases from three commercial products. Metalloproteases from these sources were purified in one step with high efficiency and purity (more than 95%). Compared with the previously reported methods, this protocol resulted in several positive features, such as fewer steps, better activity recoveries, and higher purity. Coupled with efficacy, time-saving procedure and accessible reagents, this novel affinity purification protocol represents a potential important tool for industrial application.

Supplementary Materials: The following are available online at www.mdpi.com/1660-3397/15/1/5/s1. Figure S1: The ESI-MS analysis of the affinity ligand. The possible structures of the chemicals in principal peaks are shown. The ESI-MS cone voltage (170 V) was selected. Scanning was performed from m/z 100 to 1000 in 10 s, and several scans were summed to obtain the final spectrum, Figure S2: SDS-PAGE analysis of three purified and crude commercial metalloproteases. (A) Analysis of crude (Line 1) and purified (Line 2) MP; (B) Analysis of crude (Line 2) and purified (Line 1) DENIE-B LPS-P; (C) Analysis of crude (Line 2) and purified (Line 1) ViscozymeL.

Acknowledgments: This project was funded by National Science Foundation-Shandong province Joint Fund (U1406402-5); International Science and Technology Cooperation and Exchanges (2014DFG30890); Postdoctoral Science Foundation of China (2016M590673 and 2015M582170); Postdoctoral Science Foundation of Shandong; Postdoctoral Researcher Applied Research Project of Qingdao (Q51201601 and Q51201613); National Natural Science Foundation of China (41376175); The Scientific and Technological Innovation Project Financially Supported by Qingdao National Laboratory for Marine Science and Technology (2015ASKJ02); National Hi-tech R&D Program (2014AA093516); Science and Technique Plan of Qingdao (14-2-4-11-jch). International collaboration (PSR-SIIRI) supported by the Ministry of Science and Technology of China and the Ministry of Economy, Science and Innovation of Quebec, Canada (MESI).

Author Contributions: S.L., S.L. and M.S. conceived and designed the experiments. S.L., L.W., J.L., Y.W. and J.H. performed the experiments. S.L., L.W., Y.W. and X.X. analyzed the data. S.L., X.X. and M.S. wrote the main manuscript text. All authors reviewed the manuscript.

Conflicts of Interest: We declare that we have no competing financial interests.

References

1. Li, Q.; Yi, L.; Marek, P.; Iverson, B.L. Commercial proteases: Present and future. *FEBS Lett.* **2014**, *587*, 1155–1163. [CrossRef] [PubMed]
2. Gupta, R.; Beg, Q.; Khan, S.; Chauhan, B. An overview on fermentation, downstream processing and properties of microbial alkaline proteases. *Appl. Microbiol. Biotechnol.* **2002**, *60*, 381–395. [PubMed]
3. Wang, F.; Hao, J.H.; Yang, C.Y.; Sun, M. Cloning, expression, and identification of a novel extracellular cold-adapted alkaline protease gene of the marine bacterium strain YS-80-122. *Appl. Biochem. Biotechnol.* **2010**, *162*, 1497–1505. [CrossRef] [PubMed]
4. Hao, J.H.; Sun, M. Purification and characterization of a cold alkaline protease from a psychrophilic *Pseudomonas aeruginosa* HY1215. *Appl. Biochem. Biotechnol.* **2015**, *175*, 715–722. [CrossRef] [PubMed]
5. Kunamneni, A.; Durvasula, R. Streptokinase—A drug for thrombolytic therapy: A patent review. *Recent Adv. Cardiovasc. Drug Discov.* **2014**, *9*, 106–121. [CrossRef] [PubMed]
6. Craik, C.S.; Page, M.J.; Madison, E.L. Proteases as therapeutics. *Biochem. J.* **2011**, *435*, 1–16. [CrossRef] [PubMed]
7. Wang, W.J.; Yu, X.H.; Wang, C.; Yang, W.; He, W.S.; Zhang, S.J.; Yan, Y.G.; Zhang, J. MMPs and ADAMTSs in intervertebral disc degeneration. *Clin. Chim. Acta* **2015**, *448*, 238–246. [CrossRef] [PubMed]
8. Erban, T. Purification of tropomyosin, paramyosin, actin, tubulin, troponin and kinases for chemiproteomics and its application to different scientific fields. *PLoS ONE* **2011**, *6*, e22860. [CrossRef] [PubMed]
9. Nakamura, K.; Suzuki, T.; Hasegawa, M.; Kato, Y.; Sasaki, H.; Inouye, K. Characterization of p-aminobenzamidine-based sorbent and its use for high-performance affinity chromatography of trypsin-like proteases. *J. Chromatogr. A* **2003**, *1009*, 133–139. [CrossRef]

10. Braganza, V.J.; Simmons, W.H. Tryptase from rat skin: Purification and properties. *Biochemistry* **1991**, *30*, 4997–5007. [CrossRef] [PubMed]
11. Burchacka, E.; Witkowska, D. The role of serine proteases in the pathogenesis of bacterial infections. *Postep. Hig. Med. Doświadczalnej* **2016**, *70*, 678–694. [CrossRef] [PubMed]
12. Jean, M.; Raghavan, A.; Charles, M.L.; Robbins, M.S.; Wagner, E.; Rivard, G.É.; Charest-Morin, X.; Marceau, F. The isolated human umbilical vein as a bioassay for kinin-generating proteases: An in vitro model for therapeutic angioedema agents. *Life Sci.* **2016**, *155*, 180–188. [CrossRef] [PubMed]
13. McMurray, J.J.; Dickstein, K.; Køber, L.V. Aliskiren, enalapril, or aliskiren and enalapril in heart failure. *N. Engl. J. Med.* **2016**, *374*, 1521–1532. [CrossRef] [PubMed]
14. Häse, C.C.; Finkelstein, R.A. Bacterial extracellular zinc-containing metalloproteases. *Microbiol. Rev.* **1993**, *12*, 823–837.
15. Adekoya, O.A.; Sylte, I. The thermolysin family (M4) of enzymes: Therapeutic and biotechnological potential. *Chem. Biol. Drug Des.* **2009**, *73*, 7–16. [CrossRef] [PubMed]
16. Gowda, C.D.; Shivaprasad, H.V.; Kumar, R.V.; Rajesh, R.; Saikumari, Y.K.; Frey, B.M.; Frey, F.J.; Sharath, B.K.; Vishwanath, B.S. Characterization of major zinc containing myonecrotic and procoagulant metalloprotease 'malabarin' from non lethal trimeresurus malabaricus snake venom with thrombin like activity: Its neutralization by chelating agents. *Curr. Top. Med. Chem.* **2011**, *11*, 2578–2588. [CrossRef] [PubMed]
17. Lei, F.F.; Cui, C.; Zhao, H.F.; Tang, X.L.; Zhao, M.M. Purification and characterization of a new neutral metalloprotease from marine *Exiguobacterium* sp. SWJS2. *Biotechnol. Appl. Biochem.* **2016**, *63*, 238–248. [CrossRef] [PubMed]
18. Majumder, R.; Banik, S.P.; Khowala, S. Purification and characterisation of κ-casein specific milk-clotting metalloprotease from *Termitomyces clypeatus* MTCC 5091. *Food Chem.* **2015**, *173*, 441–448. [CrossRef] [PubMed]
19. Zhao, H.L.; Yang, J.; Chen, X.L.; Su, H.N.; Zhang, X.Y.; Huang, F.; Zhou, B.C.; Xie, B.B. Optimization of fermentation conditions for the production of the M23 protease Pseudoalterin by deep-sea *Pseudoalteromonas* sp. CF6-2 with artery powder as an inducer. *Molecules* **2014**, *19*, 4779–4790. [CrossRef] [PubMed]
20. Shao, X.; Ran, L.Y.; Liu, C.; Chen, X.L.; Zhang, X.Y.; Qin, Q.L.; Zhou, B.C.; Zhang, Y.Z. Culture condition optimization and pilot scale production of the M12 metalloprotease myroilysin produced by the deep-sea bacterium *Myroides profundi* D25. *Molecules* **2015**, *20*, 11891–11901. [CrossRef] [PubMed]
21. Li, S.Y.; Wang, L.N.; Yang, J.; Liu, J.Z.; Lin, S.X.; Hao, J.H.; Sun, M. Affinity purification of metalloprotease from marine bacterium using immobilized metal affinity chromatography. *J. Sep. Sci.* **2016**, *39*, 2050–2056. [CrossRef] [PubMed]
22. Chessa, J.P.; Petrescu, I.; Bentahir, M.; Van Beeumen, J.; Gerday, C. Purification, physico-chemical characterization and sequence of a heat labile alkaline metalloprotease isolated from a psychrophilic *Pseudomonas* species. *Biochim. Biophys. Acta* **2000**, *1479*, 265–274. [CrossRef]
23. Martínez, C.A.; Seidel-Morgenstern, A. Purification of single-chain antibody fragments exploiting pH-gradients in simulated moving bed chromatography. *J. Chromatogr. A* **2016**, *1434*, 29–38. [CrossRef] [PubMed]
24. Chen, B.; Li, R.; Li, S.Y.; Chen, X.L.; Yang, K.D.; Chen, G.L.; Ma, X.X. Evaluation and optimization of the metal-binding properties of a complex ligand for immobilized metal affinity chromatography. *J. Sep. Sci.* **2016**, *39*, 518–524. [CrossRef] [PubMed]
25. Cheung, R.C.; Wong, J.H.; Ng, T.B. Immobilized metal ion affinity chromatography: A review on its applications. *Appl. Microbiol. Biotechnol.* **2012**, *96*, 1411–1420. [CrossRef] [PubMed]
26. Labrou, N.E. Design and selection of ligands for affinity chromatography. *J. Chromatogr. B* **2003**, *790*, 67–78. [CrossRef]
27. Ye, L.; Xu, A.Z.; Cheng, C.; Zhang, L.; Huo, C.X.; Huang, F.Y.; Xu, H.; Li, R.Y. Design and synthesis of affinity ligands and relation of their structure with adsorption of proteins. *J. Sep. Sci.* **2011**, *34*, 3145–3150. [CrossRef] [PubMed]
28. Havlicek, V.; Lemr, K.; Schug, K.A. Current trends in microbial diagnostics based on mass spectrometry. *Anal. Chem.* **2013**, *85*, 790–797. [CrossRef] [PubMed]
29. Zhang, S.C.; Sun, M.; Li, T.; Wang, Q.H.; Hao, J.H.; Han, Y.; Hu, X.J.; Zhou, M.; Lin, S.X. Structure analysis of a new psychrophilic marine protease. *PLoS ONE* **2011**, *6*, e26939. [CrossRef] [PubMed]
30. Ji, X.F.; Zheng, Y.; Wang, W.; Sheng, J.; Hao, J.H.; Sun, M. Virtual screening of novel reversible inhibitors for marine alkaline protease MP. *J. Mol. Graph. Model.* **2013**, *46*, 125–131. [CrossRef] [PubMed]

31. Li, D.J.; Chen, Y.; Liu, Z. Boronate affinity materials for separation and molecular recognition: Structure, properties and applications. *Chem. Soc. Rev.* **2015**, *44*, 8097–8123. [CrossRef] [PubMed]
32. Wang, S.; Ye, J.; Li, X.; Liu, Z. Boronate affinity fluorescent nanoparticles for förster resonance energy transfer inhibition assay of *cis-diol* biomolecules. *Anal. Chem.* **2016**, *88*, 5088–5096. [CrossRef] [PubMed]
33. Xue, Y.; Shi, W.; Zhu, B.; Gu, X.; Wang, Y.; Yan, C. Polyethyleneimine-grafted boronate affinity materials for selective enrichment of *cis-diol*-containing compounds. *Talanta* **2015**, *140*, 1–9. [CrossRef] [PubMed]
34. Toprak, A.; Görgün, C.; Kuru, C.İ.; Türkcan, C.; Uygun, M.; Akgöl, S. Boronate affinity nanoparticles for RNA isolation. *Mater. Sci. Eng. C Mater. Biol. Appl.* **2015**, *50*, 251–256. [CrossRef] [PubMed]
35. Jiang, H.P.; Qi, C.B.; Chu, J.M.; Yuan, B.F.; Feng, Y.Q. Profiling of cis-diol-containing nucleosides and ribosylated metabolites by boronate-affinity organic-silica hybrid monolithic capillary liquid chromatography/mass spectrometry. *Sci. Rep.* **2015**, *5*, 7785. [CrossRef] [PubMed]
36. Fasoli, E.; Reyes, Y.R.; Guzman, O.M.; Rosado, A.; Cruz, V.R.; Borges, A.; Martinez, E.; Vibha Bansal, V. Para-aminobenzamidine linked regenerated cellulose membranes for plasminogen activator purification: Effect of spacer arm length and ligand density. *J. Chromatogr. B* **2013**, *930*, 13–21. [CrossRef] [PubMed]
37. Xin, X.; Dong, D.X.; Wang, T.; Li, R.X. Affinity purification of serine proteinase from *Deinagkistrodon acutus* venom. *J. Chromatogr. B* **2007**, *859*, 111–118. [CrossRef] [PubMed]
38. Van Ness, J.; Kalbfleisch, S.; Petrie, C.R.; Reed, M.W.; Tabone, J.C.; Vermeulen, N.M. A versatile solid support system for oligodeoxynucleotide probe-based hybridization assays. *Nucleic Acids. Res.* **1991**, *19*, 3345–3350. [CrossRef] [PubMed]
39. Batra, S.; Bhushan, R. Amino acids as chiral auxiliaries in cyanuric chloride-based chiral derivatizing agents for enantioseparation by liquid chromatography. *Biomed. Chromatogr.* **2014**, *28*, 1532–1546. [CrossRef] [PubMed]
40. Chien, T.E.; Li, K.L.; Lin, P.Y.; Lin, J.L. Infrared spectroscopic study of the adsorption forms of cyanuric acid and cyanuric chloride on TiO_2. *Langmuir* **2016**, *32*, 5306–5313. [CrossRef] [PubMed]
41. Mountford, S.J.; Daly, R.; Robinson, A.J.; Hearn, M.T. Design, synthesis and evaluation of pyridine-based chromatographic adsorbents for antibody purification. *J. Chromatogr. A* **2014**, *1355*, 15–25. [CrossRef] [PubMed]
42. Smoum, R.; Rubinstein, A.; Dembitsky, V.M.; Srebnik, M. Boron containing compounds as protease inhibitors. *Chem. Rev.* **2012**, *112*, 4156–4220. [CrossRef] [PubMed]
43. Huang, S.H.; Pan, S.H.; Chen, G.G.; Huang, S.; Zhang, Z.F.; Li, Y.; Liang, Z.Q. Biochemical characteristics of a fibrinolytic enzyme purified from a marine bacterium, *Bacillus subtilis* HQS-3. *Int. J. Biol. Macromol.* **2013**, *62*, 124–130. [CrossRef] [PubMed]
44. Huang, W.Q.; Zhong, L.F.; Meng, Z.Z.; You, Z.J.; Li, J.Z.; Luo, X.C. The structure and enzyme characteristics of a recombinant leucine aminopeptidase rlap1 from *Aspergillus sojae* and its application in debittering. *Appl. Biochem. Biotechnol.* **2015**, *177*, 190–206. [CrossRef] [PubMed]
45. Guo, W.; Ruckenstein, E. A new matrix for membrane affinity chromatography and its application to the purification of concanavalin A. *J. Membr. Sci.* **2001**, *182*, 227–234. [CrossRef]
46. Dong, D.X.; Liu, H.R.; Xiao, Q.S.; Li, R.X. Affinity purification of egg yolk immunoglobulins (IgY) with a stable synthetic ligand. *J. Chromatogr. B* **2008**, *870*, 51–54. [CrossRef] [PubMed]
47. Xin, Y.; Yang, H.L.; Xiao, X.L.; Zhang, L.; Zhang, Y.R.; Tong, Y.J.; Chen, Y.; Wang, W. Affinity purification of urinary trypsin inhibitor from human urine. *J. Sep. Sci.* **2012**, *35*, 1–6. [CrossRef] [PubMed]
48. Santos-Martins, D.; Forli, S.; Ramos, M.J.; Olson, A.J. AutoDock4(Zn): An improved AutoDock force field for small-molecule docking to zinc metalloproteins. *J. Chem. Inf. Model.* **2014**, *54*, 2371–2379. [CrossRef] [PubMed]
49. Humphrey, W.; Dalke, A.; Schulten, K. VMD: Visual molecular dynamics. *J. Mol. Graph. Model.* **1996**, *14*, 33–38. [CrossRef]

© 2016 by the authors. Licensee MDPI, Basel, Switzerland. This article is an open access article distributed under the terms and conditions of the Creative Commons Attribution (CC BY) license (http://creativecommons.org/licenses/by/4.0/).

Article

Activity Improvement and Vital Amino Acid Identification on the Marine-Derived Quorum Quenching Enzyme MomL by Protein Engineering

Jiayi Wang [1,†], Jing Lin [1,†], Yunhui Zhang [1], Jingjing Zhang [1], Tao Feng [1], Hui Li [1], Xianghong Wang [1], Qingyang Sun [1], Xiaohua Zhang [1,2,3] and Yan Wang [1,2,3,*]

[1] College of Marine Life Sciences, Ocean University of China, Qingdao 266003, China; wangjiayi109911@163.com (J.W.); lynn44944@163.com (J.L.); yhzhang2011@163.com (Y.Z.); jingjingzhangnn@163.com (J.Z.); fengtao246@163.com (T.F.); l56021831@163.com (H.L.); xhwang@ouc.edu.cn (X.W.); lilysun1012@126.com (Q.S.); xhzhang@ouc.edu.cn (X.Z.)
[2] Laboratory for Marine Ecology and Environmental Science, Qingdao National Laboratory for Marine Science and Technology, Qingdao 266071, China
[3] Institute of Evolution & Marine Biodiversity, Ocean University of China, Qingdao 266003, China
* Correspondence: wangy12@ouc.edu.cn
† These authors contributed equally to this work.

Received: 4 April 2019; Accepted: 17 May 2019; Published: 21 May 2019

Abstract: MomL is a marine-derived quorum-quenching (QQ) lactonase which can degrade various N-acyl homoserine lactones (AHLs). Intentional modification of MomL may lead to a highly efficient QQ enzyme with broad application potential. In this study, we used a rapid and efficient method combining error-prone polymerase chain reaction (epPCR), high-throughput screening and site-directed mutagenesis to identify highly active MomL mutants. In this way, we obtained two candidate mutants, MomL$_{I144V}$ and MomL$_{V149A}$. These two mutants exhibited enhanced activities and blocked the production of pathogenic factors of *Pectobacterium carotovorum* subsp. *carotovorum (Pcc)*. Besides, seven amino acids which are vital for MomL enzyme activity were identified. Substitutions of these amino acids (E238G/K205E/L254R) in MomL led to almost complete loss of its QQ activity. We then tested the effect of MomL and its mutants on *Pcc*-infected Chinese cabbage. The results indicated that MomL and its mutants (MomL$_{L254R}$, MomL$_{I144V}$, MomL$_{V149A}$) significantly decreased the pathogenicity of *Pcc*. This study provides an efficient method for QQ enzyme modification and gives us new clues for further investigation on the catalytic mechanism of QQ lactonase.

Keywords: quorum quenching enzyme; error prone PCR; high-throughput screening; site-directed mutagenesis; catalytic ability; *Pectobacterium carotovorum* subsp. *carotovorum (Pcc)*

1. Introduction

Quorum sensing (QS) is a communication system that many bacteria aggregates used to regulate their aggregate size via small molecules called autoinducers [1,2]. The process of interfering with QS through degradation of signals is termed as quorum quenching (QQ). N-acyl homoserine lactones (AHLs) are QS signals used by a wide range of Gram-negative bacteria. AHL lactonase is one major type of AHL-degrading enzymes, which hydrolyses the lactone ring of AHL molecule to produce corresponding N-acyl-homoserine. Since some bacteria use QS to mediate virulence factors and antimicrobial resistance, QQ is considered to be a promising alternative for bacterial disease control, which can attenuate QS-regulated virulence factors production in many bacterial pathogens without any lethal effect and impart less-selective pressures for resistant mutants than conventional antibiotics [3–5]. QQ enzyme is one of the most well-studied methods of QQ. AiiA, the earliest identified QQ enzyme, can decrease extracellular pectolytic enzyme activity and attenuate pathogenicity of *Erwinia carotovora* [6]. A recent study shows that QQ enzyme (AiiA) and QS inhibitor (G1) demonstrated enhanced QS inhibiting effects on reducing AHL concentration when applied together [7].

MomL, a novel AHL lactonase, was isolated from *Muricauda olearia* Th120 [8,9]. This protein consists of 294 amino acids and has a molecular weight of 32.8 kDa. MomL belongs to the metallo-β-lactamase superfamily, and shows the highest identity of 56.8% with protein Aii20J, which belongs to *Tenacibaculum* sp. 20J [10]. Moreover, MomL shares 54.4% and 24.5% identity with FiaL from *Flaviramulus ichthyoenteri* T78T and AiiA from *Bacillus* sp. 240B1 [10–14]. The wide-ranging substrate properties of MomL confer great advantages in disease prevention because different pathogenic bacteria produce AHL molecules with different chain lengths. For example, AHL produced by *Burkholderia* is C8-HSL [15–17], while that of *Vibrio harveyi* is 3OC4-HSL [18]. Moreover, the ability of MomL to degrade C6-HSL is approximately 10 times higher than that of AiiA [14]. MomL exhibited degradative activity on both short and long-chain AHLs and inhibited the pathogenicity of different pathogenic bacteria [9,19]. In order to investigate its application value, MomL was heterologously expressed by *Bacillus brevis*, and the recombinant strain showed a broad antibacterial spectrum than original strain [20]. Although MomL shares the "HXHXDH~H~D" motif with other AHL lactonases in the metallo-β-lactamase superfamily, this motif of MomL performs different functions from AiiA [14,21,22]. Furthermore, little is known about its catalytic mechanism and other amino acids that are involved in the active site remain unclear. Therefore, elucidating the action mechanism helps to expand the application of MomL and paves a way for marine-derived QQ enzyme research.

Pectobacterium carotovorum subsp. *carotovorum* (Pcc) is a bacterial pathogen that can cause severe soft rot of cabbage [23–25]. Extracellular enzymes such as pectate lyases, pectinases, cellulases and proteases produced by *Pcc* are main causes for tissue maceration [26]. Disease factors produced by *Pcc* can be induced by the AHL-based QS system [27]. Thus, as an environmentally friendly biocontrol strategy, QQ can be used to prevent or alleviate symptoms caused by such infections.

Protein engineering is a multi-faceted field that can create desired protein properties via various approaches including protein structure prediction to protein selection from random mutagenesis library [28]. As an early example, the *ebgA* gene of *E. coli* K12, was deleted to lead to the synthesis of ebg enzyme and show enhanced activity toward lactose [29]. The catalytic function of cytochrome c from *Rhodothermus marinus* was enhanced more than 15-fold than industrial catalysts in forming carbon-silicon bonds [30,31]. Building high-quality mutant libraries and high efficiency screening system are crucial steps for selecting functional proteins. Site-directed mutagenesis is a valuable tool for understanding the relationship between enzyme activity and amino acids.

In this study, we improved the efficiency of mutant library establishment using a combination method of error-prone polymerase chain reaction (epPCR) and seamless cloning. In addition, an IPTG in situ photocopying technology was used to perform high-throughput screening of random mutagenesis library. We rapidly obtained two high-activity mutant proteins and identified seven amino acids which are vital for QQ ability of MomL. Furthermore, we investigated the ability of MomL and its mutants to inhibit the agricultural pathogenic bacterium *Pcc* virulence factors and the formation of soft rot on Chinese cabbage.

2. Results

2.1. Overview of the High-Efficiency Strategy of Constructing and Screening a Random Mutagenesis Library

In this study, we built a highly efficient and rapid method to obtain the required variants. This method mainly combined three types of technology, specifically epPCR, seamless cloning and isopropyl-β-D-thiogalactoside (IPTG) in situ photocopying. We selected an appropriate amino acid mutation rate and generated PCR products containing randomly mutated amino acids by performing optimized epPCR of three rounds. The PCR products were cloned into pET-24a(+) vectors via seamless cloning, and the recombinant plasmids were transformed into *E. coli* BL21(DE3). *Chromobacterium violaceum* CV026 can produce violacein in the presence of AHLs with N-acyl side chains from C4 to C8 in length. When QQ substances were added, the production of violacein was inhibited. Therefore, in the screening plate containing exogenous C6-HSL and the indicator CV026, C6-HSL can be degraded and the plate will not turn violet when the imprinted *E. coli* BL21 colonies of the random mutagenesis library produced active MomL enzyme. Single colonies were imprinted on the screening plates containing IPTG and indicator CV026. The QQ ability of MomL was estimated by either the white halo or the halo diameter produced in the screening plate and positive mutants were selected. The method used in this study was highly efficient and faster than the traditional method (Figure 1). The analyzation for the efficiency and feasibility of this method were performed using MomL protein as an example.

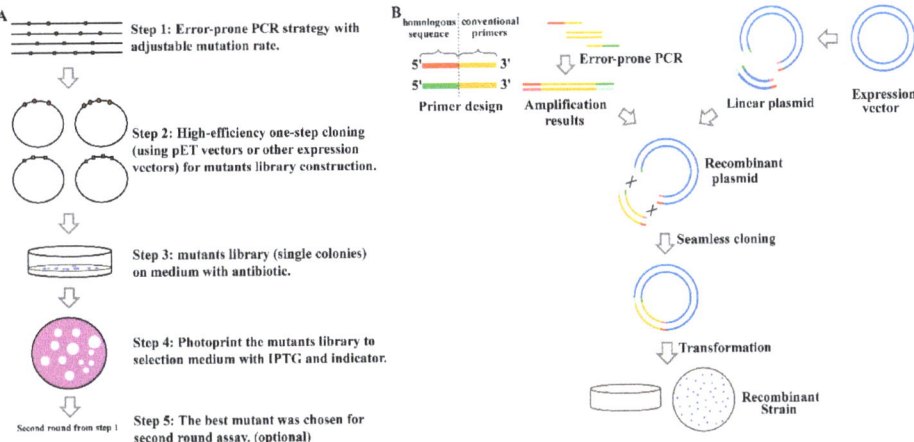

Figure 1. The schematic diagram of high efficiency strategy of constructing and screening random mutagenesis library (**A**) and the process of error-prone polymerase chain reaction (epPCR) and seamless cloning (**B**).

2.2. Error-Prone Polymerase Chain Reaction (EpPCR) Condition Optimization with Suitable Mutation Efficiency

EpPCR randomly introduces mutant sites, and the mismatch rate is related to the magnesium and manganese ion contents [32,33]. In order to build a more efficient mutant library, 1% were selected as the optimal amino acid mutation rate. To determine the appropriate mismatch rate, Mg^{2+} concentration gradient ranging from 1 to 8 mM and Mn^{2+} gradient ranging from 0 to 0.6 mM were detected respectively. As shown in Figure S1A,B, specific DNA bands were observed following PCR in different Mg^{2+} or Mn^{2+} concentration gradient. Next, orthogonal test of the two factors (Mg^{2+} and Mn^{2+}) was conducted based on the results of the single factor experiment. Appropriate DNA bands were obtained under the 10 orthogonal test conditions (Figure S1C). We randomly selected 100 single colonies of each condition for sequencing. The results demonstrated that under 1 mM Mg^{2+} and 0.2 mM Mn^{2+} conditions, 70% of single colonies were suffered in 1–3 mutant sites while no mutation was presented in the other 23% of colonies. Fifty percent of the mutant proteins contained no more than 2 mutant sites under the 2 mM Mg^{2+} and 0.1 mM Mn^{2+} conditions. The average mutation frequency distribution was under 2 mM Mg^{2+} and 0.15 mM Mn^{2+}, including 2-3 mutant sites in 50% of mutant proteins and 2-5 mutant sites in 83% of mutant proteins. Besides, under this condition, 100 detected mutant libraries existed at least one mutant site. Under 2 mM Mg^{2+} and 0.2 mM Mn^{2+}, premature translational termination occurred in 30% of mutant proteins, and there were 7–11 mutant sites per protein. Ultimately, based on these results, 2 mM Mg^{2+} and 0.15 mM Mn^{2+} conditions were selected to maintain the amino acid mutation rate at approximately 1% (Table S2).

2.3. Screening of a Mutation Library Based on a Seamless and EpPCR Strategy

A random mutagenesis library was built using the selected epPCR conditions. Next, the traditional cloning method (TA cloning) and seamless cloning were separately employed to ligate random mutant fragments and vectors. There were obvious differences in mutation library abundance between the two methods. The number of mutant proteins obtained via seamless cloning was 15–20 times higher than that obtained using TA cloning (Figure 2). Besides, randomly sequencing results showed that 92% of single colonies contained the *momL* gene while only 8% were false positive colonies with self-ligation plasmids.

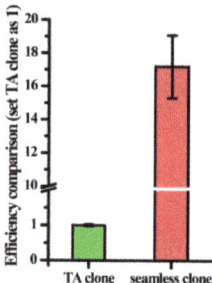

Figure 2. Efficiency comparison between seamless cloning and traditional cloning (TA). All data are presented as mean ± standard deviation (SD, $n = 3$).

We obtained more than 5000 mutant strains for random mutagenesis library. Subsequently, IPTG in situ photocopying technology was utilized to efficiently screen mutant proteins. In the prescreening step, QQ ability of MomL was estimated by whether the visual white halo showed in screening plate. In this step, approximately 3000 strains were screened; 10% of strains that produced larger halo diameters were chosen for second-round screening. In second round screening step, mutants were screened using crude enzyme supernatant in CV026-loaded screening plate. Single colonies M1–M8 were selected from the area with large white halos while M9–M10 were identified from the region lacking white halos (Figure 3 and Figure S2). Two high-activity mutant proteins, M2 and M3, and the mutant proteins M9 and M10, which lacked activity, were selected for sequencing. The results indicated that Ile144 in M2 was mutated to Val (I144V), and Val149 in M3 was mutated to Ala (V149A). In addition, four amino acids in M9 were mutated, namely, E238G, N179S, N51Y and K82R, and four amino acids in M10 were mutated, namely, M228V, T84A, K205E and L254R.

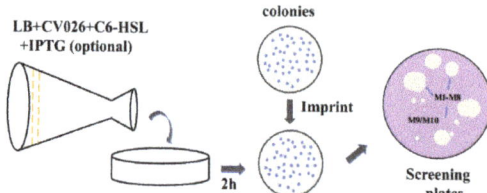

Figure 3. Screening target proteins by isopropyl-β-D-thiogalactoside (IPTG) *in situ* photocopying. M1–M8 are single colonies with highly activity; M9–M10 indicate inactive proteins.

2.4. Analysis of Amino Acids in Mutant Proteins

To further analyze the functions of single amino acids, we mutated the above amino acids loci and constructed 10 single amino acid mutants: $MomL_{I144V}$, $MomL_{V149A}$, $MomL_{N51Y}$, $MomL_{N179S}$, $MomL_{M228V}$, $MomL_{K205E}$, $MomL_{E238G}$, $MomL_{L254R}$, $MomL_{T84A}$, and $MomL_{K82R}$ (Figure 4). Biochemical test indicated that the activities of $MomL_{I144V}$ and $MomL_{V149A}$ were 1.3 and 1.8 times higher, respectively, than that of wild-type MomL (Figure 5A). Furthermore, MomL$_{E238G}$ was inactive, and the activities of MomL$_{K205E}$ and MomL$_{L254R}$ were reduced by 80%–90% compared to MomL. The activities of MomL$_{N179S}$/MomL$_{N51Y}$/MomL$_{K82R}$/MomL$_{M228V}$/MomL$_{T84A}$ also decreased, ranging from 40–80% of wild-type MomL activity (Figure 5B). The results indicated that Glu238, Lys205, Leu254, Thr84 and Asn179 are related to hydrolysis reaction of C6-HSL. Changes in every single site can reduce the enzyme activity to 50% or more.

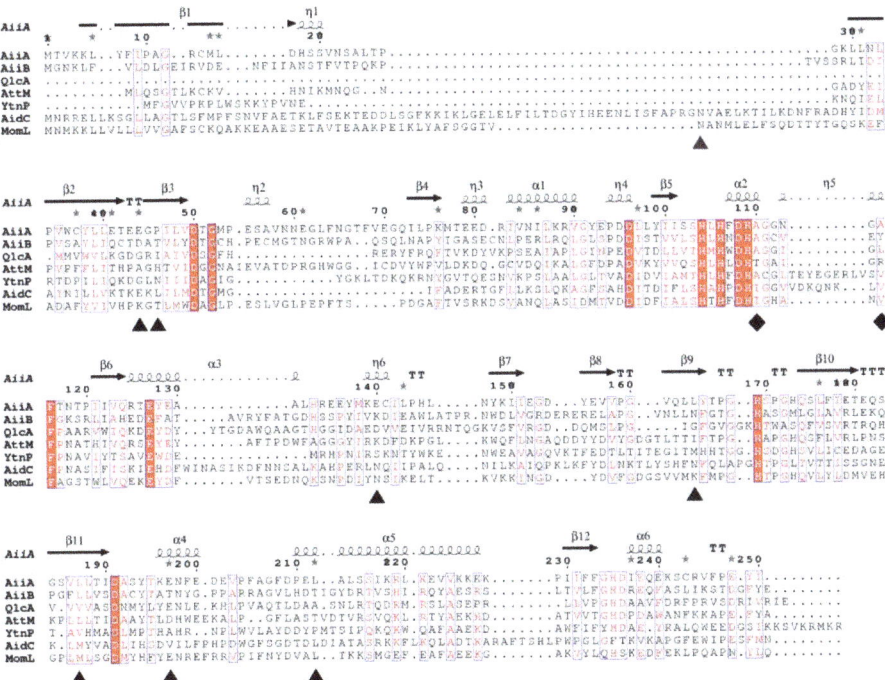

Figure 4. Multiple sequence alignment of amino acid sequences of MomL, putative homologues, and other representative *N*-acyl homoserine lactone (AHL) lactonases. Sequence alignment was performed by the MUSCLE program in the MEGA software package and enhanced by ESPript 3.0. MomL homologue from *Eudoraea adriatica* (WP_019670967) showed the highest score when BLASTP searching nonredundant (NR) databases. Other sequences of AHL lactonase are AiiA from *Bacillus* sp. strain 240B1 (AAF62398), AidC from *Chryseobacterium* sp. strain StRB126 (BAM28988), QlcA from unculturable soil bacteria, and AttM (AAD43990), AiiB (NP 396590) from *Agrobacterium fabrum* C58 and YtnP from *Bacillus*. Filled triangles show amino acids which are essential for MomL activity. Filled rhombuses show amino acids, the mutation of which increased MomL activity.

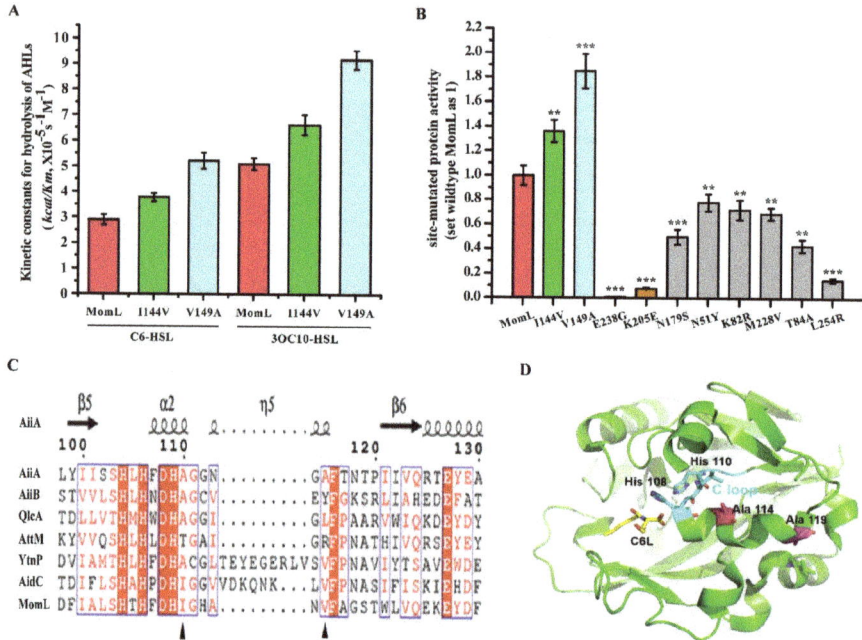

Figure 5. (**A**) Enzyme kinetics experiments of MomL and mutant proteins on different substrates C6-HSL and 3OC10-HSL. (**B**) Protein activity test of mutant proteins. All data are presented as mean ± standard deviation (SD, n = 3). An unpaired t-test was performed for testing significant differences between groups (*** $P < 0.001$, ** $P < 0.01$, * $P < 0.05$). (**C**) Multiple-sequence alignment of the amino acid sequences of MomL, putative homologues, and other representative AHL lactonases. The multiple-sequence alignment procedure is the same as described in Figure 4. (**D**) The structure and active site of AiiA, the homologous protein of MomL. A114 and A119 in AiiA are located near the C-loop.

By screening mutant proteins, we rapidly obtained two live mutant proteins and identified seven amino acids that are involved in QQ ability of MomL. By multiple sequence alignment of MomL and other AHL lactonases belonging to the metallo-β-lactamase superfamily, we found that Ile144, Val149, Asn179, Lys205 were variable amino acids in the conserved domain "HXHXDH ~ 60aa ~ H", and may be directly related to the catalytic reaction; while Thr84, Glu238 and Leu254 were amino acids outside the conserved domain, and may be related to maintaining protein stability. In addition, by analyzing the structure of AiiA, the homologous protein of MomL, we found that I144 and V149 in MomL (A114 and A119 in AiiA) are located near the catalytic ring of the active center (C-loop in Figure 5C,D). We speculated that the mutation of I144V and V149A may affect the enzyme activity by affecting the conformation of the C-loop.

2.5. The Effect of Mutant Proteins on the Virulence Factors and Survival of Pectobacterium Carotovorum Subsp. Carotovorum (Pcc)

The inhibitory effects of mutant proteins on pectate lyase, the virulence factor of the plant pathogenic bacterium *Pcc*, was analyzed. Wild-type MomL, MomL$_{I144V}$ and MomL$_{V149A}$ inhibited the expression of the pectate lyase gene, and the inhibitory effect of MomL$_{V149A}$ was slightly higher than the wild-type MomL. We also analyzed the gene expression of pectate lyase when treated by MomL$_{E238G}$, MomL$_{K205E}$ and MomL$_{L254R}$. The mutation of these three amino acids resulted in the inability to inhibit pectate lyase gene expression (Figure 6A). In addition, the yield of pectate lyase was determined and the results were consistent with the transcriptional analysis. MomL$_{I144V}$ and MomL$_{V149A}$ greatly reduced pectate lyase yield, while MomL$_{E238G}$, MomL$_{K205E}$ and MomL$_{L254R}$ did not (Figure 6B). Besides, the presence of MomL$_{I144V}$ and MomL$_{V149A}$ reduced the *Pcc* survival rate under stress conditions to 30%–45% of the survival of *Pcc* alone. The presence of MomL$_{E238G}$ did not affect *Pcc* survival. Furthermore, the boiled MomL did not affect the *Pcc* survival rate (Figure 6C). We speculated that site-directed mutagenesis of *momL* led to changes in other function of the mutant proteins, such as the fold of the enzyme, stability, substrate interaction and many other performance parameters, and thus resulted in reduced survival of *Pcc*. However, the specific mechanism needs to be studied further.

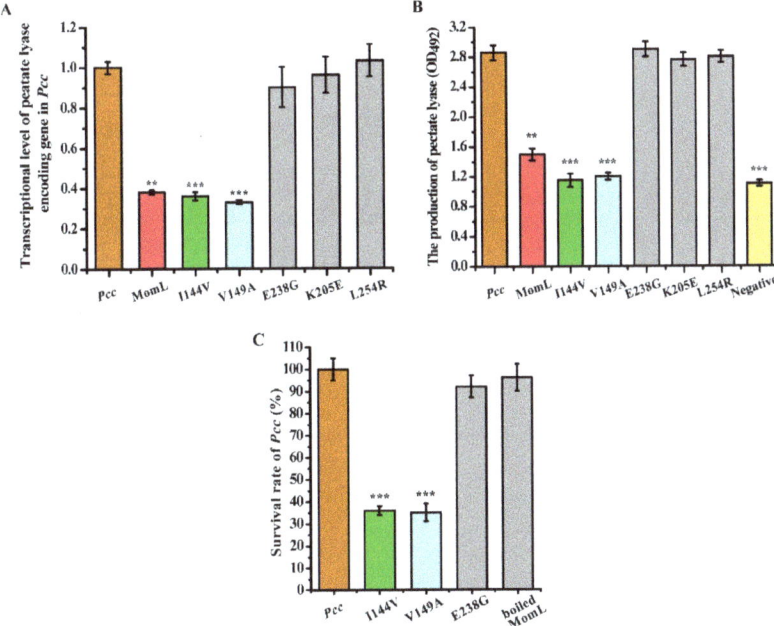

Figure 6. Transcriptional level of pectate lyase encoding gene in *Pectobacterium carotovorum* subsp. *carotovorum* (*Pcc*) (**A**) and the production of pectate lyase (OD$_{492}$). (**B**). Effects of MomL and mutant proteins towards the *Pcc* survival rate (**C**). All data are presented as mean ± standard deviation (SD, *n* = 3). An unpaired t-test was performed for testing significant differences between groups (*** $P < 0.001$, ** $P < 0.01$, * $P < 0.05$).

2.6. Effects of MomL and Mutant Proteins on Pcc Infection of Chinese Cabbage

To further analyze MomL and its mutants, their treatment effect towards soft rot of Chinese cabbage was tested. When treated by *Pcc* alone, approximately 2/3 of the cabbage leaf area was infected and decomposed. Following infection with *Pcc* and treatment with MomL, only a small percentage of tissue was infected. However, inactivated MomL applied in combination with *Pcc* did not reduce the degree of decay in Chinese cabbage. The decay areas of the cabbage after treatment with MomL$_{E238G}$, MomL$_{K205E}$ and *Pcc* were comparable to those obtained with *Pcc* infection alone. The application of MomL$_{L254R}$ and *Pcc* together reduced the decay area by approximately 50% compared with that treated by *Pcc* alone. The treatment effects of MomL$_{I144V}$ and MomL$_{V149A}$ were the most significant. After the application of MomL$_{I144V}$ and MomL$_{V149A}$, the *Pcc* infection rate on Chinese cabbage decreased obviously (Figure 7). Overall, in infection experiments, the bacterial survival rate significantly decreased by more than 50% after adding MomL or active mutant proteins. The results indicated that co-culture with MomL or mutant proteins can relieve the symptoms caused by *Pcc*, and this may be due to the decrease of virulence factors such as pectate lyase.

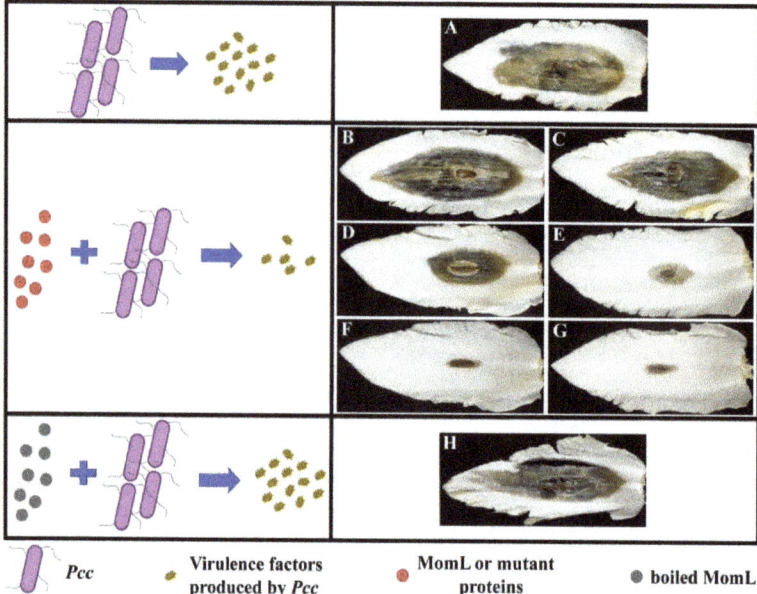

Figure 7. Effects of MomL and mutant proteins on *Pcc* infection of Chinese cabbage. (**A**) *Pcc*; (**B**) *Pcc* with MomL$_{E238G}$; (**C**) *Pcc* with MomL$_{K205E}$; (**D**) *Pcc* with MomL$_{L254R}$; (**E**) *Pcc* with MomL; (**F**) *Pcc* with MomL$_{I144V}$; (**G**) *Pcc* with MomL$_{V149A}$; (**H**) *Pcc* with boiled MomL. The results shown are representative of biological duplicates.

3. Discussion

Marine metagenomic data revealed that QQ is a common activity in marine bacteria [34]. Many QQ enzymes have been identified from marine species, such as Aii20J from *Tenacibaculum* sp. strain 20J, QsdH from *Pseudoalteromonas byunsanensis* strain 1A01261 and AiiC from *Anabaena* sp. PCC 7120 [35,36]. QQ enzymes have broad application prospects in aquaculture disease control, biofouling prevention and drugs development [37,38]. Improving the degrading ability of QQ enzymes will lead to highly stable and efficient proteins for industrial use. Thus, further studies about marine aquatic QQ can expand marine QQ bioresource application and pave a way to solve problems related to aquaculture and agriculture that is conducted in a saline environment [39]. The marine-derived QQ enzyme MomL, a novel type of AHL lactonase with an unknown action mechanism, was investigated in this study. MomL demonstrates a wide antimicrobial spectrum and provides a promising alternative for disease control due to its ability to inhibit the pathogenicity induced by the AHL QS system. The amino acids and active site in MomL have not previously been explored, except for the "HXHXDH~H~D" motif. Hence, we focused on MomL to improve its bacteriostatic activity, explore its highly active mutant proteins, and identify amino acids involved in enzyme activity via site-directed mutagenesis, thus providing a theoretical basis for its mechanism of action.

Among protein engineering strategies, random mutagenesis methods are usually applied to study properties that are not understood rationally. EpPCR is standard method for random mutagenesis due to its robustness and simplicity in use [40]. A seamless cloning technique is used to insert a targeted fragment into any location in the vector without relying on an enzymatic site. The main factor affecting epPCR was the concentration of Mn^{2+}, which can result in higher mutation frequencies at higher concentrations. Other influential factors including the concentration of Mg^{2+}, the proportion of deoxyribonucleoside triphosphates (dNTPs), and even the PCR reaction cycles. In this study, mutation frequencies were controlled at 1–3%. Thus, each protein contained 3-5 mutations. After multiple analyses, we ultimately determined the concentrations of dNTP, Mg^{2+} and Mn^{2+} for the use in next step. We screened amplification enzymes using epPCR to identify high-fidelity enzymes with improved cloning efficiency, but unsatisfactory mutation rates resulted in the low diversity of the random mutant library. Ultimately, Taq enzyme was chosen for epPCR. At the beginning of each reaction, the Taq enzyme produced higher mutant library diversity with 2–5 mutations per protein but achieved low seamless cloning efficiency, thus limiting the number of transformants. Presumably, the A-end of the Taq enzyme affected the efficiency of seamless cloning, which was optimized in our study. We removed the A-end using the HS DNA polymerase (Takara Primer STAR®). The entire experimental time was shortened to one-fifth of the time required for traditional experiment, and the efficiency of mutant library establishment was nearly 10 times higher than that achieved previously. Furthermore, the efficiency of positive cloning during mutant library construction was as high as 92%. Thus, our strategy demonstrated wide applications for establishing protein mutant libraries, and greatly improved the efficiency of seamless cloning.

The first two approaches involve large high-throughput selection, and only 10%–20% of bacteria on a parent plate can be transferred to a sub-plate in the traditional method. But IPTG in situ photocopying is a high-throughput screening system. By performing single colony dilution and counting the number of single colonies, we increased the transferred number of bacteria to 50%. This type of screening method holds great applicable value for other QQ enzymes' screening. By screening mutant proteins, we rapidly obtained two highly active mutants of MomL and identified seven amino acids which are involved in enzyme activity. However, given the lack of MomL crystals structure, the deep catalytic mechanism remains to be characterized. In infection experiments, the bacterial survival rate significantly decreased by more than 50% after adding highly active mutant proteins to *Pcc*. MomL and its mutant proteins also reduced the virulence factor pectate lyase produced by *Pcc*. We applied these proteins to infect Chinese cabbage and found that the infection symptoms were alleviated after adding MomL or its mutants, indicating MomL and its mutants can be an alternative strategy for disease control. We are currently characterizing the minimum concentration and maximum time required for MomL treatment to facilitate the application of MomL alike to actual utilization.

4. Materials and Methods

4.1. Bacterial Strains, Plasmids, Media, Growth Conditions, and Chemicals

Pectobacterium carotovorum subsp. *carotovorum* (*Pcc*) was purchased from the CGMCC (China General Microbiological Culture Collection, Beijing, China) [41]. *E. coli* strain AHL882-5 was used to express the MomL protein. *E. coli* strain BL21(DE3) was used as a host for protein expression. Proteins were expressed following the cloning of random mutants of the *momL* gene into pET-24a(+). The strain *Chromobacterium violaceum* CV026 was used as an indicator in the AHL activity bioassay [42]. C6-HSL was purchased from the Cayman Chemical Company and prepared in dimethyl sulfoxide (DMSO). *M. olearia* Th120, CV026 and *Pcc* were routinely cultured on Luria-Bertani (LB) agar at 28 °C. *E. coli* strain AHL882-5 was cultured in LB medium at 37 °C. When required, 25 μg/mL kanamycin was added to the solid or liquid media.

4.2. Random Mutant Library Construction and Identification of High-Activity Mutants

The mutant library of the AHL lactonase MomL was constructed using error prone PCR (epPCR). The primers for epPCR are listed in Table S1. Each 100-μL epPCR reaction contained 10 μL of 10× PCR buffer (Takara, Shiga, Japan), 8 μL of dNTP mixture (2.5 mM dATP, 2.5 mM dGTP, 10 mM dCTP, and 10 mM dTTP), 1 μL of the primer *momL*-F (20 μM), 1 μL of the primer *momL*-R (20 μM), 1 μL of template plasmid from strain AHL882-5, 1 μL of Taq DNA Polymerase (Takara, 5 U/μL), appropriate metal ions, and deionized water to a final volume of 100 µL. PCR was conducted using the following conditions: denaturation at 94 °C for 10 min, followed by 30 cycles of denaturation at 94 °C for 30 s, annealing at 55 °C for 30 s, extension at 72 °C for 60 s, and a final incubation at 72 °C for 10 min. The resulting PCR products were digested with Prime STAR® HS DNA Polymerase (Takara) to improve the ligation efficiency. They were then further digested with DpnI (NEB, Ipswich, MA, USA) to remove template plasmids and were finally purified using a PCR product purification kit (Biomed, Beijing, China) according to the manufacturer's instructions. Purified mutant *momL* genes were ligated into the linear vector pET-24a(+) via seamless cloning. Recombinant plasmids were transformed into *E. coli* BL21(DE3), diluted with fresh LB medium, plated on LB agar containing 25 µg/mL kanamycin, and cultured at 37 °C overnight.

4.3. High-Throughput Screening of High-Activity Mutants

We added 1 mL of overnight cultured CV026, 7.5 µl C6-HSL (DMSO, 1 mM) and 0.5 mM IPTG (final concentration) to 15 mL of molten semisolid LB agar (1%, w/v) before the agar was poured into the plates. When the agar solidified, colonies growing on LB agar containing 25 µg/mL kanamycin were imprinted on the selection plate using sterile toothpicks. After the prescreening step, choosing mutants that produced a white halo for the next round screening. In second-round screening, the mutants were induced to expression in 0.5 mM IPTG condition, the supernatant of the cultures were collected after centrifugation at 12,000 rpm for 10 min at 4 °C and filtered through 0.22-µm-pore-size filter to test the AHL lactonase activity. The CV026 screening plate was prepared as described above without adding with 0.5 mM IPTG. The medium was punched using a sterile tip and the crude enzyme supernatant were added into the hole (with MomL crude enzyme supernatant as positive control and LB medium as negative control).

4.4. Expression and Purification of Mutant Proteins

Colonies surrounded by white halos on the purple background of the plate were picked and cultured in LB medium (with 25 µg/mL kanamycin) in a shaking incubator at 37 °C. Protein expression was induced with 0.5 mM IPTG at an original OD_{600} of 0.5–0.7 at 16 °C for 12 h. Cells were harvested via centrifugation at 12,000 rpm for 10 min at 4 °C. Cell pellets were resuspended gently in binding buffer (20 mM Tris-HCl with 10 mM imidazole, 0.5 M NaCl, pH 8) and disrupted by sonication on ice. The cell debris was removed via centrifugation at 12,000 rpm for 10 min at 4 °C, and the supernatants were filtered through a 0.22 µm pore-size filter. Before they were loaded onto NTA-Ni (Qiagen) columns, the supernatants were confirmed using a CV026 plate assay previously described by McClean [42]. Then, the mutant proteins were eluted using a specific wash buffer (20 mM Tris-HCl containing different amounts of imidazole, 0.5 M NaCl, pH 8) from the NTA-Ni columns and evaluated by sodium dodecyl sulfate polyacrylamide gel electrophoresis (SDS-PAGE).

4.5. N-Acyl Homoserine Lactone (AHL) Lactonases Activity Assay

The relative activity of mutant proteins was measured using a pH-sensitive colorimetric assay previously described by K Tang [9]. The test system consisted of a reaction buffer, morpholinepropanesulfonic acid (MOPS), a pH indicator, bromothymol blue (BTB), AHL substrate C6-HSL and the enzyme undergoing measurement. When AHL molecules were degraded to donate protons, the pH was weakly altered. Color changes due to BTB were measured using a microplate reader.

4.6. Site-Directed Mutagenesis of MomL

To study multi-site mutant proteins, we constructed site-directed single mutant via site-directed mutagenesis [43]. Mutation sites and primers [44] are listed in Table S1. The mutated genes were amplified using Primer STAR GXL DNA polymerase (Takara) and cyclized after phosphorylation. Recombinant plasmids were expressed in *E. coli* BL21(DE3) and mutant proteins were purified as previously described. The enzyme activity of each mutant protein was measured as described above.

4.7. Kinetic Assay of MomL and Mutant Proteins Activities

The catalytic activities of MomL and mutant proteins were measured by a pH sensitive colorimetric assay [45]. Briefly, 3.5nM enzyme, C6-HSL/3OC10-HSL (0.156 to 5 mM) were added to a MOPS (5 mM, pH 7.1)/BTB (1 mM) system of total 100 uL. Due to the pH-sensitive dye (BTB) mediated color change, OD_{630} was continuously measured using a microplate absorbance reader [9]. Initial rates were calculated and a GraphPad Prism software was used for calculating K_m and k_{cat} values. A standard curve using HCl was constructed to reflect the relationship between the absorbance change and the proton concentration, the value of OD_{630} would decrease by 0.193 after adding with 100 nmol of HCl.

4.8. Effects of MomL and Mutant Proteins on the Pathogenicity of Pcc

Pcc was cultured to the exponential phase and inoculated in 5 mL of LB medium containing equal amounts of enzymes (MomL, $MomL_{I144V}$ and $MomL_{V149A}$), and ddH_2O was added as a positive control. The cultures were grown on a shaker (170 rpm) at 28 °C for 24 h. Extracellular pectate lyase activity was determined using a DNS assay. First, 0.2% polygalacturonic acid reaction buffer (0.2% polygalacturonic acid; 0.2 M NaCl in 0.05 M sodium acetate buffer at pH 5.2) was prepared. Then, to obtain culture samples for enzyme assays, cultures were centrifuged at 4000 rpm for 5 min, and the supernatants were filter-sterilized through a 0.22 μm filter at 4 °C or on ice. Two hundred microliters of enzyme and 400 μL of 0.2% polygalacturonic acid reaction buffer were blended and immersed in a tube maintained at a constant temperature of 48 °C for 30 min. Four hundred microliters of DNS were added to the 600 μL reaction system, and the mixture was incubated in boiling water for 5 min and centrifuged at 12,000 rpm for 1 min; the precipitate was then discarded. The supernatant was diluted three times, and the absorbance was measured at 492 nm.

4.9. Pcc Survival Rate Assay

Pcc was cultured as mentioned above. A bacterial suspension was diluted with fresh LB and plated on LB agar; after 15 h of growth at 28 °C, colonies were counted. The colonies on Pcc plates were set to 100%.

4.10. Pcc Infection Experiment

Leaves were selected from the same Chinese cabbage, and 10 μL of Pcc bacterial suspension with enzyme (MomL and mutant proteins) were added to a cut surface. The inoculated leaves were incubated in sterile dishes at 28 °C for 24–48 h. Pcc alone was used as the positive control.

Supplementary Materials: The following are available online at http://www.mdpi.com/1660-3397/17/5/300/s1. Figure S1: The detection of gel electrophoresis of *momL* fragment with series of epPCR condition. Figure S2: The protein activity test. Table S1: Mutation sites and mutation primers of MomL. Table S2: The mutation rate comparison of epPCR products with different conditions.

Author Contributions: Conceptualization, J.W. and Y.W.; Data curation, Y.Z., J.Z., T.F., H.L., X.W. and Q.S.; Formal analysis, J.W. and J.L.; Methodology, J.W. and J.L.; Software, J.L.; Supervision, X.Z. and Y.W.; Writing—original draft, J.W.; Writing—review and editing, X.Z. and Y.W.

Funding: This work was supported by the Fundamental Research Funds for the Central Universities (No. 201941009), the National Natural Science Foundation of China (No. 31870023, 31571970 and 41506160), the Young Elite Scientists Sponsorship Program by CAST (No. YESS20160009).

Conflicts of Interest: The authors declare no competing interests.

References

1. Mion, S.; Rémy, B.; Plener, L.; Chabrière, É.; Daudé, D. Quorum sensing and quorum quenching: How to disrupt bacterial communication to inhibit virulence? *Med. Sci. (Paris)* **2019**, *35*, 31–38. [CrossRef]
2. Miller, M.B.; Bassler, B.L. Quorum sensing in bacteria. *Ann. Rev. Microbiol.* **2000**, *55*, 165–199. [CrossRef] [PubMed]

3. Williams, P.; Winzer, K.; Chan, W.C.; Cámara, M. Look who's talking: Communication and quorum sensing in the bacterial world. *Philos. Trans. R. Soc. Lond. B. Biol. Sci.* **2007**, *362*, 1119–1134. [CrossRef] [PubMed]
4. Allen, R.C.; Popat, R.; Diggle, S.P.; Brown, S.P. Targeting virulence: Can we make evolution-proof drugs? *Nat. Rev. Microbiol.* **2014**, *12*, 300. [CrossRef]
5. Whiteley, M.; Diggle, S.P.; Greenberg, E.P. Progress in and promise of bacterial quorum sensing research. *Nature* **2017**, *551*, 313–320. [CrossRef] [PubMed]
6. Dong, Y.H.; Xu, J.L.; Li, X.Z.; Zhang, L.H. AiiA, an enzyme that inactivates the acylhomoserine lactone quorum-sensing signal and attenuates the virulence of *Erwinia carotovora*. *Proc. Natl. Acad. Sci. USA* **2000**, *97*, 3526–3531. [CrossRef] [PubMed]
7. Fong, J.; Zhang, C.; Yang, R.; Boo, Z.Z.; Tan, S.K.; Nielsen, T.E.; Givskov, M.; Liu, X.W.; Bin, W.; Su, H. Combination Therapy Strategy of Quorum Quenching Enzyme and Quorum Sensing Inhibitor in Suppressing Multiple Quorum Sensing Pathways of *P. aeruginosa*. *Sci. Rep.* **2018**, *8*, 1155. [CrossRef] [PubMed]
8. Hwang, C.Y.; Kim, M.H.; Bae, G.D.; Zhang, G.I.; Kim, Y.H.; Cho, B.C. *Muricauda olearia* sp. nov., isolated from crude-oil-contaminated seawater, and emended description of the genus *Muricauda*. *Int. J. Syst. Evol. Microbiol.* **2009**, *59*, 1856–1861. [CrossRef]
9. Tang, K.; Su, Y.; Brackman, G.; Cui, F.; Zhang, Y.; Shi, X.; Coenye, T.; Zhang, X.H. MomL, a novel marine-derived N-acyl homoserine lactonase from *Muricauda olearia*. *Appl. Environ. Microbiol.* **2015**, *81*, 774–782. [CrossRef]
10. Mayer, C.; Romero, M.; Muras, A.; Otero, A. Aii20J, a wide-spectrum thermostable N-acylhomoserine lactonase from the marine bacterium *Tenacibaculum* sp. 20J, can quench AHL-mediated acid resistance in *Escherichia coli*. *Appl. Microbiol. Biotechnol.* **2015**, *99*, 9523–9539. [CrossRef]
11. Liu, D.; Momb, J.; Thomas, P.W.; Moulin, A.; Petsko, G.A.; Fast, W.; Ringe, D. Mechanism of the quorum-quenching lactonase (AiiA) from *Bacillus thuringiensis*. 1. Product-bound structures. *Biochemistry* **2008**, *47*, 7706–7714. [CrossRef] [PubMed]
12. Momb, J.; Wang, C.; Liu, D.; Thomas, P.W.; Petsko, G.A.; Guo, H.; Ringe, D.; Fast, W. Mechanism of the quorum-quenching lactonase (AiiA) from *Bacillus thuringiensis*. 2. Substrate modeling and active site mutations. *Biochemistry* **2008**, *47*, 7715–7725. [CrossRef] [PubMed]
13. Zhang, Y.; Liu, J.; Tang, K.; Yu, M.; Coenye, T.; Zhang, X.H. Genome analysis of *Flaviramulus ichthyoenteri* Th78^T in the family *Flavobacteriaceae*: Insights into its quorum quenching property and potential roles in fish intestine. *BMC Gen.* **2015**, *16*, 1–10. [CrossRef] [PubMed]
14. Dong, Y.H.; Wang, L.H.; Xu, J.L.; Zhang, H.B.; Zhang, X.F.; Zhang, L.H. Quenching quorum-sensing-dependent bacterial infection by an N-acyl homoserine lactonase. *Nature* **2001**, *411*, 813–817. [CrossRef] [PubMed]
15. Chan, K.G.; Atkinson, S.; Mathee, K.; Sam, C.K.; Chhabra, S.R.; Cámara, M.; Koh, C.L.; Williams, P. Characterization of N-acylhomoserine lactone-degrading bacteria associated with the *Zingiber officinale* (ginger) rhizosphere: Co-existence of quorum quenching and quorum sensing in *Acinetobacter* and *Burkholderia*. *BMC Microbiol.* **2011**, *11*, 51. [CrossRef] [PubMed]
16. Hong, K.W.; Koh, C.L.; Sam, C.K.; Yin, W.F.; Chan, K.G. Complete genome sequence of *Burkholderia* sp. Strain GG4, a betaproteobacterium that reduces 3-oxo-N-acylhomoserine lactones and produces different N-acylhomoserine lactones. *J. Bacteriol.* **2012**, *194*, 6317. [CrossRef] [PubMed]
17. Uroz, S.; Chhabra, S.R.; Cámara, M.; Williams, P.; Oger, P.; Dessaux, Y. N-Acylhomoserine lactone quorum-sensing molecules are modified and degraded by *Rhodococcus erythropolis* W2 by both amidolytic and novel oxidoreductase activities. *Microbiology* **2005**, *151*, 3313–3322. [CrossRef]
18. Gooding, J.R.; May, A.L.; Hilliard, K.R.; Campagna, S.R. Establishing a quantitative definition of quorum sensing provides insight into the information content of the autoinducer signals in *Vibrio harveyi* and *Escherichia coli*. *Biochemistry* **2010**, *49*, 5621–5623. [CrossRef]
19. Wang, Y.; Li, H.; Cui, X.; Zhang, X.H. A novel stress response mechanism, triggered by indole, involved in quorum quenching enzyme MomL and iron-sulfur cluster in *Muricauda olearia* Th120. *Sci. Rep.* **2017**, *7*, 4252. [CrossRef]
20. Zhang, J.; Wang, J.; Feng, T.; Du, R.; Tian, X.; Wang, Y.; Zhang, X.-H. Heterologous Expression of the Marine-Derived Quorum Quenching Enzyme MomL Can Expand the Antibacterial Spectrum of *Bacillus brevis*. *Mar. Drugs* **2019**, *17*, 128. [CrossRef]

21. Kiran, M.D.; Adikesavan, N.V.; Cirioni, O.; Giacometti, A.; Silvestri, C.; Scalise, G.; Ghiselli, R.; Saba, V.; Orlando, F.; Shoham, M.; et al. Discovery of a quorum-sensing inhibitor of drug-resistant staphylococcal infections by structure-based virtual screening. *Mol. Pharmacol.* **2008**, *73*, 1578–1586. [CrossRef] [PubMed]
22. Roy, V.; Fernandes, R.; Tsao, C.; Bentley, W. Cross Species Quorum Quenching Using a Native AI-2 Processing Enzyme. *ACS Chem. Biol.* **2010**, *5*, 223. [CrossRef]
23. Corbett, M.; Virtue, S.; Bell, K.; Birch, P.; Burr, T.; Hyman, L.; Lilley, K.; Poock, S.; Toth, I.; Salmond, G. Identification of a new quorum-sensing-controlled virulence factor in *Erwinia carotovora* subsp. *atroseptica* secreted via the type II targeting pathway. *Mol. Plant Microbe Interact.* **2005**, *18*, 334–342. [CrossRef]
24. Burr, T.; Barnard, A.M.L.; Corbett, M.J.; Pemberton, C.L.; Simpson, N.J.L.; Salmond, G.P.C. Identification of the central quorum sensing regulator of virulence in the enteric phytopathogen, *Erwinia carotovora*: The VirR repressor. *Mol. Microbiol.* **2006**, *59*, 113–125. [CrossRef]
25. Lee, D.H.; Lim, J.A.; Lee, J.; Roh, E.; Jung, K.; Choi, M.; Oh, C.; Ryu, S.; Yun, J.; Heu, S. Characterization of genes required for the pathogenicity of *Pectobacterium carotovorum* subsp. *carotovorum* Pcc21 in Chinese cabbage. *Microbiology* **2013**, *159*, 1487–1496. [CrossRef]
26. Kotoujansky, A. Molecular Genetics of Pathogenesis by Soft-Rot *Erwinias*. *Annu. Rev. Phytopathol.* **1987**, *25*, 405–430. [CrossRef]
27. Jafra, S.; Jalink, H.; Schoor, R.V.D.; Wolf, J.M.V.D. *Pectobacterium carotovorum* subsp. *carotovorum* Strains Show Diversity in Production of and Response to N-acyl Homoserine Lactones. *J. Phytopathol.* **2010**, *154*, 729–739. [CrossRef]
28. Lane, M.D.; Seelig, B. Advances in the directed evolution of proteins. *Curr. Opin. Chem. Biol.* **2014**, *22*, 129–136. [CrossRef] [PubMed]
29. Campbell, J.H.; Lengyel, J.A.; Langridge, J. Evolution of a Second Gene for β-Galactosidase in *Escherichia coli*. *Proc. Natl. Acad. Sci. USA* **1973**, *70*, 1841–1845. [CrossRef]
30. Kan, S.B.J.; Lewis, R.D.; Chen, K.; Arnold, F.H. Directed evolution of cytochrome c for carbon-silicon bond formation: Bringing silicon to life. *Science* **2016**, *354*, 1048–1051. [CrossRef]
31. Kan, S.B.J.; Huang, X.Y.; Gumulya, Y.; Chen, K.; Arnold, F.H. Genetically programmed chiral organoborane synthesis. *Nature* **2017**, *552*, 132. [CrossRef] [PubMed]
32. Drummond, D.A.; Iverson, B.L.; Georgiou, G.; Arnold, F.H. Why high-error-rate random mutagenesis libraries are enriched in functional and improved proteins. *J. Mol. Biol.* **2005**, *350*, 806–816. [CrossRef] [PubMed]
33. Hartwig, A. Role of magnesium in genomic stability. *Mut. Res.* **2001**, *475*, 113–121. [CrossRef]
34. Romero, M.; Martin-Cuadrado, A.B.; Otero, A. Determination of Whether Quorum Quenching Is a Common Activity in Marine Bacteria by Analysis of Cultivable Bacteria and Metagenomic Sequences. *Appl. Environ. Microbiol.* **2012**, *78*, 6345. [CrossRef] [PubMed]
35. Romero, M.; Diggle, S.P.; Heeb, S.; Cámara, M.; Otero, A. Quorum quenching activity in *Anabaena* sp. PCC 7120: Identification of AiiC, a novel AHL-acylase. *FEMS Microbiol. Lett.* **2008**, *280*, 73–80. [CrossRef]
36. Wei, H.; Lin, Y.; Yi, S.; Liu, P.; Jie, S.; Shao, Z.; Liu, Z. QsdH, a Novel AHL Lactonase in the RND-Type Inner Membrane of Marine *Pseudoalteromonas byunsanensis* Strain 1A01261. *PLoS ONE* **2012**, *7*, e46587.
37. Kim, J.H.; Choi, D.C.; Yeon, K.M.; Kim, S.R.; Lee, C.H. Enzyme-immobilized nanofiltration membrane to mitigate biofouling based on quorum quenching. *Environ. Sci. Technol.* **2011**, *45*, 1601–1607. [CrossRef]
38. Torres, M.; Rubio-Portillo, E.; Anton, J.; Ramos-Espla, A.A.; Quesada, E.; Llamas, I. Selection of the N-Acylhomoserine Lactone-Degrading Bacterium *Alteromonas stellipolaris* PQQ-42 and of Its Potential for Biocontrol in Aquaculture. *Front. Microbiol.* **2016**, *7*, 646. [CrossRef]
39. Torres, M.; Dessaux, Y.; Llamas, I. Saline Environments as a Source of Potential Quorum Sensing Disruptors to Control Bacterial Infections: A Review. *Mar. Drugs* **2019**, *17*, 191. [CrossRef] [PubMed]
40. Ruff, A.J.; Dennig, A.; Schwaneberg, U. To get what we aim for—Progress in diversity generation methods. *FEBS J.* **2013**, *280*, 2961–2978. [CrossRef] [PubMed]
41. Pennypacker, B.W.; Dickey, R.S.; Nelson, P.E. A histological comparison of the response of a chrysanthemum cultivar susceptible to *Erwinia chrysanthemi* and *Erwinia* subsp. *carotovora*. *Phytopathology* **1981**, *71*.
42. Mcclean, K.H.; Winson, M.K.; Fish, L.; Taylor, A.; Chhabra, S.R.; Camara, M.; Daykin, M.; Lamb, J.H.; Swift, S.; Bycroft, B.W.; et al. Quorum sensing and *Chromobacterium violaceum*: Exploitation of violacein production and inhibition for the detection of N-acylhomoserine lactones. *Microbiology* **1997**, *143*, 3703. [CrossRef] [PubMed]

43. Song, Q.; Wang, Y.; Yin, C.; Zhang, X.H. LaaA, a novel high-active alkalophilic alpha-amylase from deep-sea bacterium *Luteimonas abyssi* XH031 T. *En. Microbial. Technol.* **2016**, *90*, 83–92. [CrossRef] [PubMed]
44. Primrose, S.B.; Twyman, R.M. *Principles of Gene Manipulation and Genomics*, 7th ed.; Blackwell: Oxford, UK, 2006.
45. Chapman, E.; Wong, C.H. A pH sensitive colorometric assay for the high-Throughput screening of enzyme inhibitors and substrates: A case study using kinases. *Bioorg. Med. Chem.* **2002**, *10*, 551–555. [CrossRef]

© 2019 by the authors. Licensee MDPI, Basel, Switzerland. This article is an open access article distributed under the terms and conditions of the Creative Commons Attribution (CC BY) license (http://creativecommons.org/licenses/by/4.0/).

Article

Characteristics of a Novel Manganese Superoxide Dismutase of a Hadal Sea Cucumber (*Paelopatides* sp.) from the Mariana Trench

Yanan Li [1,2], Xue Kong [1,2] and Haibin Zhang [1,*]

1. Institute of Deep-Sea Science and Engineering, Chinese Academy of Sciences, Sanya 572000, China; liyn@idsse.ac.cn (Y.L.); kongx@sidsse.ac.cn (X.K.)
2. College of Earth and Planetary Sciences, University of Chinese Academy of Sciences, Beijing 100039, China
* Correspondence: hzhang@idsse.ac.cn; Tel./Fax: +86-0898-8838-0935

Received: 17 December 2018; Accepted: 15 January 2019; Published: 1 February 2019

Abstract: A novel, cold-adapted, and acid-base stable manganese superoxide dismutase (Ps-Mn-SOD) was cloned from hadal sea cucumber *Paelopatides* sp. The dimeric recombinant enzyme exhibited approximately 60 kDa in molecular weight, expressed activity from 0 °C to 70 °C with an optimal temperature of 0 °C, and resisted wide pH values from 2.2–13.0 with optimal activity (> 70%) at pH 5.0–12.0. The Km and Vmax of Ps-Mn-SOD were 0.0329 ± 0.0040 mM and 9112 ± 248 U/mg, respectively. At tested conditions, Ps-Mn-SOD was relatively stable in divalent metal ion and other chemicals, such as β-mercaptoethanol, dithiothreitol, Tween 20, Triton X-100, and Chaps. Furthermore, the enzyme showed striking stability in 5 M urea or 4 M guanidine hydrochloride, resisted digestion by proteases, and tolerated a high hydrostatic pressure of 100 MPa. The resistance of Ps-Mn-SOD against low temperature, extreme acidity and alkalinity, chemicals, proteases, and high pressure make it a potential candidate in biopharmaceutical and nutraceutical fields.

Keywords: expression; purification; deep-sea enzyme; pCold vector

1. Introduction

Reactive oxygen species (ROS) are necessary for various physiological functions, such as signaling pathways and immune responses; the mass accumulation of ROS will damage bio-macromolecules, leading to cell death and various diseases [1,2]. Superoxide dismutases (SODs, EC 1.15.1.1) are one of the most important antioxidant enzymes that clear ROS by converting them into oxygen and hydrogen peroxide. According to the different metal cofactors, several types, such as Cu,Zn-SOD, Mn-SOD, Fe-SOD, cambialistic SOD (activated with either Fe or Mn), Ni-SOD, and Fe,Zn-SOD, have been reported in many species [3–7].

Studies have shown that SODs are related to immune reactions in invertebrates, as exemplified by bacterial and viral invasion [4,8], environmental pollution [9,10], and temperature stimulation [11]. Recently, Xie et al. indicates that antioxidant is related to the deep-sea environmental adaptability [12]. On the other hand, point mutations and activity loss of SODs lead to serious diseases and death in vertebrates. For example, the mice model of mitochondria SOD-deficiency is characterized by neurodegeneration, myocardial injury, and perinatal death [13,14]. A strong link is observed between Alzheimer's disease, tumor, amyotrophic lateral sclerosis, and SODs [15,16]. Hence, the physiological significance of SODs allows their application in the therapeutic and nutraceutical fields. To date, SODs have been reported to exhibit positive effects on inflammatory diseases, arthritis tumor, and promotion [17–19]. An orally effective form of SOD (glisodin) has been developed by Isocell Pharma, and it showed cosmetic and health benefits in human subjects [20,21]. Producing SOD using engineered bacteria is one of the most promising methods to obtain high yield and inexpensive SODs for application. Therefore, the development of SODs with remarkable characteristics is particularly urgent.

Sea cucumbers are highly important commercial sea foods owing to their high nutritional value, and they are distributed from shallow water to the deep sea [22]. Although deep sea is an extremely low-temperature and high hydrostatic-pressured environment for most living organisms, holothurians dominate benthic megafaunal communities in hadal trenches and form "the kingdom of Holothuroidea" when food is abundant [23]. Extreme environments, such as the deep sea, are ideal for the development of new enzymes; numerous novel enzymes with unique activities, such as proteases and lipases, have been identified from the deep sea [24,25]. Considering the promising applications of SODs in therapeutic and nutraceutical fields, relationship with the adaptability of the deep-sea environment and limited studies in extreme organisms, especially in hadal sea cucumbers, we report a novel manganese superoxide dismutase from hadal sea cucumber *Paelopatides* sp. (Ps-Mn-SOD), which inhabits a depth of 6500 m in the Mariana Trench, analyzed its biochemical characteristics, and evaluated its stability for potential use in the food and preliminarily nutraceutical fields.

2. Results

2.1. Sequence Characteristics

The ORF of Ps-Mn-SOD is 768 bp long, encoding 255 amino acids. A signal peptide was detected at the N-terminal of deduced amino acid sequence. The N- and C-terminal domains spanned from Lys-34 to Ser-127 and Pro-137 to Leu-242, respectively. Four conserved amino acid residues, namely, His-63, His-119, Asp-209, His-213 are responsible for manganese coordination. A conserved residue of Tyr-35 is responsible for the second coordination sphere of the metal [26]. A highly conserved Mn-SOD signature sequence with the pattern D-x-[WF]-E-H-[STA]-[FY] existed in Ps-Mn-SOD (DVWEHAYY). The predicted secondary structure contained 13 α-helices and 4 β-strands. The deduced theoretical isoelectric point was 5.05, and the molecular weight was 29.29 kDa. The instability index of 36.97 classified the protein as stable. The 3D model of Ps-Mn-SOD was predicted using the x-ray template of *Bacillus subtilis*, which shared 45.27% sequence identity (PDB ID: 2RCV) [27]. This model shows that Ps-Mn-SOD is presented as a homodimer, and each subunit embraces one manganese ion. The global and per-residue model qualities were assessed using the QMEAN scoring function [28]. GMQE and QMEAN4 Z-scores reached 0.64 and −2.63, respectively, suggesting the accuracy of predicted 3D model of Ps-Mn-SOD. Figure 1 and Supplementary Figure S1 provide the related structural information of Ps-Mn-SOD.

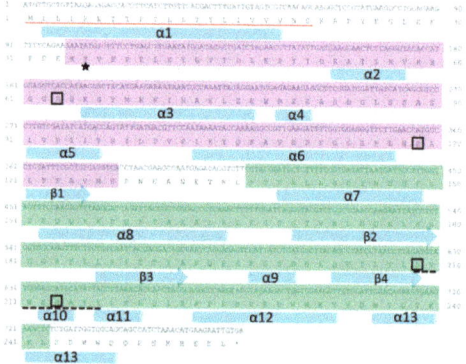

Figure 1. Nucleotide and corresponding amino acid sequences of Ps-Mn-SOD. The signal peptide is drawn with a red line. The signature sequence DVWEHAYY is underlined with dotted line. N- and C-terminal domains are marked with purple and green shades, respectively. Four conserved amino acid residues for manganese coordination are boxed. Asterisk points to the highly conserved Tyr-35 residue. Cylinders and arrows represent helices and strands, respectively.

2.2. Homology and Phylogenetic Analysis

Multiple alignment and pairwise homology analysis between Ps-Mn-SOD and other invertebrates were performed, and the results are shown in Figure 2 and Supplementary Table S1. Multiple alignment of Ps-Mn-SOD with other invertebrates indicated that four amino acids were responsible for manganese binding, and the signature sequences are highly conserved in different Mn-SOD sources and were also identified in Ps-Mn-SOD (Figure 2). The highest similarity and identity were shared with *Apostichopus japonicus* (83.9% and 78.0%), followed by *Capitella teleta* (66.9% and 47.9%), *Exaiptasia pallida* (66.3% and 47.7%), *Strongylocentrotus purpuratus* (65.1% and 47.0%), *Mizuhopecten yessoensis* (64.4% and 46.7%), and *Stylophora pistillata* (63.1% and 45.8%). To determine the type of SOD present, we performed phylogenetic analysis based on the amino acid sequences of the determined SOD types in Genebank (Figure 3). The results showed that the present SOD clustered with *A. japonicus* and evidently a Mn-SOD type with high bootstrap values.

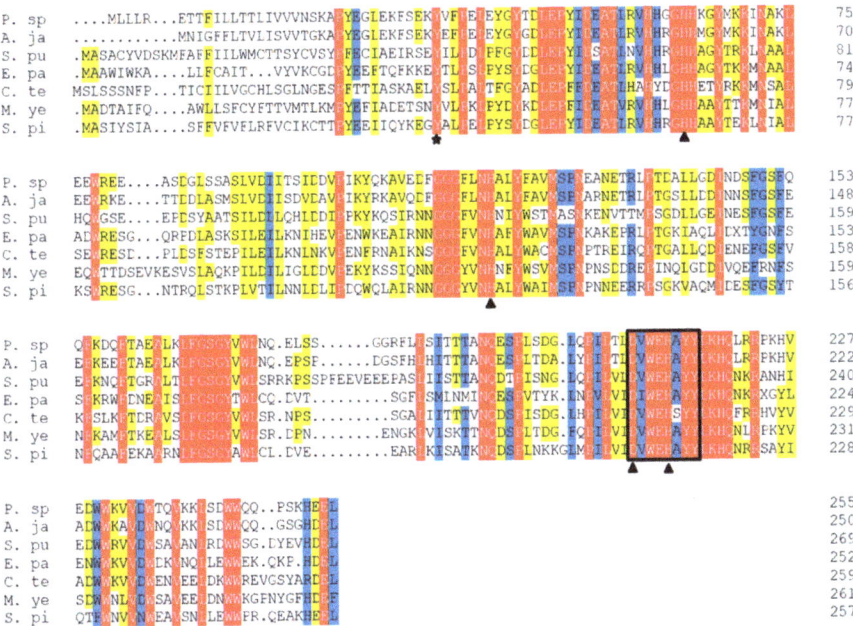

Figure 2. Multiple alignment of Ps-Mn-SOD with other invertebrates. Mn-SOD signature sequence is boxed. Triangles point to the active sites for manganese coordination. Asterisk points to the highly conserved Tyr-35 residue.

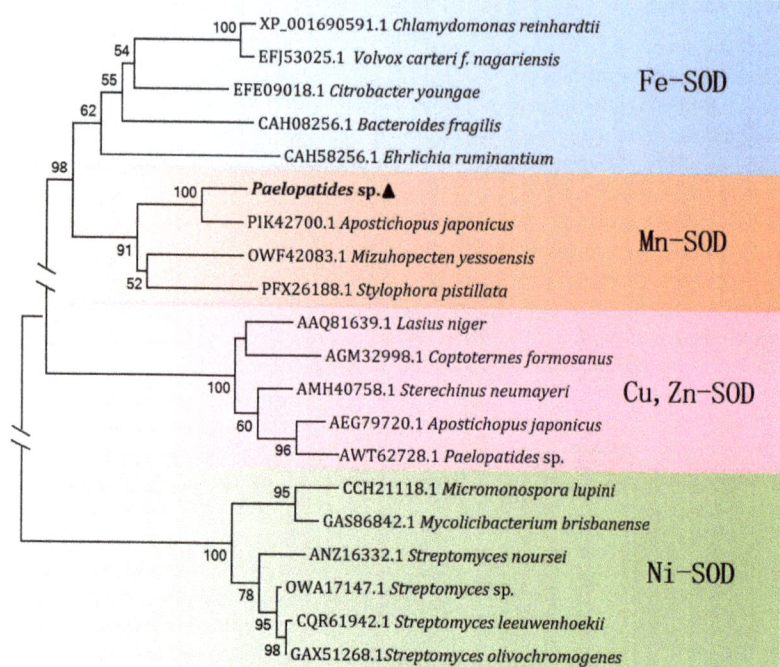

Figure 3. Neighbor-joining phylogenetic tree of SODs based on amino acid sequence homology. Bootstrap values below 50 are cut off. Ps-Mn-SOD is displayed in bold.

2.3. Expression, Purification, and Validation of Ps-Mn-SOD

The Ps-Mn-SOD gene was expressed with a His-tag in *E. coli*. Supplementary Figure S2 shows the SDS-PAGE analysis results. Recombinant Ps-Mn-SOD was expressed under 0.1 mM IPTG at 15 °C for 24 h and produced a distinct band at approximately 30 kDa, consistent with the previously estimated molecular weight (Supplementary Figure S2, lanes 1 and 2). The protein was purified under native conditions due to its highly soluble expression in the supernatant (Supplementary Figure S2, lanes 3 and 4). The maximum protein yield approximated 4.39 mg/L culture. Western blot analysis was performed to verify its successful expression (Supplementary Figure S2, lanes 5 and 6).

2.4. Characterizations of Ps-Mn-SOD

2.4.1. Effects of Temperature on Ps-Mn-SOD

The activity of Ps-Mn-SOD was determined from 0 °C to 80 °C, with the optimum temperature observed at 0 °C. A stable activity was observed at low temperatures, with > 70% activity highlighted from 0 °C to 60 °C. The activity was maintained at 2.53% at 70 °C and lost at 80 °C (Figure 4A).

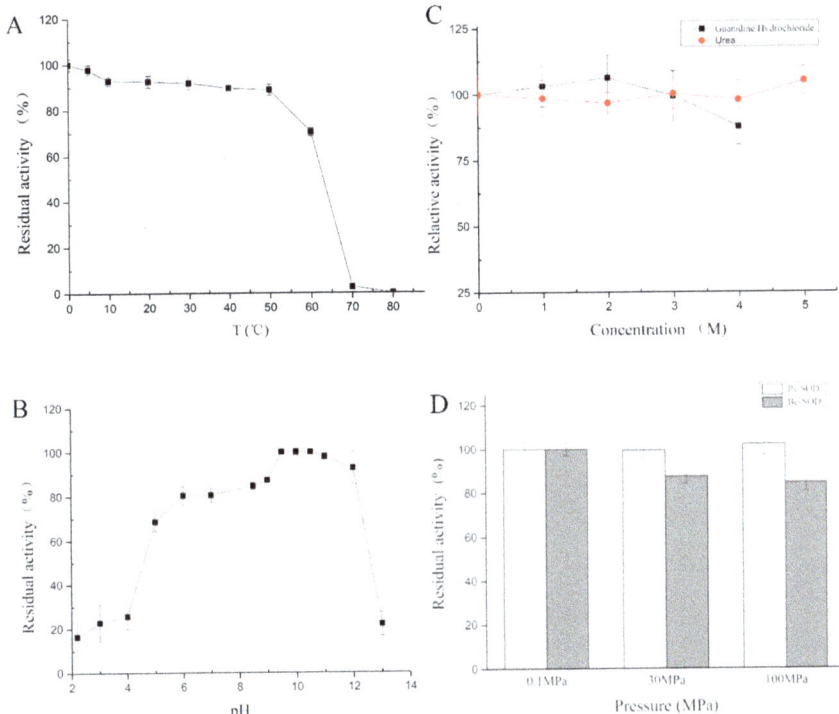

Figure 4. Effects of temperature (**A**), pH (**B**), urea and guanidine hydrochloride (**C**), and high hydrostatic pressure (**D**) on Ps-Mn-SOD. Ps-SOD and Be-SOD represent SOD from *Paelopatides* sp. and bovine erythrocytes, respectively.

2.4.2. Effects of pH on Ps-Mn-SOD

The activity of recombinant Ps-Mn-SOD was measured under pH 2.2–13.0, with an optimum pH observed at 10.5 (Figure 4B). Ps-Mn-SOD could resist extreme pH values (> 20% at pH 3.0–13.0) and showed optimal activity (> 70%) at pH 5.0–12.0.

2.4.3. Effects of Chemicals on Ps-Mn-SOD

The effects of metal ions on Ps-Mn-SOD activity were determined at 0.1 or 1 mM final concentration (Table 1). Ps-Mn-SOD activity was inhibited by Mn^{2+}, Co^{2+}, Ni^{2+}, Zn^{2+}, and 1 mM Cu^{2+} and Ba^{2+}. In particular, Co^{2+} showed more significant inhibition effect on Ps-Mn-SOD activity. Mg^{2+} and Ca^{2+} showed minimal effects.

Table 2 provides the effects of inhibitors, detergents, and denaturants on Ps-Mn-SOD activity. Ps-Mn-SOD activity was strongly inhibited by ethylene diamine tetraacetic acid (EDTA) and SDS and especially sensitive to SDS. Reductant dithiothreitol (DTT) and β-mercaptoethanol (β-ME) minimally affected enzyme activity. Detergents of Tween 20, Triton X-100, and Chaps slightly enhanced enzyme activity at 0.1% concentration.

Table 1. Effects of metal ions on Ps-Mn-SOD. ** $p < 0.01$.

Divalent Metal Ions	Concentration/mmol·L^{-1}	Relative Activity/%
Control	—	100 ± 2.39
Mn^{2+}	0.1	92.89 ± 1.53 **
	1	84.99 ± 2.77 **
Co^{2+}	0.1	80.13 ± 1.23 **
	1	61.49 ± 1.54 **
Ni^{2+}	0.1	95.70 ± 2.38 **
	1	94.42 ± 2.92 **
Zn^{2+}	0.1	90.25 ± 1.76 **
	1	90.99 ± 4.63 **
Cu^{2+}	0.1	98.39 ± 3.97
	1	88.99 ± 5.44 **
Ba^{2+}	0.1	99.16 ± 2.18
	1	95.94 ± 2.40 **
Mg^{2+}	0.1	100.68 ± 3.27
	1	100.61 ± 2.16
Ca^{2+}	0.1	99.71 ± 1.13
	1	100.39 ± 4.48

Table 2. Effects of inhibitors, reductant, and detergents. * $p < 0.05$; ** $p < 0.01$.

Divalent Metal Ions	Concentration	Relative Activity/%
Control	—	100 ± 2.84
EDTA	1 mmol·L^{-1}	64.33 ± 3.08 **
	10 mmol·L^{-1}	58.03 ± 2.59 **
DTT	1 mmol·L^{-1}	96.36 ± 4.65
	10 mmol·L^{-1}	97.00 ± 5.46
β-ME	1 mmol·L^{-1}	96.53 ± 4.47
	10 mmol·L^{-1}	101.85 ± 3.72
Tween 20	0.1%	109.96 ± 6.62 **
	1%	105.01 ± 3.28 **
Chaps	0.1%	103.13 ± 2.32 *
	1%	99.32 ± 3.66
Triton X-100	0.1%	105.02 ± 3.29 **
	1%	99.22 ± 3.79
SDS	0.1%	5.21 ± 3.45 **
	1%	6.11 ± 4.15 **

The enzyme could resist the strong denaturation of urea and guanidine hydrochloride (Figure 4C) and maintain an almost full activity after 1 h treatment in 5 M urea or 4 M guanidine hydrochloride.

Hydrogen peroxide and sodium azide were used to determine the SOD type (Figure 5 and Supplementary Figure S4). After treatment of the recombinant Ps-Mn-SOD using 10 mM hydrogen peroxide and sodium azide at 25 °C for 1 h, the relative activities were 7.73% and 90.39%, respectively. This showed that the SOD from *Paelopatides* sp. belongs to Fe/Mn-SOD family, in accordance with previous phylogenetic analysis and 3D structure prediction.

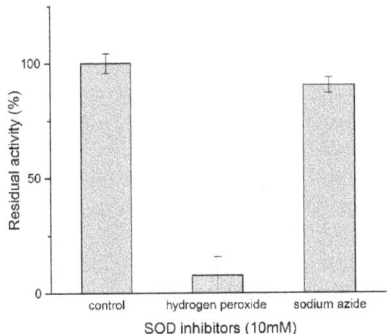

Figure 5. SOD type assay.

2.4.4. Effects of Digestive Enzymes on Ps-Mn-SOD

Digestion experiment was performed to test the stability of recombinant Ps-Mn-SOD in digestive fluid. Residual enzyme activity was measured after different incubation times for 0–4 h at 37 °C and pH 7.4. As shown in Table 3 and Supplementary Table S2, although the Ps-Mn-SOD sequence putatively contains 30 chymotrypsin and 23 trypsin cleavage sites, the enzyme could still maintain intact activity after 4 h treatment at an enzyme/substrate (w/w) ratio of 1/100.

Table 3. Cleavage effect of digestive enzyme on Ps-Mn-SOD at different time periods. Results are shown as mean ($n = 3$) ± SD. ** $p < 0.01$.

Time (h)	Relative Activity (%)
0	100 ± 2.21
1	108.66 ± 5.70
2	104.46 ± 4.54
3	103.73 ± 3.24
4	106.72 ± 4.80

2.4.5. Effects of High Hydrostatic Pressure on Ps-Mn-SOD

As shown in Figure 4D, the recombinant Ps-Mn-SOD could maintain full activity with increasing hydrostatic pressure until 100 MPa. By contrast, the SOD from bovine erythrocytes exhibited reduced activity of 84.57% when the pressure reached 100 MPa.

2.4.6. Kinetic Parameters

The kinetic parameters of recombinant Ps-Mn-SOD were determined using a series of xanthine (0.006–0.6 mM) concentrations at 37 °C and pH 8.2 (Supplementary Figure S3) based on the Michaelis–Menten equation. The K_m and V_{max} values of Ps-Mn-SOD were 0.0329 ± 0.0040 mM and 9112 ± 248 U/mg, respectively. The R^2 value of the curve fitting was 0.9815.

3. Discussion

Mn-SODs are predominantly found in mitochondria, as the first line of antioxidant defense, which are involved in cellular physiology, such as cell impairment and immune-responsive [8]. The important biological functions of Mn-SODs have attracted increasing attention among researchers. Novel Mn-SODs with remarkable characteristics will have great applications in food, cosmetic, and pharmaceutical industries. In the present study, a novel and kinetically stable Mn-SOD derived from hadal sea cucumber was cloned, expressed, and characterized.

Based on preliminary data, the Ps-Mn-SOD is frigostabile, consistent with the fact that the protein was derived from hadal area, which maintained > 90% activity below 20 °C with the optimum temperature observed at 0 °C. In contrast, Mn-SOD from ark shell, *Scapharca broughtonii*, showed <40% activity below 20 °C [4]. Mn-SOD from seahorse, *Hippocampus abdominalis*, showed <80% and continuously reduced activity below 20 °C [8].

Mn-SODs in several sources have been found to function at wide pH values. For example, a hyperthermostable Mn-SOD from *Thermus thermophilus* HB27 maintained >70% activity at pH 4.0–8.0 [29]; Mn-SOD from deep-sea thermophile *Geobacillus* sp. EPT3 maintained >70% activity at pH 7.0–9.0 [30]; and Mn-SOD from *Thermoascus aurantiacus* var. *levisporus* only maintained >40% activity at pH 6.0–9.0 [31]. In contrast, the present Ps-Mn-SOD could maintain >70% activity at pH 5.0–12.0, showing remarkably wide pH values adaptation. Furthermore, after 1 h treatment in extremely acidic (pH 2.2) or alkaline (pH 13.0) conditions, Ps-Mn-SOD still maintained ~20% activity, showing remarkable stability to extreme pH values. The pH assays also showed that Ps-Mn-SOD is more stable under alkaline (pH 8.5–12.0) than acidic (pH 2.2–5.0) conditions. Metal ligands may undergo protonation at low pH but exhibit stability in alkaline conditions [32]. Similar studies on seahorse and bay scallop SODs were also reported [8,33].

Ps-Mn-SOD is relatively stable in chemicals, such as urea, guanidine hydrochloride, β-ME, DTT, etc. It maintained almost 100% activity after 1 h treatment of 5 M urea or 4 M guanidine hydrochloride at 25 °C, showing excellent resistance to strong protein denaturants. By comparison, the Mn-SOD from deep-sea thermophile *Geobacillus* sp. EPT3 maintained > 70% residual activity in 2.5 M urea or guanidine hydrochloride after 30 min treatment [30]. Fe-SOD from Antarctic yeast *Rhodotorula mucilaginosa* showed relatively low tolerance to urea [34]. However, based on our obtained data (unpublished and [35]), SODs from hadal sea cucumbers constantly exhibited excellent resistance to perturbation of denaturants. In addition, Ps-Mn-SOD maintained 97.00% and 99.22% residual activity after 1 h treatment of 10 mM DTT and 1% Triton X-100, respectively. While Mn-SOD from deep-sea thermophile *Geobacillus* sp. EPT3 only maintained 84.10% and 70.30% activity after 30 min treatment of corresponding chemicals [30].

As expected, Ps-Mn-SOD could also resist the perturbation by high hydrostatic pressure compared to the homolog from atmospheric pressure organism, because it was derived from a hadal field. Given the limitations of our equipment, the experiment was not performed at pressure more than 100 MPa. In fact, Ps-Mn-SOD might resist >100 MPa hydrostatic pressure. Similar results have been reported in other deep-sea enzymes, such as RNA polymerase from *Shewanella violacea* [36], N-acetylneuraminate lyase from *Mycoplasma* sp. [37], and lactate dehydrogenase b from *Corphaenoides armatus* [38]. Nonetheless, the sensitivity of enzymes to high hydrostatic pressure is not always related to the depth where the organisms lived. For example, two polygalacturonases from the hadal yeast *Cryptococcus liquefaciens* strain N6 exhibited an almost constant activity from 0.1 to 100 MPa. While, at the same pressure, polygalacturonase from *Aspergillus japonicus*, which lives under atmospheric pressure, increased by approximately 50% [39]. However, limited studies reported in detail the pressure assays of SODs, proving the difficulty in the interpretation of their pressure tolerance mechanism.

Altogether, these features render Ps-Mn-SOD a potential candidate in the biopharmaceutical and nutraceutical fields.

4. Materials and Methods

4.1. Material and Reagents

Hadal sea cucumber was collected at the depth of 6500 m in the Mariana Trench (10° 57.1693' N 141° 56.1719' E). Total RNA was extracted using RNeasy Plus Universal Kits from Qiagen, Hilden, Germany, and reverse-transcribed to cDNA. The transcriptome was obtained by sequencing assembly and annotation by Novogene Company (Tianjin, China). The following reagents were purchased from Takara, Tokyo, Japan: PrimeScript™ II 1st strand cDNA Synthesis Kit, PrimeSTAR® GXL DNA Polymerase, *E. coli* DH5α, and pG-KJE8/BL21 competent cells, pCold II vector, restriction enzymes *BamH* I, and *Pst* I, T4-DNA ligase, and DNA and protein markers. The 1 mL Ni-NTA affinity column, BCA protein assay kit, primers, and trypsin/chymotrypsin complex (2400:400) were obtained from Sangon Biotech Company, Shanghai, China. Polyvinylidene difluoride (PVDF) membrane was obtained from Millipore Company, USA. The primary (ab18184) and secondary antibodies (ab6789) were obtained from Abcam, Cambridge, UK. Pierce™ ECL Plus Western blot analysis substrate was obtained from ThermoFisher, Waltham, MA, USA.

4.2. Cloning and Recombinant

For the manganese SOD (Ps-Mn-SOD) gene, the Mn-SOD sequences of Holothuroidea in GenBank were submitted to the transcriptome database of *Paelopatides* sp. to run a local blast using Bioedit 7.0 software. The open reading frame (ORF) of Ps-Mn-SOD (deleted signal peptide) was amplified by primers Ps-Mn-SOD-S: CGGGATCCAAGGCTCCGTATGAAGGCCTGGAGA and Ps-Mn-SOD-A: AACTGCAGTCACAATTCTTCATGTTTAGATGGC using the cDNA as template (the underlined restriction enzyme sites). The sequence was submitted to GenBank database with accession numbers MK182093. The purified and digested PCR product was ligated with pCold II vector. The recombinant plasmids, that is, pCold II-Ps-Mn-SOD, were transformed into *E. coli* DH5α, and positive clones were verified by sequencing.

4.3. Protein Overproduction, Purification, and Confirmation

The recombinant plasmids were transformed into *E. coli* chaperone competent cells pG-KJE8/BL21, which were inoculated in liquid Luria-Bertani medium (containing 100 µg/mL ampicillin, 20 µg/mL chloramphenicol, 0.5 mg/mL L-arabinose, and 2 ng/mL tetracycline), proliferated at 37 °C until the OD_{600} reached 0.4–0.6, cooled on an ice–water mixture for 40 min, added isopropyl β-D-1-thiogalactopyranoside (IPTG) with a final concentration of 0.1 mM, and then incubated for 24 h at 15 °C to produce the recombinant protein. Cells were harvested, washed with $1 \times$ phosphate-buffered saline, resuspended in binding buffer (50 mM Na_3PO_4, 300 mM NaCl, and 20 mM imidazole, pH 7.4), and then sonicated on ice. The supernatant harboring the recombinant protein was separated from cell debris by centrifugation at 12000 g and 4 °C for 20 min and then applied to 1 mL Ni-NTA column for purification of the target protein based on its 6× His-tag, according to the manufacturer's instructions. The harvested target protein was dialyzed with $1 \times$ tris buffered saline (TBS) at 4 °C for 24 h against three changes of $1 \times$ TBS and finally stored at −80 °C for further experiments. The expression condition was analyzed on 12% sodium dodecyl sulfate polyacrylamide gel electrophoresis (SDS-PAGE) and confirmed using Western blot analysis. The recombinant protein on 12% SDS-PAGE gel was transferred to a PVDF membrane, which was successively incubated with primary (diluted 1:5000) and secondary antibodies (diluted 1:10000), dyed with Pierce™ ECL Plus Western blot analysis substrate, and detected under chemiluminescent imaging system. Additional details were as described by Li et al. [35].

4.4. Bioinformatics Analyses

The amino acid sequence of Ps-Mn-SOD was translated using ExPASy translation tool (http://web.expasy.org/translate/). The signal peptide, secondary structure, motif sequences, and 3D homology model were predicted by SignalP 4.1 Server (http://www.cbs.dtu.dk/services/SignalP/), Scratch Protein Predictor (http://scratch.proteomics.ics.uci.edu/), InterPro Scan (http://www.ebi.ac.uk/InterProScan/), and Swiss model server (http://swissmodel.expasy.org/) [40], respectively. The physicochemical properties of Ps-Mn-SOD were predicted using ExPASy ProtParam tool (http://web.expasy.org/protparam/). The possible cleavage sites of trypsin and chymotrypsin on Ps-Mn-SOD were predicted using the peptide cutter software (http://web.expasy.org/peptide_cutter/). Multiple alignments of Ps-Mn-SOD were processed using DNAMAN 7.0.2 software. Homology analysis was constructed by pairwise alignment tool (https://www.ebi.ac.uk/Tools/psa/emboss_needle/). The neighbor-joining phylogenic tree was generated in MEGA 7.0 with bootstrap values 1000.

4.5. Enzyme Assays

SOD activity was determined via spectrophotometric method using the SOD assay kit from Nanjing Jiancheng Institute of Biology and Engineering (Code No. A001-1-1, Nanjing, China). Each measurement point contained three replicates, and the results are shown as mean (n = 3) ± standard deviation (SD). The 1 × TBS was used as the blank control. One unit of SOD activity was defined as the amount of enzyme that inhibited 50% of chromogen production at 550 nm.

The purified Ps-Mn-SOD was quantified, and residual activities were determined after incubation under different variables, including temperature, pH, chemicals, digestive enzymes, and high hydrostatic pressure. Considering temperature, proteins were treated from 0 °C to 80 °C for 15 min with an interval of 10 °C [33,34]. Proteins were treated at pH 2.2–13 for 1 h at 25 °C [4,34]. The enzymatic activity at optimum temperature and pH was set as 100%. With regard to chemicals, the proteins were mixed with an equal treatment solution at different final concentrations for 40 min at 25 °C [34,41]. The incubation time of urea, guanidine hydrochloride, hydrogen peroxide, and sodium azide was expanded to 1 h. The enzyme activity without chemicals was set as 100%. For proteolytic susceptibility assay, the mass ratio of recombinant Ps-Mn-SOD and trypsin/chymotrypsin complex was 1:100, and the group incubated for 0 h was considered with 100% enzyme activity [34,42]. For high hydrostatic pressure, proteins were treated at 0.1, 30, and 100 MPa for 2 h at 5 °C. The enzyme activity at 0.1 MPa was set as 100%, and bovine erythrocyte SOD was selected for comparison from atmospheric organism. Kinetics of Ps-Mn-SOD were measured as previously described by Li et al. [35].

4.6. Statistical Analysis

Independent sample T-test was used for statistical analysis for each of the two groups using SPSS 21.0 (IBM Company, Armonk, NY, USA); $p < 0.05$ was considered statistically significant.

Supplementary Materials: The following are available online at http://www.mdpi.com/1660-3397/17/2/84/s1, Figure S1: The predicted 3D model of Ps-Mn-SOD. Red spheres represent manganese ions. (**A**) homodimer. (**B**) close-up of the manganese ion binding site. Figure S2: Analysis of SDS-PAGE. M: protein marker, Lane 1: total proteins before induction, Lane 2: total proteins after induction, Lane 3: inclusion body after ultrasonication, Lane 4: supernatant after ultrasonication, Lane 5: Western blot of recombinant protein, Lane 6: purified protein. Figure S3: The curve of kinetic parameters of Ps-Mn-SOD. Figure S4: SOD type assay. The result is expressed using specific activity. Table S1. Pairwise alignment analysis between Ps-Mn-SOD and other species. Table S2. The prediction of cleavage site of Ps-Mn-SOD.

Author Contributions: H.Z. collected the sample, Y.L. designed and performed the experiments, Y.L. and H.Z. prepared the manuscript, and X.K. gave advice during the experiments.

Funding: We thank J.C. for total RNA extraction. We also thank for language modification given by L.Y. in South China Sea Fisheries Research Institute, Chinese Academy of Fishery Sciences. This work was We changed "supported by The National Key Research and Development Program of China (2017YFC0306600, 2018YFC0309804), Hundred Talents Program of CAS (SIDSSE–BR–201401).

Conflicts of Interest: The authors declare no conflict of interest.

References

1. Torres, M.A. ROS in biotic interactions. *Physiol. Plant.* **2010**, *138*, 414–429. [CrossRef] [PubMed]
2. Kawanishi, S.; Inoue, S. Damage to DNA by reactive oxygen and nitrogen species. *Seikagaku* **1997**, *69*, 1014–1017. [PubMed]
3. Kim, F.J.; Kim, H.P.; Hah, Y.C.; Roe, J.H. Differential expression of superoxide dismutases containing Ni and Fe/Zn in *Streptomyces coelicolor*. *Eur. J. Biochem.* **1996**, *241*, 178–185. [CrossRef] [PubMed]
4. Zheng, L.; Wu, B.; Liu, Z.; Tian, J.; Yu, T.; Zhou, L.; Sun, X.; Yang, A.G. A manganese superoxide dismutase (MnSOD) from ark shell, *Scapharca broughtonii*: Molecular characterization, expression and immune activity analysis. *Fish Shellfish Immunol.* **2015**, *45*, 656–665. [CrossRef] [PubMed]
5. Zheng, Z.; Jiang, Y.H.; Miao, J.L.; Wang, Q.F.; Zhang, B.T.; Li, G.Y. Purification and characterization of a cold-active iron superoxide dismutase from a psychrophilic bacterium, *Marinomonas* sp. NJ522. *Biotechnol. Lett.* **2006**, *28*, 85–88. [CrossRef] [PubMed]

6. Krauss, I.R.; Merlino, A.; Pica, A.; Rullo, R.; Bertoni, A.; Capasso, A.; Amato, M.; Riccitiello, F.; De Vendittis, E.; Sica, F. Fine tuning of metal-specific activity in the Mn-like group of cambialistic superoxide dismutases. *RSC Adv.* **2015**, *5*, 87876–87887. [CrossRef]
7. Sheng, Y.; Abreu, I.A.; Cabelli, D.E.; Maroney, M.J.; Miller, A.-F.; Teixeira, M.; Valentine, J.S. Superoxide Dismutases and Superoxide Reductases. *Chem. Rev.* **2014**, *114*, 3854–3918. [CrossRef] [PubMed]
8. Ncn, P.; Godahewa, G.I.; Lee, S.; Kim, M.J.; Hwang, J.Y.; Kwon, M.G.; Hwang, S.D.; Lee, J. Manganese-superoxide dismutase (MnSOD), a role player in seahorse (*Hippocampus abdominalis*) antioxidant defense system and adaptive immune system. *Fish Shellfish Immunol.* **2017**, *68*, 435–442.
9. Kim, B.M.; Rhee, J.S.; Park, G.S.; Lee, J.; Lee, Y.M.; Lee, J.S. Cu/Zn- and Mn-superoxide dismutase (SOD) from the copepod *Tigriopus japonicus*: Molecular cloning and expression in response to environmental pollutants. *Chemosphere* **2011**, *84*, 1467–1475. [CrossRef]
10. Li, C.; He, J.; Su, X.; Li, T. A manganese superoxide dismutase in blood clam *Tegillarca granosa*: Molecular cloning, tissue distribution and expression analysis. *Comp. Biochem. Physiol. B Biochem. Mol. Biol.* **2011**, *159*, 64–70. [CrossRef]
11. Wang, H.; Yang, H.; Liu, J.; Yanhong, L.I.; Liu, Z. Combined effects of temperature and copper ion concentration on the superoxide dismutase activity in *Crassostrea ariakensis*. *Acta Oceanol. Sin.* **2016**, *35*, 51–57. [CrossRef]
12. Xie, Z.; Jian, H.; Jin, Z.; Xiao, X. Enhancing the adaptability of the deep-sea bacterium Shewanella piezotolerans WP3 to high pressure and low temperature by experimental evolution under H_2O_2 stress. *Appl. Environ. Microbiol.* **2017**, *84*, e02342-17. [CrossRef] [PubMed]
13. Lebovitz, R.M.; Zhang, H.; Vogel, H.; Cartwright, J.; Dionne, L.; Lu, N.; Huang, S.; Matzuk, M.M. Neurodegeneration, myocardial injury, and perinatal death in mitochondrial superoxide dismutase-deficient mice. *Proc. Natl. Acad. Sci. USA* **1996**, *93*, 9782–9787. [CrossRef] [PubMed]
14. Li, Y.; Huang, T.T.; Carlson, E.J.; Melov, S.; Ursell, P.C.; Olson, J.L.; Noble, L.J.; Yoshimura, M.P.; Berger, C.; Chan, P.H. Dilated cardiomyopathy and neonatal lethality in mutant mice lacking manganese superoxide dismutase. *Nat. Genet.* **1995**, *11*, 376–381. [CrossRef] [PubMed]
15. Delacourte, A.; Defossez, A.; Ceballos, I.; Nicole, A.; Sinet, P.M. Preferential localization of copper zinc superoxide dismutase in the vulnerable cortical neurons in Alzheimer's disease. *Neurosci. Lett.* **1988**, *92*, 247–253. [CrossRef]
16. Kruman, I.I.; Pedersen, W.A.; Springer, J.E.; Mattson, M.P. ALS-linked Cu/Zn-SOD mutation increases vulnerability of motor neurons to excitotoxicity by a mechanism involving increased oxidative stress and perturbed calcium homeostasis. *Exp. Neurol.* **1999**, *160*, 28–39. [CrossRef] [PubMed]
17. Cullen, J.J.; Weydert, C.; Hinkhouse, M.M.; Ritchie, J.; Domann, F.E.; Spitz, D.; Oberley, L.W. The role of manganese superoxide dismutase in the growth of pancreatic adenocarcinoma. *Cancer Res.* **2003**, *63*, 1297–1303.
18. Zhang, Y.; Wang, J.Z.; Wu, Y.J.; Li, W.G. Anti-inflammatory effect of recombinant human superoxide dismutase in rats and mice and its mechanism. *Acta Pharmacol. Sin.* **2002**, *23*, 439–444.
19. Luisa, C.M.; Jorge, J.C.; Van'T, H.R.; Cruz, M.E.; Crommelin, D.J.; Storm, G. Superoxide dismutase entrapped in long-circulating liposomes: Formulation design and therapeutic activity in rat adjuvant arthritis. *BBA Biomembr.* **2002**, *1564*, 227–236. [CrossRef]
20. Cloarec, M.; Caillard, P.; Provost, J.C.; Dever, J.M.; Elbeze, Y.; Zamaria, N. GliSODin, a vegetal sod with gliadin, as preventative agent vs. atherosclerosis, as confirmed with carotid ultrasound-B imaging. *Eur. Ann. Allergy Clin. Immunol.* **2007**, *39*, 45–50.
21. Muth, C.M.; Glenz, Y.; Klaus, M.; Radermacher, P.; Speit, G.; Leverve, X. Influence of an orally effective SOD on hyperbaric oxygen-related cell damage. *Free Radic. Res.* **2004**, *38*, 927–932. [CrossRef] [PubMed]
22. Liu, Y.-X.; Zhou, D.-Y.; Liu, Z.-Q.; Lu, T.; Song, L.; Li, D.-M.; Dong, X.-P.; Qi, H.; Zhu, B.-W.; Shahidi, F. Structural and biochemical changes in dermis of sea cucumber (Stichopus japonicus) during autolysis in response to cutting the body wall. *Food Chem.* **2018**, *240*, 1254–1261. [CrossRef] [PubMed]
23. Jamieson, A.J.; Gebruk, A.; Fujii, T.; Solan, M. Functional effects of the hadal sea cucumber Elpidia atakama (Echinodermata: Holothuroidea, Elasipodida) reflect small-scale patterns of resource availability. *Mar. Biol.* **2011**, *158*, 2695–2703. [CrossRef]
24. Zhang, J.; Lin, S.; Zeng, R. Cloning, expression, and characterization of a cold-adapted lipase gene from an antarctic deep-sea psychrotrophic bacterium, *Psychrobacter* sp. 7195. *J. Microbiol. Biotechnol.* **2007**, *17*, 604. [PubMed]

25. Michels, P.C.; Clark, D.S. Pressure-enhanced activity and stability of a hyperthermophilic protease from a deep-sea methanogen. *Appl. Environ. Microbiol.* **1997**, *63*, 3985–3991. [PubMed]
26. Borgstahl, G.E.O.; Parge, H.E.; Hickey, M.J.; Beyer, W.F., Jr.; Hallewell, R.A.; Tainer, J.A. The structure of human mitochondrial manganese superoxide dismutase reveals a novel tetrameric interface of two 4-helix bundles. *Cell* **1992**, *71*, 107–118. [CrossRef]
27. Liu, P.; Ewis, H.E.; Huang, Y.J.; Lu, C.D.; Tai, P.C.; Weber, I.T. Crystal Structure of the Bacillus subtilis Superoxide Dismutase. *Acta Crystallogr.* **2008**, *63 Pt 12*, 1003–1007.
28. Benkert, P.; Biasini, M.; Schwede, T. Toward the estimation of the absolute quality of individual protein structure models. *Bioinformatics* **2011**, *27*, 343–350. [CrossRef]
29. Liu, J.; Yin, M.; Hu, Z.; Lu, J.; Cui, Z. Purification and characterization of a hyperthermostable Mn-superoxide dismutase from *Thermus thermophilus* HB27. *Extremophiles* **2011**, *15*, 221–226. [CrossRef]
30. Zhu, Y.B.; Wang, G.H.; Ni, H.; Xiao, A.F.; Cai, H.N. Cloning and characterization of a new manganese superoxide dismutase from deep-sea thermophile *Geobacillus* sp. EPT3. *World J. Microbiol. Biotechnol.* **2014**, *30*, 1347–1357. [CrossRef]
31. Song, N.N.; Zheng, Y.; Shi-Jin, E.; Li, D.C. Cloning, expression, and characterization of thermostable Manganese superoxide dismutase from *Thermoascus aurantiacus* var. *levisporus*. *J. Microbiol.* **2009**, *47*, 123–130. [CrossRef] [PubMed]
32. Dolashki, A.; Abrashev, R.; Stevanovic, S.; Stefanova, L.; Ali, S.A.; Velkova, L.; Hristova, R.; Angelova, M.; Voelter, W.; Devreese, B. Biochemical properties of Cu/Zn-superoxide dismutase from fungal strain *Aspergillus niger* 26. *Spectrochim. Acta Part A Mol. Biomol. Spectrosc.* **2009**, *71*, 975–983. [CrossRef] [PubMed]
33. Bao, Y.; Li, L.; Xu, F.; Zhang, G. Intracellular copper/zinc superoxide dismutase from bay scallop *Argopecten irradians*: Its gene structure, mRNA expression and recombinant protein. *Fish Shellfish Immunol.* **2009**, *27*, 210–220. [CrossRef] [PubMed]
34. Kan, G.; Wen, H.; Wang, X.; Zhou, T.; Shi, C. Cloning and characterization of iron-superoxide dismutase in Antarctic yeast strain Rhodotorula mucilaginosa AN5. *J. Basic Microbiol.* **2017**, *57*, 680–690. [CrossRef] [PubMed]
35. Li, Y.; Kong, X.; Chen, J.; Liu, H.; Zhang, H. Characteristics of the Copper, Zinc Superoxide Dismutase of a Hadal Sea Cucumber (*Paelopatides* sp.) from the Mariana Trench. *Mar. Drugs* **2018**, *16*, 169. [CrossRef] [PubMed]
36. Kawano, H.; Nakasone, K.; Matsumoto, M.; Yoshida, Y.; Usami, R.; Kato, C.; Abe, F. Differential pressure resistance in the activity of RNA polymerase isolated from *Shewanella violacea* and *Escherichia coli*. *Extremophiles* **2004**, *8*, 367–375. [CrossRef]
37. Wang, S.L.; Li, Y.L.; Han, Z.; Chen, X.; Chen, Q.J.; Wang, Y.; He, L.S. Molecular Characterization of a NovelN-Acetylneuraminate Lyase from a Deep-Sea Symbiotic Mycoplasma. *Mar. Drugs* **2018**, *16*, 80. [CrossRef]
38. Brindley, A.A.; Pickersgill, R.W.; Partridge, J.C.; Dunstan, D.J.; Hunt, D.M.; Warren, M.J. Enzyme sequence and its relationship to hyperbaric stability of artificial and natural fish lactate dehydrogenases. *PLoS ONE* **2008**, *3*, e2042. [CrossRef]
39. Abe, F.; Minegishi, H.; Miura, T.; Nagahama, T.; Usami, R.; Horikoshi, K. Characterization of cold- and high-pressure-active polygalacturonases from a deep-sea yeast, *Cryptococcus liquefaciens* strain N6. *J. Agric. Chem. Soc.* **2006**, *70*, 296–299.
40. Sujiwattanarat, P.; Pongsanarakul, P.; Temsiripong, Y.; Temsiripong, T.; Thawornkuno, C.; Uno, Y.; Unajak, S.; Matsuda, Y.; Choowongkomon, K.; Srikulnath, K. Molecular cloning and characterization of Siamese crocodile (*Crocodylus siamensis*) copper, zinc superoxide dismutase (CSI-Cu,Zn-SOD) gene. *Comp. Biochem. Physiol. A Mol. Integr. Physiol.* **2016**, *191*, 187–195. [CrossRef]
41. Zhu, Y.; Li, H.; Ni, H.; Liu, J.; Xiao, A.; Cai, H. Purification and biochemical characterization of manganesecontaining superoxide dismutase from deep-sea thermophile *Geobacillus* sp. EPT3. *Acta Oceanol. Sin.* **2014**, *33*, 163–169. [CrossRef]
42. Ken, C.F.; Hsiung, T.M.; Huang, Z.X.; Juang, R.H.; Lin, C.T. Characterization of Fe/Mn−Superoxide Dismutase from Diatom *Thallassiosira weissflogii*: Cloning, Expression, and Property. *J. Agric. Food Chem.* **2005**, *53*, 1470–1474. [CrossRef] [PubMed]

© 2019 by the authors. Licensee MDPI, Basel, Switzerland. This article is an open access article distributed under the terms and conditions of the Creative Commons Attribution (CC BY) license (http://creativecommons.org/licenses/by/4.0/).

Article

A Novel Cold-Adapted Leucine Dehydrogenase from Antarctic Sea-Ice Bacterium *Pseudoalteromonas* sp. ANT178

Yatong Wang, Yanhua Hou, Yifan Wang, Lu Zheng, Xianlei Xu, Kang Pan, Rongqi Li and Quanfu Wang *

School of Marine Science and Technology, Harbin Institute of Technology, Weihai 264209, China; wangyatong199311@163.com (Y.W.); marry7718@163.com (Y.H.); daid01@126.com (Y.W.); zhenglu0206@126.com (L.Z.); 17863108956@163.com (X.X.); m15662319228@163.com (K.P.); l1263769417@163.com (R.L.)
* Correspondence: wangquanfuhit@hit.edu.cn; Tel./Fax: +86-631-568-7240

Received: 13 September 2018; Accepted: 27 September 2018; Published: 1 October 2018

Abstract: L-*tert*-leucine and its derivatives are useful as pharmaceutical active ingredients, in which leucine dehydrogenase (LeuDH) is the key enzyme in their enzymatic conversions. In the present study, a novel cold-adapted LeuDH, *psleudh*, was cloned from psychrotrophic bacteria *Pseudoalteromonas* sp. ANT178, which was isolated from Antarctic sea-ice. Bioinformatics analysis of the gene *psleudh* showed that the gene was 1209 bp in length and coded for a 42.6 kDa protein containing 402 amino acids. PsLeuDH had conserved Phe binding site and NAD$^+$ binding site, and belonged to a member of the Glu/Leu/Phe/Val dehydrogenase family. Homology modeling analysis results suggested that PsLeuDH exhibited more glycine residues, reduced proline residues, and arginine residues, which might be responsible for its catalytic efficiency at low temperature. The recombinant PsLeuDH (rPsLeuDH) was purified a major band with the high specific activity of 275.13 U/mg using a Ni-NTA affinity chromatography. The optimum temperature and pH for rPsLeuDH activity were 30 °C and pH 9.0, respectively. Importantly, rPsLeuDH retained at least 40% of its maximum activity even at 0 °C. Moreover, the activity of rPsLeuDH was the highest in the presence of 2.0 M NaCl. Substrate specificity and kinetic studies of rPsLeuDH demonstrated that L-leucine was the most suitable substrate, and the catalytic activity at low temperatures was ensured by maintaining a high k_{cat} value. The results of the current study would provide insight into Antarctic sea-ice bacterium LeuDH, and the unique properties of rPsLeuDH make it a promising candidate as a biocatalyst in medical and pharmaceutical industries.

Keywords: leucine dehydrogenase; cold-adapted; Antarctic bacterium; sea-ice; homology modeling

1. Introduction

Leucine dehydrogenase (LeuDH; EC 1.4.1.9), a NAD$^+$ dependent oxidoreductase, which catalyzes reversible L-leucine and other branched chain L-amino acids deamination reaction to the formation of the corresponding α-keto acid [1]. The enzyme was first identified in *Bacillus cereus* [2], and then was found in some microorganisms *Bacillus licheniformis* [3], *Bacillus sphaericus* [4], *Citrobacter freundii* [5], and *Laceyella sacchari* [6]. Moreover, crystal structures of the LeuDH from *Sporosarcina psychrophila* [7] and *Bacillus sphaericus* have been described [8].

LeuDH is used as a biocatalyst to format amino acids for using in the pharmaceutical industry by catalyzing the corresponding α-keto acids [9]. However, some of α-keto acids are unstable and degraded during prolonged incubation at moderate temperatures, such as 37 °C [10]. Importantly, cold-adapted enzymes that exhibit high levels of activity at room temperature (20–25 °C) should be useful for converting such unstable α-keto acids. What is more, cold-adapted enzymes have better conversion rates, the specificity of substrate and product, fewer by-products, which are required in the modern industry [11]. Although many LeuDHs have already been characterized, only a few cold-adapted LeuDH have been reported, such as LeuDH from *Alcanivorax dieselolei* [12] and *Sporosarcina psychrophila* [7].

Antarctic sea-ice, due to its specific geographical location and climate, is considered as an extreme environment on the earth. To develop the ability to withstand the extreme environment, sea-ice microorganisms have evolved several adaptive strategies and would be the new and promising microbial sources of cold-adapted enzymes. In our previous studies, some cold-adapted enzymes were isolated from Antarctic sea-ice bacteria and had become interesting for industrial applications [13,14]. It is well-known that L-*tert*-leucine and its derivatives are useful as pharmaceutical active ingredients and chiral auxiliaries, while LeuDH is a key enzyme for the enzymatic production of L-*tert*-leucine. Here, we briefly describe the homology modeling, expression, and characterization of cold-adapted LeuDH from Antarctic sea-ice bacterium. This LeuDH had unique properties make it good candidate for future medical and pharmaceutical industry applications.

2. Results and Discussion

2.1. Gene Cloning and Sequence Analysis

The *psleudh* gene was amplified from genomic DNA of the strain ANT178. It consisted of an ORF of 1209 bp, encoded a protein of 402 amino acid resides with a theoretical p*I* of 5.08. Furthermore, the DNA sequence of *psleudh* was submitted to the GenBank database with the accession number of MH322031. Based on sequences alignment, PsLeuDH showed the highest sequence similarity (88.0%) with LeuDH from *Pseudoalteromonas nigrifaciens* (ASM53600), followed by a sequence similarity of 65.0% with LeuDH from *Colwellia piezophila* (WP_019029130). More importantly, PsLeuDH had a conserved Phe binding site (I344) and NAD^+ binding sites (G233, G235, T236, V237, D256, I257, A261, C290, A291, C312, and N314). The coenzyme binding domain of NAD^+ in LeuDH was capable of catalyzing the reversible oxidative deamination of L-leucine and several other branched chain amino acids to form the corresponding 2-oxo acid derivatives. This domain could be classified as a member of the Rossmann fold superfamily, comprising a plurality of different dehydrogenases, wherein the amino acid dehydrogenase family comprises a common feature: a beta-sheet-alpha helix-beta sheet conformation [15]. PsLeuDH had this structural feature from Figure 1, further demonstrating that PsLeuDH was a member of the Glu/Leu/Phe/Val dehydrogenase family.

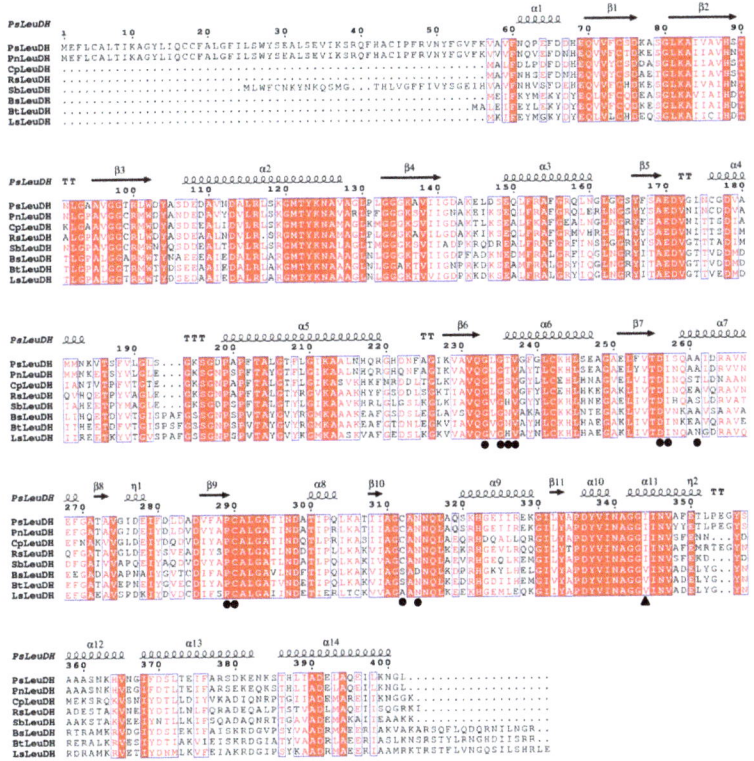

Figure 1. Amino acid sequence alignment of PsLeuDH and related LeuDH. PsLeuDH, *Pseudoalteromonas* sp. ANT178 LeuDH (MH322031); PnLeuDH, *Pseudoalteromonas nigrifaciens* (ASM53600); CpLeuDH, *Colwellia piezophila* (WP_019029130); RsLeuDH, *Rheinheimera salexigens* (WP_070050751); SbLeuDH, *Shewanella baltica* BA175 (AEG11165); BsLeuDH, *Bacillus sphaericus* ATCC4525 (PDB ID:1LEH); BtLeuDH, *Bacillus thuringiensis* (WP_001162678); and LsLeuDH, *Laceyella sacchari* (KR065697). ●, NAD binding site; ▲, Phe binding site.

2.2. Homology Modeling and Analysis of PsLeuDH

BsLeuDH (PDB ID:1LEH), encoded 364 amino acids, was isolated from mesophilic bacteria *Bacillus sphaericus* ATCC4525 [16], which exhibited the highest sequence identity (51%) to PsLeuDH using DALI server. The comparative analysis of the 3D structure of PsLeuDH and the mesophilic enzyme Bs-LeuDH was shown in Figure 2. It could be seen that two LeuDHs had a similar NAD^+ binding site and Phe binding site.

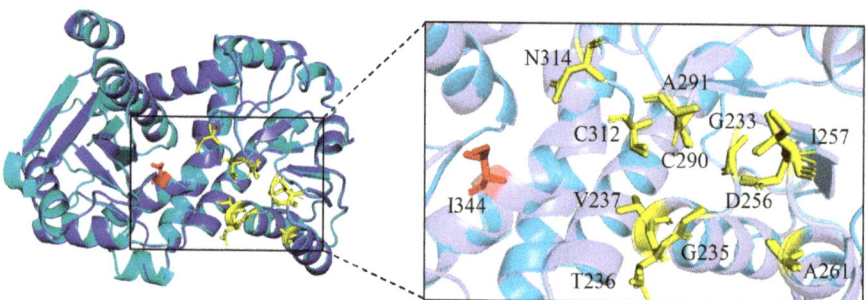

Figure 2. Three-dimensional structure comparison of PsLeuDH and BsLeuDH model. PsLeuDH, tv-blue; BsLeuDH, cyan; NAD$^+$ binding site, yellow ball stick model; Phe binding site, red ball stick model.

Comparison of structural adaptation characteristics and amino acid substitutions between PsLeuDH and BsLeuDH was shown in Table 1. It can be seen that PsLeuDH exhibited several cold-adapted features. Firstly, the number of electrostatic interactions of PsLeuDH was less than BsLeuDH, which might make the structure of PsLeuDH more flexible [17]. PsLeuDH also had less hydrophobic interactions compared to BsLeuDH, it might make PsLeuDH less rigid and contributed to decrease in structural stability [18]. Secondly, PsLeuDH revealed higher glycine residues and fewer proline and arginine residues that could affect the cold-adapted proteins properties which might offer higher flexibility to proteins [19]. Several amino acid residues in BsLeuDH were replaced by glycine residues in PsLeuDH. The glycine residues might improve the flexibility of the active site, and regulate the entropy of protein unfolding [10], thus probably improving the catalytic efficiency of the enzyme at low temperature. Additionally, proline might reduce the configuration entropy of the unfolding of protein molecules [20] and reduce the stability of enzyme molecules. Additionally, the stability of enzyme was also a significant factor to determine its catalytic characteristics. Some arginine residues in PsLeuDH were replaced by other residues at the same position in BsLeuDH. One of the stability factors in protein structure referred to salt bridges formed by arginine residues [19], arginine might make protein molecules more stable through ionic interaction. Compared with mesophilic enzyme BsLeuDH, PsLeuDH had higher flexibility and lower thermal stability, resulting in higher catalytic efficiency at low temperature [21].

Table 1. Comparison of structural adaption features and amino acid substitutions between PsLeuDH and its homolog (BsLeuDH).

Parameters	PsLeuDH	Bs-LeuDH	Expected Effect on PsLeuDH
Electrostatic interactions			
Salt Bridge (2.5 to 4.0)	17	22	
Hydrogen Bonds ($\leq$3.3 Å)	368	403	Protein stability
Cation-pi interactions	3	11	
Aromatic interactions	6	8	
Hydrophobic interactions	227	318	Thermolability
Glycine residues	42	36	
Proline residues	9	11	
Arginine residues	10	17	Flexibility
Glycine substitution (PsLeuDH → BsLeuDH)	G163 → N107, G177 → D121, G238 → A185, G240 → A187, G275 → A222, G401 →V346		
Proline substitution (PsLeuDH → BsLeuDH)	A94 → P38, A143 →P87, S320 → P267, S385 → P330		
Proline substitution (PsLeuDH → BsLeuDH)	P131 → N75, P63 → M7		Stability
Arginine substitution (PsLeuDH → BsLeuDH)	R219 → F166, R264 → A211, R327 → H274, R378 → I323		

2.3. Expression and Purification of the rPsLeuDH

The gene coding for the PsLeuDH was cloned into the pET-28a (+) vector and expressed in *E. coli* BL21 (DE3) under IPTG induction (Figure 3, Lane 3). rPsLeuDH was purified in a single step using His-tag affinity chromatography. A major band was observed on SDS-PAGE with about the molecular weight 44.4 kDa (Figure 3, Lane 4, 5). It is noteworthy that the last purified rPsLeuDH exhibited the highest specific activity of 275.13 U/mg.

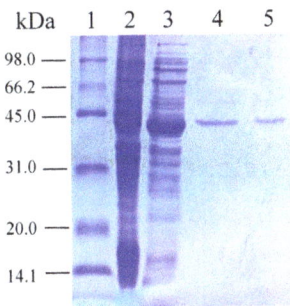

Figure 3. Expression and purification analysis of PsLeuDH. Lane 1: molecular weight standard marker; Lane 2: crude extract from the BL21/pET-28a (+); Lane 3: crude extract from the BL21/pET-28a (+)-PsLeuDH with IPTG induction; Lane 4: rPsLeuDH eluted with 50 mM imidazole; Lane 5: rPsLeuDH eluted with 100 mM imidazole.

2.4. Effects of Temperature and pH on Activity and Stability of rPsLeuDH

The temperature characteristic of rPsLeuDH was shown in Figure 4a. It exhibited the highest activity at 30 °C, and that of a cold-adapted LeuDH was 30 °C [12], whereas thermophilic LeuDH was approximately 40–65 °C [6,22], or (60–75 °C) [5]. It is worth pointing out that rPsLeuDH retained 40% of the highest activity at 0 °C, suggested that the enzyme is a cold-adapted enzyme [23]. Furthermore, the thermostability of rPsLeuDH was assessed in Figure 4b. It was stable and retained 85% of its initial activity after incubating at 30 °C after 120 min. While, after incubating at 50 °C for 20 min, it was only 30% of its activity lower than other cold-adapted LeuDHs from *Alcanivorax dieselolei* [12] and *Sporosarcina psychrophila* [7]. However, thermostable LeuDH could retain full activity after incubation at 65 °C for 10 min [24]. The above results indicated that rPsLeuDH had thermal instability, which was another significant feature of cold-adapted enzyme [25]. The effect of pH on rPsLeuDH activity was shown in Figure 4c. The activity of rPsLeuDH was higher under alkaline conditions (pH 7.0–10.0), with the highest activity at pH 9.0. Similar results were described in other LeuDHs such as *Sporosarcina psychrophile* (pH 8.5–11.0) [7], *Laceyella sacchari* (pH 9.5–11) [6] and *Citrobacter freundii* (pH 9.0 to 11.0) [5]. After 30 min of exposure to pH 6.0–10.0, the stability of rPsLeuDH showed a similar pattern with that of the activity response to pH (Figure 4d). This broad range of pH dependence for the activity and stability made the rPsLeuDH probably useful for medical industrial applications.

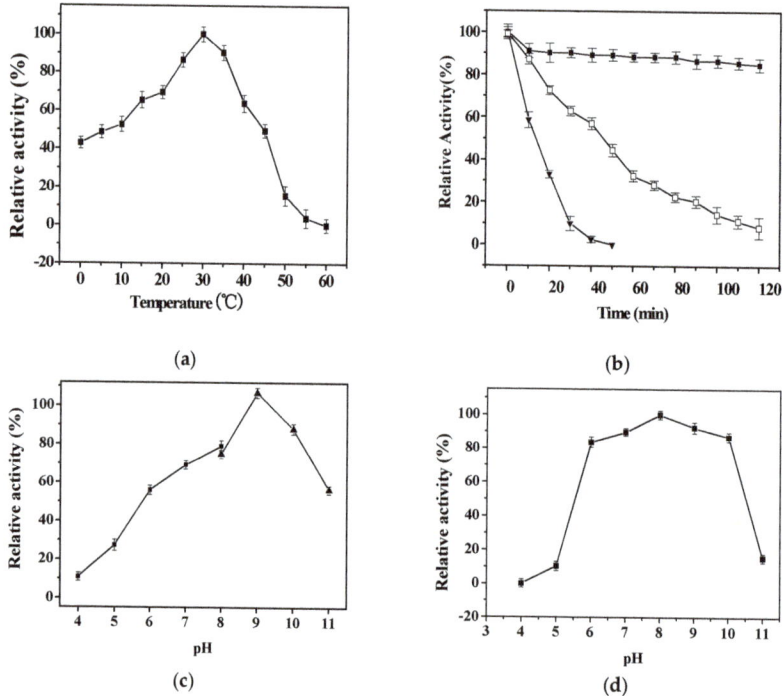

Figure 4. Effects of temperature and pH on the activity and stability of rPsLeuDH. (**a**) Effect of temperature on the activity of rPsLeuDH. (**b**) Effect of temperature on the stability of rPsLeuDH. (■) 30 °C, (□) 40 °C, (▼) 50 °C. (**c**) Effect of pH on the activity of rPsLeuDH. (**d**) Effect of pH on the stability of rPsLeuDH. Data are presented as mean ± SD (n = 3).

2.5. Effects of NaCl Concentration and Different Reagents on the Activity of PsLeuDH

The effect of NaCl concentration on the rPsLeuDH activity was shown in Figure 5. It could be seen that rPsLeuDH was stable at 0–3.0 M NaCl, with the highest activity at 2.0 M NaCl, which may be related to high salinity in the Antarctic sea ice environment. The similar result was also found in LeuDH from *Bacillus licheniformis* [3] and *Thermoactinomyces intermedius* [24] after high salt concentration treatment. The effect of various reagents on the rPsLeuDH activity was listed in Table 2. rPsLeuDH was completely inhibited by 1 mM $Pb(NO_3)_2$ and $BaCl_2$. Inhibitions by 1 mM $CrCl_2$ and $CdCl_2$ were 86.7% and 92.4%, respectively, while only partially inhibited by other metals salt in some extent. In addition, rPsLeuDH was sensitive to Thiourea and ethanol, but Triton X-100 kept the enzyme activity.

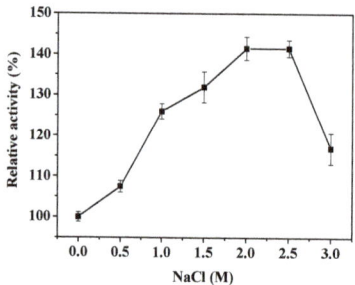

Figure 5. Effect of salt concentration on the activity of rPsLeuDH.

Table 2. Effects of different reagents on the activity of rPsLeuDH.

Reagent	Concentration	Relative Activity (%)	Reagent	Concentration	Relative Activity (%)
None		100 ± 0.0			
KCl	1 mM	99.7 ± 1.6	KCl	5 mM	40.0 ± 1.9
CoCl$_2$	1 mM	90.1 ± 1.7	CoCl$_2$	5 mM	70.0 ± 2.0
MgCl$_2$	1 mM	87.9 ± 0.8	MgCl$_2$	5 mM	65.8 ± 1.2
CaCl$_2$	1 mM	87.9 ± 0.4	CaCl$_2$	5 mM	68.1 ± 0.9
ZnCl$_2$	1 mM	80.0 ± 2.5	ZnCl$_2$	5 mM	72.2 ± 2.0
FeCl$_2$	1 mM	75.1 ± 2.2	FeCl$_2$	5 mM	62.4 ± 1.7
CuCl$_2$	1 mM	61.0 ± 2.2	CuCl$_2$	5 mM	41.0 ± 1.5
HgCl$_2$	1 mM	29.2 ± 0.3	HgCl$_2$	5 mM	12.3 ± 1.9
CrCl$_2$	1 mM	13.3 ± 0.3	CrCl$_2$	5 mM	5.8 ± 2.9
CdCl$_2$	1 mM	7.6 ± 0.5	CdCl$_2$	5 mM	0.0 ± 0.0
Pb(NO$_3$)$_2$	1 mM	0.0 ± 0.0	Pb(NO$_3$)$_2$	5 mM	0.0 ± 0.0
BaCl$_2$	1 mM	0.0 ± 0.0	BaCl$_2$	5 mM	0.0 ± 0.0
EDTA	1 mM	91.8 ± 2.7	EDTA	5 mM	84.2 ± 2.1
Thiourea	1 mM	51.5 ± 4.0	Thiourea	5 mM	34.3 ± 2.6
Triton X-100	0.2%	102.7 ± 1.4	Ethanol	25%	67.5 ± 1.4

2.6. The Substrate Specificity Analysis and Kinetic Parameters of rPsLeuDH

The substrate specificity analysis of rPsLeuDH was listed in Table 3. It could catalyze and utilize five substrates, indicating that rPsLeuDH possessed a broad spectrum of substrates in catalytic oxidation reaction. L-leucine was the most suitable substrate for rPsLeuDH, which was the similar with other microbial LeuDH [6,22]. The kinetic parameters of rPsLeuDH were determined. K_m and V_m of L-leucine were calculated as 0.33 mM and 15.24 μmol/min·mg, respectively. Besides, the k_{cat} value of L-leucine was 30.13/s, demonstrating that rPsLeuDH had a high affinity to substrates and was conducive to improving catalytic efficiency at low temperature.

Table 3. Substrate specificity analysis of rPsLeuDH.

Substrate	V_m (μmol/min·mg)	K_m (mM)	k_{cat} (1/s)	k_{cat}/K_m (mM^{-1} s^{-1})
L-lecine	15.24	0.33	30.13	91.30
L-tyrosine	13.35	0.48	26.39	54.98
L-proline	10.52	0.64	20.80	32.50
DL-methionine	8.38	0.75	16.57	22.09
L-arginine	7.13	0.84	14.09	16.77

2.7. The Thermodynamic Parameters of rPsLeuDH

Thermodynamic parameters such as ΔH, ΔS and ΔG at different temperature (0–30 °C) were calculated and listed in Table 4. At 0, 10, 20, and 30 °C, the k_{cat} value of rPsLeuDH were 12.25, 14.96, 20.20 and 30.13/s, respectively, indicating that the k_{cat} value increased with increasing temperature, which was similar to the k_{cat} change trend of cold-adapted β-D-galactosidase at different temperatures [26]. rPsLeuDH also exhibited lower ΔH, ΔS and ΔG and higher k_{cat} at low temperature, as compared to mesophilic enzyme, which may be mainly related to the conformation of cold adapted protein [27]. On the other hand, it may also be related to increasing the efficiency of binding of the substrate to the catalytic site [28].

Table 4. Thermodynamic parameter of the rPsLeuDH.

Temperature (°C)	ΔH (KJ/mol)	ΔS (J/mol K)	ΔG (KJ/mol)	k_{cat} (1/s)
0	18.27	−156.45	61.01	12.25
10	18.19	−157.75	62.90	14.96
20	18.11	−158.02	64.43	20.20
30	18.02	−157.28	65.70	30.13

3. Materials and Methods

3.1. Microorganisms and Growth Conditions

The strain *Pseudoalteromonas* sp. ANT178, isolated from sea ice in Antarctica (68°30′ E, 65°00′ S), was used as a source of *psleudh* gene. The strain ANT178 was cultivated in the 2216E sea water medium (initial pH 7.5, 5 g/L peptone, and 1 g/L yeast extract) for 96 h at 12 °C. *E. coli* BL21 (DE3) was used as the plasmid host.

3.2. Sequence Analysis of LeuDH Gene

The open reading frame and amino acid sequences of *psleudh* gene were computed (https://www.ncbi.nlm.nih.gov/orffinder/). The theoretical molecular weight and p*I* were also analyzed using the ExPASy Compute pI/Mw tool (http://web.expasy.org/computepi). Multiple sequence alignment of the amino acids of PsLeuDH was performed using Bioedit 7.2 and ESPript 3.0 [29].

3.3. Protein Homology Modeling

A homology model of LeuDH was built with SWISS-MODEL. LeuDH from mesophilic bacteria *Bacillus sphaericus* ATCC4525 (PDB ID:1LEH) [16] was selected as the template. The structure figures were created with PyMOL software (DeLano Scientific LLC, San Carlos, CA, USA). Salt bridges were carried out using VMD 1.9.3. (University of Illinois Urbana-Champaign, Champaign, IL, USA). For the hydrogen bonds, a cut-off distance of 3.3 Å was set. Cation-pi interactions, aromatic interactions, ionic interactions, and hydrogen bonds were predicted by the Protein Interactions Calculator program (http://pic.mbu.iisc.ernet.in).

3.4. Molecular Cloning, Expression and Purification of rPsLeuDH

The genome of *Pseudoalteromonas* sp. ANT178 was sequenced and annotated using high-throughput technologies (data not shown). The full-length gene of *psleudh* was amplified by PCR using the primers 5′-GATGGATCCATGGAATTT TTATGTG-3′ (*Bam*HI site underlined) and 5′-CAGAAGCTTGAAGACCGTTTT TAAG-3′ (*Hin*dIII site underlined) according to its genome sequence. PCR was performed with Taq DNA polymerase (TaKaRa Bio, Dalian, China). The product was then directly cloned into the corresponding sites of the pET-28a (+) vector and transformed into *E. coli* BL21. The transformants with the *psleudh* gene were grown in Luria-Bertani (LB) medium supplemented with 100 mg/L kanamycin and cultured by shaking at 37 °C until the OD_{600} reached 0.6–0.8. Then, 1.0 mM sopropyl-β-D-thiogalactopyranoside (IPTG) was added for induction. The bacterial cells were cultured at 37 °C for 2–3 h, and then the culture temperature was shifted to 28 °C to induce the protein expression for 6 h. The induced cells centrifuged at 4 °C and 7500× g for 15 min and subjected to ultrasonic disruption with 150 W (JY96-IIN, Shanghai, China). The insoluble debris was removed by centrifuged at 4 °C and 7500× g for 15 min, and the supernatant was harvested as crude protein (21.99 mg). Purification of rPsLeuDH with the His-tagged was purified using Ni-NTA affinity chromatography. The purified protein (1.11 mg) was eluted with 10, 50, 100 and 250 mM imidazole buffer (20 mM Tris-HCl, 500 mM NaCl, pH 8.0) at a flow rate of 1.0 mL/min. The purity and the molecular mass of the rPsLeuDH were determined by SDS-PAGE, using 12.0% polyacrylamide gels.

3.5. Assay of rPsLeuDH Activity

The standard enzyme assay were based on traditional method and modified on basis [1,30]. The oxidation reaction activity assay was determined by 200 μL reaction system. It contained 0.1 M Glycine-NaOH (pH 10.4) buffer, 10 mM L-leucine and 10 μL purified enzyme (0.62 μg), which incubated at 30 °C for 2 min. After adding 1 mM NAD^+, the changes of absorbance at 340 nm within 1 min were detected. Futhermore, the reductive amination reaction system containing (200 μL) 0.2 M NH_4Cl-NH_4OH buffer (pH 9.0), 5 mM TMP and 10 μL purified enzyme at 30 °C for 2 min, after adding 0.2 mM NADH, changes in absorbance at 340 nm within 1 min were measured. One unit of LeuDH activity was defined as the amount of enzyme catalyzed the formation or reduction of 1 µmoL NADH/min at 30 °C.

3.6. Characterization of the Purified rPsLeuDH

The optimal temperature of the purified rPsLeuDH was determined with the standard assay at temperatures from 0 °C to 60 °C. To evaluate the thermostability, the purified enzyme was incubated at three different temperatures (30, 40, and 50 °C) for 120 min, and the residual activity was measured by the standard enzyme assays. The optimal pH of the purified enzyme was determined at 30 °C using Citric acid/Na_2HPO_4 buffer (0.2 M) and NH_4Cl-NH_4OH buffer (0.2 M) for pH ranges 4.0–8.0 and 8.0–10.0, respectively. To assess pH stability, the rPsLeuDH was pretreated at pH 4.0–11.0 in the absence of substrate at 30 °C for 30 min, and the residual activity was measured by the standard enzyme assays. The purified rPsLeuDH was incubated at 0–3.0 M NaCl at 30 °C for 30 min, and remaining activity was assayed with the standard enzyme assays. The effects of different reagents on the rPsLeuDH activity were assayed with the standard enzyme assay after pre-incubating enzyme in different metal ions at 30 °C for 30 min. Enzyme activity assayed without any reagent was defined as control (100%).

3.7. Kinetic Parameter of the rPsLeuDH

To assess the kinetics parameters, the Lineweaver-Burk plot method was used to calculate the K_m and V_m of rPsLeuDH [31]. The kinetic constants of NADH (0.025 mM–0.4 mM), L-leucine (0.05 mM–2 mM), L-tyrosine (0.05 mM–2 mM), L-proline (0.05 mM–2 mM), DL-methionine (0.05 mM–2 mM), L-arginine (0.05 mM–2 mM), TMP (0.05 mM–2 mM), and NAD^+ (0.025 mM–0.4 mM) were determined by the above method in rPsLeuDH.

3.8. Thermodynamic Parameter of the rPsLeuDH

The k_{cat} parameter is the reaction rate constant for the enzymatic-substrate complex chemical conversion into the enzyme and the product. k_{cat} was calculated based on kinetics experiments, and the thermodynamic related parameters were assayed by the modification method of Feller [27] as follows:

$$k_{cat} = Ae^{-\frac{E_a}{RT}} \tag{1}$$

$$\Delta H = E_a - RT \tag{2}$$

$$\Delta S = R\left(Ink_{cat} - 24.76 - InT + \frac{E_a}{RT}\right) \tag{3}$$

$$\Delta G = \Delta H - T\Delta S \tag{4}$$

where A is the constant, E_a is the activation energy of the reaction, R is the gas constant (8.314 J mol^{-1} K^{-1}), ΔH is the enthalpy of activation, ΔS is the entropy of activation, and ΔG is the free energy of activation.

4. Conclusions

A novel cold-adapted leucine dehydrogenase gene (*psleudh*) was cloned from Antarctic sea-ice bacterium and expressed in *E. coli* (DE3). Through homology modeling and comparison with its homologous enzyme (BsLeuDH), it was suggested that more glycine residues, reduced proline residues and arginine residues might be responsible for its catalytic efficiency at low temperature. rPsLeuDH was purified and characterized with higher activity at 30 °C, high salt (3.0 M), remarkable pH stability (pH 6.0–10.0), and higher specific activity (275.13 U/mg). These unique properties of rPsLeuDH make it a promising candidate as a biocatalyst in the enzymatic production of L-*tert*-leucine at room temperature.

Author Contributions: Y.W., Y.H. and Q.W. took charge of the research and designed the experiments; Y.W., L.Z., X.X., K.P., R.L., Y.W. and Y.H. performed the experiments and analyzed the data; Y.W. and Q.W. wrote the paper.

Funding: This research was funded by the National Natural Science Foundation of China (41876149), the Natural Science Foundation of Shandong Province (ZR2017MC046) and the Key Research and Development Plan of Shandong Province (2018GHY115021).

Conflicts of Interest: The authors declare no conflict of interest.

References

1. Ohshima, T.; Wandrey, C.; Sugiura, M.; Soda, K. Screening of thermostable leucine and alanine dehydrogenases in thermophilic *Bacillus* strains. *Biotechnol. Lett.* **1985**, *7*, 871–876. [CrossRef]
2. Sanwal, B.D.; Zink, M.W. L-leucine dehydrogenase of *Bacillus cereus*. *Arch. Biochem. Biophys.* **1961**, *94*, 430–435. [CrossRef]
3. Nagata, S.; Bakthavatsalam, S.; Galkin, A.G.; Asada, H.; Sakai, S.; Esaki, N.; Soda, K.; Ohshima, T.; Nagasaki, S.; Misono, H. Gene cloning, purification, and characterization of thermostable and halophilic leucine dehydrogenase from a halophilic thermophile, *Bacillus licheniformis* TSN9. *Appl. Microbiol. Biotechnol.* **1995**, *44*, 432–438. [CrossRef] [PubMed]
4. Katoh, R.; Nagata, S.; Misono, H. Cloning and sequencing of the leucine dehydrogenase gene from *Bacillus sphaericus*, IFO 3525 and importance of the C-terminal region for the enzyme activity. *J. Mol. Catal. B Enzym.* **2003**, *23*, 239–247. [CrossRef]
5. Mahdizadehdehosta, R.; Kianmehr, A.; Khalili, A. Isolation and characterization of leucine dehydrogenase from a thermophilic *Citrobacter freundii* JK-91 strain isolated from Jask Port. *Iran. J. Microbiol.* **2013**, *5*, 278–284. [PubMed]
6. Zhu, W.J.; Li, Y.; Jia, H.H.; Wei, P.; Zhou, H.; Jiang, M. Expression, purification and characterization of a thermostable leucine dehydrogenase from the halophilic thermophile *Laceyella sacchari*. *Biotechnol. Lett.* **2016**, *38*, 855–861. [CrossRef] [PubMed]
7. Zhao, Y.; Wakamatsu, T.; Doi, K.; Sakuraba, H.; Ohshima, T. A psychrophilic leucine dehydrogenase from *Sporosarcina psychrophila*: Purification, characterization, gene sequencing and crystal structure analysis. *J. Mol. Catal. B Enzym.* **2012**, *83*, 65–72. [CrossRef]
8. Turnbull, A.P.; Ashford, S.R.; Baker, P.J.; Rice, D.W.; Rodgers, F.H.; Stillman, T.J.; Hanson, R.L. Crystallization and quaternary structure analysis of the NAD(+)-dependent leucine dehydrogenase from *Bacillus sphaericus*. *J. Mol. Biol.* **1994**, *236*, 663–665. [CrossRef] [PubMed]
9. Zhu, L.; Wu, Z.; Jin, J.M.; Tang, S.Y. Directed evolution of leucine dehydrogenase for improved efficiency of L-*tert*-leucine synthesis. *Appl. Microbiol. Biotechnol.* **2016**, *100*, 5805–5813. [CrossRef] [PubMed]
10. Galkin, A.; Kulakova, L.; Ashida, H.; Sawa, Y.; Esaki, N. Cold-adapted alanine dehydrogenases from two Antarctic bacterial strains: Gene cloning, protein characterization, and comparison with mesophilic and thermophilic counterparts. *Appl. Environ. Microb.* **1999**, *65*, 4014–4020.
11. Gerday, C.; Aittaleb, M.; Bentahir, M.; Chessa, J.P.; Claverie, P.; Collins, T.; D'Amico, S.; Dumont, J.; Garsoux, G.; Georlette, D.; et al. Cold-adapted enzymes: From fundamentals to biotechnology. *Trend Biotechnol.* **2000**, *18*, 103–107. [CrossRef]
12. Jiang, W.; Sun, D.F.; Lu, J.X.; Wang, Y.L.; Wang, S.Z.; Zhang, Y.H.; Fang, B.S. A cold-adapted leucine dehydrogenase from marine bacterium *Alcanivorax dieselolei*: Characterization and L-*tert*-leucine production. *Eng. Life Sci.* **2016**, *16*, 283–289. [CrossRef]
13. Shi, Y.L.; Wang, Q.F.; Hou, Y.H.; Hong, Y.Y.; Han, X.; Yi, J.L.; Qu, J.J.; Lu, Y. Molecular cloning, expression and enzymatic characterization of glutathione s-transferase from Antarctic sea-ice bacteria *Pseudoalteromonas* sp. ANT506. *Microbiol. Res.* **2014**, *169*, 179–184. [CrossRef] [PubMed]
14. Wang, Y.T.; Han, H.; Cui, B.Q.; Hou, Y.H.; Wang, Y.F.; Wang, Q.F. A glutathione peroxidase from Antarctic psychrotrophic bacterium *Pseudoalteromonas* sp. ANT506: Cloning and heterologous expression of the gene and characterization of recombinant enzyme. *Bioengineered* **2017**, *8*, 742–749. [CrossRef] [PubMed]
15. Kuroda, S.I.; Tanizawa, K.; Sakamoto, Y.; Tanaka, H.; Soda, K. Alanine dehydrogenases from two Bacillus species with distinct thermostabilities: Molecular cloning, DNA and protein sequence determination, and structural comparison with other NAD(P)(+)-dependent dehydrogenases. *Biochemistry* **1990**, *29*, 1009–1015. [CrossRef] [PubMed]

16. Baker, P.J.; Turnbull, A.P.; Sedelnikova, S.E.; Stillman, T.J.; Rice, D.W. A role for quaternary structure in the substrate specificity of leucine dehydrogenase. *Structure* **1995**, *3*, 693–705. [CrossRef]
17. Paredes, D.I.; Watters, K.; Pitman, D.J.; Bystroff, C.; Dordick, J.S. Comparative void-volume analysis of psychrophilic and mesophilic enzymes: Structural bioinformatics of psychrophilic enzymes reveals sources of core flexibility. *BMC Struct. Biol.* **2011**, *11*, 42–50. [CrossRef] [PubMed]
18. Li, F.L.; Shi, Y.; Zhang, J.X.; Gao, J.; Zhang, Y.W. Cloning, expression, characterization and homology modeling of a novel water-forming NADH oxidase from *Streptococcus* mutans ATCC 25175. *Int. J. Biol. Macromol.* **2018**, *113*, 1073–1079. [CrossRef] [PubMed]
19. Mohammadi, S.; Parvizpour, S.; Razmara, J.; Abu Bakar, F.D.; Illias, R.M.; Mahadi, N.M.; Murad, A.M. Structure prediction of a novel Exo-β-1,3-Glucanase: Insights into the cold adaptation of psychrophilic yeast *Glaciozyma antarctica* PI12. *Interdiscip. Sci. Comput. Life Sci.* **2016**, *10*, 157–168. [CrossRef] [PubMed]
20. Herning, T.; Yutani, K.; Inaka, K.; Kuroki, R.; Matsushima, M.; Kikuchi, M. Role of proline residues in human lysozyme stability: A scanning calorimetric study combined with X-ray structure analysis of proline mutants. *Biochemistry* **1992**, *31*, 7077–7085. [CrossRef] [PubMed]
21. Siglioccolo, A.; Gerace, R.; Pascarella, S. "Cold spots" in protein cold adaptation: Insights from normalized atomic displacement parameters (B'-factors). *Biophys. Chem.* **2010**, *153*, 104–114. [CrossRef] [PubMed]
22. Li, J.; Pan, J.; Zhang, J.; Xu, J.H. Stereoselective synthesis of L-*tert*-leucine by a newly cloned leucine dehydrogenase from *Exiguobacterium sibiricum*. *J. Mol. Catal. B Enzym.* **2014**, *105*, 11–17. [CrossRef]
23. Feller, G.; Narinx, E.; Arpigny, J.L.; Aittaleb, M.; Baise, E.; Genicot, S.; Gerday, C. Enzymes from psychrophilic organisms. *FEMS Microbiol. Rev.* **1996**, *18*, 189–202. [CrossRef]
24. Ohshima, T.; Nishida, N.; Bakthavatsalam, S.; Kataoka, K.; Takada, H.; Yoshimura, T.; Soda, K.; Esaki, N. The purification, characterization, cloning and sequencing of the gene for a halostable and thermostable leucine dehydrogenase from *Thermoactinomyces intermedius*. *Eur. J. Biochem.* **1994**, *222*, 305–312. [CrossRef] [PubMed]
25. Michetti, D.; Brandsdal, B.O.; Bon, D.; Isaksen, G.V.; Tiberti, M.; Papaleo, E. A comparative study of cold- and warm-adapted endonucleases a using sequence analyses and molecular dynamics simulations. *PLoS ONE* **2017**, *12*, e0169586. [CrossRef] [PubMed]
26. Pawlak-Szukalska, A.; Wanarska, M.; Popinigis, A.T.; Kur, J. A novel cold-active β-D-galactosidase with transglycosylation activity from the Antarctic *Arthrobacter* sp. 32cB-Gene cloning, purification and characterization. *Process Biochem.* **2014**, *49*, 2122–2133. [CrossRef]
27. Lonhienne, T.; Gerday, C.; Feller, G. Psychrophilic enzymes: Revisiting the thermodynamic parameters of activation may explain local flexibility. *Biochim. Biophys. Acta* **2000**, *1543*, 1–10. [CrossRef]
28. Khrapunov, S.; Chang, E.; Callender, R.H. Thermodynamic and structural adaptation differences between the mesophilic and psychrophilic lactate dehydrogenases. *Biochemistry* **2017**, *56*, 3587–3595. [CrossRef] [PubMed]
29. Robert, X.; Gouet, P. Deciphering key features in protein structures with the new ENDscript server. *Nucleic Acids Res.* **2014**, *42*, W320–W324. [CrossRef] [PubMed]
30. Ohshima, T.; Nagata, S.; Soda, K. Purification and characterization of thermostable leucine dehydrogenase from *Bacillus* stearothermophilus. *Arch. Microbiol.* **1985**, *141*, 407–411. [CrossRef]
31. Shang, Z.C.; Zhang, L.L.; Wu, Z.J.; Gong, P.; Li, D.P.; Zhu, P.; Gao, H.J. The activity and kinetic parameters of oxidoreductases in phaeozem in response to long-term fertiliser management. *J. Soil Sci. Plant Nutr.* **2012**, *12*, 597–607. [CrossRef]

© 2018 by the authors. Licensee MDPI, Basel, Switzerland. This article is an open access article distributed under the terms and conditions of the Creative Commons Attribution (CC BY) license (http://creativecommons.org/licenses/by/4.0/).

MDPI
St. Alban-Anlage 66
4052 Basel
Switzerland
Tel. +41 61 683 77 34
Fax +41 61 302 89 18
www.mdpi.com

Marine Drugs Editorial Office
E-mail: marinedrugs@mdpi.com
www.mdpi.com/journal/marinedrugs

MDPI
St. Alban-Anlage 66
4052 Basel
Switzerland
Tel. +41 61 683 77 34
Fax +41 61 302 89 18
www.mdpi.com

Marine Drugs Editorial Office
E-mail: marinedrugs@mdpi.com
www.mdpi.com/journal/marinedrugs

www.ingramcontent.com/pod-product-compliance
Lightning Source LLC
LaVergne TN
LVHW070243100526
838202LV00015B/2170